Graduate Texts in Mathematics 302

Graduate Texts in Mathematics

Series Editors:
Patricia Hersh, *University of Oregon*
Ravi Vakil, *Stanford University*
Jared Wunsch, *Northwestern University*

Associate Editors:
Alexei Borodin, *Massachusetts Institute of Technology*
Richard D. Canary, *University of Michigan*
David Eisenbud, *University of California, Berkeley & SLMath*
Brian C. Hall, *University of Notre Dame*
June Huh, *Princeton University*
Eugenia Malinnikova, *Stanford University*
Akhil Mathew, *University of Chicago*
Peter J. Olver, *University of Minnesota*
John Pardon, *State University of New York*
Jeremy Quastel, *University of Toronto*
Wilhelm Schlag, *Yale University*
Barry Simon, *California Institute of Technology*
Melanie Matchett Wood, *Harvard University*
Yufei Zhao, *Massachusetts Institute of Technology*

Graduate Texts in Mathematics bridge the gap between passive study and creative understanding, offering graduate-level introductions to advanced topics in mathematics. The volumes are carefully written as teaching aids and highlight characteristic features of the theory. Although these books are frequently used as textbooks in graduate courses, they are also suitable for individual study.

Loukas Grafakos

Fundamentals of Fourier Analysis

Loukas Grafakos
Department of Mathematics
University of Missouri
Columbia, MO, USA

ISSN 0072-5285 ISSN 2197-5612 (electronic)
Graduate Texts in Mathematics
ISBN 978-3-031-56499-4 ISBN 978-3-031-56500-7 (eBook)
https://doi.org/10.1007/978-3-031-56500-7

Mathematics Subject Classification: 42AXX, 42BXX

© The Editor(s) (if applicable) and The Author(s), under exclusive license to Springer Nature Switzerland AG 2024

This work is subject to copyright. All rights are solely and exclusively licensed by the Publisher, whether the whole or part of the material is concerned, specifically the rights of translation, reprinting, reuse of illustrations, recitation, broadcasting, reproduction on microfilms or in any other physical way, and transmission or information storage and retrieval, electronic adaptation, computer software, or by similar or dissimilar methodology now known or hereafter developed.
The use of general descriptive names, registered names, trademarks, service marks, etc. in this publication does not imply, even in the absence of a specific statement, that such names are exempt from the relevant protective laws and regulations and therefore free for general use.
The publisher, the authors and the editors are safe to assume that the advice and information in this book are believed to be true and accurate at the date of publication. Neither the publisher nor the authors or the editors give a warranty, expressed or implied, with respect to the material contained herein or for any errors or omissions that may have been made. The publisher remains neutral with regard to jurisdictional claims in published maps and institutional affiliations.

This Springer imprint is published by the registered company Springer Nature Switzerland AG
The registered company address is: Gewerbestrasse 11, 6330 Cham, Switzerland

Paper in this product is recyclable.

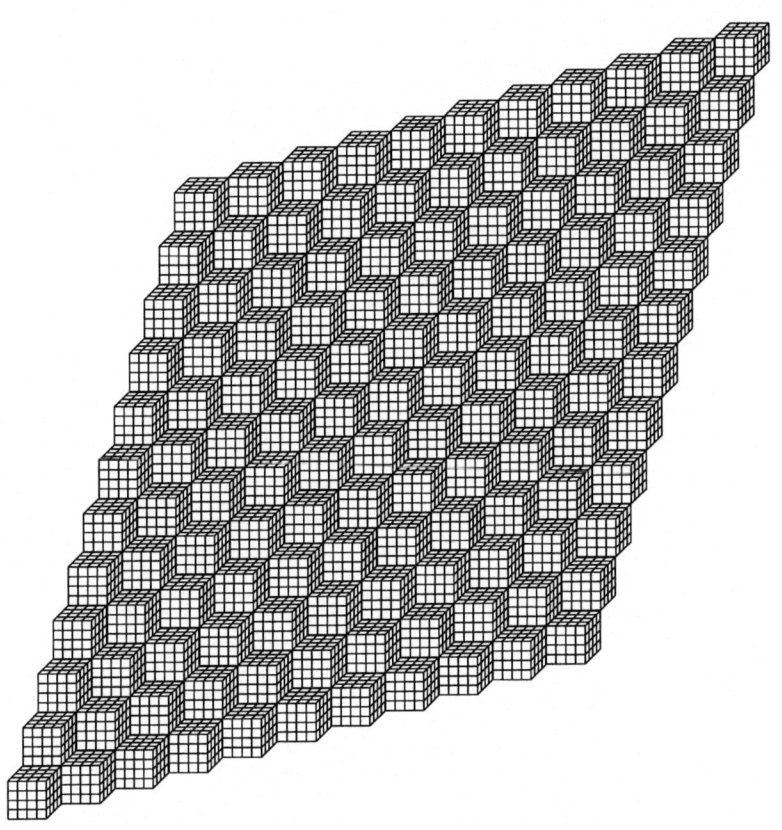

To my students and collaborators

Preface

I am truly indebted to the authors of the graduate texts that taught me mathematics through the years. It has long been my wish to express this gratitude by writing a graduate textbook that would benefit generations of younger mathematicians. This was my intention when I started working on the volumes *Classical Fourier Analysis* (GTM 249) and *Modern Fourier Analysis* (GTM 250). But for some reason the end result was a bit different; these books grew too big and are mostly used today as references. Nonetheless, I hope this monograph achieves my original goal. The present text is designed to introduce Euclidean Fourier Analysis to students who have successfully completed first-year graduate courses in Real Analysis and Complex Variables. The material is self-contained, thoughtfully planned out, and presented with the intention of building a solid foundation in Fourier Analysis within two semesters. Each section is complemented by up to a dozen exercises that range in difficulty from "straightforward" to "fairly challenging," in no particular order. Better comprehension is definitely achieved by solving these exercises, and readers are urged to do so in order to test their understanding. I hope that students will profit from this text, and instructors will enjoy teaching from it. Above all else, I hope this book will inspire many people to study harmonic analysis.

This book is designed to be self-contained and for this reason, many peripheral tools needed are included in Appendices A–G; however, in a few instances, results from Appendices A–D of [31] are also used. While this book may omit some popular themes, it is only intended to serve the purposes of a two-semester course and is written to contain as many topics of general interest as possible. The outline below and the interdependence chart will assist instructors in deciding which sections/chapters they wish to omit and/or replace with topics of their own interest without affecting the logical flow of the exposition.

Chapter 1 begins with a section containing a summary of important results on measure theory and Lebesgue spaces, with selected proofs. This section could be omitted as it is usually covered in courses on real variables. The remaining sections of this chapter focus on weak L^p spaces, interpolation, maximal functions, the Lebesgue differentiation theorem, convolutions, smooth functions, and approximate identities.

These fundamental notions build an arsenal of indispensable tools in analysis. I suggest they be covered in detail.

The Fourier transform is introduced in the first section of Chapter 2 as an operator on L^1 and in the third section as an operator on L^2, after Fourier inversion is settled in the second section. The Hausdorff–Young inequality is obtained by means of complex interpolation. The latter topic is presented in the entire range of Lebesgue spaces and is based on the results in Appendices B and C. Case II in the proof of Theorem 2.4.1 is not needed in the sequel and could be excluded. Theorem 2.5.7 in Section 2.5 could also be skipped. Sections 2.6 and 2.7 provide the necessary tools from distribution theory that a harmonic analyst needs to know. These sections are sine qua non to audiences without background on distributions. Section 2.8 introduces L^p Fourier multipliers and is important in this book. Finally, Section 2.9 is optional and is recommended to audiences with interest in oscillatory integrals.

Chapter 3 is concerned with the theory of singular integrals. The theory is intuitively built starting with the one-dimensional Hilbert transform (first section). The following two sections contain higher-dimensional analogs. The L^2 boundedness of singular integrals is the topic of Section 3.4 and the L^p boundedness is the topic of Section 3.6, which is established in terms of the Calderón–Zygmund decomposition (Section 3.5). The last section of this chapter deals with maximal singular integrals, while Section 3.7 could be skipped if there is time pressure.

The first three sections of the fourth chapter are concerned with vector-valued extensions of the L^p boundedness results of singular integrals. To simplify the presentation, vectors are restricted to the finite-dimensional case while analogous results for infinite vectors are obtained by a limiting process. The results in these sections are obtained as in the previous chapter, the only difference being that the functions involved here take values in a finite-dimensional Banach space. These ideas find fruitful applications in Littlewood–Paley theory, which is studied in Sections 4.4 and 4.5. The last section of this chapter (Section 4.6) contains product-type variants of these results that are subsequently needed only in the proof of the Marcinkiewicz multiplier theorem (Section 5.7).

Chapter 5 is concerned with fractional integration and differentiation. The first two sections focus on the Riesz and Bessel potentials and some of their basic properties. A version of the Miklhin–Hörmander multiplier theorem is the focus of Section 5.3. A short exposition of Sobolev spaces, including the Sobolev embedding theorem, is given in Section 5.4. Stein's interpolation theorem for analytic families (Section 5.5) furnishes an elegant way to obtain estimates for fractional derivatives. The Calderón–Torchinsky multiplier theorem provides an extension of Miklhin–Hörmander's theorem and is presented in Section 5.6. The chapter ends with the study of the Marcinkiewicz multiplier theorem (Section 5.7); this requires the results of Section 4.6.

The space of functions of bounded mean oscillation (BMO) is studied in Chapter 6. The most important theorem in this chapter is the John–Nirenberg theorem, which is proved in Section 6.2. A version of BMO called dyadic BMO is investigated in Section 6.3. A useful tool in the study of BMO is the sharp maximal function, which is discussed in Section 6.4. This is used in interpolation via BMO, a topic examined in the last section of this chapter (Section 6.5).

Chapter 7 focuses on Hardy spaces H^p. The first section is concerned with the interplay of smoothness and cancellation, which is explicitly manifested in many estimates. Although emphasis is placed on the Hardy space H^1, the basic definitions, examples, and properties of Hardy spaces are discussed for all $p > 0$; these include properties of the grand maximal function. An important topic discussed in this chapter is the atomic decomposition of H^1 (Section 7.6). This is based on the Whitney decomposition of open sets in $\mathbf{R}^n$, contained in the previous section. The action of singular integrals on the Hardy space H^1 is examined in Section 7.7. The final topic of this section is the duality between H^1 and BMO.

The final chapter of this book deals with A_p weights. The appearance of the A_p condition is motivated in Sections 8.1 and 8.2. Basic properties of A_p weights are discussed in Section 8.3, while weighted L^p estimates are obtained in Section 8.4. Other topics studied in this chapter are factorization (Section 8.5), reverse Hölder property (Section 8.6), and weighted estimates for singular integrals (Section 8.7).

The following chart concisely displays the interdependence of the chapters:

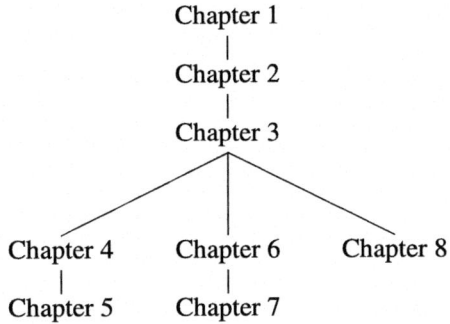

My intention is to maintain an errata website, which can be accessed via the link:

https://grafakos.missouri.edu

I am extremely thankful to all the people who have assisted me in the preparation of this textbook. First, I would like to express my gratitude to Jan Bouwe van den Berg, Georgios Dosidis, and Lenka Slavíková, who used a preliminary version of the book in the classroom and provided me with valuable feedback.

Many thanks to Nathan Bushman, Aniruddha Deshmukh, Xinyu Gao, Luigi Fontana, Steve Goldschmidt, Wyatt Gregory, Lixin He, Kristen Kaliski, Carlo Morpurgo, Felipe Noguera Rodriguez, Aritro Pathak, Dani Rozenbroek, Daniel Sinambela, Gregory Slease, Derek Sparrius, Arun Suresh, Konstantinos Tselios, and James Warta for suggesting improvements to the exposition and content.

I am deeply indebted to John Lucas for the multitude of corrections he provided me with. I must admit that I have never encountered a more careful reader in my career. His feedback was significant and greatly improved this book. A very special acknowledgment goes to Sean Douglas who drew almost all the pictures in the text and provided me with an abundance of corrections and suggestions. I am also very thankful to Springer's Mathematics executive editor Elizabeth Loew for her support and assistance throughout the preparation of this book. Finally, I gratefully acknowledge the support of the Simons Foundation Fellowship No. 819503 which provided me with a teaching release during the academic year 2021–2022 and precious time to work on the manuscript.

I wish to dedicate this book to all my graduate students and collaborators who have enriched my life throughout the years.

Columbia, Missouri Loukas Grafakos
February 2024

Contents

1 **Introductory Material** 1
   1.1 A Review of Lebesgue Spaces 1
   Exercises 8
   1.2 The Distribution Function and Weak L^p Spaces 10
   Exercises 14
   1.3 Real Interpolation 16
   Exercises 20
   1.4 The Hardy–Littlewood Maximal Operator 22
   Exercises 27
   1.5 The Lebesgue Differentiation Theorem 28
   Exercises 31
   1.6 Convolution 32
   Exercises 35
   1.7 Smoothness and Smooth Functions with Compact Support 37
   Exercises 41
   1.8 Schwartz Functions 42
   Exercises 44
   1.9 Approximate Identities 45
   Exercises 48

2 **Fourier Transforms, Tempered Distributions, Approximate Identities** 51
   2.1 The Fourier Transform on L^1 51
   Exercises 55
   2.2 Fourier Inversion 56
   Exercises 59
   2.3 The Fourier Transform on L^2 61
   Exercises 68
   2.4 Complex Interpolation and the Hausdorff–Young Inequality 69
   Exercises 74

	2.5 Approximate Identities and Almost Everywhere Convergence	75
	Exercises	81
	2.6 Tempered Distributions	82
	Exercises	89
	2.7 Basic Operations with Tempered Distributions	89
	Exercises	95
	2.8 L^p Fourier Multipliers	96
	Exercises	101
	2.9 Van der Corput Lemma	102
	Exercises	105
3	**Singular Integrals**	107
	3.1 The Hilbert Transform	107
	Exercises	111
	3.2 Homogeneous Singular Integrals and Riesz Transforms	112
	Exercises	118
	3.3 Calderón–Zygmund Singular Integrals	119
	Exercises	122
	3.4 L^2 Boundedness of Calderón–Zygmund Operators	123
	Exercises	127
	3.5 The Calderón–Zygmund Decomposition	128
	Exercises	132
	3.6 L^2 Boundedness Implies L^p Boundedness	133
	Exercises	138
	3.7 The Hilbert Transform and the Poisson Kernel	139
	Exercises	143
	3.8 Maximal Singular Integrals	145
	Exercises	148
4	**Vector-Valued Singular Integrals and Littlewood–Paley Theory**	151
	4.1 The Vector-Valued Calderón–Zygmund Theorem	151
	Exercises	157
	4.2 Applications of Vector-Valued Inequalities	158
	Exercises	162
	4.3 A Matrix-Valued Calderón–Zygmund Theorem and Its Applications	164
	Exercises	170
	4.4 Littlewood–Paley Theory	172
	Exercises	177
	4.5 Reverse Littlewood–Paley Inequalities	179
	Exercises	185
	4.6 Littlewood–Paley Theory of Product Type	187
	Exercises	193

5 Fractional Integrability or Differentiability and Multiplier Theorems ... 195
  5.1 Powers of the Laplacian and Riesz Potentials ... 195
  Exercises ... 199
  5.2 Bessel Potentials ... 201
  Exercises ... 206
  5.3 The Mikhlin and Hörmander Multiplier Theorems ... 207
  Exercises ... 213
  5.4 Sobolev Spaces ... 214
  Exercises ... 219
  5.5 Interpolation of Analytic Families of Operators ... 220
  Exercises ... 229
  5.6 The Calderón–Torchinsky Multiplier Theorem ... 230
  Exercises ... 238
  5.7 The Marcinkiewicz Multiplier Theorem ... 239
  Exercises ... 245

6 Bounded Mean Oscillation ... 247
  6.1 Basic Properties of Functions of Bounded Mean Oscillation ... 247
  Exercises ... 252
  6.2 The John–Nirenberg Theorem ... 254
  Exercises ... 259
  6.3 Dyadic Maximal Functions and Dyadic BMO ... 260
  Exercises ... 265
  6.4 The Sharp Maximal Function ... 266
  Exercises ... 271
  6.5 Interpolation Using BMO ... 272
  Exercises ... 277

7 Hardy Spaces ... 279
  7.1 Smoothness and Cancellation ... 279
  Exercises ... 283
  7.2 Definition of Hardy Spaces and Preliminary Estimates ... 283
  Exercises ... 288
  7.3 H^p Atoms ... 289
  Exercises ... 293
  7.4 Grand Maximal Function ... 293
  Exercises ... 302
  7.5 The Whitney Decomposition of Open Sets ... 303
  Exercises ... 310
  7.6 Atomic Decomposition of H^1 ... 311
  Exercises ... 317

7.7 Singular Integrals on the Hardy Space H^1 319
  Exercises 322
  7.8 Duality Between H^1 and *BMO* 323
  Exercises 328

8 Weighted Inequalities 329
  8.1 Appearance of Weights 329
  Exercises 333
  8.2 The A_p Condition 334
  Exercises 339
  8.3 Properties of A_p Weights 340
  Exercises 343
  8.4 Strong-Type A_p Estimates 345
  Exercises 349
  8.5 The Jones Factorization of Weights 349
  Exercises 353
  8.6 Reverse Hölder Property of A_p Weights 354
  Exercises 361
  8.7 Weighted Estimates for Singular Integral Operators 362
  Exercises 367

Historical Notes 369

Appendix A Orthogonal Matrices 371

Appendix B Subharmonic Functions 373

Appendix C Poisson Kernel on the Unit Strip 377

Appendix D Density for Subadditive Operators 383

Appendix E Transposes and Adjoints of Linear Operators ... 385

Appendix F Faà di Bruno Formula 387

Appendix G Besicovitch Covering Lemma 389

Glossary 393

References 397

Index 403

Chapter 1
Introductory Material

1.1 A Review of Lebesgue Spaces

In this section we review some basic facts concerning the Lebesgue spaces L^p. We state a variety of results concerning these spaces but only provide proofs selectively, as the material can be found in many textbooks on real variables

A *measure space* is a set X equipped with a σ-algebra $\mathscr{F}$ of subsets of X, called *measurable sets*, and a function μ from the measurable subsets to $[0,\infty]$, called a *positive measure*, that satisfies $\mu(\emptyset) = 0$ and

$$\mu\Big(\bigcup_{j=1}^{\infty} B_j\Big) = \sum_{j=1}^{\infty} \mu(B_j)$$

for any sequence B_j of pairwise disjoint elements of $\mathscr{F}$. In this situation we refer to the triple $(X, \mathscr{F}, \mu)$ as a measure space, although we often do not indicate the dependence on the σ-algebra $\mathscr{F}$ if this is evident. Measurable sets of measure zero are called *null sets*. Measure spaces X in this text are assumed to be *complete*, which means that subsets of null sets in $\mathscr{F}$ also belong to $\mathscr{F}$. A measure space X is called σ-*finite* if there is a sequence of sets $X_n \in \mathscr{F}$ with $\mu(X_n) < \infty$ such that

$$X = \bigcup_{n=1}^{\infty} X_n.$$

The *Borel sets* form the smallest σ-algebra that contains the open sets in a topological space. A *positive Borel measure* is a positive measure on the Borel sets.

A real-valued function f on a measure space $(X, \mathscr{F}, \mu)$ is called *measurable* if for all real numbers λ we have $\{x \in X : f(x) > \lambda\} \in \mathscr{F}$. A characteristic function χ_B of a subset B of X is measurable if and only if $B \in \mathscr{F}$. A complex-valued function g is measurable if and only if both its real part $\operatorname{Re} g$ and its imaginary part $\operatorname{Im} g$ are measurable functions. In measure spaces we identify two functions that are equal except on a set of measure zero; we indicate this fact by saying that the

© The Author(s), under exclusive license to Springer Nature Switzerland AG 2024
L. Grafakos, *Fundamentals of Fourier Analysis*, Graduate Texts in Mathematics 302,
https://doi.org/10.1007/978-3-031-56500-7_1

functions are equal a.e. A *simple function* is a finite linear combination of characteristic functions of sets in $\mathscr{F}$; that is $f = \sum_{j=1}^{N} c_j \chi_{B_j}$, $c_j \in \mathbf{C}$, $N \in \mathbf{Z}^+$. If $\mu(B_j) < \infty$ for all j, then f is called *finitely simple*. The sets $B_j \in \mathscr{F}$ can be chosen to be pairwise disjoint, in which case $\sum_{j=1}^{N} c_j \chi_{B_j}$ is the *standard representation* of f. Every nonnegative measurable function is the pointwise limit of an increasing sequence of simple functions; if the space is σ-finite, these simple functions can be chosen to be finitely simple.

Definition 1.1.1. For $0 < p < \infty$, we define the L^p norm (or quasi-norm if $p < 1$) of a complex-valued function f on a measure space (X, μ) by

$$\|f\|_{L^p(X,\mu)} = \left(\int_X |f(x)|^p \, d\mu(x) \right)^{\frac{1}{p}} = \left(\int_X |f|^p \, d\mu \right)^{\frac{1}{p}}. \tag{1.1.1}$$

For $p = \infty$ this norm is defined by

$$\|f\|_{L^\infty(X,\mu)} = \operatorname*{ess.sup}_{x \in X} |f(x)| = \inf \{ a > 0 : \mu(\{x \in X : |f(x)| > a\}) = 0 \}. \tag{1.1.2}$$

For any $0 < p \leq \infty$ we define $L^p(X, \mu)$ to be the space of all complex-valued measurable functions f with $\|f\|_{L^p} < \infty$. A function is called *integrable* if it lies in L^1.

The preceding definition implies that $\mu(\{x : |f(x)| > \|f\|_{L^\infty}\}) = 0$.

Definition 1.1.2. For any $1 < p < \infty$ we use the notation $p' = \frac{p}{p-1}$ to indicate the *dual exponent* of p. Moreover, we set $1' = \infty$ and $\infty' = 1$.

Note that the definition of the dual exponent implies that $p'' = p$ for all $p \in [1, \infty]$.

Theorem 1.1.3. *(Hölder's inequality) Let $f, g, f_1, \ldots, f_m$ ($m \geq 2$) be nonzero measurable functions on a measure space (X, μ). Then*
(a) If $1 \leq q \leq \infty$, then we have

$$\|fg\|_{L^1} \leq \|f\|_{L^q} \|g\|_{L^{q'}}. \tag{1.1.3}$$

(b) If $0 < p_1, \ldots, p_m, p \leq \infty$ and $1/p = 1/p_1 + \cdots + 1/p_m$, then

$$\|f_1 \cdots f_m\|_{L^p} \leq \|f_1\|_{L^{p_1}} \cdots \|f_m\|_{L^{p_m}}. \tag{1.1.4}$$

Thus $f_j \in L^{p_j}$ for all $j = 1, \ldots, m$ implies that $f_1 \cdots f_m \in L^p$.
(c) Suppose that no f_i is identically equal to zero a.e. and that $p < \infty$. If equality holds in (1.1.4) then $\left(|f_j|/\|f_j\|_{L^{p_j}} \right)^{p_j} = \left(|f_k|/\|f_k\|_{L^{p_k}} \right)^{p_k}$ a.e. for all $p_j, p_k < \infty$ and $|f_i| = \|f_i\|_{L^\infty}$ at almost all points for which $f_1 \cdots f_m$ is not zero, when $p_i = \infty$.

Proof. (a) If $q = \infty$, restrict the integral of fg over the set $\{|f| \leq \|f\|_{L^\infty}\}$ (as its complement has measure zero) to obtain the conclusion. When $q = 1$ reverse the roles of f and g. For $1 < q < \infty$ consider the function $\psi(t) = \frac{1}{q} t^q + \frac{1}{q'} b^{q'} - tb$ on $[0, \infty)$ for some fixed $b > 0$. This function is decreasing on $[0, b^{q'/q}]$, increasing on $[b^{q'/q}, \infty)$,

1.1 A Review of Lebesgue Spaces

and satisfies $\psi(t) = 0$ if and only if $t = b^{q'/q}$. Thus $\psi(t) \geq 0$. For $b = |g(x)|/\|g\|_{L^{q'}}$ the integral of $\psi(|f(x)|/\|f\|_{L^q})$ over X is nonnegative, yielding (1.1.3). Moreover, equality holds in (1.1.3) if and only if $(|f|/\|f\|_{L^q})^q = (|g|/\|g\|_{L^{q'}})^{q'}$ a.e.

(b) When $m = 2$ assume that $\max(p_1, p_2) < \infty$; otherwise the assertion is straightforward. Using (1.1.3) with $f = |f_1|^p$, $g = |f_2|^p$, $q = p_1/p$, and $q' = p_2/p$ we obtain (1.1.4) when $m = 2$. Additionally, a previous observation yields that equality holds in this case if and only if $\big(|f_1|/\|f_1\|_{L^{p_1}}\big)^{p_1} = \big(|f_2|/\|f_2\|_{L^{p_2}}\big)^{p_2}$ a.e. The case of $m \geq 3$ follows by induction, by applying the case for $m-1$ functions with exponents $1/q = 1/p_1 + \cdots + 1/p_{m-1}$ and the case $m = 2$ with exponents $1/p = 1/q + 1/p_m$ to the functions $f_1 \cdots f_{m-1}$ and f_m, respectively.

(c) As $p < \infty$ there is at least one index $p_i < \infty$. Let us reindex the p_j such that $p_j < \infty$ for $1 \leq j \leq \kappa$ and $p_j = \infty$ when $j \geq \kappa+1$; note $\kappa \geq 1$. We have

$$\|f_1 \cdots f_m\|_{L^p} \leq \|f_1 \cdots f_\kappa\|_{L^p} \prod_{j=\kappa+1}^m \|f_j\|_{L^\infty} \leq \prod_{j=1}^\kappa \|f_j\|_{L^{p_j}} \prod_{j=\kappa+1}^m \|f_j\|_{L^\infty}. \quad (1.1.5)$$

If equality holds in (1.1.4) then equality holds in the second inequality in (1.1.5), so we must have

$$\|f_1 \cdots f_\kappa\|_{L^p} = \|f_1\|_{L^{p_1}} \cdots \|f_\kappa\|_{L^{p_\kappa}}.$$

We may assume $\kappa \geq 2$ as the assertion about the f_j when $p_j < \infty$ is straightforward when $\kappa = 1$. For $i \in \{2,\ldots,\kappa\}$ set $1/q_i = 1/p_1 + 1/p_i$ and notice that equality must hold in the second inequality below (all norms are nonvanishing by assumption)

$$\Big\|\prod_{j=1}^\kappa f_j\Big\|_{L^p} \leq \|f_1 f_i\|_{L^{q_i}} \prod_{2 \leq j \neq i \leq \kappa} \|f_j\|_{L^{p_j}} \leq \prod_{j=1}^\kappa \|f_j\|_{L^{p_j}}.$$

This reduces matters to the case of equality when $m = 2$, from which we obtain $\big(|f_1|/\|f_1\|_{L^{p_1}}\big)^{p_1} = \big(|f_i|/\|f_i\|_{L^{p_i}}\big)^{p_i}$ a.e. for all $1 \leq i \leq \kappa$ by part (a). Now set

$$F = f_1 \cdots f_\kappa \quad \text{and} \quad H = \prod_{j=\kappa+1}^m (f_j \|f_j\|_{L^\infty}^{-1});$$

then H is well defined as no f_j vanishes identically. Since equality holds in (1.1.4), then equality must also hold in the first inequality in (1.1.5); this gives $\|FH\|_{L^p}^p = \|F\|_{L^p}^p$ and as $|H| \leq 1$ a.e. it follows that $|F|^p(1-|H|^p) = 0$ a.e. Consequently on the set $\{x \in X : f_1(x) \cdots f_m(x) \neq 0\}$ which is contained in $\{x \in X : F(x) \neq 0\}$, we must have $|f_i| = \|f_i\|_{L^\infty}$ a.e. when $i \geq \kappa+1$ (that is, when $p_i = \infty$). $\square$

The assertion in (c) may fail if $p = \infty$; indeed, the functions $f_1(x) = f_2(x) = e^{-|x|}$ on $\mathbf{R}$ satisfy $\|f_1 f_2\|_{L^\infty} = 1 = \|f_1\|_{L^\infty}\|f_2\|_{L^\infty}$ but $|f_1(x)| < 1 = \|f_1\|_{L^\infty}$ when $x \neq 0$.

Proposition 1.1.4. *Let (X,μ) be a measure space. For any q, let $\mathscr{D}^q$ be a dense subspace of $L^q(X,\mu)$. If $1 \leq p < \infty$ and $f \in L^p$ then*

$$\|f\|_{L^p} = \sup\Big\{\Big|\int_X fg\,d\mu\Big| : g \in \mathscr{D}^{p'} \text{ with } \|g\|_{L^{p'}} = 1\Big\}. \quad (1.1.6)$$

Also, if (X,μ) is a σ-finite measure space, then for any $f \in L^\infty$ we have

$$\|f\|_{L^\infty} = \sup\left\{\left|\int_X fg\,d\mu\right| : g \in \mathcal{D}^1 \text{ with } \|g\|_{L^1} = 1\right\}. \tag{1.1.7}$$

Proof. Indeed, the direction $\geq$ in (1.1.6) follows by Hölder's inequality. For the converse direction, if f is not the zero function in L^p, we denote its complex conjugate by $\overline{f}$. For $1 \leq p < \infty$ we choose

$$g = \frac{\overline{f}|f|^{p-2}}{\|f\|_{L^p}^{p-1}}$$

and we note that $\|g\|_{L^{p'}} = 1$ and that for this choice of g the integral in (1.1.6) is actually equal to $\|f\|_{L^p}$. Then (1.1.6) is obtained by the density of $\mathcal{D}^{p'}$ in $L^{p'}$.

Notice that the direction $\geq$ in (1.1.7) is immediate. For the converse direction, let $X = \bigcup_{m=1}^\infty X_m$, where X_m is an increasing sequence of measurable subsets of X with $\mu(X_m) < \infty$. Given a nonzero function f in $L^\infty(X,\mu)$ and $0 < \delta < \|f\|_{L^\infty}$, the set $B_\delta = \{|f| > \|f\|_{L^\infty} - \delta\}$ has positive μ measure (which could be infinite); then $0 < \mu(B_\delta \cap X_m) < \infty$ for some $m \in \mathbf{Z}^+$ and we define

$$g_\delta = \frac{\overline{f}}{|f|} \frac{\chi_{B_\delta \cap X_m}}{\mu(B_\delta \cap X_m)}$$

recalling that $|f|$ does not vanish on B_δ. Notice that $\|g_\delta\|_{L^1} = 1$ and

$$\sup_{\|g\|_{L^1}=1}\left|\int_X fg\,d\mu\right| \geq \left|\int_X f g_\delta\,d\mu\right| = \frac{1}{\mu(B_\delta \cap X_m)}\int_{B_\delta \cap X_m}|f|\,d\mu \geq \|f\|_{L^\infty} - \delta.$$

We now find $h_\delta \in \mathcal{D}^1$ such that $\|h_\delta - g_\delta\|_{L^1} \leq \delta$. Letting $\delta \to 0$, we obtain the direction $\leq$ in (1.1.7). $\square$

Proposition 1.1.5. *Let $0 < p \leq \infty$ and f, g be functions in $L^p = L^p(X,\mu)$.*
(a) *(Minkowski's inequality) If $1 \leq p \leq \infty$, then*

$$\|f+g\|_{L^p} \leq \|f\|_{L^p} + \|g\|_{L^p}. \tag{1.1.8}$$

(b) *When $0 < p < 1$ we have*

$$\|f+g\|_{L^p} \leq 2^{\frac{1-p}{p}}\left(\|f\|_{L^p} + \|g\|_{L^p}\right). \tag{1.1.9}$$

(c) *For all $0 < p \leq \infty$ we have*

$$\|f+g\|_{L^p}^{\min(1,p)} \leq \|f\|_{L^p}^{\min(1,p)} + \|g\|_{L^p}^{\min(1,p)}. \tag{1.1.10}$$

Proof. Part (a) is straightforward when $p = \infty$ and can be derived from Proposition 1.1.4 when $1 \leq p < \infty$. Part (b) is based on the inequalities $(a+b)^p \leq a^p + b^p$ and

1.1 A Review of Lebesgue Spaces

$(A+B)^{\frac{1}{p}} \leq 2^{\frac{1-p}{p}}(A^{\frac{1}{p}}+B^{\frac{1}{p}})$, which are valid for $a,b,A,B \geq 0$ when $0 < p < 1$; the second inequality is a consequence of Hölder's inequality with exponents $1/p$ and $1/(1-p)$ applied to the functions φ and ψ defined on $X = \{1,2\}$ (equipped with discrete measure) by $\varphi(1) = \varphi(2) = 1$ and $\psi(1) = A$, $\psi(2) = B$. Part (c) is contained in (a) for $p \geq 1$ and uses $(a+b)^p \leq a^p + b^p$ for $p < 1$. □

Additionally, $\|\lambda f\|_{L^p(X,\mu)} = |\lambda| \|f\|_{L^p(X,\mu)}$ and $\|f\|_{L^p(X,\mu)} = 0$ implies that $f = 0$ (μ-a.e.), and thus the L^p spaces are normed linear spaces for $1 \leq p \leq \infty$ and quasi-normed linear spaces when $p < 1$. Moreover, these spaces are complete.

Theorem 1.1.6. *(Fatou's lemma)* Let g, h, f_n, $n = 1, 2, \ldots$, be real-valued measurable functions on a measure space (X, μ).
(a) If $f_n \geq 0$ for all n then

$$\int_X \liminf_{n \to \infty} f_n \, d\mu \leq \liminf_{n \to \infty} \int_X f_n \, d\mu. \tag{1.1.11}$$

(b) *The conclusion in (1.1.11) also holds if $f_n \geq g$ a.e. for all n where $\int_X g\, d\mu > -\infty$.*
(c) *Suppose that if $f_n \leq h$ a.e. for all $n = 1, 2, \ldots$ with $\int_X h\, d\mu < +\infty$. Then*

$$\limsup_{n \to \infty} \int_X f_n \, d\mu \leq \int_X \limsup_{n \to \infty} f_n \, d\mu. \tag{1.1.12}$$

Proof. The cases $\int_X g\, d\mu = \infty$ and $\int_X h\, d\mu = -\infty$ are trivial. So we may assume that

$$-\infty < \int_X g\, d\mu \quad \text{and} \quad \int_X h\, d\mu < \infty.$$

Then parts (b) and (c) follow by applying (a) to $f_n - g$ and $h - f_n$, respectively. □

Theorem 1.1.7. *(Lebesgue monotone convergence theorem[1])* Let f be a measurable function on a measure space (X, μ) and let $\{f_n\}_{n=1}^{\infty}$ be a sequence of real-valued measurable functions that converges to f.
(a) *If $0 \leq f_1 \leq f_2 \leq f_3 \leq \cdots$ a.e., then*

$$\int_X f_n\, d\mu \uparrow \int_X f\, d\mu. \tag{1.1.13}$$

(b) *The conclusion in (1.1.13) holds if $f_1 \leq f_2 \leq f_3 \leq \cdots$ a.e. and $\int_X f_1\, d\mu > -\infty$.*
(c) *If $f_1 \geq f_2 \geq f_3 \geq \cdots$ a.e. and $\int_X f_1\, d\mu < +\infty$, then*

$$\int_X f_n\, d\mu \downarrow \int_X f\, d\mu. \tag{1.1.14}$$

Proof. Part (b) follows by applying part (a) to $f_n - f_1$ when $\int_X f_1\, d\mu < +\infty$; the case $\int_X f_1\, d\mu = \infty$ is trivial. Likewise, part (c) follows by applying part (a) to $f_1 - f_n$ when $\int_X f_1\, d\mu > -\infty$, while (1.1.14) is immediate when $\int_X f_1\, d\mu = -\infty$. □

---
[1] Often abbreviated as LMCT.

Theorem 1.1.8. *Let f, f_n, $n = 1, 2, \ldots$, be complex-valued measurable functions on a measure space (X, μ) such that $f_n \to f$ a.e. and g, g_n, $n = 1, 2, \ldots$, are nonnegative integrable functions on X.*
(a) (***Lebesgue dominated convergence theorem***[2]) *If $|f_n| \leq g$ a.e. for all n, then*

$$\lim_{n \to \infty} \int_X |f_n - f| \, d\mu = 0. \tag{1.1.15}$$

(b) (***Generalized dominated convergence theorem***) *If $|f_n| \leq g_n$ for all n, $g_n \to g$ a.e. and $\int_X g_n \, d\mu \to \int_X g \, d\mu$, then (1.1.15) holds.*

Proof. To prove part (b) consider the sequence $|f_n - f| - g_n$ which satisfies $|f_n - f| - g_n \leq g$ a.e. As $0 \leq \int_X g \, d\mu < \infty$, applying part (c) of Fatou's lemma we obtain

$$\limsup_{n \to \infty} \int_X |f_n - f| - g_n \, d\mu \leq \int_X \limsup_{n \to \infty} (|f_n - f| - g_n) \, d\mu = -\int_X g \, d\mu. \tag{1.1.16}$$

But as the limit of $\int_X g_n \, d\mu$ exists, we have

$$\limsup_{n \to \infty} \int_X |f_n - f| - g_n \, d\mu = \limsup_{n \to \infty} \int_X |f_n - f| \, d\mu - \int_X g \, d\mu.$$

Adding $\int_X g \, d\mu$ to both sides in (1.1.16) we obtain $\limsup_{n \to \infty} \int_X |f_n - f| \, d\mu = 0$. □

Theorem 1.1.9. *Let $0 < p \leq \infty$ and let (X, μ) be a measure space. Every Cauchy sequence in $L^p(X, \mu)$ is convergent in L^p and has a subsequence that converges a.e.*

Proof. Let $\{f_n\}_{n=1}^{\infty}$ be a Cauchy sequence in $L^p(X, \mu)$. We can find a sequence of natural numbers $n_1 < n_2 < \cdots$ such that $\|f_{n_{k+1}} - f_{n_k}\|_{L^p}^{\min(1,p)} < 2^{-k}$ for all $k = 1, 2, \ldots$. By the Lebesgue monotone convergence theorem the function

$$G = \sum_{j=1}^{\infty} |f_{n_{j+1}} - f_{n_j}| = \lim_{k \to \infty} \sum_{j=1}^{k-1} |f_{n_{j+1}} - f_{n_j}|$$

lies in L^p and thus it is finite a.e. This implies that the series defining G converges a.e. and so there is a measurable function f on X such that

$$f_{n_k} - f_{n_1} = \sum_{j=1}^{k-1} (f_{n_{j+1}} - f_{n_j}) \to f - f_{n_1} \quad \text{a.e.}$$

as $k \to \infty$. But $|f - f_{n_1}| \leq \sup_k |f_{n_k} - f_{n_1}| \leq G$, and as $G \in L^p$, we conclude that $f - f_{n_1} \in L^p$, hence $f \in L^p$. Now $|f_{n_k} - f| \leq G + |f - f_{n_1}| \in L^p$; thus the LDCT yields $\|f_{n_k} - f\|_{L^p} \to 0$. To complete the proof, given $\varepsilon > 0$, find N such that for

$$m, n \geq N \implies \|f_n - f_m\|_{L^p}^{\min(1,p)} < \frac{\varepsilon}{2}.$$

---

[2] Often abbreviated as LDCT.

1.1 A Review of Lebesgue Spaces

Also, as $\|f_{n_k} - f\|_{L^p} \to 0$, we can find k_0 such that

$$k \geq k_0 \implies \|f_{n_k} - f\|_{L^p}^{\min(1,p)} < \frac{\varepsilon}{2}.$$

We pick $k_1 \geq k_0$ such that $n_{k_1} \geq N$. Combining these implications, we obtain

$$n \geq N \implies \|f_n - f\|_{L^p}^{\min(1,p)} \leq \|f_n - f_{n_{k_1}}\|_{L^p}^{\min(1,p)} + \|f_{n_{k_1}} - f\|_{L^p}^{\min(1,p)} < \frac{\varepsilon}{2} + \frac{\varepsilon}{2}.$$

These facts show that $f_{n_k} \to f$ a.e. and that $f_n \to f$ in L^p. □

Theorem 1.1.10. *(Riesz representation theorem) Let (X, μ) be a measure space.*
(a) Given a nonzero complex-valued bounded linear functional T on $L^p(X)$, where $1 < p < \infty$, there is a unique (a.e.) function $h \in L^{p'}$ such that

$$T(f) = \int_X f h \, d\mu \qquad \text{for all } f \in L^p(X),$$

and moreover the norm of T on $L^p(X)$ equals $\|h\|_{L^{p'}}$.
(b) The assertion in part (a) is also valid if $p = 1$ and (X, μ) is σ-finite.

Theorem 1.1.11. *Let (X, μ) and (Y, ν) be two σ-finite measure spaces and let F be a measurable function on $X \times Y$, equipped with product measure.*
(a) (Tonelli's theorem) We have

$$\int_X \left(\int_Y |F(x,y)| \, d\nu(y) \right) d\mu(x) = \int_Y \left(\int_X |F(x,y)| \, d\mu(x) \right) d\nu(y). \qquad (1.1.17)$$

(b) (Fubini's theorem) If either expression in (1.1.17) is finite, then

$$\int_X \left(\int_Y F(x,y) \, d\nu(y) \right) d\mu(x) = \int_Y \left(\int_X F(x,y) \, d\mu(x) \right) d\nu(y). \qquad (1.1.18)$$

Theorem 1.1.12. *(Minkowski integral inequality) Let (X, μ) and (Y, ν) be two σ-finite measure spaces and let $1 \leq p \leq \infty$. Then for every nonnegative measurable function F on the product space $(X, \mu) \times (Y, \nu)$ we have*

$$\left[\int_Y \left(\int_X F(x,y) \, d\mu(x) \right)^p d\nu(y) \right]^{\frac{1}{p}} \leq \int_X \left[\int_Y F(x,y)^p \, d\nu(y) \right]^{\frac{1}{p}} d\mu(x),$$

with a suitable modification when $p = \infty$.

In this text, Lebesgue measure is denoted by dx, dy, etc. depending on the variable of integration. *Lebesgue measurable* subsets of $\mathbf{R}^n$ are simply called *measurable*. The *Lebesgue measure* of a Lebesgue measurable subset E of the real line is denoted by $|E| = \int_E 1 \, dx$. For a measurable subset E of $\mathbf{R}^n$, $L^p(E)$ denotes $L^p(E, |\cdot|)$, i.e., the set E equipped with the restriction of the Lebesgue measure on it. Subsets of $\mathbf{R}^n$ are assumed to be equipped with Lebesgue measure, unless indicated otherwise. Finite linear combinations of characteristic functions of cubes with sides parallel to the axes on $\mathbf{R}^n$ are called *step functions*. It is known that step functions are dense in $L^p(\mathbf{R}^n)$ for every $p < \infty$.

A natural measure on the space $\mathbf{Z}$ is *counting measure* ν defined on subsets A of $\mathbf{Z}$ by $\nu(A) =$ cardinality of A. For $0 < p \le \infty$, $\ell^p(\mathbf{Z})$ denotes the space $L^p(\mathbf{Z}, \nu)$.

Example 1.1.13. The spaces $L^p([0,1])$ decrease and the spaces $\ell^p(\mathbf{Z})$ increase as p increases. Indeed for $0 < p < q \le \infty$, by Hölder's inequality, we have

$$\|f\|_{L^p([0,1])} \le |[0,1]|^{\frac{1}{p}-\frac{1}{q}} \|f\|_{L^q([0,1])} = \|f\|_{L^q([0,1])},$$

so $L^q([0,1])$ is contained in $L^p([0,1])$.

Now $\|\{b(k)\}_k\|_{\ell^p(\mathbf{Z})} = \left(\int_{\mathbf{Z}} |b(k)|^p d\nu\right)^{\frac{1}{p}} = \left(\sum_{k\in\mathbf{Z}} |b(k)|^p\right)^{\frac{1}{p}}$; hence

$$\|\{b(k)\}_k\|_{\ell^q} = \left(\sum_{k\in\mathbf{Z}} |b(k)|^{q-p+p}\right)^{\frac{1}{q}} \le \|\{b(k)\}_k\|_{\ell^\infty}^{\frac{q-p}{q}} \|\{b(k)\}_k\|_{\ell^p}^{\frac{p}{q}} \le \|\{b(k)\}_k\|_{\ell^p}$$

as clearly $\|\{b(k)\}_k\|_{\ell^\infty} \le \|\{b(k)\}_k\|_{\ell^p(\mathbf{Z})}$. This shows that $\ell^p(\mathbf{Z})$ embeds in $\ell^q(\mathbf{Z})$.

Exercises

1.1.1. Let $\Phi : [0,\infty) \to [0,\infty)$ be a continuous increasing function with $\Phi(0) = 0$ such that for some $K > 0$ and all $t, s \in [0,\infty)$ we have $\Phi(t+s) \le K(\Phi(t) + \Phi(s))$ (*quasi-subadditivity*). Let $f, f_n, n = 1, 2, \ldots$, be measurable functions on (X, μ) that satisfy $f_n \to f$ a.e. and $\int_X \Phi(|f|) d\mu < \infty$. Prove that

$$\int_X \Phi(|f_n - f|) d\mu \to 0 \iff \int_X \Phi(|f_n|) d\mu \to \int_X \Phi(|f|) d\mu$$

as $n \to \infty$. Apply this result to $\Phi(t) = |t|^p$ and $\Phi(t) = |t|^p \ln(|t|+1)^q$, $0 < p, q < \infty$. [*Hint:* Apply Theorem 1.1.8 (b) with $g_n = K(\Phi(|f_n|) + \Phi(|f|))$.]

1.1.2. Suppose that $1 \le p < \infty$ and f is a measurable function on a measure space (X, μ) such that fg lies in L^1 for any $g \in L^{p'}$. Prove that $f \in L^p$. Derive the same conclusion if $p = \infty$ and (X, μ) is σ-finite. [*Hint:* Show by contradiction that there is a positive finite constant C such that $\|fg\|_{L^1} \le C\|g\|_{L^{p'}}$ for all $g \in L^{p'}$. Then use Proposition 1.1.4.]

1.1.3. Let $0 < p_0 \le \infty$. Show that the function $h(t) = |t|^{-\frac{1}{p_0}} \left(1 + |\log|t||\right)^{-\frac{1}{p_0} - 1}$ lies in $L^p(\mathbf{R})$ if and only if $p = p_0$.

1.1.4. Let f_j, $1 \le j \le n$, be real-valued measurable functions on a measure space (X, μ) and suppose that

1.1 A Review of Lebesgue Spaces

$$\|f_1\|_{L^p} = \|f_2\|_{L^p} = \cdots = \|f_n\|_{L^p} = \left\|\frac{f_1 + \cdots + f_n}{n}\right\|_{L^p}$$

for some $1 < p < \infty$. Prove that $f_1 = \cdots = f_n \geq 0$ a.e. or $f_1 = \cdots = f_n \leq 0$ a.e.
[*Hint:* Using the case of equality in Hölder's inequality, first show that if $a_i \in \mathbf{R}$ and $\frac{1}{n}\sum_{i=1}^n |a_i|^p = |\frac{1}{n}\sum_{i=1}^n a_i|^p$, then $a_i = a_j \geq 0$ for all i,j or $a_i = a_j \leq 0$ for all i,j.]

1.1.5. (Jensen's inequality) A differentiable real-valued function φ is called *convex* on an open interval I if and only if the function lies above all of its tangents, i.e., $\varphi(t) \geq \varphi(t_0) + \varphi'(t_0)(t - t_0)$ for all $t, t_0 \in I$. Suppose that (X, μ) is a measure space with $0 < \mu(X) < \infty$, g is a real-valued function on X whose range lies in an open interval I, and φ is a convex function on I. Prove that

$$\varphi\left(\frac{1}{\mu(X)}\int_X g\, d\mu\right) \leq \frac{1}{\mu(X)}\int_X \varphi(g)\, d\mu.$$

[*Hint:* Start with $\varphi(t) \geq \varphi(t_0) + \varphi'(t_0)(t - t_0)$ taking $t = g(x)$ and $t_0 = \frac{1}{\mu(X)}\int_X g\, d\mu$.]

1.1.6. Let $0 < p_0 \leq p \leq p_1 \leq \infty$ and let $\frac{1}{p} = \frac{1-\theta}{p_0} + \frac{\theta}{p_1}$ for some $\theta \in [0,1]$. Prove

$$\|f\|_{L^p} \leq \|f\|_{L^{p_0}}^{1-\theta} \|f\|_{L^{p_1}}^{\theta}.$$

[*Hint:* For $p < \infty$ and $\theta \in (0,1)$ use Hölder's inequality with exponents $\frac{p_0}{(1-\theta)p}, \frac{p_1}{\theta p}$.]

1.1.7. Let (X, μ) be a measure space with $\mu(X) < \infty$. Show that for any measurable function f on X we have

$$\lim_{p \to \infty} \|f\|_{L^p} = \|f\|_{L^\infty}.$$

[*Hint:* We may assume that f is nonzero and lies in L^{p_0} for some $p_0 < \infty$. One direction is a consequence of the inequality $\|f\|_{L^p} \leq \|f\|_{L^{p_0}}^{\frac{p_0}{p}} \|f\|_{L^\infty}^{1 - \frac{p_0}{p}}$. For $0 < \delta < \|f\|_{L^\infty}$ use that

$$\|f\|_{L^p} \geq (\|f\|_{L^\infty} - \delta)\mu(\{|f| > \|f\|_{L^\infty} - \delta\})^{\frac{1}{p}}$$

for the other inequality.]

1.1.8. Let (X, μ) be a measure space with $0 < \mu(X) < \infty$ and suppose f is measurable function on X which satisfies $0 < \|f\|_{L^\infty(X)} < \infty$. Prove that

$$\lim_{p \to \infty} \frac{\|f\|_{L^{p+1}}^{p+1}}{\|f\|_{L^p}^p} = \|f\|_{L^\infty} \quad \text{and} \quad \lim_{p \to \infty} \frac{\|f\|_{L^{2p}}^2}{\|f\|_{L^p}} = \|f\|_{L^\infty}.$$

[*Hint:* Show that both ratios lie between $\mu(X)^{-\frac{1}{p}}\|f\|_{L^p}$ and $\|f\|_{L^\infty}$.]

1.1.9. Let g be a measurable function on a measure space (X,μ) that satisfies:
$$\int_{|g|<n} \frac{|g|^2}{\frac{1}{n}+|g|}\,d\mu \leq 1$$
for all $n = 1, 2, \ldots$. Prove that
$$\lim_{n\to\infty} \int_{|g|\geq n} \frac{|g|^2}{\frac{1}{n}+|g|}\,d\mu = 0.$$

1.1.10. For $x,t \geq 0$ let $\varphi(t,x) = t\sin(x) + \cos(x)$. Using that $1/x = \int_0^\infty e^{-xt}\,dt$ for $x > 0$, show
$$\int_{1/n}^n \frac{\sin x}{x}\,dx = \int_0^\infty \frac{e^{-t/n}\varphi(t,1/n) - e^{-nt}\varphi(t,n)}{t^2+1}\,dt.$$
Prove that $e^{-tx}|\varphi(t,x)| \leq 2$ for $x,t \geq 0$ and conclude via the LDCT that
$$\int_0^\infty \frac{\sin x}{x}\,dx = \frac{\pi}{2}.$$

1.1.11. Let f be a continuous and integrable function on $[0,\infty)$. Show that
$$\lim_{\varepsilon \to 0} \frac{1}{\varepsilon} \int_0^\infty f(t)e^{-t/\varepsilon}\,dt = f(0).$$

1.1.12. Let $0 < p < \infty$ and let f be a measurable function on a measure space (X,μ). If
$$E_k(f) = \{x \in X : |f(x)| > 2^k\},$$
show that
$$(1 - 2^{-p}) \sum_{k \in \mathbf{Z}} 2^{kp}\mu(E_k(f)) \leq \|f\|_{L^p}^p \leq 2^p \sum_{k \in \mathbf{Z}} 2^{kp}\mu(E_k(f)).$$

1.1.13. Let f_j be functions on $L^p(X,\mu)$ for some $0 < p < 1$ and let $s > 0$. Show that for any $0 < \varepsilon < s$ we have
$$\Big\|\sum_{j=0}^\infty 2^{-js} f_j\Big\|_{L^p} \leq \Big(1 - 2^{-\frac{\varepsilon p}{1-p}}\Big)^{\frac{p-1}{p}} \sum_{j=0}^\infty 2^{-j(s-\varepsilon)} \|f_j\|_{L^p}.$$

1.2 The Distribution Function and Weak L^p Spaces

The spaces studied in this section provide an alternative way to quantitatively measure a function. A good tool to achieve this purpose is the distribution function.

1.2 The Distribution Function and Weak L^p Spaces

Definition 1.2.1. The *distribution function* of a measurable function f on a measure space (X, μ) is the function D_f defined on $[0, \infty)$ as follows:

$$D_f(\lambda) = \mu(\{x \in X : |f(x)| > \lambda\}). \tag{1.2.1}$$

As λ increases the set in (1.2.1) decreases, so the distribution function D_f is decreasing.

Example 1.2.2. In Figure 1.1 the function $f : [-\pi, 0] \to [0, 1]$ given by $f(x) = |\sin(x)|$ is plotted on $[-\pi, 0]$ and its distribution function

$$D_f(\lambda) = \pi - 2\arcsin(\lambda)$$

is plotted on $[0, 1]$. Notice that $D_f(0) = \pi$ is equal to the measure of the support of f. Also note that $\max(f) = 1$ and $D_f(1) = 0$.

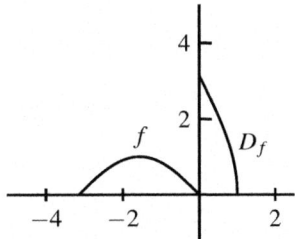

Fig. 1.1 The function f of Example 1.2.2 and its distribution function D_f.

The distribution function D_f can be used to precisely evaluate the L^p norm of a function. The following proposition contains the relevant identity.

Proposition 1.2.3. *Let (X, μ) be a σ-finite measure space. For any increasing, continuously differentiable function φ on $[0, \infty)$ with $\varphi(0) = 0$ and every measurable function f on X with $\varphi(|f|)$ integrable on X, we have*

$$\int_X \varphi(|f|) \, d\mu = \int_0^\infty \varphi'(\lambda) D_f(\lambda) \, d\lambda. \tag{1.2.2}$$

Proof. As X is a σ-finite measure space and $\varphi' \geq 0$, we can apply Tonelli's theorem. We have

$$\int_0^\infty \varphi'(\lambda) D_f(\lambda) \, d\lambda = \int_0^\infty \varphi'(\lambda) \int_X \chi_{\{x: |f(x)| > \lambda\}} \, d\mu \, d\lambda$$

$$= \int_X \int_0^{|f(x)|} \varphi'(\lambda) \, d\lambda \, d\mu$$

$$= \int_X \varphi(|f(x)|) - \varphi(0) \, d\mu.$$

This proves (1.2.2) since $\varphi(0) = 0$. □

Corollary 1.2.4. *Given $0 < p < \infty$ and a measurable function f on a σ-finite measure space (X, μ), we have*

$$\|f\|_{L^p}^p = p \int_0^\infty \lambda^{p-1} D_f(\lambda) \, d\lambda. \tag{1.2.3}$$

Definition 1.2.5. Given a measurable function f on a measure space (X,μ), we define the weak L^p (quasi-)norm of f as follows:

$$\|f\|_{L^{p,\infty}} = \sup_{\lambda>0} \lambda \, D_f(\lambda)^{\frac{1}{p}}. \tag{1.2.4}$$

For $0 < p < \infty$, the space *weak $L^p(X,\mu)$*, denoted by $L^{p,\infty}(X,\mu)$, is defined as the set of all μ-measurable functions f such that $\|f\|_{L^{p,\infty}} < \infty$. We also define $L^{\infty,\infty}(X,\mu) = L^\infty(X,\mu)$. The notation $L^{p,\infty}(\mathbf{R}^n)$ is reserved for $L^{p,\infty}(\mathbf{R}^n, |\cdot|)$.

We now explain why $\|f\|_{L^{p,\infty}}$ is a quasi-norm.[3] First one notes that

$$\|f\|_{L^{p,\infty}} = 0 \Rightarrow \mu(\{x \in X : |f(x)| > \lambda\}) = 0 \;\forall\, \lambda > 0 \Rightarrow f = 0 \quad \mu\text{-a.e.} \tag{1.2.5}$$

Also, we have

$$D_{f+g}(\lambda) \le D_f\!\left(\tfrac{\lambda}{2}\right) + D_g\!\left(\tfrac{\lambda}{2}\right) \;\Longrightarrow\; D_{f+g}(\lambda)^{\frac{1}{p}} \le c_p\Big[D_f\!\left(\tfrac{\lambda}{2}\right)^{\frac{1}{p}} + D_g\!\left(\tfrac{\lambda}{2}\right)^{\frac{1}{p}}\Big]$$

where $c_p = \max(1, 2^{1/p-1})$. This implies

$$\|f+g\|_{L^{p,\infty}} \le 2c_p\!\left(\|f\|_{L^{p,\infty}} + \|g\|_{L^{p,\infty}}\right). \tag{1.2.6}$$

Combining (1.2.5) with (1.2.6) and the simple fact that for any $c \in \mathbf{C}$ we have $\|cf\|_{L^{p,\infty}} = |c|\,\|f\|_{L^{p,\infty}}$, we obtain that $L^{p,\infty}(X,\mu)$ is a *quasi-normed space*.

The weak L^p spaces are larger than the usual L^p spaces. We have the following:

Proposition 1.2.6. *For any $0 < p < \infty$ and any f in $L^p(X,\mu)$ we have*

$$\|f\|_{L^{p,\infty}} \le \|f\|_{L^p}.$$

Hence the embedding $L^p(X,\mu) \subseteq L^{p,\infty}(X,\mu)$ holds.

Proof. Using Chebyshev's inequality:

$$\lambda^p D_f(\lambda) \le \int_{\{x:|f(x)|>\lambda\}} |f(x)|^p\, d\mu \le \|f\|_{L^p}^p$$

we obtain $\|f\|_{L^{p,\infty}} \le \|f\|_{L^p}$. $\square$

The inclusion $L^p \subseteq L^{p,\infty}$ is strict as the function $h(x) = |x|^{-n/p}$ obviously does not lie in $L^p(\mathbf{R}^n)$ but belongs to $L^{p,\infty}(\mathbf{R}^n)$. To see this, we note that

$$\big|\{x \in \mathbf{R}^n : |x|^{-\frac{n}{p}} > \lambda\}\big| = \big|\{x \in \mathbf{R}^n : |x| < \lambda^{-\frac{p}{n}}\}\big| = |B(0,1)|\lambda^{-p},$$

which implies that $\|h\|_{L^{p,\infty}} = |B(0,1)|^{1/p} < \infty$. Here $B(0,1) = \{x \in \mathbf{R}^n : |x| < 1\}$.

---

[3] A quasi-norm on a vector space Z is a nonnegative function $\|\cdot\|_Z$ with the properties of a norm, but with a relaxed triangle inequality $\|z+z'\|_Z \le B(\|z\|_Z + \|z'\|_Z)$, for some $B \ge 1$ and all $z, z' \in Z$.

1.2 The Distribution Function and Weak L^p Spaces

Remark 1.2.7. Note that for $0 < r < \infty$, $\||f|^r\|_{L^p} = \|f\|_{L^{pr}}^r$. One can verify that the same property holds for the weak L^p quasi-norm, i.e., $\||f|^r\|_{L^{p,\infty}} = \|f\|_{L^{pr,\infty}}^r$.

Definition 1.2.8. The space $L^1_{\text{loc}}(\mathbf{R}^n, |\cdot|)$ of *locally integrable* functions is the set of all Lebesgue-measurable functions f on $\mathbf{R}^n$ that satisfy

$$\int_K |f(x)|\,dx < \infty \tag{1.2.7}$$

for any compact subset K of $\mathbf{R}^n$.

Example 1.2.9. (a) $L^p(\mathbf{R}^n)$ is contained in $L^1_{\text{loc}}(\mathbf{R}^n)$ for $1 \leq p \leq \infty$ (Theorem 1.1.3).
(b) For $0 < p < 1$, $L^p(\mathbf{R}^n) \setminus L^1_{\text{loc}}(\mathbf{R}^n) \neq \emptyset$; i.e., it contains $|x|^{-\frac{n}{p}}\left(\log\frac{1}{|x|}\right)^{-\frac{2}{p}}\chi_{|x|\leq 1}$.
(c) There exist functions in $L^{1,\infty}(\mathbf{R}^n) \setminus L^1_{\text{loc}}(\mathbf{R}^n)$, such as $|x|^{-n}$.
(d) The function $e^{e^{|x|}}$ lies in $L^1_{\text{loc}}(\mathbf{R}^n)$ but not in $L^p(\mathbf{R}^n)$ for any $p > 0$.

Theorem 1.2.10. *For a measurable function f on a σ-finite measure space (X,μ) define*

$$\||f\||_{L^{p,\infty}} = \sup_{0 < \mu(E) < \infty} \mu(E)^{-1+\frac{1}{p}} \int_E |f|\,d\mu.$$

Let $1 < p < \infty$. Then $\||\cdot\||_{L^{p,\infty}}$ is a norm on $L^{p,\infty}$ that satisfies

$$\|f\|_{L^{p,\infty}} \leq \||f\||_{L^{p,\infty}} \leq \frac{p}{p-1}\|f\|_{L^{p,\infty}}.$$

Proof. Let $E \subseteq X$ such that $0 < \mu(E) < \infty$ and let $f \in L^{p,\infty}(X,\mu)$. By Proposition 1.2.3 we write

$$\begin{aligned}\int_E |f|\,d\mu &= \int_0^\infty \mu(\{|f| > \lambda\} \cap E)\,d\lambda \\ &\leq \int_0^\infty \min\left(\mu(E), \frac{\|f\|_{L^{p,\infty}}^p}{\lambda^p}\right)d\lambda \\ &= \frac{p}{p-1}\mu(E)^{1-\frac{1}{p}}\|f\|_{L^{p,\infty}},\end{aligned} \tag{1.2.8}$$

which follows by splitting the integral at $\lambda = \|f\|_{L^{p,\infty}}\mu(E)^{-\frac{1}{p}}$. Therefore,

$$\||f\||_{L^{p,\infty}} \leq \frac{p}{p-1}\|f\|_{L^{p,\infty}}.$$

As X is σ-finite, we can write $X = \bigcup_{N=1}^\infty X_N$ with $X_1 \subseteq X_2 \subseteq \cdots$ and $\mu(X_N) < \infty$. For $\lambda > 0$ let $E_\lambda = \{|f| > \lambda\}$ and $E_\lambda^N = \{|f| > \lambda\} \cap X_N$. Then we clearly have $\mu(E_\lambda^N) \leq \mu(X_N) < \infty$ and $\int_{E_\lambda^N} |f|\,d\mu \geq \lambda\mu(E_\lambda^N)$. Let us fix $\lambda > 0$ and $N \in \mathbf{Z}^+$. If $\mu(E_\lambda^N) > 0$, then

$$\||f\||_{L^{p,\infty}} = \sup_{0 < \mu(E) < \infty} \mu(E)^{-1+\frac{1}{p}} \int_E |f|\,d\mu \geq \mu(E_\lambda^N)^{-1+\frac{1}{p}}\lambda\mu(E_\lambda^N) = \lambda\mu(E_\lambda^N)^{\frac{1}{p}},$$

but this inequality is also trivially valid when $\mu(E_\lambda^N) = 0$. Letting $N \to \infty$ we deduce

$$|||f|||_{L^{p,\infty}} \geq \lambda \mu(E_\lambda)^{\frac{1}{p}}$$

for every $\lambda > 0$, so taking the supremum over $\lambda > 0$ we obtain $|||f|||_{L^{p,\infty}} \geq \|f\|_{L^{p,\infty}}$. The fact that $||| \cdot |||_{L^{p,\infty}}$ is a norm on $L^{p,\infty}$ is straightforward and is omitted. $\square$

Taking E to be a compact subset of $\mathbf{R}^n$ in (1.2.8), we obtain the following.

Corollary 1.2.11. *For $1 < p < \infty$ we have that $L^{p,\infty}(\mathbf{R}^n) \subseteq L^1_{\text{loc}}(\mathbf{R}^n)$.*

Proposition 1.2.12. *For $0 < p < \infty$, $L^{p,\infty}(\mathbf{R}^n)$ is a complete quasi-normed space.*

Proof. Let $\{f_k\}_k$ be a Cauchy sequence in $L^{p,\infty}$. Then $\{f_k\}_k$ is Cauchy in $L^q(E_N)$, where $E_N = B(0,N)$, $N = 1, 2, \ldots$, and $q < p$, by Exercise 1.2.6 (a). By Theorem 1.1.9 for each $N = 1, 2, \ldots$ there is a subsequence that converges a.e. on E_N. As $\mathbf{R}^n = \cup_{N=1}^\infty E_N$, the diagonal subsequence $\{f_{n_k}\}_k$ converges a.e. to a measurable function f on $\mathbf{R}^n$. Given $\varepsilon > 0$ there is an $l_0 \in \mathbf{Z}^+$ such that when $l, k \geq l_0$ we have $\|f_{n_l} - f_{n_k}\|_{L^{p,\infty}} \leq \varepsilon$. As $f_{n_l} - f_{n_k} \to f - f_{n_k}$ a.e. as $l \to \infty$, it follows that $|f - f_{n_k}| = \liminf_{l \to \infty} |f_{n_l} - f_{n_k}|$ a.e. and also $|f| = \liminf_{l \to \infty} |f_{n_l}|$ a.e. Exercise 1.2.4 gives $D_f(\lambda) \leq \liminf_{l \to \infty} D_{f_{n_l}}(\lambda)$ and $D_{f - f_{n_k}}(\lambda) \leq \liminf_{l \to \infty} D_{f_{n_l} - f_{n_k}}(\lambda)$ for any $\lambda > 0$. These yield

$$\sup_{\lambda > 0} \lambda D_f(\lambda)^{\frac{1}{p}} \leq \sup_{\lambda > 0} \liminf_{l \to \infty} \lambda D_{f_{n_l}}(\lambda)^{\frac{1}{p}} \leq \liminf_{l \to \infty} \sup_{\lambda > 0} \lambda D_{f_{n_l}}(\lambda)^{\frac{1}{p}} < \infty,$$

as every Cauchy sequence is bounded. Thus f lies in $L^{p,\infty}(\mathbf{R}^n)$. Also, for $k \geq l_0$,

$$\sup_{\lambda > 0} \lambda D_{f - f_{n_k}}(\lambda)^{\frac{1}{p}} \leq \sup_{\lambda > 0} \liminf_{l \to \infty} \lambda D_{f_{n_l} - f_{n_k}}(\lambda)^{\frac{1}{p}} \leq \liminf_{l \to \infty} \sup_{\lambda > 0} \lambda D_{f_{n_l} - f_{n_k}}(\lambda)^{\frac{1}{p}} \leq \varepsilon,$$

hence $f_{n_k} \to f$ in $L^{p,\infty}$ as $k \to \infty$. This implies $f_k \to f$ in $L^{p,\infty}$, as in quasi-normed spaces, Cauchy sequences with convergent subsequences are also convergent. $\square$

Exercises

1.2.1. Suppose $f_k \geq 0$ are measurable functions on (X, μ). Prove that if $f_k \uparrow f$ as $k \to \infty$ μ-a.e., then $D_{f_k} \uparrow D_f$ as $k \to \infty$.

1.2.2. (Lebesgue monotone convergence theorem for weak L^p spaces) Let $f_k \geq 0$ be measurable functions on a measure space (X, μ) and $0 < p < \infty$. Suppose $f_k \uparrow f$ as $k \to \infty$. Show that

$$\lim_{k \to \infty} \|f_k\|_{L^{p,\infty}} = \|f\|_{L^{p,\infty}}.$$

[*Hint:* Consider the cases $\|f\|_{L^{p,\infty}} < \infty$ and $\|f\|_{L^{p,\infty}} = \infty$. Use the previous exercise.]

1.2.3. (Fatou's lemma for weak L^p spaces) Let $f_k \geq 0$ be measurable functions on a measure space (X,μ) and $0 < p < \infty$. Prove that

$$\left\|\liminf_{k\to\infty} f_k\right\|_{L^{p,\infty}} \leq \liminf_{k\to\infty} \|f_k\|_{L^{p,\infty}}.$$

[*Hint:* Set $g_k = \inf\{f_l : l \geq k\}$ and use the previous exercise.]

1.2.4. Suppose f and f_k are measurable functions on $\mathbf{R}^n$. Prove that if $|f| \leq \liminf_{k\to\infty} |f_k|$ a.e., then $D_f \leq \liminf_{k\to\infty} D_{f_k}$.

1.2.5. Let $0 < p_0 < p < p_1 \leq \infty$ and let $\frac{1}{p} = \frac{1-\theta}{p_0} + \frac{\theta}{p_1}$ for some $\theta \in (0,1)$. Prove

$$\|f\|_{L^{p,\infty}} \leq \|f\|_{L^{p_0,\infty}}^{1-\theta} \|f\|_{L^{p_1,\infty}}^{\theta}.$$

1.2.6. Let (X,μ) be a measure space and let E be a subset of X with $\mu(E) < \infty$. Assume that f is in $L^{p,\infty}(X,\mu)$ for some $0 < p < \infty$.
(a) Show that for $0 < q < p$ we have

$$\int_E |f(x)|^q \, d\mu(x) \leq \frac{p}{p-q} \mu(E)^{1-\frac{q}{p}} \|f\|_{L^{p,\infty}}^q.$$

(b) Prove that if $\mu(X) < \infty$ and $0 < q < p < \infty$, then

$$L^p(X,\mu) \subsetneq L^{p,\infty}(X,\mu) \subsetneq L^q(X,\mu).$$

(c) Conclude that $L^{p,\infty}(\mathbf{R}^n)$ is contained in $L^1_{\text{loc}}(\mathbf{R}^n)$ when $p > 1$.

1.2.7. (Hölder's inequality for weak spaces) Let f_1 be in $L^{p_1,\infty}$ and f_2 be in $L^{p_2,\infty}$ of a measure space X where $0 < p_1, p_2 < \infty$. Given $\frac{1}{p} = \frac{1}{p_1} + \frac{1}{p_2}$, prove that

$$\|f_1 f_2\|_{L^{p,\infty}} \leq \left[(p_2/p_1)^{\frac{p_1}{p_1+p_2}} + (p_1/p_2)^{\frac{p_2}{p_1+p_2}}\right] \|f_1\|_{L^{p_1,\infty}} \|f_2\|_{L^{p_2,\infty}}.$$

Observe that the preceding inequality also extends to the case where p_1, p_2 equal ∞.
[*Hint:* For $\|f_j\|_{L^{p_j,\infty}} = 1$, $j = 1,2$, use $D_{f_1 f_2}(\lambda) \leq \mu(\{|f_1| > \lambda/s\}) + \mu(\{|f_2| > s\}) \leq (s/\lambda)^{p_1} + (1/s)^{p_2}$ and minimize over $s > 0$.]

1.2.8. Let $f \in L^1([0,\infty))$ and $g \in L^1((-\infty,0])$. Prove that the function

$$x \mapsto \int_\mathbf{R} f(x+t)g(x-t) \frac{dt}{t}$$

lies in $L^{1/2,\infty}(\mathbf{R})$ with quasi-norm bounded by $4\|f\|_{L^1}\|g\|_{L^1}$. [*Hint:* Control this function pointwise by $|x|^{-1}G(x)$, for some $G \geq 0$ with $\|G\|_{L^1} \leq \frac{1}{2}\|f\|_{L^1}\|g\|_{L^1}$.]

1.3 Real Interpolation

Let $0 < p < \infty$. Given an L^p function f on a measure space (X,μ) and $\lambda > 0$ we consider the splitting

$$f = \overbrace{f\chi_{|f|>\lambda}}^{f_0} + \overbrace{f\chi_{|f|\leq\lambda}}^{f_1}.$$

Let us pick p_0, p_1 such that $0 < p_0 < p < p_1 \leq \infty$. We note that f_0 lies in L^{p_0} and that f_1 lies in L^{p_1}. Indeed, since $p_0 - p < 0$, we have

$$\|f_0\|_{L^{p_0}}^{p_0} = \int_{|f|>\lambda} |f(x)|^p |f(x)|^{p_0-p}\,d\mu \leq \lambda^{p_0-p}\|f\|_{L^p}^p,$$

and likewise, as $p_1 - p > 0$, we obtain

$$\|f_1\|_{L^{p_1}} \leq \lambda^{1-\frac{p}{p_1}}\|f\|_{L^p}^{\frac{p}{p_1}}.$$

Thus $L^p(X)$ is contained in $L^{p_0}(X) + L^{p_1}(X) = \{f_0 + f_1 : f_j \in L^{p_j}(X), j = 0, 1\}$.

Definition 1.3.1. Let T be an operator defined on a linear space of complex-valued measurable functions on a measure space (X,μ) and taking values in the set of all complex-valued finite almost everywhere measurable functions on a measure space (Y,ν). T is called *subadditive* if for all f, g in the domain of T we have

$$|T(f+g)| \leq |T(f)| + |T(g)| \qquad \nu\text{-a.e.} \qquad (1.3.1)$$

T is called *quasi-subadditive*, that is, if there is a constant $K > 0$ such that for all f, g in the domain of T we have

$$|T(f+g)| \leq K(|T(f)| + |T(g)|) \qquad \nu\text{-a.e.} \qquad (1.3.2)$$

Definition 1.3.2. An operator that maps L^p to $L^{p,\infty}$ is called of weak type (p,p).

The next result, known as the *Marcinkiewicz interpolation theorem*, claims that if a subadditive operator is of weak types (p_0, p_0) and (p_1, p_1), then it maps L^p to L^p for p between p_0 and p_1.

Theorem 1.3.3. *(Marcinkiewicz interpolation theorem) Let (X,μ), (Y,ν) be σ-finite measure spaces and $0 < p_0 < p_1 \leq \infty$. Suppose that T is an operator defined on $L^{p_0}(X) + L^{p_1}(X)$, taking values in the space of measurable functions on Y, which is quasi-subadditive, i.e., it satisfies (1.3.2). Assume there exist $A_0, A_1 < \infty$ such that*

$$\|T(f)\|_{L^{p_0,\infty}(Y)} \leq A_0 \|f\|_{L^{p_0}(X)} \qquad \text{for all } f \in L^{p_0}(X), \qquad (1.3.3)$$

$$\|T(f)\|_{L^{p_1,\infty}(Y)} \leq A_1 \|f\|_{L^{p_1}(X)} \qquad \text{for all } f \in L^{p_1}(X). \qquad (1.3.4)$$

Then for all $p_0 < p < p_1$ and for all f in $L^p(X)$ we have

$$\|T(f)\|_{L^p(Y)} \leq A\|f\|_{L^p(X)}, \qquad (1.3.5)$$

1.3 Real Interpolation

where

$$A = 2K \left(\frac{p}{p-p_0} + \frac{p}{p_1-p} \right)^{\frac{1}{p}} A_0^{\frac{\frac{1}{p}-\frac{1}{p_1}}{\frac{1}{p_0}-\frac{1}{p_1}}} A_1^{\frac{\frac{1}{p_0}-\frac{1}{p}}{\frac{1}{p_0}-\frac{1}{p_1}}}. \tag{1.3.6}$$

Proof. Consider first the case $p_1 = \infty$. Write $f = f_0^\lambda + f_1^\lambda$, where

$$f_0^\lambda(x) = \begin{cases} f(x) & \text{for } |f(x)| > \gamma\lambda, \\ 0 & \text{for } |f(x)| \leq \gamma\lambda, \end{cases}$$

$$f_1^\lambda(x) = \begin{cases} f(x) & \text{for } |f(x)| \leq \gamma\lambda, \\ 0 & \text{for } |f(x)| > \gamma\lambda, \end{cases}$$

and $\gamma = (2A_1K)^{-1}$; then for almost all $y \in Y$ we have

$$|T(f_1^\lambda)(y)| \leq \|T(f_1^\lambda)\|_{L^\infty} \leq A_1 \|f_1^\lambda\|_{L^\infty} \leq A_1\gamma\lambda = \lambda/2K.$$

It follows that $v(\{y \in Y : |T(f_1^\lambda)(y)| > \frac{\lambda}{2K}\}) = 0 = D_{T(f_1^\lambda)}(\lambda/2K)$, hence

$$D_{T(f)}(\lambda) \leq D_{T(f_0^\lambda)}(\lambda/2K) + D_{T(f_1^\lambda)}(\lambda/2K) = D_{T(f_0^\lambda)}(\lambda/2K).$$

Since T maps L^{p_0} to $L^{p_0,\infty}$ with norm at most A_0, we write

$$D_{T(f_0^\lambda)}(\lambda/2K) \leq \frac{(2A_0K)^{p_0} \|f_0^\lambda\|_{L^{p_0}}^{p_0}}{\lambda^{p_0}} = \frac{(2A_0K)^{p_0}}{\lambda^{p_0}} \int_{|f|>\gamma\lambda} |f(x)|^{p_0} d\mu. \tag{1.3.7}$$

Using (1.3.7) and Proposition 1.2.3 [which applies as (Y,v) is σ-finite], we obtain

$$\begin{aligned}
\|T(f)\|_{L^p}^p &= p \int_0^\infty \lambda^{p-1} D_{T(f)}(\lambda) d\lambda \\
&\leq p \int_0^\infty \lambda^{p-1} D_{T(f_0^\lambda)}(\lambda/2K) d\lambda \\
&\leq p \int_0^\infty \lambda^{p-1} \frac{(2A_0K)^{p_0}}{\lambda^{p_0}} \int_{|f|>\frac{\lambda}{2A_1K}} |f(x)|^{p_0} d\mu \, d\lambda \\
&= p(2A_0K)^{p_0} \int_X |f(x)|^{p_0} \int_0^{2A_1K|f(x)|} \lambda^{p-p_0-1} d\lambda \, d\mu \\
&= \frac{p(2A_1K)^{p-p_0}(2A_0K)^{p_0}}{p-p_0} \int_X |f(x)|^p d\mu,
\end{aligned}$$

having used Tonelli's theorem, as (X,μ) is σ-finite. This proves the theorem with

$$A = 2K \left(\frac{p}{p-p_0} \right)^{\frac{1}{p}} A_1^{1-\frac{p_0}{p}} A_0^{\frac{p_0}{p}}. \tag{1.3.8}$$

Observe that the constant in (1.3.8) coincides with that in (1.3.6) when $p_1 = \infty$.

Assume now that $p_1 < \infty$. Fix a function f in $L^p(X)$ and $\lambda > 0$. We split $f = f_0^\lambda + f_1^\lambda$, where f_0^λ is in L^{p_0} and f_1^λ is in L^{p_1}. The splitting is obtained by cutting $|f|$ at height $\delta\lambda$ for some $\delta > 0$ to be determined later. Set

$$f_0^\lambda(x) = \begin{cases} f(x) & \text{for } |f(x)| > \delta\lambda, \\ 0 & \text{for } |f(x)| \leq \delta\lambda, \end{cases}$$

$$f_1^\lambda(x) = \begin{cases} f(x) & \text{for } |f(x)| \leq \delta\lambda, \\ 0 & \text{for } |f(x)| > \delta\lambda. \end{cases}$$

As noted earlier, the (potentially) unbounded part f_0^λ is an L^{p_0} function and the bounded part of f_1^λ is an L^{p_1} function. Using the quasi-subadditivity property (1.3.2) of T we obtain that

$$|T(f)| \leq K|T(f_0^\lambda)| + K|T(f_1^\lambda)|, \qquad \nu\text{-a.e.,}$$

which implies the ν-a.e. inclusion

$$\{y \in Y : |T(f)(y)| > \lambda\} \subseteq \{y \in Y : |T(f_0^\lambda)(y)| > \tfrac{\lambda}{2K}\} \cup \{y \in Y : |T(f_1^\lambda)(y)| > \tfrac{\lambda}{2K}\},$$

and therefore

$$D_{T(f)}(\lambda) \leq D_{T(f_0^\lambda)}(\lambda/2K) + D_{T(f_1^\lambda)}(\lambda/2K). \tag{1.3.9}$$

Hypotheses (1.3.3) and (1.3.4) together with (1.3.9) now give

$$D_{T(f)}(\lambda) \leq \frac{A_0^{p_0}}{(\lambda/2K)^{p_0}} \int_{|f|>\delta\lambda} |f(x)|^{p_0} d\mu + \frac{A_1^{p_1}}{(\lambda/2K)^{p_1}} \int_{|f|\leq\delta\lambda} |f(x)|^{p_1} d\mu.$$

In view of the last estimate and Proposition 1.2.3 (which can be used since Y is a σ-finite measure space), we obtain that

$$\|T(f)\|_{L^p}^p \leq p(2A_0K)^{p_0} \int_0^\infty \lambda^{p-1} \lambda^{-p_0} \int_{|f|>\delta\lambda} |f(x)|^{p_0} d\mu\, d\lambda$$

$$+ p(2A_1K)^{p_1} \int_0^\infty \lambda^{p-1} \lambda^{-p_1} \int_{|f|\leq\delta\lambda} |f(x)|^{p_1} d\mu\, d\lambda$$

$$= p(2A_0K)^{p_0} \int_X |f(x)|^{p_0} \int_0^{\frac{1}{\delta}|f(x)|} \lambda^{p-1-p_0} d\lambda\, d\mu$$

$$+ p(2A_1K)^{p_1} \int_X |f(x)|^{p_1} \int_{\frac{1}{\delta}|f(x)|}^\infty \lambda^{p-1-p_1} d\lambda\, d\mu$$

$$= \frac{p(2A_0K)^{p_0}}{p-p_0} \frac{1}{\delta^{p-p_0}} \int_X |f(x)|^{p_0} |f(x)|^{p-p_0} d\mu$$

$$+ \frac{p(2A_1K)^{p_1}}{p_1-p} \frac{1}{\delta^{p-p_1}} \int_X |f(x)|^{p_1} |f(x)|^{p-p_1} d\mu$$

$$= \left(\frac{p}{p-p_0} \frac{(2A_0K)^{p_0}}{\delta^{p-p_0}} + \frac{p}{p_1-p} (2A_1K)^{p_1} \delta^{p_1-p}\right) \|f\|_{L^p}^p, \tag{1.3.10}$$

1.3 Real Interpolation

and the convergence of the integrals in λ is justified from $p_0 < p < p_1$, while the interchange of the integrals (Tonelli's theorem) uses the hypothesis that (X, μ) is a σ-finite measure space. We pick $\delta > 0$ such that

$$(2A_0K)^{p_0}\frac{1}{\delta^{p-p_0}} = (2A_1K)^{p_1}\delta^{p_1-p},$$

and observe that the constant in the parentheses in (1.3.10) is equal to the pth power of the constant in (1.3.6). We have therefore proved the theorem when $p_1 < \infty$. □

In some applications the operator T is not a priori defined on the entire space $L^{p_0}(X) + L^{p_1}(X)$, but on a subspace of it. Let us suppose below that $\mathscr{F}$ is a subset of $L^{p_0}(X) + L^{p_1}(X)$ with the property that $f \in \mathscr{F} \implies f\chi_E \in \mathscr{F}$, where E is a measurable subset of X of finite measure. For instance, $\mathscr{F}$ could be the space of all finitely simple[4] functions. We note that in the proof of Theorem 1.3.3 if $f \in \mathscr{F}$, then so are f_0^λ and f_1^λ. Thus the proof given provides the following result.

Theorem 1.3.4. *Let (X, μ), (Y, ν) be σ-finite measure spaces and $0 < p_0 < p_1 \leq \infty$. Suppose that T is defined on $\mathscr{F}$, takes values in the space of measurable functions on Y, and satisfies (1.3.2). Assume there exist $A_0, A_1 < \infty$ such that for $\kappa = 0, 1$*

$$\big\|T(f)\big\|_{L^{p_\kappa,\infty}(Y)} \leq A_\kappa \big\|f\big\|_{L^{p_\kappa}(X)} \qquad \text{for all } f \in \mathscr{F}.$$

Then (1.3.5) holds for $p_0 < p < p_1$ and all $f \in \mathscr{F}$. Thus, if $\mathscr{F}$ is dense in $L^p(X)$ and T is linear (or positive symmetric subadditive operator, see Appendix D), then T admits a unique bounded extension on $L^p(X)$ that also satisfies (1.3.5).

The proof of Theorem 1.3.3 essentially contains the following result.

Proposition 1.3.5. *Let (X, μ) be a σ-finite measure space and let $0 < p_0 < p < p_1 \leq \infty$ be related as in $1/p = (1-\theta)/p_0 + \theta/p_1$ for some $\theta \in (0,1)$. Then for any $f \in L^{p_0,\infty}(X) \cap L^{p_1,\infty}(X)$ we have*

$$\|f\|_{L^p} \leq \left(\frac{p}{p-p_0} + \frac{p}{p_1-p}\right)^{\frac{1}{p}} \|f\|_{L^{p_0,\infty}}^{1-\theta} \|f\|_{L^{p_1,\infty}}^{\theta}. \qquad (1.3.11)$$

Proof. Set

$$B = \|f\|_{L^{p_1,\infty}}^{\frac{p_1}{p_1-p_0}} / \|f\|_{L^{p_0,\infty}}^{\frac{p_0}{p_1-p_0}}, \qquad (1.3.12)$$

which equals $B = \|f\|_{L^\infty}$ when $p_1 = \infty$. Assume first that $p_1 < \infty$. Then we have

$$\|f\|_{L^p(X,\mu)}^p = p\int_0^\infty \lambda^{p-1} D_f(\lambda)\,d\lambda$$

$$\leq p\int_0^\infty \lambda^{p-1} \min\left(\frac{\|f\|_{L^{p_0,\infty}}^{p_0}}{\lambda^{p_0}}, \frac{\|f\|_{L^{p_1,\infty}}^{p_1}}{\lambda^{p_1}}\right)d\lambda$$

---
[4] Finite linear combinations of characteristic functions of sets of finite measure.

$$= p \int_0^B \lambda^{p-1-p_0} \|f\|_{L^{p_0,\infty}}^{p_0} d\lambda + p \int_B^\infty \lambda^{p-1-p_1} \|f\|_{L^{p_1,\infty}}^{p_1} d\lambda \quad (1.3.13)$$

$$= \frac{p}{p-p_0} \|f\|_{L^{p_0,\infty}}^{p_0} B^{p-p_0} + \frac{p}{p_1-p} \|f\|_{L^{p_1,\infty}}^{p_1} B^{p-p_1}$$

$$= \left(\frac{p}{p-p_0} + \frac{p}{p_1-p}\right) (\|f\|_{L^{p_0,\infty}}^{p_0})^{\frac{p_1-p}{p_1-p_0}} (\|f\|_{L^{p_1,\infty}}^{p_1})^{\frac{p-p_0}{p_1-p_0}},$$

where we use that $p - p_1 < 0 < p - p_0$. When $p_1 = \infty$, we have $D_f(\lambda) = 0$ for $\lambda > B = \|f\|_{L^\infty}$ and thus the second integral in (1.3.13) does not appear. We obtain

$$\|f\|_{L^p}^p \leq \frac{p}{p-p_0} \|f\|_{L^{p_0,\infty}}^{p_0} \|f\|_{L^\infty}^{p-p_0},$$

which is a restatement of (1.3.11) when $p_1 = \infty$. □

Example 1.3.6. Let $0 < \alpha < \beta < \infty$. Then $|x|^{-\alpha} \in L^{n/\alpha,\infty}(\mathbf{R}^n)$, $|x|^{-\beta} \in L^{n/\beta,\infty}(\mathbf{R}^n)$. Let $h(x) = \min(|x|^{-\alpha}, |x|^{-\beta})$, $x \neq 0$. Then $h \in L^p(\mathbf{R}^n)$ for $n/\beta < p < n/\alpha$.

Exercises

1.3.1. Let $0 < p, q < \infty$. Verify that Theorem 1.3.3 applies to operators of the form $T(f) = |S(f)|^p \log(1 + |S(f)|)^q$, where S is a linear operator (or another quasi-subadditive operator) on $L^p(X) + L^q(X)$. [*Hint:* See Exercise 1.1.1.]

1.3.2. Let $A_0, A_1 > 0$, $0 < p, q_0, q_1 \leq \infty$, and let (X, μ), (Y, ν) be σ-finite measure spaces. Suppose that T is a mapping defined on $L^p(X, \mu)$ that satisfies

$$\|T(f)\|_{L^{q_0}(Y,\nu)} \leq A_0 \|f\|_{L^p} \quad \text{and} \quad \|T(f)\|_{L^{q_1}(Y,\nu)} \leq A_1 \|f\|_{L^p}$$

for all $f \in L^p(X, \mu)$. Let $0 \leq \theta \leq 1$. Prove that

$$\|T(f)\|_{L^q(Y,\nu)} \leq A_0^{1-\theta} A_1^\theta \|f\|_{L^p}$$

for all $f \in L^p(X, \mu)$, where $1/q = (1-\theta)/q_0 + \theta/q_1$. [*Hint:* Use Exercise 1.1.6.]

1.3.3. Let $A_0, A_1 > 0$, $0 < p, q_0, q_1 \leq \infty$, $q_0 \neq q_1$, and let (X, μ), (Y, ν) be σ-finite measure spaces. Suppose that T is a mapping defined on $L^p(X, \mu)$ that satisfies

$$\|T(f)\|_{L^{q_0,\infty}(Y,\nu)} \leq A_0 \|f\|_{L^p} \quad \text{and} \quad \|T(f)\|_{L^{q_1,\infty}(Y,\nu)} \leq A_1 \|f\|_{L^p}$$

for all $f \in L^p(X, \mu)$. Let $\theta \in (0,1)$ and $1/q = (1-\theta)/q_0 + \theta/q_1$. Show that there is a constant $C = C(q_0, q_1, q)$ such that

$$\|T(f)\|_{L^q(Y,\nu)} \leq C A_0^{1-\theta} A_1^\theta \|f\|_{L^p},$$

for all $f \in L^p(X, \mu)$. [*Hint:* Use Proposition 1.3.5.]

1.3 Real Interpolation

1.3.4. Let T be a subadditive operator defined on $L^1(X,\mu) + L^\infty(X,\mu)$ that takes values in the space of measurable functions on (Y,ν) and that satisfies $|T(f)| \le T(|f|)$ for all $f \in L^1 + L^\infty$. Suppose that T maps L^1 to $L^{1,\infty}$ with bound A_0 and L^∞ to itself with bound $A_1 > 0$. Given $1 < p < \infty$, prove that T maps $L^p(X)$ to $L^p(Y)$ with norm at most
$$\frac{p}{p-1} A_0^{\frac{1}{p}} A_1^{1-\frac{1}{p}}.$$
[*Hint:* Given $\lambda > 0$, $\gamma \in (0,1)$, and f measurable, write $|f| = f_0 + f_1$, where $f_0 = \max\left(|f| - \frac{\gamma\lambda}{A_1}, 0\right)$ and $f_1 = \min\left(|f|, \frac{\gamma\lambda}{A_1}\right)$. Then $\{T(|f|) > \lambda\} \subseteq \{|T(f_0)| > (1-\gamma)\lambda\}$. Then choose a suitable γ.]

1.3.5. Let (X,μ), (Y,ν) be σ-finite measure spaces, and let $0 < p_0 < p_1 \le \infty$. Define p via $\frac{1-\theta}{p_0} + \frac{\theta}{p_1} = \frac{1}{p}$, where $0 < \theta < 1$. Let T be a subadditive operator defined on $L^{p_0}(X) + L^{p_1}(X)$ and taking values in the space of measurable functions on Y. Suppose T maps L^{p_0} to L^∞ with norm A_0 and L^{p_1} to L^∞ with norm A_1. Prove that T maps L^p to L^∞ with norm at most $2 A_0^{1-\theta} A_1^\theta$.

1.3.6. Let (X,μ) and (Y,ν) be σ-finite measure spaces. Let $0 < p < p_1 \le \infty$, $0 < B < \infty$, and let $\Phi : [0,\infty) \to [0,\infty)$ be a measurable function such that
$$A = \int_0^1 \lambda^{p-1} \Phi(1/\lambda)\,d\lambda < \infty.$$
Let T be a linear operator that maps $L^{p_1}(X)$ to $L^{p_1,\infty}(Y)$ with norm B that satisfies
$$\nu(\{y \in Y : |T(f)(y)| > \lambda\}) \le A \int_X \Phi\left(\frac{|f(x)|}{\lambda}\right) d\mu$$
for all finite simple functions f on X and all $\lambda > 0$. Prove that T has a bounded extension from $L^p(X)$ to itself. [*Hint:* Set $f^\lambda = f\chi_{|f|>\lambda}$ and $f_\lambda = f\chi_{|f|\le\lambda}$. When $p_1 < \infty$, add the estimates
$$p\lambda^{p-1} \nu(\{|T(f^\lambda)| > \lambda\}) \le Ap\lambda^{p-1} \int_{|f|>\lambda} \Phi\left(\frac{|f(x)|}{\lambda}\right) d\mu$$
and
$$p\lambda^{p-1} \nu(\{|T(f_\lambda)| > \lambda\}) \le p\lambda^{p-1} B^{p_1} \int_{|f|\le\lambda} \frac{|f(x)|^{p_1}}{\lambda^{p_1}} d\mu,$$
and integrate over λ to estimate $\frac{1}{2^p}\|T(f)\|_{L^p}^p$. In the case where $p_1 = \infty$, use
$$\nu(\{|T(f)| > 2B\lambda\}) \le \nu(\{|T(f^\lambda)| > B\lambda\})$$
to complete the proof.]

1.3.7. (Vector-valued Marcinkiewicz interpolation) Let (X,μ), (Y,ν) be σ-finite measure spaces and $0 < p_0 < p_1 \le \infty$. Fix quasi-normed spaces Z, W. Define

$L^{p_0}(X,Z)$ to be the space of all functions $\vec{f}: X \to Z$ such that $x \mapsto \|\vec{f}(x)\|_Z$ are μ-measurable and belong to $L^{p_0}(X,\mu)$, and analogously define $L^{p_0,\infty}(X,Z)$. Let $\vec{T}$ be a mapping from $L^{p_0}(X,Z) + L^{p_1}(X,Z)$ to $\{\vec{h}: Y \to W,\text{ with } \|\vec{h}\|_W \; \nu\text{-measurable}\}$. Suppose $\vec{T}$ satisfies the following quasi-subadditivity property: for some $K > 0$ and all $\vec{f}, \vec{g}$ in $L^{p_0}(X,Z) + L^{p_1}(X,Z)$ we have $\|\vec{T}(\vec{f}+\vec{g})\|_W \leq K\|\vec{T}(\vec{f})\|_W + K\|\vec{T}(\vec{g})\|_W$ ν-a.e. Assume there exist $A_0, A_1 < \infty$ such that

$$\|\vec{T}(\vec{f})\|_{L^{p_0,\infty}(Y,W)} \leq A_0 \|\vec{f}\|_{L^{p_0}(X,Z)} \qquad \text{for all } \vec{f} \in L^{p_0}(X),$$

$$\|\vec{T}(\vec{f})\|_{L^{p_1,\infty}(Y,W)} \leq A_1 \|\vec{f}\|_{L^{p_1}(X,Z)} \qquad \text{for all } \vec{f} \in L^{p_1}(X).$$

Let A be as in (1.3.6). Prove that for all $p_0 < p < p_1$ and all $\vec{f}$ in $L^p(X,Z)$ we have

$$\|\vec{T}(\vec{f})\|_{L^p(Y,W)} \leq A \|\vec{f}\|_{L^p(X,Z)}.$$

[*Hint:* Adapt the proof of Theorem 1.3.3 by cutting $\|\vec{f}\|_Z$ at height $\delta\lambda$.]

1.4 The Hardy–Littlewood Maximal Operator

Let v_n be the volume of the unit ball $B(0,1)$ in $\mathbf{R}^n$.

Definition 1.4.1. Let f be a measurable function on $\mathbf{R}^n$. The function

$$\mathcal{M}(f)(x) = \sup_{\delta > 0} \operatorname*{Avg}_{B(x,\delta)} |f| = \sup_{\delta > 0} \frac{1}{v_n \delta^n} \int_{|y|<\delta} |f(x-y)|\, dy$$

is called the *centered Hardy–Littlewood maximal function* of f.

A few remarks are in order. Suppose that f is not locally integrable. Then there exists a compact set L such that $\int_L |f|\, dx = \infty$. Then for any $x \in \mathbf{R}^n$ there is a ball $B(x,R)$ that contains L. It follows that $\mathcal{M}(f)(x) \geq \frac{1}{v_n R^n} \int_L |f|\, dy = \infty$. Thus $\mathcal{M}(f)$ is interesting only when f is a locally integrable function.

Secondly, the definition of $\mathcal{M}$ remains unchanged if the open ball $B(x,\delta)$ is replaced by the closed ball $\overline{B(x,\delta)}$. Indeed, the balls $\overline{B(x,\delta)}$ and $B(x,\delta)$ have the same measure, and so f has vanishing integral over the null set $\overline{B(x,\delta)} \setminus B(x,\delta)$.

Obviously we have $\mathcal{M}(f) = \mathcal{M}(|f|) \geq 0$; thus the maximal function is a positive operator. We show later that $\mathcal{M}(f)$ pointwise controls f (i.e., $\mathcal{M}(f) \geq |f|$ a.e.). Note that $\mathcal{M}$ maps L^∞ to itself; that is, we have

$$\|\mathcal{M}(f)\|_{L^\infty} \leq \|f\|_{L^\infty}.$$

$\mathcal{M}$ is a sublinear operator; i.e., it satisfies

$$\mathcal{M}(f+g) \leq \mathcal{M}(f) + \mathcal{M}(g) \quad \text{and} \quad \mathcal{M}(\lambda f) = |\lambda|\mathcal{M}(f)$$

1.4 The Hardy–Littlewood Maximal Operator

for all locally integrable functions f and g and all complex constants λ.

Let us compute the Hardy–Littlewood maximal function of a specific function.

Example 1.4.2. On $\mathbf{R}$, let f be the characteristic function of the interval $[a,b]$. For $x \in (a,b)$, clearly $\mathcal{M}(f) = 1$. For $x \geq b$, a simple calculation shows that the largest average of f over all intervals $(x-\delta, x+\delta)$ is obtained when $\delta = x-a$. Similarly, when $x \leq a$, the largest average is obtained when $\delta = b-x$. Therefore,

$$\mathcal{M}(f)(x) = \begin{cases} (b-a)/2|x-b| & \text{when } x \leq a, \\ 1 & \text{when } x \in (a,b), \\ (b-a)/2|x-a| & \text{when } x \geq b. \end{cases}$$

Observe that $\mathcal{M}(f)$ has a jump at $x=a$ and $x=b$ equal to one-half that of f. See Figure 1.2.

If f is locally integrable and $R > 0$ is given, then by considering the average of f over the ball $B(x, |x|+R)$, which contains the ball $B(0,R)$, we obtain

$$\mathcal{M}(f)(x) \geq \frac{\int_{B(0,R)} |f(y)|\,dy}{v_n(|x|+R)^n} \tag{1.4.1}$$

for all $x \in \mathbf{R}^n$. An interesting consequence of (1.4.1) is the following: suppose that $f \neq 0$ on a set of positive measure E, then $\mathcal{M}(f)$ is not in $L^1(\mathbf{R}^n)$. In other words, if f is in $L^1_{\text{loc}}(\mathbf{R}^n)$ and $\mathcal{M}(f)$ is in $L^1(\mathbf{R}^n)$, then $f = 0$ a.e. To see this, integrate (1.4.1) over the ball $\mathbf{R}^n$ to deduce that $\|f\chi_{B(0,R)}\|_{L^1} = 0$ and thus $f(x) = 0$ for almost all x in the ball $B(0,R)$. Since this is valid for all $R = 1, 2, 3, \ldots$, it follows that $f = 0$ a.e. in $\mathbf{R}^n$.

Another remarkable locality property of $\mathcal{M}$ is that if $\mathcal{M}(f)(x_0) = 0$ for some x_0 in $\mathbf{R}^n$, then $f = 0$ a.e. To see this, we take $x = x_0$ in (1.4.1) to deduce that $\|f\chi_{B(0,R)}\|_{L^1} = 0$ and as before we have that $f = 0$ a.e. on every ball centered at the origin, i.e., $f = 0$ a.e. in $\mathbf{R}^n$.

A related analog of $\mathcal{M}(f)$ is its uncentered version $M(f)$, defined as the supremum of all averages of f over all open balls containing a given point.

Definition 1.4.3. The *uncentered Hardy–Littlewood maximal function* $M(f)$ of a measurable function f is the supremum of the averages of $|f|$ over all open balls that contain a given point, i.e.,

$$M(f)(x) = \sup_{\substack{\delta > 0 \\ y:\, |y-x| < \delta}} \frac{1}{|B(y,\delta)|} \int_{B(y,\delta)} |f(z)|\,dz.$$

Clearly we have $\mathcal{M}(f) \leq M(f)$. Now, if $x \in B(x_0, R)$ then $B(x_0, R) \subseteq B(x, 2R)$, so

$$\frac{1}{v_n R^n} \int_{B(x_0,R)} |f(y)|\,dy \leq \frac{2^n}{v_n(2R)^n} \int_{B(x,2R)} |f(y)|\,dy \leq 2^n \mathcal{M}(f)(x);$$

hence taking the supremum over $B(x_0, R)$ containing x yields $M(f)(x) \leq 2^n \mathcal{M}(f)(x)$. Thus the boundedness properties of M are identical to those of $\mathcal{M}$.

Example 1.4.4. On $\mathbf{R}$, let f be the characteristic function of the interval $I = [a,b]$. For $x \in (a,b)$, clearly $M(f)(x) = 1$. For $x > b$, a calculation shows that the largest average of f over all intervals $(y - \delta, y + \delta)$ that contain x is obtained when $\delta = \frac{1}{2}(x - a)$ and $y = \frac{1}{2}(x + a)$. Similarly, when $x < a$, the largest average is obtained when $\delta = \frac{1}{2}(b - x)$ and $y = \frac{1}{2}(b + x)$. We conclude that

$$M(f)(x) = \begin{cases} (b-a)/|x-b| & \text{when } x \leq a, \\ 1 & \text{when } x \in (a,b), \\ (b-a)/|x-a| & \text{when } x \geq b. \end{cases}$$

Observe that $M(f)(x)$ does not have a jump at $x = a$ and $x = b$ and in fact we have

$$M(f)(x) = \left(1 + \frac{\operatorname{dist}(x, I)}{|I|}\right)^{-1}.$$

This function is shown in Figure 1.2 when $a = -1$ and $b = 1$.

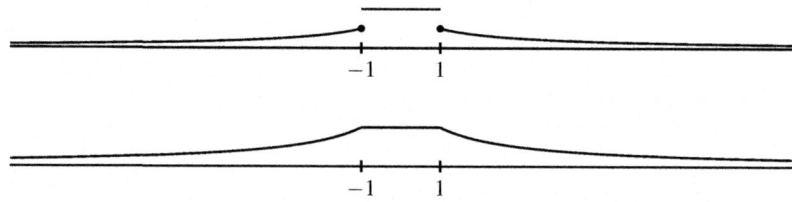

Fig. 1.2 The centered (top) and uncentered Hardy–Littlewood maximal functions of $\chi_{[-1,1]}$. The maximum value of both functions is 1 and is attained on $(-1,1)$.

We are now ready to obtain some basic properties of maximal functions. We need the following covering lemma.

Lemma 1.4.5. *Let $\{B_1, B_2, \ldots, B_N\}$ be a finite collection of open balls in $\mathbf{R}^n$. Then there exists a finite subcollection $\{B_{j_1}, \ldots, B_{j_L}\}$ of pairwise disjoint balls such that*

$$\sum_{r=1}^{L} |B_{j_r}| \geq \frac{1}{3^n} \left| \bigcup_{i=1}^{N} B_i \right|. \tag{1.4.2}$$

Proof. Let us reindex the balls so that

$$|B_1| \geq |B_2| \geq \cdots \geq |B_N|.$$

Set $j_1 = 1$. Let j_2 be the least index $s > j_1$ such that B_s is disjoint from B_{j_1}. This means that for all indices m (if any) with $j_1 < m < j_2$, B_m intersects B_{j_1}. Choose j_3

1.4 The Hardy–Littlewood Maximal Operator

to be the least index $s > j_2$ such that B_s is disjoint from $B_{j_1} \cup B_{j_2}$. This means that for all indices m (if any) with $j_2 < m < j_3$, B_m intersects B_{j_1} or B_{j_2}.

Having chosen $j_1, j_2, \ldots, j_i$, let j_{i+1} be the least index $s > j_i$ such that B_s is disjoint from $B_{j_1} \cup \cdots \cup B_{j_i}$. As before the selection procedure ensures that if the index m satisfies $j_i < m < j_{i+1}$, then B_m intersects one of the balls $B_{j_1}, \ldots, B_{j_i}$. Since we have a finite number of balls, this process will terminate when an index j_L is found such that for all $m > j_L$ the balls B_m intersect one of the $B_{j_1}, \ldots, B_{j_L}$.

We have now selected pairwise disjoint balls $B_{j_1}, \ldots, B_{j_L}$. If some B_m was not selected, that is, $m \notin \{j_1, \ldots, j_L\}$, as observed before, then B_m must intersect a selected ball B_{j_r} for some $j_r < m$. Then B_m has smaller size than B_{j_r} and we must have $B_m \subset 3B_{j_r}$. This shows that the union of the unselected balls is contained in the union of the triples of the selected balls. Therefore, the union of all balls is contained in the union of the triples of the selected balls. Thus

$$\left| \bigcup_{i=1}^{N} B_i \right| \leq \left| \bigcup_{r=1}^{L} 3B_{j_r} \right| \leq \sum_{r=1}^{L} |3B_{j_r}| = 3^n \sum_{r=1}^{L} |B_{j_r}|,$$

and the required conclusion follows. □

It was noted earlier that $\mathcal{M}(f)$ and $M(f)$ are never in L^1 if f is not zero a.e. However, it is true that these functions lie in $L^{1,\infty}$ when f is in L^1.

Theorem 1.4.6. *For any measurable function f we have*

$$|\{M(f) > \lambda\}| \leq \frac{3^n}{\lambda} \int_{\{M(f) > \lambda\}} |f(y)| \, dy. \tag{1.4.3}$$

Consequently, the uncentered Hardy–Littlewood maximal operator M maps $L^1(\mathbf{R}^n)$ to $L^{1,\infty}(\mathbf{R}^n)$ with constant at most 3^n.

Proof. We claim that the set $E_\lambda = \{x \in \mathbf{R}^n : M(f)(x) > \lambda\}$ is open. Indeed, for $x \in E_\lambda$, there is an open ball B_x that contains x such that $|B_x|^{-1} \int_{B_x} |f(y)| \, dy > \lambda$. Then $M(f)(y) > \lambda$ for any $y \in B_x$, and thus $B_x \subseteq E_\lambda$. This proves that E_λ is open.

Let K be a compact subset of E_λ. For each $x \in K$ there exists an open ball B_x containing the point x such that

$$\int_{B_x} |f(y)| \, dy > \lambda |B_x|. \tag{1.4.4}$$

Observe that $B_x \subset E_\lambda$ for all x. By compactness there exists a finite subcover $\{B_{x_1}, \ldots, B_{x_N}\}$ of K. Using Lemma 1.4.5 we find a subcollection of pairwise disjoint balls $B_{x_{j_1}}, \ldots, B_{x_{j_L}}$ such that (1.4.2) holds. Using (1.4.4) and (1.4.2) we obtain

$$|K| \leq \left| \bigcup_{i=1}^{N} B_{x_i} \right| \leq 3^n \sum_{i=1}^{L} |B_{x_{j_i}}| \leq \frac{3^n}{\lambda} \sum_{i=1}^{L} \int_{B_{x_{j_i}}} |f(y)| \, dy \leq \frac{3^n}{\lambda} \int_{E_\lambda} |f(y)| \, dy,$$

since all the balls $B_{x_{j_i}}$ are disjoint and contained in E_λ. Taking the supremum over all compact $K \subseteq E_\lambda$ and using the inner regularity of Lebesgue measure, we deduce (1.4.3). We have now proved that M maps $L^1 \to L^{1,\infty}$ with bound 3^n. □

Corollary 1.4.7. *Let $1 < p < \infty$. The uncentered Hardy–Littlewood maximal operator maps $L^p(\mathbf{R}^n)$ to $L^p(\mathbf{R}^n)$ with constant at most $3^{n/p} 2p(p-1)^{-1}$.*

Proof. It is straightforward that M maps $L^\infty \to L^\infty$ with constant 1 and is well defined on $L^1 + L^\infty$. Applying Theorem 1.3.3 we obtain

$$\|M\|_{L^p \to L^p} \leq 2\left(\frac{p}{p-1}\right)^{\frac{1}{p}} 3^{\frac{n}{p}} \leq \frac{2p}{p-1} 3^{\frac{n}{p}}, \qquad (1.4.5)$$

and this completes the proof. □

We note that Exercise 1.3.4 yields the slightly better bound for $\|M\|_{L^p \to L^p}$ in (1.4.5) without the factor of 2 on the right.

Example 1.4.8. Let $R > 0$ and $x_0 \in \mathbf{R}^n$. Then we have

$$\frac{R^n}{(|x - x_0| + R)^n} \leq \mathcal{M}(\chi_{B(x_0,R)})(x) \leq 3^n \frac{R^n}{(|x - x_0| + R)^n}. \qquad (1.4.6)$$

The lower estimate in (1.4.6) is an easy consequence of the fact that the ball $B(x, |x - x_0| + R)$ contains the ball $B(x_0, R)$. For the upper estimate, we first consider $|x - x_0| \leq 2R$, in which case we have

$$\mathcal{M}(\chi_{B(x_0,R)})(x) \leq 1 \leq \frac{3^n R^n}{(|x - x_0| + R)^n}.$$

In the case where $|x - x_0| > 2R$, if the balls $B(x,r)$ and $B(x_0, R)$ intersect, we must have that $r > |x - x_0| - R$. But note that $|x - x_0| - R > \frac{1}{3}(|x - x_0| + R)$, since $|x - x_0| > 2R$. We conclude that for $|x - x_0| > 2R$ we have

$$\mathcal{M}(\chi_{B(x_0,R)})(x) \leq \sup_{r > 0} \frac{|B(x,r) \cap B(x_0, R)|}{|B(x,r)|} \leq \sup_{r > |x-x_0|-R} \frac{v_n R^n}{v_n r^n} \leq \frac{R^n}{\left(\frac{1}{3}(|x-x_0|+R)\right)^n},$$

and thus the upper estimate in (1.4.6) holds. An analogous estimate is valid for $M(\chi_{B(x_0,R)})$.

Before ending this section we discuss an analogous situation when balls are replaced by cubes. Let f be a measurable function. Define the *uncentered maximal function with respect to cubes* by

$$M_c(f)(x) = \sup_{\substack{Q \ni x \\ Q \text{ cube}}} \frac{1}{|Q|} \int_Q |f(y)| \, dy,$$

where the cubes are assumed to have sides parallel to the coordinate planes. Note that it does not matter if the cubes Q in the supremum are taken to be open or closed.

1.4 The Hardy–Littlewood Maximal Operator

Then Lemma 1.4.5, Theorem 1.4.6, and Corollary 1.4.7 are also valid if we replace balls by open cubes. The main feature of Lemma 1.4.5 is that if two cubes with sides parallel to the axes intersect, then the smaller one is contained in the triple of the larger one. Repeating the reasoning leading to (1.4.5) we obtain the following analogous estimates for M_c:

$$\|M_c\|_{L^1 \to L^{1,\infty}} \leq 3^n, \qquad \|M_c\|_{L^p \to L^p} \leq 2\frac{3^{\frac{n}{p}} p}{p-1} \qquad (1.4.7)$$

when $1 < p < \infty$. Again by Exercise 1.3.4 the factor of 2 on the right can be removed.

Exercises

1.4.1. Show that for any measurable function f on $\mathbf{R}^n$ we have

$$\mathcal{M}(f)(x) = \sup_{\substack{r>0 \\ r \in \mathbf{Q}}} \frac{1}{v_n r^n} \int_{|x-y| \leq r} |f(y)| dy.$$

1.4.2. Let $f_k \geq 0$ be measurable functions on $\mathbf{R}^n$ such that $f_k \uparrow f$ pointwise a.e. as $k \to \infty$. Show that $M(f_k)$ increases pointwise to $M(f)$ and likewise for $\mathcal{M}$.

1.4.3. Let $f_k \geq 0$ be measurable functions on $\mathbf{R}^n$. Show that

$$M(\liminf_{k \to \infty} f_k) \leq \liminf_{k \to \infty} M(f_k).$$

1.4.4. Show that the set $\{x \in \mathbf{R}^n : \mathcal{M}(f)(x) > \lambda\}$ is open for any $f \in L^1_{\text{loc}}(\mathbf{R}^n)$. [*Hint:* Show that averages over small shifted balls of a given ball B are close to the average over B.]

1.4.5. For a ball B with radius $R > 0$, let B^* be any concentric multiple of B.
(a) Prove that[5] $\mathcal{M}(\chi_B) \approx \mathcal{M}(\chi_{B^*})$.
(b) Show that for any $x \in \mathbf{R}^n$ we have

$$\mathcal{M}(\chi_B)(x) \approx \frac{R^n}{(R + \text{dist}(x, B))^n},$$

where $\text{dist}(x, B)$ is the distance from x to B. [*Hint:* Use (1.4.6).]

1.4.6. Let f be a measurable function on $\mathbf{R}^n$. Define the *centered maximal function of f with respect to cubes*, by

$$\mathcal{M}_c(f)(x) = \sup_{\varepsilon > 0} \frac{1}{|Q(x,\varepsilon)|} \int_{Q(x,\varepsilon)} |f(y)| dy,$$

---

[5] $A \approx B$ means that for some $c, c' > 0$ we have $c < \frac{A}{B} < c'$ uniformly in all parameters involved.

where $Q(x,\varepsilon)$ is a cube parallel to the axes with side length 2ε centered at x.
(a) Prove that $\mathcal{M}_c(f) \approx \mathcal{M}(f)$. [*Hint:* See Figure 8.1.]
(b) Prove that $\mathcal{M}_c(f) \approx M_c(f)$.
(c) Show that for any cube Q with side length $\ell(Q)$ we have

$$M_c(\chi_Q)(x) \approx \frac{\ell(Q)^n}{(\ell(Q)+d_x)^n},$$

where d_x is the distance from x to Q (which is zero if $x \in \overline{Q}$).
(d) Conclude that for any concentric multiple Q^* of Q we have $M_c(\chi_Q) \approx M_c(\chi_{Q^*})$.

1.4.7. Let $h(t) = \frac{1}{t}(\log\frac{1}{t})^{-2}\chi_{(0,1/2)}$. Prove that $h \in L^1(\mathbf{R})$ but $\mathcal{M}(h) \notin L^1_{\text{loc}}(\mathbf{R})$.

1.4.8. Show that for every $1 < p < \infty$ and for any f in $L^{p,\infty}(\mathbf{R}^n)$, $M(f)$ lies in $L^{p,\infty}(\mathbf{R}^n)$ and we have

$$\|M(f)\|_{L^{p,\infty}} \leq \frac{3^n p}{p-1}\|f\|_{L^{p,\infty}}.$$

[*Hint:* First take $f \in L^p(\mathbf{R}^n) \cap L^1(\mathbf{R}^n)$. Prove that

$$\lambda |\{M(f) > \lambda\}|^{\frac{1}{p}} \leq 3^n |\{M(f) > \lambda\}|^{\frac{1}{p}-1} \int_{\{M(f)>\lambda\}} |f(y)|\,dy$$

and deduce from this the claimed inequality for $f \in L^p$. For a general $f \in L^{p,\infty}(\mathbf{R}^n)$, write $f_k = |f|\chi_{|f|\leq k}\chi_{B(0,k)}$ and use Exercises 1.2.2 and 1.4.2.]

1.5 The Lebesgue Differentiation Theorem

Recall that $L^1_{\text{loc}}(\mathbf{R}^n)$ consists of all measurable functions on $\mathbf{R}^n$ that are integrable over every compact set.

Theorem 1.5.1. *Let f be a function in $L^1_{\text{loc}}(\mathbf{R}^n)$. Then*

$$\lim_{\delta \to 0} \frac{1}{|B(x,\delta)|}\int_{B(x,\delta)} |f(y)-f(x)|\,dy = 0 \quad \text{for almost all } x \in \mathbf{R}^n. \tag{1.5.1}$$

Proof. We tile $\mathbf{R}^n$ as the union of cubes $Q_k = k + [0,1)^n$ where $k \in \mathbf{Z}^n$. It suffices to show the claimed almost everywhere convergence for any such cube Q_k. Let us fix such a cube $Q_k = k + [0,1)^n$. In proving (1.5.1) for $x \in Q_k$, we can replace f by $f_k = f\chi_{3Q_k}$, as we may assume that $\delta < 1/4$ when taking the limit in (1.5.1).

Define the *oscillation* of an integrable function g on $\mathbf{R}^n$ by

$$\mathscr{O}_g(x) = \limsup_{\delta \downarrow 0} \sup_{\delta' < \delta} \frac{1}{|B(x,\delta')|}\int_{B(x,\delta')} |g(y)-g(x)|\,dy. \tag{1.5.2}$$

1.5 The Lebesgue Differentiation Theorem

As g may only be defined almost everywhere, so is $\mathscr{O}_g$. Let φ be a continuous function with compact support on $\mathbf{R}^n$. Then φ is uniformly continuous; hence, given $\varepsilon > 0$, there is a $\delta_0 > 0$ such that

$$|y - z| < \delta_0 \implies |\varphi(y) - \varphi(z)| \leq \varepsilon.$$

This implies that for $\delta' < \delta < \delta_0$ we have

$$\frac{1}{|B(x,\delta')|} \int_{B(x,\delta')} |\varphi(y) - \varphi(x)| \, dy \leq \varepsilon.$$

Taking the supremum over all $\delta' < \delta$ and then the limit as $\delta \downarrow 0$, we obtain that $\mathscr{O}_\varphi(x) \leq \varepsilon$. As $\varepsilon > 0$ is arbitrary we deduce that $\mathscr{O}_\varphi(x) = 0$ for all $x \in \mathbf{R}^n$.

Given our fixed compactly supported integrable function f_k and given $\varepsilon' > 0$, there is a continuous function with compact support φ such that $\|f_k - \varphi\|_{L^1} < \varepsilon'$. As the oscillation function is subadditive, it follows that

$$\mathscr{O}_{f_k} \leq \mathscr{O}_{f_k - \varphi} + \mathscr{O}_\varphi = \mathscr{O}_{f_k - \varphi} \leq \mathscr{O}_{f_k} + \mathscr{O}_\varphi = \mathscr{O}_{f_k} \qquad \text{a.e.} \qquad (1.5.3)$$

Thus $\mathscr{O}_{f_k} = \mathscr{O}_{f_k - \varphi}$ a.e. We now write

$$\{\mathscr{O}_{f_k} > 0\} = \bigcup_{m=1}^\infty \{\mathscr{O}_{f_k} > \frac{1}{m}\}. \qquad (1.5.4)$$

Noticing that

$$\mathscr{O}_{f_k - \varphi} \leq \mathcal{M}(f_k - \varphi) + |f_k - \varphi| \qquad \text{a.e.,}$$

for each $m \in \mathbf{Z}^+$, we have

$$\begin{aligned}
&|\{x \in \mathbf{R}^n : \mathscr{O}_{f_k}(x) > 1/m\}| \\
&= |\{x \in \mathbf{R}^n : \mathscr{O}_{f_k - \varphi}(x) > 1/m\}| \\
&\leq |\{x \in \mathbf{R}^n : \mathcal{M}(f_k - \varphi)(x) + |(f_k - \varphi)(x)| > 1/m\}| \\
&\leq |\{x \in \mathbf{R}^n : \mathcal{M}(f_k - \varphi)(x) > 1/2m\}| + |\{x \in \mathbf{R}^n : |f_k(x) - \varphi(x)| > 1/2m\}| \\
&\leq 3^n(2m)\|f_k - \varphi\|_{L^1} + (2m)\|f_k - \varphi\|_{L^1} \\
&\leq (3^n + 1) 2m\varepsilon',
\end{aligned}$$

having used Theorem 1.4.6 and Chebyshev's inequality. As $\varepsilon' > 0$ is arbitrary, it follows that $|\{x \in \mathbf{R}^n : \mathscr{O}_{f_k}(x) > 1/m\}| = 0$. Using (1.5.4) yields that $\mathscr{O}_{f_k} = 0$ a.e. But $\mathscr{O}_{f_k} = \mathscr{O}_f$ a.e. on Q_k, and thus $\mathscr{O}_f = 0$ a.e. on Q_k. As this is true for any $k \in \mathbf{Z}^n$, it follows that $\mathscr{O}_f = 0$ a.e. on $\mathbf{R}^n$. This proves (1.5.1). $\square$

Definition 1.5.2. Let f be a locally integrable function on $\mathbf{R}^n$. Points x in $\mathbf{R}^n$ for which $f(x)$ is defined and (1.5.1) holds are called the *Lebesgue points* of f. The set of Lebesgue points is called the *Lebesgue set* of f and is denoted by $\mathscr{L}_f$.

Theorem 1.5.1 asserts that the Lebesgue set of a locally integrable function on $\mathbf{R}^n$ is a set of full measure; i.e., its complement has measure zero.

Corollary 1.5.3. *(One-dimensional Lebesgue differentiation theorem)* Given f in $L^1(\mathbf{R})$ define $F(x) = \int_{-\infty}^{x} f(t)\,dt$ for $x \in \mathbf{R}$. Then $F' = f$ on $\mathscr{L}_f$.

Proof. For $x_0 \in \mathscr{L}_f$ we have
$$\left|\frac{F(x_0+h) - F(x_0)}{h} - f(x_0)\right| \leq \frac{1}{|h|} \int_{x_0-|h|}^{x_0+|h|} |f(t) - f(x_0)|\,dt \to 0$$
as $h \to 0$. Thus $F'(x_0)$ exists and equals $f(x_0)$ whenever $x_0 \in \mathscr{L}_f$. $\square$

Corollary 1.5.4. *(Lebesgue differentiation theorem)* Let f be in $L^1_{\text{loc}}(\mathbf{R}^n)$. Then
$$\lim_{\delta \to 0} \frac{1}{|B(x,\delta)|} \int_{B(x,\delta)} f(y)\,dy = f(x)$$
for every x in $\mathscr{L}_f$, in particular for almost all $x \in \mathbf{R}^n$.

As a consequence of the preceding corollary we obtain

Corollary 1.5.5. Let f be in $L^1_{\text{loc}}(\mathbf{R}^n)$ and let x be a Lebesgue point of f. Then
$$|f(x)| \leq \mathcal{M}(f)(x).$$
In particular, this inequality holds for almost all points $x \in \mathbf{R}^n$.

Corollary 1.5.6. Let $f \in L^1_{\text{loc}}(\mathbf{R}^n)$ and $x \in \mathscr{L}_f$. For any $\delta > 0$, let $B_{x,\delta}$ be a closed ball of radius δ that contains x and shrinks down to $\{x\}$ as $\delta \to 0$. Then
$$\lim_{\delta \to 0} \frac{1}{|B_{x,\delta}|} \int_{B_{x,\delta}} |f(y) - f(x)|\,dy = 0 \tag{1.5.5}$$
and consequently
$$\lim_{\delta \to 0} \frac{1}{|B_{x,\delta}|} \int_{B_{x,\delta}} f(y)\,dy = f(x). \tag{1.5.6}$$
The same assertions are valid if $B_{x,\delta}$ are closed cubes of side length δ that contain x and shrink down to $\{x\}$ as $\delta \to 0$.

Proof. Recall $B(y,t)$ is defined to be the open ball of radius $t > 0$ centered at y. Let $x \in \mathscr{L}_f$. Then $B_{x,\delta} \subseteq \overline{B(x,2\delta)}$ as $B_{x,\delta}$ contains x, and we write
$$\frac{1}{|B_{x,\delta}|} \int_{B_{x,\delta}} |f(y) - f(x)|\,dy \leq \frac{2^n}{|B(x,2\delta)|} \int_{\overline{B(x,2\delta)}} |f(y) - f(x)|\,dy \to 0 \tag{1.5.7}$$
as $\delta \to 0$ by (1.5.1) and Definition 1.5.2. This yields (1.5.5). If $B_{x,\delta}$ are closed cubes of side length δ that contain x, then $B_{x,\delta} \subseteq \overline{B(x,\sqrt{n}\delta)}$ and (1.5.7) holds with $|B(0,\sqrt{n})|$ on the right in place of 2^n. Thus (1.5.5) holds in this case as well. Finally (1.5.6) is an immediate consequence of (1.5.5). $\square$

Exercises

1.5.1. For a locally integrable function f on $\mathbf{R}^n$, $b \in \mathbf{C} \setminus \{0\}$, $\lambda > 0$, and $x_0 \in \mathbf{R}^n$ define the operations $f_\lambda(x) = \lambda^{-n} f(\lambda^{-1} x)$ (L^1 dilation), $\tau^{x_0} f(x) = f(x - x_0)$ (translation), and $\widetilde{f}(x) = f(-x)$ (reflection) for all $x \in \mathbf{R}^n$. Prove the following:

1. $\mathscr{L}_{bf} = \mathscr{L}_f$.
2. $\mathscr{L}_{\widetilde{f}} = -\mathscr{L}_f = \{-y : y \in \mathscr{L}_f\}$.
3. $\mathscr{L}_{\overline{f}} = \mathscr{L}_f$, $\overline{f}$ here denotes complex conjugation.
4. $\mathscr{L}_{\tau^{x_0} f} = x_0 + \mathscr{L}_f = \{x_0 + y : y \in \mathscr{L}_f\}$.
5. $\mathscr{L}_{f_\lambda} = \lambda \mathscr{L}_f = \{\lambda y : y \in \mathscr{L}_f\}$.
6. $\mathscr{L}_{f \circ A} = A^{-1} \mathscr{L}_f = \{A^{-1} y : y \in \mathscr{L}_f\}$, where A is an orthogonal matrix.

Moreover, if g is another locally integrable function, prove that $\mathscr{L}_f \cap \mathscr{L}_g \subseteq \mathscr{L}_{f+g}$.

1.5.2. Show that for every $f \in L^1_{\text{loc}}(\mathbf{R}^n)$ there is a set E_f of measure zero such that

$$\lim_{\varepsilon \to 0} \frac{1}{|B(x,\varepsilon)|} \int_{B(x,\varepsilon)} \left| f(y) - \frac{1}{|B(x,\varepsilon)|} \int_{B(x,\varepsilon)} f(z) dz \right| dy = 0$$

for all $x \in \mathbf{R}^n \setminus E_f$.

1.5.3. Let f be in $L^p(\mathbf{R}^n)$ for some p satisfying $1 \leq p < \infty$. Show that

$$\lim_{\delta \to 0} \frac{1}{|B(x,\delta)|} \int_{B(x,\delta)} |f(y) - f(x)|^p \, dy = 0 \quad \text{for almost all } x \in \mathbf{R}^n.$$

1.5.4. Let g be in $L^p(\mathbf{R}^n)$ for some p satisfying $0 < p < 1$. Show that

$$\lim_{\delta \to 0} \frac{1}{|B(x,\delta)|} \int_{B(x,\delta)} |g(y) - g(x)|^p \, dy = 0 \quad \text{for almost all } x \in \mathbf{R}^n.$$

[*Hint:* For every rational number a there is a set E_a of Lebesgue measure zero such that for $x \in \mathbf{R}^n \setminus E_a$ we have

$$\lim_{\delta \to 0} \frac{1}{|B(x,\delta)|} \int_{B(x,\delta)} |g(y) - a|^p \, dy = |g(x) - a|^p,$$

since the function $y \mapsto |f(y) - a|^p$ is in $L^1_{\text{loc}}(\mathbf{R}^n)$. By considering an enumeration of the rationals, find a set of measure zero E such for $x \notin E$ the preceding limit exists for all rationals a and by continuity for all real numbers a, in particular for $a = g(x)$.]

1.5.5. Given $N \in \mathbf{Z}^+$ and $f \in L^1_{\text{loc}}(\mathbf{R}^n)$, define the function

$$F_N(f) = \sum_{Q \in \mathscr{D}(N)} \left(\frac{1}{|Q|} \int_Q f(y) \, dy \right) \chi_Q,$$

where $\mathscr{D}(N)$ is the set of all cubes Q of the form $\prod_{i=1}^{n} \left[2^{-N}m_i, 2^{-N}(m_i+1)\right)$ with $m_i \in \mathbf{Z}$ and $N = 1,2,3,\ldots$. Prove that $F_N(f) \to f$ a.e. as $N \to \infty$. [*Hint:* Use the previous exercise.]

1.6 Convolution

Definition 1.6.1. Suppose that for given f, g in $L^1_{\text{loc}}(\mathbf{R}^n)$ we have

$$\int_{\mathbf{R}^n} |f(y)| |g(x-y)| \, dy < \infty \qquad \text{for almost all } x \in \mathbf{R}^n. \tag{1.6.1}$$

Then for almost all $x \in \mathbf{R}^n$ we define the *convolution* of f and g as

$$(f * g)(x) = \int_{\mathbf{R}^n} f(y) g(x-y) \, dy.$$

If $f, g \geq 0$ are measurable functions on $\mathbf{R}^n$, then we define $(f*g)(x)$ as the value of the integral of the nonnegative measurable function $y \mapsto f(y)g(x-y)$. Notice that in this case the integral may be infinite for many values of x; see Example 1.6.5.

Changing variables $y' = x - y$, we see that the convolution of f and g can also be written as

$$(f*g)(x) = \int_{\mathbf{R}^n} g(y') f(x-y') \, dy' = (g*f)(x)$$

whenever the integral converges absolutely. Hence convolution is a *commutative* operation. It is also *associative*, in the sense $f*(g*h) = (f*g)*h$ a.e., provided

$$\int_{\mathbf{R}^n} \int_{\mathbf{R}^n} |f(y)| |g(z-y)| |h(x-z)| \, dz \, dy < \infty \quad \text{for almost all } x \in \mathbf{R}^n. \tag{1.6.2}$$

But (1.6.2) is a consequence of

$$\int_{\mathbf{R}^n} \int_{\mathbf{R}^n} \int_{\mathbf{R}^n} |f(y)| |g(z-y)| |h(x-z)| \, dz \, dy \, dx < \infty,$$

a fact that can be verified by applying Tonelli's theorem (integrating first in x, then in z, and finally in y).

Remark 1.6.2. If both f and g are integrable functions, then (1.6.1) holds. Indeed,

$$\int_{\mathbf{R}^n} \int_{\mathbf{R}^n} |f(y)| |g(x-y)| \, dy \, dx$$
$$= \int_{\mathbf{R}^n} \int_{\mathbf{R}^n} |f(y)| |g(x-y)| \, dx \, dy$$
$$= \int_{\mathbf{R}^n} |f(y)| \int_{\mathbf{R}^n} |g(x-y)| \, dx \, dy$$

1.6 Convolution

$$= \int_{\mathbf{R}^n} |f(y)| \int_{\mathbf{R}^n} |g(x)| \, dx \, dy$$
$$= \|f\|_{L^1(\mathbf{R}^n)} \|g\|_{L^1(\mathbf{R}^n)}$$
$$< +\infty,$$

having used Tonelli's theorem. Thus (1.6.1) holds. Since

$$|f * g| \leq |f| * |g|,$$

we conclude

$$\|f * g\|_{L^1(\mathbf{R}^n)} \leq \|f\|_{L^1(\mathbf{R}^n)} \|g\|_{L^1(\mathbf{R}^n)}. \tag{1.6.3}$$

Example 1.6.3. On $\mathbf{R}$ consider the convolution of the two characteristic functions $\chi_{[-a,a]}$ and $\chi_{[-b,b]}$, where $0 < a \leq b < \infty$. Then for any $x \in \mathbf{R}$ we obtain

$$(\chi_{[-a,a]} * \chi_{[-b,b]})(x) = |[-a,a] \cap [-b+x,b+x]|$$

and a straightforward calculation yields that

$$|[-a,a] \cap [-b+x,b+x]| = \begin{cases} 2a & \text{when } |x| \leq b-a, \\ a+b-|x| & \text{when } b-a < |x| \leq a+b, \\ 0 & \text{when } |x| > a+b. \end{cases}$$

The following example indicates how the convolution improves smoothness.

Example 1.6.4. Let $h = \chi_{[-1,1]}$. A calculation gives that $(h*h)(x) = 2 - |x|$ for $|x| \leq 2$ and $(h*h)(x) = 0$ for $|x| > 2$. Also $(h*h*h)(x)$ equals

$$\begin{cases} 3 - |x|^2 & \text{if } |x| \leq 1, \\ 4 - 2|x| + \frac{(|x|-1)^2}{2} & \text{if } 1 < |x| \leq 3, \\ 0 & \text{if } 3 < |x|. \end{cases}$$

It turns out that $h*h$ is continuous (but not continuously differentiable) and $h*h*h$ lies in $\mathscr{C}^1$ but not in $\mathscr{C}^2$. The graphs of $h*h*h$ and its derivative are shown in Figure 1.3.

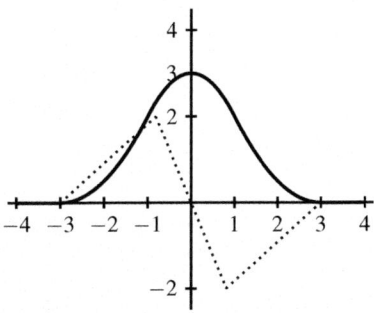

Fig. 1.3 The triple convolution $h*h*h$ and its piecewise linear derivative (dotted) are plotted.

Example 1.6.5. We compute the convolution of $\chi_{[-1,1]}$ and $|x|^{-1}$ on $\mathbf{R}$:

$$h(x) = (\chi_{[-1,1]} * |\cdot|^{-1})(x) = \int_{-1+x}^{1+x} \frac{dy}{|y|} = \begin{cases} \log \frac{x+1}{x-1} & \text{if } |x| > 1, \\ \infty & \text{if } |x| \leq 1. \end{cases}$$

Notice that $h(x)$ decays like $1/|x|$ as $|x| \to \infty$. In general, the convolution inherits the worst decay of the two functions; see Exercise 1.8.4.

The following inequality concerning convolution is of fundamental importance.

Theorem 1.6.6. *(Minkowski convolution inequality)* Let $1 \leq p \leq \infty$. *For f in $L^p(\mathbf{R}^n)$ and g in $L^1(\mathbf{R}^n)$ we have that $g*f$ exists a.e. and satisfies*

$$\|g*f\|_{L^p(\mathbf{R}^n)} \leq \|g\|_{L^1(\mathbf{R}^n)} \|f\|_{L^p(\mathbf{R}^n)}. \tag{1.6.4}$$

Proof. We may assume that $1 < p < \infty$, since the case $p = 1$ was considered in Remark 1.6.2 and the case $p = \infty$ is straightforward. We first show that

$$(|g| * |f|)(x) = \int_{\mathbf{R}^n} |f(x-y)| \, |g(y)| \, dy \tag{1.6.5}$$

exists a.e. Applying Hölder's inequality in (1.6.5) with respect to the measure $|g(y)|dy$ to the functions $y \mapsto f(x-y)$ and 1 with exponents p and p', respectively, we obtain

$$(|g| * |f|)(x) \leq \left(\int_{\mathbf{R}^n} |f(x-y)|^p |g(y)| \, dy \right)^{\frac{1}{p}} \left(\int_{\mathbf{R}^n} |g(y)| \, dy \right)^{\frac{1}{p'}}. \tag{1.6.6}$$

Taking L^p norms of both sides of (1.6.6) we deduce

$$\begin{aligned}
\big\| |g| * |f| \big\|_{L^p} &\leq \left(\|g\|_{L^1}^{p-1} \int_{\mathbf{R}^n} \int_{\mathbf{R}^n} |f(x-y)|^p |g(y)| \, dy \, dx \right)^{\frac{1}{p}} \\
&= \left(\|g\|_{L^1}^{p-1} \int_{\mathbf{R}^n} \int_{\mathbf{R}^n} |f(x-y)|^p \, dx \, |g(y)| \, dy \right)^{\frac{1}{p}} \\
&= \left(\|g\|_{L^1}^{p-1} \int_{\mathbf{R}^n} \int_{\mathbf{R}^n} |f(x)|^p \, dx \, |g(y)| \, dy \right)^{\frac{1}{p}} \\
&= \left(\|f\|_{L^p}^p \|g\|_{L^1} \|g\|_{L^1}^{p-1} \right)^{\frac{1}{p}} \\
&= \|f\|_{L^p} \|g\|_{L^1} < \infty,
\end{aligned}$$

using Tonelli's theorem. This shows that $|g| * |f|$ is finite a.e. and satisfies (1.6.4). Thus the integral in (1.6.5) converges absolutely, hence it converges. This yields that $f * g$ is well defined and satisfies (1.6.4) as $|g * f| \leq |g| * |f|$. $\square$

In proving L^p estimates, it is often convenient to work with continuous functions with compact support. As $L^p(\mathbf{R}^n)$ functions can be approximated by step functions and step functions can be approximated by trapezoidal functions in the L^p norm, we see that continuous functions with compact support are dense in $L^p(\mathbf{R}^n)$. A slightly more general fact is proven in Proposition 1.7.4.

Theorem 1.6.7. *Let $1 \leq p \leq \infty$. For given $f \in L^p(\mathbf{R}^n)$ and $g \in L^{p'}(\mathbf{R}^n)$ the function $f * g$ is uniformly continuous and bounded.*

1.6 Convolution

Proof. Let us assume $1 \leq p < \infty$. The case $p = \infty$ can be handled by reversing the roles of f and g. Given $\varepsilon > 0$, let φ be a continuous function with compact support such that $\|f - \varphi\|_{L^p} < \varepsilon$. Let us suppose that the support of φ is contained in $B(0, M)$. Then φ is uniformly continuous, so there is $\delta > 0$ such that

$$x \in \mathbf{R}^n, |h| < \delta \implies |\varphi(x+h) - \varphi(x)| < \varepsilon |B(0, M+1)|^{-\frac{1}{p}}.$$

For $|h| < \min(\delta, 1)$ Hölder's inequality yields

$$\left|(\varphi * g)(x+h) - (\varphi * g)(x)\right| \leq \left[\int_{|y| \leq M+1} |\varphi(y+h) - \varphi(y)|^p \, dy\right]^{\frac{1}{p}} \|g\|_{L^{p'}}$$

$$\leq (\varepsilon |B(0, M+1)|^{-\frac{1}{p}}) |B(0, M+1)|^{\frac{1}{p}} \|g\|_{L^{p'}}.$$

Then for $|h| < \min(\delta, 1)$ we have

$$|(f*g)(x+h) - (f*g)(x)|$$
$$\leq |(\varphi * g)(x+h) - (\varphi * g)(x)| + |((f - \varphi) * g)(x+h) - ((f - \varphi) * g)(x)|$$
$$\leq \varepsilon \|g\|_{L^{p'}} + 2 \|f - \varphi\|_{L^p} \|g\|_{L^{p'}}$$
$$\leq 3\varepsilon \|g\|_{L^{p'}}.$$

This proves the uniform continuity of $f * g$ on $\mathbf{R}^n$. Its boundedness is a consequence of Hölder's inequality. $\square$

Exercises

1.6.1. Show that the support of the convolution of two functions is contained in the algebraic sum[6] of the supports of the two functions.

1.6.2. Let f, g, h be nonnegative measurable functions on $\mathbf{R}^n$ and let $1 \leq p < \infty$. Prove that

$$((f*g)^p * h)^{\frac{1}{p}} \leq \min\left[f * (g^p * h)^{\frac{1}{p}}, g * (f^p * h)^{\frac{1}{p}}\right].$$

[*Hint:* Use the Minkowski integral inequality.]

1.6.3. Let $\alpha \in \mathbf{R}^+$ and $\beta \in \mathbf{R}$. Consider the functions $g(t) = e^{-\alpha t} \chi_{t>0}$ and $h(t) = e^{i\beta t}$ defined on the real line. Show that for any positive integer m we have

$$\underbrace{g * \cdots * g}_{m \text{ times}} * h = (\alpha + i\beta)^{-m} h.$$

1.6.4. Consider the Gaussian function $G(x) = e^{-\pi |x|^2}$ on $\mathbf{R}^n$. Show that $(G * G)(x) = G(x/\sqrt{2})/(\sqrt{2})^n$. [*Hint:* Change variables $y = y' + \frac{x}{2}$.]

---

[6] The algebraic sum of the sets A and B is the set $A + B = \{a + b : a \in A, b \in B\}$.

1.6.5. (a) Let $f \in L^1(\mathbf{R}^n)$ and $g \in L^\infty(\mathbf{R}^n)$ and suppose that g has compact support. Prove that $(f * g)(x) \to 0$ as $|x| \to \infty$.
(b) Provide examples of $f \in L^1(\mathbf{R})$ compactly supported and $g \in L^\infty(\mathbf{R})$ noncompactly supported, such that $|f * g|$ is a constant; hence the assertion in (a) fails.

1.6.6. Let K be a positive integrable function on $\mathbf{R}^n$ and let $1 \le p \le \infty$. Prove that the norm of the operator $T(f) = f * K$ from $L^p(\mathbf{R}^n)$ to itself is equal to $\|K\|_{L^1}$.
$\bigl[$*Hint:* Clearly, $\|T\|_{L^p \to L^p} \le \|K\|_{L^1}$. Conversely, fix $0 < \varepsilon < 1$ and let N be a positive integer. Let $\chi_N = \chi_{B(0,N)}$ and for any $R > 0$ let $K_R = K\chi_{B(0,R)}$, where $B(x,R)$ is the ball of radius R centered at x. Observe that for $|x| \le (1-\varepsilon)N$, we have $B(0,N\varepsilon) \subseteq B(x,N)$; thus $\int_{\mathbf{R}^n} \chi_N(x-y) K_{N\varepsilon}(y)\, dy = \int_{\mathbf{R}^n} K_{N\varepsilon}(y)\, dy = \|K_{N\varepsilon}\|_{L^1}$. Then for $p < \infty$

$$\frac{\|K * \chi_N\|_{L^p}^p}{\|\chi_N\|_{L^p}^p} \ge \frac{\|K_{N\varepsilon} * \chi_N\|_{L^p(B(0,(1-\varepsilon)N))}^p}{\|\chi_N\|_{L^p}^p} \ge \|K_{N\varepsilon}\|_{L^1}^p (1-\varepsilon)^n.$$

Let $N \to \infty$ first and then $\varepsilon \to 0$. The case $p = \infty$ is straightforward.$\bigr]$

1.6.7. Let $1 < p < \infty$. (a) Let K be an integrable function on $\mathbf{R}^n$. Show that

$$\|f * K\|_{L^{p,\infty}} \le \frac{p}{p-1} \|K\|_{L^1} \|f\|_{L^{p,\infty}}$$

for all f in $L^{p,\infty}$. Thus the operator $f \mapsto f * K$ maps $L^{p,\infty}(\mathbf{R}^n)$ to $L^{p,\infty}(\mathbf{R}^n)$.
(b) Let $K \in L^{p,\infty}(\mathbf{R}^n)$. Prove that the operator $f \mapsto f * K$ maps $L^1(\mathbf{R}^n)$ to $L^{p,\infty}(\mathbf{R}^n)$ with norm at most $\frac{p}{p-1}\|K\|_{L^{p,\infty}}$.
$\bigl[$*Hint:* Part (a): Use Theorem 1.2.10. Part (b): Reverse the roles of f and K in (a).$\bigr]$

1.6.8. Let K be a nonnegative function in $L^1_{\mathrm{loc}}(\mathbf{R}^n)$ and let $0 < p \le \infty$. Suppose that there is a positive constant C such that the inequality

$$\|f * K\|_{L^{p,\infty}} \le C \|f\|_{L^p}$$

holds for all nonnegative functions f in $L^{p,\infty}$. Prove that $K \in L^1(\mathbf{R}^n)$. Obtain the same conclusion when $\|f\|_{L^p}$ is replaced by $\|f\|_{L^{p,\infty}}$ in the hypothesis. [*Hint:* Use that

$$\chi_{B(0,2R)} * K \ge \left(\int_{B(0,R)} K(x)\, dx\right) \chi_{B(0,R)}$$

and let $R \to \infty$. Here $B(0,r) = \{x : |x| < r\}$.]

1.6.9. Let Ω be a measurable subset of $\mathbf{R}^n$ and let $K \ge 0$ be an even measurable function on $\mathbf{R}^n$. Let $T_K(f) = f * K$ for f measurable.
(a) Show that for $1 \le p \le \infty$ we have

$$\|T_K\|_{L^p(\Omega) \to L^p(\mathbf{R}^n)} \le \|F\|_{L^\infty(\Omega)}, \quad \text{where} \quad F(x) = \int_\Omega K(x-y)\, dy.$$

(b) Assume $p = 1$. Show that $\|F\|_{L^\infty(\Omega)} < \infty$ if and only if $\|T_K\|_{L^1(\Omega) \to L^1(\mathbf{R}^n)} < \infty$. [*Hint:* Part (b): Define $\Omega_m = \{x \in \mathbf{R}^n : F(x) \geq m, |x| \leq m\}$. Assuming $\|F\|_{L^\infty(\Omega)} = \infty$ and $|\Omega_m| > 0$, show that $\int_\Omega T_K(f_m)(x)dx \geq m$, where $f_m = \frac{1}{|\Omega_m|}\chi_{\Omega_m}$.]

1.7 Smoothness and Smooth Functions with Compact Support

We begin by introducing notation relevant to several variables. We denote the magnitude of $x = (x_1, \ldots, x_n) \in \mathbf{R}^n$ by $|x| = (x_1^2 + \cdots + x_n^2)^{1/2}$. The *partial derivative* of a function f on $\mathbf{R}^n$ with respect to the jth variable x_j, if it exists, is denoted by $\partial_j f$. The *gradient* of a function f is the vector $\nabla f = (\partial_1 f, \ldots, \partial_n f)$, assuming $\partial_j f$ exist for all j. Higher-order partial derivatives of a function can be obtained by multiple applications of ∂_j. In particular, the mth partial derivative of f with respect to the jth variable is denoted by $\partial_j^m f$, if it exists.

Let $N \in \mathbf{Z}^+$. The space of functions in $\mathbf{R}^n$ all of whose partial derivatives of order at most N are continuous is denoted by $\mathscr{C}^N(\mathbf{R}^n)$ and the space of all *infinitely differentiable functions* on $\mathbf{R}^n$ by $\mathscr{C}^\infty(\mathbf{R}^n)$; functions in $\mathscr{C}^\infty(\mathbf{R}^n)$ are also called *smooth*. The space of $\mathscr{C}^\infty$ functions with compact support on $\mathbf{R}^n$ is denoted by $\mathscr{C}_0^\infty(\mathbf{R}^n)$. A *multi-index* α is an ordered n-tuple of nonnegative integers. For a multi-index $\alpha = (\alpha_1, \ldots, \alpha_n)$, $|\alpha| = \alpha_1 + \cdots + \alpha_n$ denotes its total size (or magnitude) and $\alpha! = \alpha_1! \cdots \alpha_n!$ denotes the product of the factorials of its entries. Given a multi-index $\alpha = (\alpha_1, \ldots, \alpha_n)$ and f in $\mathscr{C}^{|\alpha|}(\mathbf{R}^n)$, $\partial^\alpha f$ denotes the mixed derivative $\partial_1^{\alpha_1} \cdots \partial_n^{\alpha_n} f$ which remains invariant if the partial derivatives are taken in a different order. Then $|\alpha|$ indicates the *total number of derivatives* that appear in $\partial^\alpha f$. Finally, for $1 \leq j \leq n$, the vector e_j is defined as the element of $\mathbf{R}^n$ all of whose coordinates are zero except for the jth one, which equals 1.

For $x \in \mathbf{R}^n$ and $\alpha = (\alpha_1, \ldots, \alpha_n)$ a multi-index, we set $x^\alpha = x_1^{\alpha_1} \cdots x_n^{\alpha_n}$. Multi-indices will be denoted by the letters $\alpha, \beta, \gamma, \delta, \ldots$. It is straightforward to verify that

$$|x^\alpha| \leq |x|^{|\alpha|}. \tag{1.7.1}$$

The converse inequality in (1.7.1) fails as one coordinate of x may vanish. However, for each $k = 1, 2, \ldots$ we have for some $0 < c_{n,k} < d_{n,k} < \infty$, the following inequality:

$$c_{n,k}|x|^k \leq \sum_{|\beta|=k} |x^\beta| \leq d_{n,k}|x|^k \tag{1.7.2}$$

for all $x \in \mathbf{R}^n \setminus \{0\}$. To prove (1.7.2), we notice that the function

$$y \mapsto \sum_{|\beta|=k} |y^\beta|$$

defined on $\mathbf{R}^n$ has no zeros on the unit sphere and is continuous. Then it is bounded above and below by constants on the unit sphere $\mathbf{S}^{n-1}$. These constants appear in the double inequality (1.7.2). A related inequality is

$$C_{n,k}(1+|x|)^k \leq \sum_{|\beta|\leq k} |x^\beta| \leq D_{n,k}(1+|x|)^k. \tag{1.7.3}$$

This follows from (1.7.2) for $|x| \geq 1$, while for $|x| < 1$ both terms in (1.7.3) are at least 1 as $|x^{(0,\ldots,0)}| = 1$. Here again $0 < C_{n,k} < D_{n,k} < \infty$.

We end the preliminaries by noting the validity of the one-dimensional *Leibniz rule*

$$\frac{d^m}{dt^m}(fg) = \sum_{k=0}^{m} \binom{m}{k} \frac{d^k f}{dt^k} \frac{d^{m-k}g}{dt^{m-k}}, \tag{1.7.4}$$

for all $\mathscr{C}^m$ functions f,g on $\mathbf{R}$, and its multidimensional analog

$$\partial^\alpha(fg) = \sum_{\beta \leq \alpha} \binom{\alpha_1}{\beta_1} \cdots \binom{\alpha_n}{\beta_n} (\partial^\beta f)(\partial^{\alpha-\beta}g), \tag{1.7.5}$$

for f,g in $\mathscr{C}^{|\alpha|}(\mathbf{R}^n)$ for some multi-index α, where the notation $\beta \leq \alpha$ in (1.7.5) means that β ranges over all multi-indices satisfying $0 \leq \beta_j \leq \alpha_j$ for all $1 \leq j \leq n$. We observe that identity (1.7.5) is easily deduced by repeated application of (1.7.4), which in turn is obtained by induction.

Theorem 1.7.1. *Let $m \in \mathbf{Z}^+$, $1 \leq q < \infty$, $g \in L^q(\mathbf{R}^n)$, and $\varphi \in \mathscr{C}^m(\mathbf{R}^n) \cap L^{q'}(\mathbf{R}^n)$. Moreover, assume that $\partial^\alpha \varphi$ lies in $L^{q'}(\mathbf{R}^n)$ for all multi-indices α with $|\alpha| \leq m$. Then $\varphi * g$ lies in $\mathscr{C}^m(\mathbf{R}^n)$ and $\partial^\alpha(\varphi * g) = (\partial^\alpha \varphi) * g \in L^\infty$ for all $|\alpha| \leq m$.*

Proof. Let e_j be the unit vector $(0, \ldots, 1, \ldots, 0)$ with 1 in the jth entry and zeros in all the other entries. If $q > 1$ we initially make the additional assumption that g has compact support. If $q = 1$ this initial assumption is not necessary.

We fix an arbitrary $x_0 \in \mathbf{R}^n$ and we show that $\varphi * g$ has a jth partial derivative at x_0, i.e., prove the case $\alpha = e_j$. Using the fundamental theorem of calculus write

$$\Lambda(g,\varphi)(t,x_0) = \frac{(g*\varphi)(x_0+te_j) - (g*\varphi)(x_0)}{t} - g*\partial_j\varphi(x_0)$$

$$= \int_0^1 \int_{\mathbf{R}^n} \left[(\partial_j\varphi(y+tse_j) - \partial_j\varphi(y))g(x_0-y)\right] dy\, ds. \tag{1.7.6}$$

We note that the integrand in (1.7.6) tends pointwise to zero as $t \to 0$ by the fact that $\varphi \in \mathscr{C}^1$; moreover it is bounded by $2\|\partial_j\varphi\|_{L^\infty(x_0-\mathrm{supp}g)}|g(x_0-\cdot)|$ which lies in $L^1(\mathbf{R}^n \times [0,1], dy\,ds)$. The last assertion follows by the hypotheses of the theorem if $q = 1$ and by the additional assumption that it lies in L^q and is supported in a compact set, on which $\partial_j\varphi$ is certainly bounded, if $q > 1$. The LDCT then yields that $\Lambda(g,\varphi)(t,x_0) \to 0$ as $t \to 0$ for any fixed $x_0 \in \mathbf{R}^n$ when g has compact support.

We now remove the assumption that g has compact support if $q > 1$. Given $g \in L^q$, set $g_M(x) = g(x)\chi_{|x|\leq M}$ for $M > 0$. Given a point x_0 and $\varepsilon > 0$ we find an M such that $\|g - g_M\|_{L^q} < \varepsilon$ (as $q < \infty$) and we pick a $\delta > 0$ such that for $|t| < \delta$ we have $|\Lambda(g_M,\varphi)(t,x_0)| < \varepsilon$. Additionally, by Hölder's inequality we obtain

$$\left|\Lambda(g - g_M, \varphi)(t,x_0)\right| \leq \int_0^1 \|g - g_M\|_{L^q} \|\partial_j\varphi(\cdot + tse_j) - \partial_j\varphi\|_{L^{q'}} ds \leq 2\varepsilon \|\partial_j\varphi\|_{L^{q'}}.$$

1.7 Smoothness and Smooth Functions with Compact Support

Combining these estimates, for $|t| < \delta$, we deduce

$$\left|\Lambda(g,\varphi)(t,x_0)\right| \leq \left|\Lambda(g_M,\varphi)(t,x_0)\right| + \left|\Lambda(g-g_M,\varphi)(t,x_0)\right| \leq \varepsilon + 2\left\|\partial_j\varphi\right\|_{L^{q'}}\varepsilon.$$

Thus $\Lambda(g,\varphi)(t,x_0)$ converges to zero as $t \to 0$; hence $\partial_j(\varphi*g) = (\partial_j\varphi)*g$ on $\mathbf{R}^n$.

The continuity and boundedness of $\partial_j(\varphi*g)$ is derived from that of $(\partial_j\varphi)*g$ by Theorem 1.6.7. Finally, assuming $\partial^\alpha\varphi \in L^{q'}$ for all $|\alpha| \leq m$, we obtain the existence, continuity, and boundedness of $\partial^\alpha(\varphi*g)$ via the identity $\partial^\alpha(\varphi*g) = (\partial^\alpha\varphi)*g$, which is proved by induction on m, applying the case $m=1$ repeatedly. $\square$

The following corollary shows that the convolution inherits the best degree of smoothness of the two functions.

Corollary 1.7.2. *Let $\varphi \in \mathscr{C}^\infty(\mathbf{R}^n)$, and $g \in L^1(\mathbf{R}^n)$. Assume that $\partial^\alpha\varphi$ lies in $L^\infty(\mathbf{R}^n)$ for all multi-indices α. Then $\varphi*g$ lies in $\mathscr{C}^\infty(\mathbf{R}^n)$.*

Proposition 1.7.3. (a) *There exists a nonnegative and nonzero smooth function supported in a given ball $B(x_0,R)$ in $\mathbf{R}^n$.*
(b) *Given $0 < r_1 < r_2 < \infty$ there exists a smooth function with values in $[0,1]$ supported in $\overline{B(0,r_2)}$ and equal to 1 on $\overline{B(0,r_1)}$.*

Proof. (a) On the real line define the function

$$g(t) = \begin{cases} e^{-1/t} & \text{when } t > 0, \\ 0 & \text{when } t \leq 0. \end{cases}$$

Notice that $g \in \mathscr{C}^m(\mathbf{R}\setminus\{0\})$, while for $t = 0$ we have

$$\lim_{t\to 0^+} \frac{g(t)-g(0)}{t} = \lim_{t\to 0^+} \frac{e^{-1/t}-0}{t} = 0, \quad \lim_{t\to 0^-} \frac{g(t)-g(0)}{t} = \lim_{t\to 0^-} \frac{0-0}{t} = 0,$$

so $g'(0) = 0$. In a similar way we can show $g''(0) = 0$ and in general $g^{(k)}(0) = 0$ for all k. All these assertions make use of the fact that $\lim_{t\to 0^+} t^{-L}e^{-1/t} = 0$ for any $L > 0$. Then the function $h(t) = g(1-t)g(1+t)$ is smooth, nonnegative, nonzero, and is supported in $[-1,1]$, and so does $h(t^2)$; see Figure 1.4.

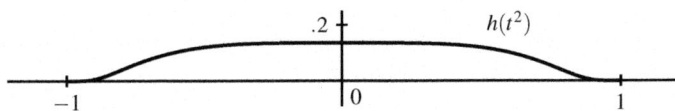

Fig. 1.4 The function $h(t^2)$.

On $\mathbf{R}^n$ consider the function $H(x) = h(|x|^2)$, which is a composition of two smooth functions, hence it is smooth, nonzero, and is supported in the unit ball. Then $H((x-x_0)/R)$ is smooth, nonzero, and supported in the ball $B(x_0,R)$.

(b) Start with a nonzero smooth function $f \geq 0$ supported in the ball $B(0, \frac{1}{2}(r_2 - r_1))$ with integral 1. Define the function $\varphi = \chi_{B(0, \frac{1}{2}(r_2+r_1))} * f$. Then φ is supported in

$$\overline{B(0, \tfrac{1}{2}(r_2 - r_1))} + \overline{B(0, \tfrac{1}{2}(r_2 + r_1))} = \overline{B(0, r_2)},$$

is nonzero, and is smooth by Theorem 1.7.1. Moreover, for $|x| \leq r_1$ we have

$$\varphi(x) = \int_{|x-y| \leq \frac{1}{2}(r_2+r_1)} f(y)\,dy = \int_{\text{supp } f} f(y)\,dy = 1,$$

as $\{y \in \mathbf{R}^n : |y - x| \leq \frac{1}{2}(r_2 + r_1)\}$ contains $\{y \in \mathbf{R}^n : |y| \leq \frac{1}{2}(r_2 - r_1)\}$, which in turn contains the support of f. Finally, notice that $0 \leq \varphi \leq \|f\|_{L^1} = 1$. $\square$

Proposition 1.7.4. *For any $0 < p < \infty$, $\mathscr{C}_0^\infty(\mathbf{R}^n)$ is dense in $L^p(\mathbf{R}^n)$.*

Proof. It suffices to prove the assertion for nonnegative functions, as a complex-valued function can be written as $f_1 - f_2 + i(f_3 - f_4)$, where $f_j \geq 0$, $j = 1, 2, 3, 4$. Given $f \geq 0$ in $L^p(\mathbf{R}^n)$ and $\varepsilon > 0$ we pick a step function $s = \sum_{j=1}^N b_j \chi_{Q_j}$ with $\|f - s\|_{L^p}^{\min(1,p)} < \varepsilon^{\min(1,p)}/2$, where $b_j > 0$ and Q_j are disjoint cubes. We claim that for each j there is a $\mathscr{C}_0^\infty$ function h_j supported in Q_j with values in $[0,1]$ such that

$$\|\chi_{Q_j} - h_j\|_{L^p}^{\min(1,p)} < \frac{\varepsilon^{\min(1,p)}}{2N b_j^{\min(1,p)}}. \tag{1.7.7}$$

To construct h_j, if Q_j has side length $\ell(Q_j)$, by Proposition 1.7.3 we find a $\mathscr{C}_0^\infty$ function φ_j supported in $[-\frac{1}{2}, \frac{1}{2}]$ with values in $[0,1]$ such that

$$\|\chi_{[-\frac{1}{2}, \frac{1}{2}]} - \varphi_j\|_{L^p(\mathbf{R})}^{\min(1,p)} < \frac{\varepsilon^{\min(1,p)}}{2N b_j^{\min(1,p)}} \frac{1}{n} \frac{1}{\ell(Q_j)^{\frac{\min(1,p)n}{p}}}. \tag{1.7.8}$$

Define $h_j(x_1, \ldots, x_n) = \prod_{k=1}^n \varphi_j\big((x_k - c_{j,k})/\ell(Q_j)\big)$, where $c_{j,k}$ is the kth coordinate of the center c_j of Q_j. Then we derive (1.7.7) using (1.7.8), the identity

$$a_1 a_2 \cdots a_n - a_1' a_2' \cdots a_n' = \sum_{k=1}^n a_1 \cdots a_{k-1} (a_k - a_k') a_{k+1}' \cdots a_n'$$

(with the obvious modifications when $k = 1, n$) and the the subadditivity of the expression $\|\cdot\|_{L^p}^{\min(1,p)}$. Finally we obtain from (1.7.7) that

$$\Big\|s - \sum_{j=1}^N b_j h_j \Big\|_{L^p}^{\min(1,p)} \leq \sum_{j=1}^N b_j^{\min(1,p)} \|\chi_{Q_j} - h_j\|_{L^p}^{\min(1,p)} < \frac{\varepsilon^{\min(1,p)}}{2},$$

which, combined with $\|f - s\|_{L^p}^{\min(1,p)} < \varepsilon^{\min(1,p)}/2$, yields $\|f - \sum_{j=1}^N b_j h_j\|_{L^p} < \varepsilon$; so the claimed density is valid. $\square$

1.7 Smoothness and Smooth Functions with Compact Support

Note that Proposition 1.7.4 fails when $p = \infty$ as the function 1 cannot be approximated in L^∞ by smooth functions with compact support.

Exercises

1.7.1. Let $a \in \mathbf{R}$. Show that $|\nabla |x|^a| = |a| |x|^{a-1}$ for all $x \in \mathbf{R}^n \setminus \{0\}$.

1.7.2. For $t \in (0,1)$ define $h(t) = e^{-1/t} e^{-1/(1-t)}$ and set $h(t) = 0$ for $t \notin (0,1)$. Show that $h \in \mathscr{C}_0^\infty(\mathbf{R})$ and that

$$x \mapsto \int_{-\infty}^x h(t)\,dt \left(\int_0^1 h(s)\,ds \right)^{-1}$$

is a smooth increasing function that vanishes for $x \leq 0$ and equals 1 for $x \geq 1$.

1.7.3. Let $0 < r_1 < r_2 < r_3 < r_4 < \infty$.
(a) Prove that there is a $\mathscr{C}^\infty(\mathbf{R}^n)$ function which vanishes when $|x| \leq r_1$ and is identically equal to 1 for $|x| \geq r_2$.
(b) Prove that there is a $\mathscr{C}_0^\infty(\mathbf{R}^n)$ function supported in the annulus $r_1 \leq |x| \leq r_4$ which is equal to 1 on the annulus $r_2 \leq |x| \leq r_3$.

1.7.4. Let $a > 1$ and $a \notin \mathbf{Z}$. Prove that $|x|^a$ lies in $\mathscr{C}^{[a]}(\mathbf{R}^n) \setminus \mathscr{C}^{[a]+1}(\mathbf{R}^n)$.

1.7.5. (Poincaré inequality) Let $-\infty < a < b < \infty$ and let u be a smooth function supported in $\Omega = (a,b) \times \mathbf{R}^{n-1}$. Prove that

$$\int_\Omega |u|^2 \, dx \leq (b-a)^2 \int_\Omega |\nabla u|^2 \, dx.$$

[*Hint:* Start with $|u(x_1, x')|^2 = \left| \int_a^{x_1} \partial_1 u(t, x')\,dt \right|^2$ for $x_1 \leq b$, apply the Cauchy–Schwarz inequality, and then integrate over $x_1 \in (a,b)$ and over $x' \in \mathbf{R}^{n-1}$.]

1.7.6. (Partitions of unity) Let K be a compact subset of $\mathbf{R}^n$ and let $\{U_\alpha\}_{\alpha \in I}$ be an open cover of K. Show that there are functions ψ_j, $j = 1, \ldots, L$, such that

(a) ψ_j are nonnegative and smooth for all $j = 1, \ldots, L$.
(b) $\sum_{j=1}^L \psi_j = 1$ on a neighborhood of K.
(c) For each j there is an index $\alpha \in I$ such that ψ_j is supported in U_α.

Such a family of functions ψ_j is called a *partition of unity subordinate to* $\{U_\alpha\}_{\alpha \in I}$. [*Hint:* For each $x \in K$ find concentric open balls B_x, B_x' such that $x \in B_x \subsetneq B_x' \subset U_\alpha$ for some α depending on x. Pass to a finite subcover $B_{x_1}, \ldots, B_{x_L}$ of K. Choose nonnegative smooth functions ϕ_j supported in B_{x_j}' and equal to 1 on B_{x_j} by Proposition 1.7.3. Then set

$$\psi_1 = \phi_1, \quad \text{and} \quad \psi_{j+1} = \phi_{j+1} \prod_{i=1}^j (1 - \phi_i)$$

for $1 \leq j < L$.]

1.8 Schwartz Functions

For a pair of multi-indices α and β and a function $f \in \mathscr{C}^\infty(\mathbf{R}^n)$ we define the $\rho_{\alpha,\beta}$ *Schwartz seminorm*[7] of f by

$$\rho_{\alpha,\beta}(f) = \sup_{x \in \mathbf{R}^n} |x^\alpha \partial^\beta f(x)|.$$

Naturally, this quantity could be infinite for certain smooth functions.

Definition 1.8.1. A $\mathscr{C}^\infty$ complex-valued function f on $\mathbf{R}^n$ is called a Schwartz function if for all multi-indices α and β we have $\rho_{\alpha,\beta}(f) < \infty$. The space of all Schwartz functions on $\mathbf{R}^n$ is denoted by $\mathscr{S}(\mathbf{R}^n)$.

Thus a $\mathscr{C}^\infty$ function is called Schwartz if and only if for every multi-index β and every $N \in \mathbf{Z}^+$ there is a constant $C_{N,\beta}$ such that for all $x \in \mathbf{R}^n$ we have

$$|\partial^\beta f(x)| \le \frac{C_{N,\beta}}{(1+|x|)^N}.$$

Obviously, every smooth function with compact support is a Schwartz function, i.e., $\mathscr{C}_0^\infty(\mathbf{R}^n)$ is contained in $\mathscr{S}(\mathbf{R}^n)$.

Example 1.8.2. The function $e^{-|x|^2}$ lies in $\mathscr{S}(\mathbf{R}^n)$ but $e^{-|x|}$ does not, since the latter fails to be differentiable at the origin. The $\mathscr{C}^\infty$ function $g(x) = (1+|x|^2)^{-10}$ is not in $\mathscr{S}(\mathbf{R}^n)$, as $\rho_{\alpha_1,0}(g) = \infty$ for $\alpha_1 = (21,0,\ldots,0)$ and $0 = (0,\ldots,0)$.

Example 1.8.3. The function $e^{-1/x} e^{-x} \chi_{[0,\infty)}$ lies in $\mathscr{S}(\mathbf{R})$ as $e^{-1/x}\chi_{[0,\infty)}$ is infinitely differentiable at the origin with vanishing derivatives of all orders.

Proposition 1.8.4. *Let f, g be in $\mathscr{S}(\mathbf{R}^n)$ and $c \in \mathbf{C}$. Then $f+g$, cf, fg, and $f*g$ lie in $\mathscr{S}(\mathbf{R}^n)$.*

Proof. The only nontrivial assertion is that $\partial^\beta(f*g)$ has rapid decay at infinity. For each $N > 0$ there are constants $C_{N,\beta}$ and $C'_{N+n+1,0}$ such that

$$\left| \int_{\mathbf{R}^n} (\partial^\beta f)(x-y) g(y)\, dy \right| \le \int_{\mathbf{R}^n} \frac{C_{N,\beta}}{(1+|x-y|)^N} \frac{C'_{N+n+1,0}}{(1+|y|)^{N+n+1}}\, dy. \quad (1.8.1)$$

Inserting the simple estimate $(1+|x-y|)^{-N} \le (1+|y|)^N (1+|x|)^{-N}$ in (1.8.1) we deduce that

$$|(\partial^\beta f * g)(x)| \le C_{N,\beta,n} (1+|x|)^{-N} \int_{\mathbf{R}^n} (1+|y|)^{-n-1} dy = C(N,\beta,n)(1+|x|)^{-N},$$

and this proves the rapid decay of $\partial^\beta f * g$ at infinity. $\square$

---

[7] The Schwartz seminorm is in fact a norm (see Exercise 1.8.2).

1.8 Schwartz Functions

Next we define convergence on the space of Schwartz functions

Definition 1.8.5. Let $\{f_j\}_{j=1}^\infty$ be a sequence of Schwartz functions. We say that f_j converges to a Schwartz function f *in the Schwartz topology*, or simply in $\mathscr{S}(\mathbf{R}^n)$, if $\rho_{\alpha,\beta}(f_j - f) \to 0$ as $j \to \infty$ for all multi-indices α, β. We then write $f_j \to f$ in $\mathscr{S}$.

In particular, if $f_j \to f$ in $\mathscr{S}$ as $j \to \infty$, then for all multi-indices β, the sequence $\partial^\beta f_j - \partial^\beta f$ tends to zero uniformly on $\mathbf{R}^n$.

Example 1.8.6. The sequence of Schwartz functions $f_j(x) = e^{-1/x}e^{-jx}\chi_{[0,\infty)}$ on the real line converges to zero in $\mathscr{S}(\mathbf{R})$ as $j \to \infty$. To verify this assertion, first we notice that for each $m \in \mathbf{Z}^+$ there is a polynomial P_m of degree $2m$ such that

$$\frac{d^m}{dx^m}(e^{-\frac{1}{x}}) = P_m(\tfrac{1}{x})e^{-\frac{1}{x}},$$

a fact that will be tacitly used in the sequel. Now for $j \geq 1$ and for nonnegative integers K, L we estimate

$$\rho_{K,L}(f_j)$$
$$\leq \sum_{l=0}^L \binom{L}{l} j^{L-l} \sup_{x \geq 1}\left|x^K e^{-jx}\frac{d^l}{dx^l}(e^{-\frac{1}{x}})\right| + \sum_{l=0}^L \binom{L}{l} j^{L-l} \sup_{0 \leq x < 1} e^{-jx}\left|\frac{d^l}{dx^l}(e^{-\frac{1}{x}})\right|.$$

The first supremum tends to zero as $j \to \infty$ since $e^{-jx} \leq e^{-j/2}e^{-x/2}$ when $j, x \geq 1$. In the second supremum notice that the lth derivative of $e^{-1/x}$ on $[0,1)$ is bounded by $C_M x^M$ for any $M \in \mathbf{Z}^+$. Choosing $M = L+1$ we bound the second term by

$$C'_L j^L e^{-jx} x^{L+1} \leq \frac{C'_L}{j} \sup_{t>0}(t^{L+1}e^{-t}),$$

which also tends to zero as $j \to \infty$.

Theorem 1.8.7. *The space $\mathscr{C}_0^\infty(\mathbf{R}^n)$ is dense in $\mathscr{S}(\mathbf{R}^n)$ in the Schwartz topology. Precisely, fix a smooth function φ with values in $[0,1]$ supported in $B(0,2)$ and equal to 1 on the unit ball $B(0,1)$. Then for any $f \in \mathscr{S}(\mathbf{R}^n)$, the sequence $f_j(x) = f(x)\varphi(x/j)$ converges to $f(x)$ in the Schwartz topology as $j \to \infty$.*

Proof. For fixed multi-indices α and β we show that $\rho_{\alpha,\beta}(f\varphi(\cdot/j) - f)$ tends to zero as $j \to \infty$. By Leibniz's rule we estimate this Schwartz seminorm by

$$\sum_{\substack{\gamma \leq \beta \\ \gamma \neq \beta}} \binom{\beta}{\gamma}\frac{1}{j^{|\beta|-|\gamma|}} \sup_{x \in \mathbf{R}^n}\left|x^\alpha(\partial^\gamma f)(x)(\partial^{\beta-\gamma}\varphi)\left(\tfrac{x}{j}\right)\right| + \sup_{x \in \mathbf{R}^n}\left|x^\alpha(\partial^\beta f)(x)\left(\varphi\left(\tfrac{x}{j}\right) - 1\right)\right|.$$

As $(\partial^{\beta-\gamma}\varphi)(x/j)$ remains bounded for all j, the first term tends to zero as $j \to \infty$, since $|\beta| - |\gamma| \geq 1$. As $\varphi(x/j) - 1 = 0$ $|x| < j$, the second supremum equals

$$\sup_{|x|\geq j}\frac{1}{|x|^2}\left|\left(\varphi\left(\frac{x}{j}\right)-1\right)\right|\left[|x|^2|x^\alpha(\partial^\beta f)(x)|\right]\leq\frac{1}{j^2}\left[\frac{1}{c_{n,2}}\sum_{k=1}^n\sum_{l=1}^n\rho_{\alpha+e_k+e_l,\beta}(f)\right],$$

which tends to 0 as $j \to \infty$. In the last inequality we made use of (1.7.2) and we set $e_k = (0,\ldots,0,1,0,\ldots,0)$, the multi-index with 1 in its kth coordinate and zero elsewhere. $\square$

Exercises

1.8.1. Which of the following functions $e^{ix}e^{-x^2}$, $e^{ie^x}e^{-x^2}$, $e^{ie^{x^2}}e^{-x^2}$, $e^{-|x|^3}$ lie in the Schwartz class $\mathscr{S}(\mathbf{R})$?

1.8.2. Suppose that a Schwartz function φ satisfies $\rho_{\alpha,\beta}(\varphi) = 0$ for some multi-indices α, β. Prove that $\varphi \equiv 0$.

1.8.3. Let $\varphi \in \mathscr{S}(\mathbf{R}^n)$ and P be a polynomial. Show that $P\varphi \in \mathscr{S}(\mathbf{R}^n)$.

1.8.4. Let $M, N > n$ and suppose that f, g are functions on $\mathbf{R}^n$ that satisfy the estimates $|f(x)| \leq A(1+|x|)^{-M}$ and $|g(x)| \leq B(1+|x|)^{-N}$ for all $x \in \mathbf{R}^n$. Prove that for some $C = C(n,N,M) > 0$ we have

$$|(f*g)(x)| \leq CAB(1+|x|)^{-\min(N,M)} \qquad \text{for all } x \in \mathbf{R}^n.$$

1.8.5. Show that the convolution of a Schwartz function with a compactly supported integrable function is another Schwartz function.

1.8.6. Suppose that $f_j \to f$ in the Schwartz topology as $j \to \infty$. Prove for all multi-indices γ and δ, $\partial^\gamma f_j \to \partial^\gamma f$ and $(\cdot)^\delta f_j \to (\cdot)^\delta f$ in the Schwartz topology, as $j \to \infty$.

1.8.7. Let f_j, f, g_j, g be Schwartz functions for $j = 1, 2, \ldots$. Suppose that $f_j \to f$ and $g_j \to g$ in $\mathscr{S}(\mathbf{R}^n)$ as $j \to \infty$. Prove that $f_j g_j \to fg$ in $\mathscr{S}(\mathbf{R}^n)$, as $j \to \infty$.

1.8.8. Let $0 < p \leq \infty$. Prove that there is a constant $C = C_{n,p}$ such that

$$\|\varphi\|_{L^p} \leq C \sum_{|\alpha|\leq[n/p]+1} \rho_{\alpha,0}(\varphi)$$

for all $\varphi \in \mathscr{S}(\mathbf{R}^n)$. Conclude that if $f_j \to f$ in the Schwartz topology, then $f_j \to f$ in L^p. [*Hint:* Use inequality (1.7.3).]

1.8.9. Let f_j, f, g_j, g be Schwartz functions for $j = 1, 2, \ldots$. Suppose that $f_j \to f$ and $g_j \to g$ in $\mathscr{S}(\mathbf{R}^n)$ as $j \to \infty$. Show that $f_j * g_j \to f * g$ in $\mathscr{S}(\mathbf{R}^n)$, as $j \to \infty$. [*Hint:* Using the inequality $|x^\alpha| \leq 2^{|\alpha|}(|x-y|^{|\alpha|}+|y|^{|\alpha|})$ and (1.7.3) prove first that

$$\rho_{\alpha,\beta}(\varphi*\psi) \leq C\left[\rho_{0,0}(\psi)\sum_{|\gamma|\leq n+1+|\alpha|}\rho_{\gamma,\beta}(\varphi)+\rho_{0,\beta}(\varphi)\sum_{|\gamma|\leq n+1+|\alpha|}\rho_{\gamma,0}(\psi)\right],$$

for some constant C that depends on n and α. Use the previous exercise.]

1.9 Approximate Identities

The futile search for a function f_0 in $L^1(\mathbf{R}^n)$ such that $f_0 * f = f$ for all $f \in L^1(\mathbf{R}^n)$ leads to the notion of approximate identities.

Definition 1.9.1. An *approximate identity* is a family of L^1 functions K_δ on $\mathbf{R}^n$ with the following three properties:

(i) There exists a constant $c > 0$ such that $\|K_\delta\|_{L^1(\mathbf{R}^n)} \leq c$ for all $\delta > 0$.
(ii) $\int_{\mathbf{R}^n} K_\delta(y)\,dy = 1$ for all $\delta > 0$.
(iii) For any $\gamma > 0$ we have that $\int_{|y|>\gamma} |K_\delta(y)|\,dy \to 0$ as $\delta \to 0$.

We also define approximate identities as sequences $\{K_m\}_{m=1}^\infty$, in which properties (i) and (ii) are valid for all $m = 1, 2, \ldots$ and property (iii) holds as $m \to \infty$.

Example 1.9.2. Let K be an integrable function on $\mathbf{R}^n$ with integral 1. Let $K_\delta(x) = \delta^{-n} K(\delta^{-1} x)$. We claim that the family $\{K_\delta\}_{\delta > 0}$ is an approximate identity. Properties (i) and (ii) are immediate, while property (iii) follows from the fact that

$$\int_{|x| \geq \gamma} |K_\delta(x)|\,dx = \int_{|x'| \geq \gamma/\delta} |K(x')|\,dx' \to 0$$

as $\delta \to 0$ for γ fixed.

Approximate identities can be thought of as families of positive functions K_δ that spike near 0 as δ becomes smaller, in such a way that the area under the graph of each function remains equal to 1, as shown in Figure 1.5.

An important example is provided by the function

$$K(x) = (\pi(x^2 + 1))^{-1},$$

which gives rise to the family

$$K_\delta(x) = \delta^{-1} K(\delta^{-1} x)$$
$$= \frac{1}{\pi} \frac{\delta}{\delta^2 + x^2}, \quad \delta > 0,$$

called the *Poisson kernel*.

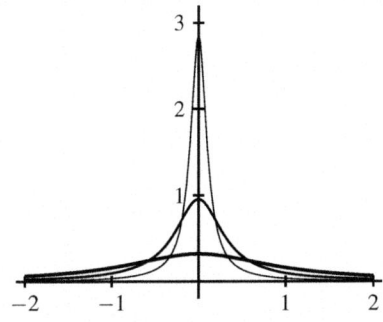

Fig. 1.5 The Poisson kernel shown for $\delta = 1$, $\delta = 1/3$, and $\delta = 1/9$ (spikiest).

Lemma 1.9.3. *Given $0 < p < \infty$ and $f \in L^p(\mathbf{R}^n)$ we have that*

$$\int_{\mathbf{R}^n} |f(x-y) - f(x)|^p\,dx \to 0 \quad \text{as } |y| \to 0.$$

Proof. Let $m_p = \min(1,p)$ for $0 < p < \infty$. The proof will be based on the subadditivity of the expression $f \mapsto \|f\|_{L^p}^{m_p}$. Given $\varepsilon > 0$ and $f \in L^p$ we find a continuous function with compact support h such that $\|f - h\|_{L^p}^{m_p} < \varepsilon^{m_p}/3$. We write

$$\|f(\cdot - y) - f\|_{L^p}^{m_p} \leq \|f(\cdot - y) - h(\cdot - y)\|_{L^p}^{m_p} + \|h(\cdot - y) - h\|_{L^p}^{m_p} + \|h - f\|_{L^p}^{m_p}.$$

Now suppose that h is supported in the compact ball $\overline{B(0,R)}$. Then for $|y| < 1$ the function $h(\cdot - y) - h$ is supported in $\overline{B(0,R+1)}$ and is bounded above by the constant $2\|h\|_{L^\infty}$. Hence $|h(x-y) - h(x)| \leq 2\|h\|_{L^\infty} \chi_{\overline{B(0,R+1)}} \in L^1(\mathbf{R}^n)$. By continuity we have that for all $x \in \mathbf{R}^n$

$$h(x-y) - h(x) \to 0$$

pointwise as $y \to 0$. The LDCT gives that $\|h(\cdot - y) - h\|_{L^p}^{m_p} \to 0$ as $|y| \to 0$. It follows that there is a $\delta > 0$ such that

$$|y| < \delta \implies \|h(\cdot - y) - h\|_{L^p}^{m_p} < \frac{\varepsilon^{m_p}}{3}.$$

Consequently, when $|y| < \min(1,\delta)$ we have $\|f(\cdot - y) - f\|_{L^p}^{m_p} < 3(\frac{\varepsilon^{m_p}}{3}) = \varepsilon^{m_p}$. $\square$

Let $\{K_\delta\}_{\delta > 0}$ be an approximate identity on $\mathbf{R}^n$. If f lies in $L^p(\mathbf{R}^n)$ for $1 \leq p < \infty$, then $K_\delta * f$ lies in $L^p(\mathbf{R}^n)$ and hence it is defined almost everywhere. However, if f lies in $L^\infty(\mathbf{R}^n)$, then the integral defining $K_\delta * f$ converges absolutely; hence $(K_\delta * f)(x)$ is well defined for every $x \in \mathbf{R}^n$.

Theorem 1.9.4. *Let $\{K_\delta\}_{\delta > 0}$ be an approximate identity.*

(a) *If f lies in $L^p(\mathbf{R}^n)$ for $1 \leq p < \infty$, then $\|K_\delta * f - f\|_{L^p(\mathbf{R}^n)} \to 0$ as $\delta \to 0$.*
(b) *If f in $L^\infty(\mathbf{R}^n)$ then $\|K_\delta * f - f\|_{L^\infty(E)} \to 0$ as $\delta \to 0$ whenever $E \subseteq \mathbf{R}^n$ and f is uniformly continuous in a neighborhood of E, in the sense that*

$$\forall \varepsilon > 0 \, \exists \delta > 0 \text{ such that } x \in E, \, |y| < \delta \implies |f(x-y) - f(x)| < \varepsilon. \quad (1.9.1)$$

Proof. (a) Let $f \in L^p(\mathbf{R}^n)$, $1 \leq p \leq \infty$. Since K_δ has integral 1, for almost all $x \in \mathbf{R}^n$ we have

$$(K_\delta * f)(x) - f(x) = \int_{\mathbf{R}^n} f(x-y) K_\delta(y) \, dy - f(x) \int_{\mathbf{R}^n} K_\delta(y) \, dy$$
$$= \int_{\mathbf{R}^n} (f(x-y) - f(x)) K_\delta(y) \, dy.$$

(Note that when $p = \infty$ the preceding identity holds for all $x \in \mathbf{R}^n$.) Taking L^p norms in x and applying Minkowski's integral inequality we obtain

$$\|K_\delta * f - f\|_{L^p(\mathbf{R}^n)} \leq \int_{\mathbf{R}^n} \|f(\cdot - y) - f\|_{L^p(\mathbf{R}^n)} |K_\delta(y)| \, dy. \quad (1.9.2)$$

Let $\varepsilon > 0$ be given. Let us first assume that $1 \leq p < \infty$. By Lemma 1.9.3, there is a $\gamma > 0$ such that

1.9 Approximate Identities

where c is the constant that appears in Definition 1.9.1 (i).

$$|y| < \gamma \implies \|f(\cdot - y) - f\|_{L^p(\mathbf{R}^n)} \leq \frac{\varepsilon}{2c} \quad (1.9.3)$$

In view of (1.9.3) it follows that

$$\int_{|y|<\gamma} \|f(\cdot-y)-f\|_{L^p(\mathbf{R}^n)} |K_\delta(y)|\,dy \leq \frac{\varepsilon}{2c} \int_{|y|<\gamma} |K_\delta(y)|\,dy \leq \frac{\varepsilon}{2c} c = \frac{\varepsilon}{2}. \quad (1.9.4)$$

Now choose δ_0 such that for $0 < \delta < \delta_0$ we have $\int_{|y|\geq\gamma} |K_\delta(y)|\,dy < \frac{\varepsilon}{4(1+\|f\|_{L^p})}$ by property (iii) in Definition 1.9.1. Then for $0 < \delta < \delta_0$ we have

$$\int_{|y|\geq\gamma} \|f(\cdot-y)-f\|_{L^p(\mathbf{R}^n)} |K_\delta(y)|\,dy \leq 2\|f\|_{L^p} \int_{|y|\geq\gamma} |K_\delta(y)|\,dy < \frac{\varepsilon}{2}. \quad (1.9.5)$$

Combining (1.9.2), (1.9.4), and (1.9.5), we deduce that for $0 < \delta < \delta_0$ we have $\|K_\delta * f - f\|_{L^p(\mathbf{R}^n)} < \varepsilon$. In other words, $K_\delta * f \to f$ in $L^p(\mathbf{R}^n)$.

(b) Now consider the case where $p = \infty$. The fact that $f \in L^\infty$ is uniformly continuous in a neighborhood of E implies that given $\varepsilon > 0$ there is a $\gamma > 0$ such that

$$x \in E, \ |y| < \gamma \implies |f(x-y) - f(x)| < \frac{\varepsilon}{2c}.$$

That is, (1.9.3) holds when $L^p(\mathbf{R}^n)$ is replaced by $L^\infty(E)$ and the proof proceeds as that in part (a). $\square$

Remark 1.9.5. Condition (1.9.1) holds in two important situations: (a) when E is compact and f is continuous on $E + \overline{B(0,\varepsilon_0)}$ for some $\varepsilon_0 > 0$, and (b) when E is a finite set and f is continuous at every point in E. In particular, if $f \in L^\infty$ is continuous at a point x_0, then $(K_\delta * f)(x_0) \to f(x_0)$ as $\delta \to 0$. Moreover, if the supports of K_δ shrink to $\{0\}$ as $\delta \to 0$, then $(K_\delta * f)(x_0)$ is well defined for δ sufficiently small without the assumption that f is globally bounded, and

$$(K_\delta * f)(x_0) - f(x_0) = \int_{\mathbf{R}^n} K_\delta(y)[f(x_0 - y) - f(x_0)]\,dy \to 0,$$

as long as f is merely continuous at x_0 (but could be unbounded near another point).

Example 1.9.6. On the line consider the approximate identity $h_\varepsilon = \frac{1}{2\varepsilon}\chi_{[-\varepsilon,\varepsilon]}$. To see how the family $h_\varepsilon * f$ tends to a function f, we take f to be the characteristic function of an interval $[a,b]$. The graph of $\chi_{[a,b]} * h_\varepsilon$ is depicted in Figure 1.6 for ε small enough. It is clear from the picture that $\chi_{[a,b]} * h_\varepsilon$ converges to $\chi_{[a,b]}$ in L^p for $p < \infty$ and also pointwise at the points of continuity of the characteristic function, i.e., all points but a,b.

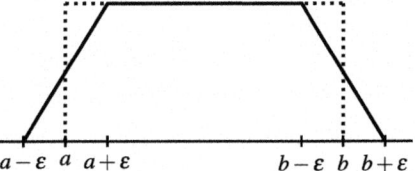

Fig. 1.6 The graph of $\chi_{[a,b]} * h_\varepsilon$ for $\varepsilon < \frac{b-a}{2}$. The dotted lines represent the function $\chi_{[a,b]}$.

A simple modification in the proof of Theorem 1.9.4 yields the following variant.

Theorem 1.9.7. *Let K_δ be a family of functions on $\mathbf{R}^n$ that satisfies properties* (i) *and* (iii) *of Definition 1.9.1 and also*

$$\int_{\mathbf{R}^n} K_\delta(y)\,dy = A \qquad (1.9.6)$$

for some fixed $A \in \mathbf{C}$ and for all $\delta > 0$.
(a) *If $f \in L^p(\mathbf{R}^n)$ for some $1 \le p < \infty$, then $\|K_\delta * f - Af\|_{L^p(\mathbf{R}^n)} \to 0$ as $\delta \to 0$.*
(b) *If f in $L^\infty(\mathbf{R}^n)$, then $\|K_\delta * f - Af\|_{L^\infty(E)} \to 0$ as $\delta \to 0$, provided f is uniformly continuous in a neighborhood of a subset E of $\mathbf{R}^n$ in the sense of* (1.9.1).

A family of functions $\{K_\delta\}_{\delta>0}$ that satisfies properties (i) and (iii) of Definition 1.9.1 and also (1.9.6) for some $A \neq 0$ is called an A-multiple of an approximate identity. In the case where $A = 0$, it is called *an approximate zero family*.

As an application of the notion of approximate identities we show that $\mathscr{C}_0^\infty(\mathbf{R}^n)$ is a dense subspace of $L^p(\mathbf{R}^n)$ for all $1 \le p < \infty$.

Example 1.9.8. Given $f \in L^p(\mathbf{R}^n)$ and $\varepsilon > 0$ we find a compactly supported function h such that $\|f - h\|_{L^p(\mathbf{R}^n)} < \varepsilon/2$. In fact such an h can be chosen to be $f\chi_{|f|<M}$ for some large M (since $f\chi_{|f|<M} \to f$ in $L^p(\mathbf{R}^n)$ as $M \to \infty$ by the LDCT). Next we find a compactly supported smooth function K on $\mathbf{R}^n$ with integral 1 and we consider the approximate identity $\{K_\delta\}_{\delta>0}$. Then in view of Theorem 1.9.4, there is a $\delta > 0$ such that $\|K_\delta * h - h\|_{L^p(\mathbf{R}^n)} < \varepsilon/2$. It follows that $\|K_\delta * h - f\|_{L^p(\mathbf{R}^n)} < \varepsilon$ and notice that $K_\delta * h$ is both smooth and compactly supported.

Exercises

1.9.1. Show that for all $x \in \mathbf{R}$ we have

$$\lim_{\varepsilon \to 0^+} \varepsilon \int_{\mathbf{R}} \frac{y\cos(\sin(x-y))}{(y^2+\varepsilon^2)^{3/2}}\,dy = 0.$$

Moreover, the convergence is uniform in x when $|x| \le M$, for any $M > 0$.

1.9.2. For $m = 1, 2, \dots$ let B_m be balls in $\mathbf{R}^n$ that contain the origin and whose measures shrink to 0 as $m \to \infty$. Prove that the family of functions $|B_m|^{-1}\chi_{B_m}$ is an approximate identity. Write $B_m = B_m^+ \cup B_m^-$, where B_m^+, B_m^- are disjoint and equimeasurable. Show that the sequence $|B_m|^{-1}\chi_{B_m^+} - |B_m|^{-1}\chi_{B_m^-}$ is an approximate zero family as $m \to \infty$.

1.9.3. Let $Q_m(t) = c_m(1-t^2)^m$ for $t \in [-1,1]$ and zero elsewhere, where c_m is a constant chosen such that $\int_{-1}^1 Q_m(t)\,dt = 1$ for all $m = 1, 2, \dots$.
(a) Prove that $c_m \le (m+1)/2$. [*Hint:* Use $(1-t^2)^m \ge (1-|t|)^m$ when $|t| \le 1$.]

(b) Prove that $\{Q_m\}_{m=1}^\infty$ is an approximate identity on $\mathbf{R}$ as $m \to \infty$.

(c) Given a continuous function f on $\mathbf{R}$ supported in $[0,1]$, show that

$$c_m \int_0^1 f(t)(1-(x-t)^2)^m \, dt = (f * Q_m)(x), \qquad x \in [0,1],$$

is a sequence of polynomials that converges to f uniformly on $[0,1]$ as $m \to \infty$.

1.9.4. (Fejér kernel) Show that the sequence of functions $\{F_m\}_{m=1}^\infty$ defined on the real line by

$$F_m(t) = \frac{1}{m+1} \left| \frac{\sin(\pi(m+1)t)}{\sin(\pi t)} \right|^2 \chi_{|t| \leq \frac{1}{2}} = \frac{1}{m+1} \left| \sum_{k=0}^m e^{2\pi i k t} \right|^2 \chi_{|t| \leq \frac{1}{2}}$$

is an approximate identity as $m \to \infty$. Also verify the preceding identity.

1.9.5. (Continuous Fejér kernel) Show that the family $\{F_\varepsilon\}_{\varepsilon > 0}$ defined on $\mathbf{R}^n$ by

$$F_\varepsilon(x_1, \ldots, x_n) = \frac{1}{\varepsilon^n} \prod_{j=1}^n \left(\frac{\sin(\pi x_j / \varepsilon)}{\pi x_j / \varepsilon} \right)^2$$

forms an approximate identity. [*Hint:* Use Exercise 2.3.3.]

1.9.6. Let $0 < b < 1$ and z be a complex number with $|z| < 1/b$. Let f be a continuous function on the line. Prove that

$$\lim_{m \to \infty} \frac{1}{b^m} \sum_{j=m}^\infty z^{j-m} \int_{b^{j+1}}^{b^j} f(t) \, dt = \frac{1-b}{1-bz} f(0).$$

[*Hint:* Apply Theorem 1.9.7 for a suitable approximate identity sequence $\{K_m\}_{m=1}^\infty$.]

1.9.7. Let Ω be an open subset of $\mathbf{R}^n$. Denote by $\mathscr{C}_0^\infty(\Omega)$ the space of all smooth functions with compact support contained in Ω. Suppose that f, g are locally integrable functions on Ω (i.e., they lie in $L^1(K)$ for any compact subset K of Ω). Assume that

$$\int f(x) \varphi(x) \, dx = \int g(x) \varphi(x) \, dx \qquad \text{for all } \varphi \text{ in } \mathscr{C}_0^\infty(\Omega).$$

Prove that $f = g$ a.e. on Ω. [*Hint:* It will suffice to prove $f = g$ a.e. on every ball $B(x_0, r)$ contained in Ω. Let $\varepsilon_0 = \text{dist}(\overline{B(x_0, r)}, \Omega^c)$ (or $\varepsilon_0 = 1$ if $\Omega = \mathbf{R}^n$). Pick a smooth function with integral 1 supported in the unit ball of $\mathbf{R}^n$ and define $h = f - g$ on $\overline{B(x_0, r + \frac{\varepsilon_0}{3})}$ and zero elsewhere. Then $h \in L^1(\mathbf{R}^n)$ and $\varphi_\varepsilon * h$ is supported in $\overline{B(x_0, r + \frac{2\varepsilon_0}{3})}$ and vanishes on $\overline{B(x_0, r)}$ for $\varepsilon < \varepsilon_0/3$. Apply Theorem 1.9.4 (a).]

Chapter 2
Fourier Transforms, Tempered Distributions, Approximate Identities

2.1 The Fourier Transform on L^1

Given the 1-periodic function

$$h(x) = -5e^{-(2\pi i)x} - 3 + 2e^{(2\pi i)x} + 17e^{3(2\pi i)x} + 5e^{4(2\pi i)x} - 2e^{6(2\pi i)x}$$

on the line, the sequence $\{a_k\}_{k=-\infty}^{+\infty}$ of coefficients

...	a_{-3}	a_{-2}	a_{-1}	a_0	a_1	a_2	a_3	a_4	a_5	a_6	a_7	a_8	...
...	0	0	-5	-3	2	0	17	5	0	-2	0	0	...

lists the number of times the exponentials $e^{k(2\pi i)x}$ appear in the function. In other words, the constant a_k "quantifies the presence" of the term $e^{k(2\pi i)x}$. Examining the magnitude of these terms, we can determine, for instance, which frequency is the most dominant (in our example it is the third). The number a_k is called the *Fourier coefficient* of h and can be isolated from the function h by integrating against the conjugate exponential of $e^{k(2\pi i)x}$, i.e.,

$$a_k = \int_0^1 h(x)e^{-2\pi i k x}dx.$$

Analogous examples can be written for functions of two or more variables that are 1-periodic in each variable. Motivated by the periodic situation, we extend the notion of "Fourier coefficient sequence" to nonperiodic functions on Euclidean spaces.

We denote the inner product of two points $x = (x_1, \ldots, x_n)$, $y = (y_1, \ldots, y_n)$ in $\mathbf{R}^n$ by $x \cdot y = \sum_{j=1}^n x_j y_j$.

Definition 2.1.1. Given f in $L^1(\mathbf{R}^n)$ we define its *Fourier transform* by

$$\widehat{f}(\xi) = \int_{\mathbf{R}^n} f(x)e^{-2\pi i x \cdot \xi}dx. \tag{2.1.1}$$

Example 2.1.2. If $\varphi(x) = e^{-\pi x^2}$ defined on $\mathbf{R}$, then $\widehat{\varphi}(\xi) = \varphi(\xi)$ for all real ξ. To prove this, consider the function

$$F(s) = \int_{-\infty}^{+\infty} e^{-\pi(t+is)^2} dt, \qquad s \in \mathbf{R}.$$

For h real in $[-1,1]$ by the fundamental theorem of calculus we write

$$\frac{F(s+h) - F(s)}{h} = -2\pi i \int_{-\infty}^{+\infty} \int_0^1 e^{-\pi(t+is+ih\theta)^2} (t+is+ih\theta) d\theta dt,$$

and by the LDCT we can pass the limit in $h \to 0$ inside both integrals. We obtain that F is differentiable and

$$F'(s) = \int_{-\infty}^{+\infty} -2\pi i (t+is) e^{-\pi(t+is)^2} dt = \int_{-\infty}^{+\infty} i \frac{d}{dt} \left(e^{-\pi(t+is)^2} \right) dt = 0.$$

Thus F is constant and equal to $F(0) = 1$. Using this fact, we calculate the Fourier transform of the function $\varphi(t) = e^{-\pi t^2}$ on $\mathbf{R}$ as follows:

$$1 = F(0) = F(s) = \int_{-\infty}^{+\infty} e^{-\pi t^2} e^{-2\pi i t s} e^{\pi s^2} dt = e^{\pi s^2} \widehat{\varphi}(s),$$

from which we conclude that $\widehat{\varphi}(s) = e^{-\pi s^2} = \varphi(s)$ for any $s \in \mathbf{R}$.

Remark 2.1.3. Notice that if $F(x_1, x_2) = f_1(x_1) f_2(x_2)$ where $f_j \in L^1(\mathbf{R})$, then

$$\widehat{F}(\xi_1, \xi_2) = \int_{\mathbf{R}} \int_{\mathbf{R}} f_1(x_1) f_2(x_2) e^{-2\pi i (x_1 \xi_1 + x_2 \xi_2)} dx_1 dx_2 = \widehat{f_1}(\xi_1) \widehat{f_2}(\xi_2).$$

An analogous calculation holds in $\mathbf{R}^n$. Thus the function $x \mapsto e^{-\pi |x|^2}$ on $\mathbf{R}^n$ equals its Fourier transform, i.e., Example 2.1.2 can be extended to all dimensions.

Example 2.1.4. The Fourier transform of the function $\chi_{[a,b]}$ on the real line is

$$\int_a^b e^{-2\pi i x \xi} dx = \begin{cases} \frac{e^{-2\pi i b \xi} - e^{-2\pi i a \xi}}{-2\pi i \xi} & \xi \neq 0, \\ b - a & \xi = 0. \end{cases}$$

This can be expressed as $\sin(2\pi b \xi)/\pi \xi$ when $b = -a > 0$.

Proposition 2.1.5. Let f be in $L^1(\mathbf{R}^n)$. Then $\widehat{f}$ is uniformly continuous on $\mathbf{R}^n$ and

$$\|\widehat{f}\|_{L^\infty} \leq \|f\|_{L^1}. \tag{2.1.2}$$

Moreover, if g also lies in $L^1(\mathbf{R}^n)$, then

$$\widehat{f * g} = \widehat{f} \widehat{g}. \tag{2.1.3}$$

2.1 The Fourier Transform on L^1

Proof. Inequality (2.1.2) is a direct consequence of Definition 2.1.1. To prove the uniform continuity of $\widehat{f}$ let $\varepsilon > 0$ be given. Notice that $f(x)(e^{-2\pi i x \cdot h} - 1) \to 0$ as $h \to 0$ for all $x \in \mathbf{R}^n$; hence by the LDCT, there is a $\delta > 0$ such that

$$|h| < \delta \implies \int_{\mathbf{R}^n} |f(x)| |e^{-2\pi i x \cdot h} - 1| dx < \varepsilon. \tag{2.1.4}$$

But $|\widehat{f}(\xi) - \widehat{f}(\xi')|$ is bounded by the integral in (2.1.4) with $h = \xi - \xi'$; thus when $|\xi - \xi'| < \delta$ we have $|\widehat{f}(\xi) - \widehat{f}(\xi')| < \varepsilon$, which gives that $\widehat{f}$ is uniformly continuous.

Identity (2.1.3) is proved via the following calculation:

$$\begin{aligned}
\widehat{f * g}(\xi) &= \int_{\mathbf{R}^n} \int_{\mathbf{R}^n} f(x-y) g(y) e^{-2\pi i x \cdot \xi} \, dy \, dx \\
&= \int_{\mathbf{R}^n} \int_{\mathbf{R}^n} f(x-y) g(y) e^{-2\pi i (x-y) \cdot \xi} e^{-2\pi i y \cdot \xi} \, dy \, dx \\
&= \int_{\mathbf{R}^n} g(y) \int_{\mathbf{R}^n} f(x-y) e^{-2\pi i (x-y) \cdot \xi} dx \, e^{-2\pi i y \cdot \xi} \, dy \\
&= \widehat{f}(\xi) \widehat{g}(\xi),
\end{aligned}$$

where the application of Fubini's theorem is justified by the absolute convergence of the preceding double integral. $\square$

We now continue with some properties of the Fourier transform. Before we do this we introduce some notation. For a measurable function f on $\mathbf{R}^n$ and $y \in \mathbf{R}^n$ we respectively define the *translation* and *reflection* of f by

$$(\tau^y f)(x) = f(x-y) \quad \text{and} \quad \widetilde{f}(x) = f(-x). \tag{2.1.5}$$

Also recall the L^1 dilation $f_\lambda(x) = \lambda^{-n} f(x/\lambda)$, $\lambda > 0$, that preserves the L^1 norm.

Proposition 2.1.6. *Given f, g in $L^1(\mathbf{R}^n)$, $y \in \mathbf{R}^n$, $b \in \mathbf{C}$, and $\lambda > 0$, we have*

(1) $\widehat{f+g} = \widehat{f} + \widehat{g}$,

(2) $\widehat{bf} = b\widehat{f}$,

(3) $\widehat{\widetilde{f}} = \widetilde{\widehat{f}}$,

(4) $\overline{\widehat{f}} = \widetilde{\widehat{\overline{f}}}$,

(5) $\widehat{\tau^y f} = e^{-2\pi i y \cdot (\cdot)} \widehat{f}$,

(6) $(e^{2\pi i (\cdot) \cdot y} f)\widehat{} = \tau^y(\widehat{f})$,

(7) $[f(\lambda \cdot)]\widehat{} = \lambda^{-n} \widehat{f}(\lambda^{-1} \cdot) = (\widehat{f})_\lambda$,

(8) $\widehat{f \circ A^t} = \dfrac{1}{|\det A|} \widehat{f} \circ A^{-1}$, *if A is a matrix with nonzero determinant,*

(9) $\widehat{f \circ A} = \widehat{f} \circ A$, *where A is an orthogonal matrix.*

Proof. Properties (1)–(4) are straightforward. Properties (5)–(7) require a suitable change of variables but they are omitted. Property (9) follows from (8) as orthogonal matrices satisfy $A^{-1} = A^t$ and $|\det A| = 1$. Finally, we prove (8). Viewing elements of $\mathbf{R}^n$ as column vectors, we set $y = A^t x$. We have $dy = |\det A| dx$ and thus we write

$$\widehat{f \circ A^t}(\xi) = \int_{\mathbf{R}^n} f(A^t x) e^{-2\pi i x \cdot \xi} dx$$

$$= \frac{1}{|\det A^t|} \int_{\mathbf{R}^n} f(y) e^{-2\pi i (A^t)^{-1} y \cdot \xi} dy$$

$$= \frac{1}{|\det A|} \int_{\mathbf{R}^n} f(y) e^{-2\pi i (A^{-1})^t y \cdot \xi} dy$$

$$= \frac{1}{|\det A|} \int_{\mathbf{R}^n} f(y) e^{-2\pi i y \cdot A^{-1} \xi} dy$$

$$= \frac{1}{|\det A|} \widehat{f}(A^{-1} \xi),$$

where we used that $(A^{-1})^t = (A^t)^{-1}$. This proves the identity in (8). □

A measurable function is called *radial* if $f \circ A = f$ a.e. for every orthogonal matrix A. For every radial function f on $\mathbf{R}^n$ there is a function f_0 on the line such that $f(x) = f_0(|x|)$ for almost all $x \in \mathbf{R}^n$.

Corollary 2.1.7. *The Fourier transform of an integrable radial function is radial.*

Proof. Let A be an orthogonal matrix. Since f is radial, we have $f = f \circ A$ a.e. This implies $\widehat{f \circ A} = \widehat{f}$. But in Proposition 2.1.6 (9) gives that $\widehat{f \circ A} = \widehat{f} \circ A$. Thus $\widehat{f} \circ A = \widehat{f}$ and this shows that $\widehat{f}$ is radial. □

Proposition 2.1.8. *Let $\varphi \in \mathscr{C}^\infty(\mathbf{R}^n)$, $\xi \in \mathbf{R}^n$, and α be a multi-index. Then*

(1) $\sup_{|\beta| \le |\alpha|} \sup_{x \in \mathbf{R}^n} (1 + |x|)^{n + \frac{1}{2}} |\partial^\beta \varphi(x)| < \infty$ *implies* $(\partial^\alpha \varphi)\widehat{\ }(\xi) = (2\pi i \xi)^\alpha \widehat{\varphi}(\xi)$.

(2) *If $(1 + |\cdot|)^{|\alpha|} \varphi \in L^1(\mathbf{R}^n)$ then $\partial^\alpha \widehat{\varphi}$ exists and equals the Fourier transform of the function $x \mapsto (-2\pi i x)^\alpha \varphi(x)$.*

(3) $\varphi \in \mathscr{S}(\mathbf{R}^n)$ *implies that* $\widehat{\varphi} \in \mathscr{S}(\mathbf{R}^n)$.

Proof. Property (1) is proved by integrating by parts $|\alpha|$ times, which is justified by the hypothesis that for all $|\beta| \le |\alpha|$, $\partial^\beta \varphi$ are integrable and vanish at ∞. We have

$$(\partial^\alpha \varphi)\widehat{\ }(\xi) = \int_{\mathbf{R}^n} (\partial^\alpha \varphi)(x) e^{-2\pi i x \cdot \xi} dx = \int_{\mathbf{R}^n} \varphi(x)(2\pi i \xi)^\alpha e^{-2\pi i x \cdot \xi} dx = (2\pi i \xi)^\alpha \widehat{\varphi}(\xi)$$

as $(-1)^{|\alpha|}(-2\pi i \xi)^\alpha = (2\pi i \xi)^\alpha$.

To prove (2), let $e_j = (0, \ldots, 1, \ldots, 0)$ have 1 at its jth entry, and zero elsewhere. Since

$$\frac{e^{-2\pi i x \cdot (\xi + h e_j)} - e^{-2\pi i x \cdot \xi}}{h} - (-2\pi i x_j) e^{-2\pi i x \cdot \xi} \to 0, \quad \text{as } h \to 0, \quad (2.1.6)$$

2.1 The Fourier Transform on L^1

and the preceding function is bounded by $4\pi|x|$ for all h and ξ, the Lebesgue dominated convergence theorem implies that the integral of the function in (2.1.6) with respect to the measure $\varphi(x)dx$ converges to zero. This proves (2) for $\alpha = e_j$. Assuming that $\partial^\beta \widehat{\varphi} = ((-2\pi i \cdot)^\beta \varphi)\widehat{}$ for some $|\beta| < |\alpha|$, applying this procedure to $(-2\pi i(\cdot))^\beta \varphi$ in place of φ, we obtain $\partial^\gamma \widehat{\varphi} = ((-2\pi i \cdot)^\gamma \varphi)\widehat{}$ where $\gamma = \beta + e_j$. This process ends when $|\gamma|$ reaches $|\alpha|$ in view of the hypothesis that $(1+|x|)^{|\alpha|}\varphi(x)$ lies in $L^1(\mathbf{R}^n)$. This inductive procedure yields (2) for the given multi-index α.

To prove (3) we employ both (1) and (2) as follows:

$$\|(\cdot)^\alpha \partial^\beta \widehat{\varphi}\|_{L^\infty} = \frac{(2\pi)^{|\beta|}}{(2\pi)^{|\alpha|}}\|[\partial^\alpha((\cdot)^\beta \varphi)]\widehat{}\|_{L^\infty} \le \frac{(2\pi)^{|\beta|}}{(2\pi)^{|\alpha|}}\|\partial^\alpha((\cdot)^\beta \varphi)\|_{L^1} < \infty,$$

where the first inequality uses (2.1.2) [Proposition 2.1.5]. $\square$

Exercises

2.1.1. Let f, g, h be integrable functions on the line. Prove the following:
(a) The Fourier transform of $(x_1, x_2) \mapsto f(x_1 - x_2)g(x_2)$ is $\widehat{f}(\xi_1)\widehat{g}(\xi_1 + \xi_2)$.
(b) The Fourier transform of $(x_1, x_2) \mapsto f(x_1 + x_2)g(x_1 - x_2)$ is

$$(\xi_1, \xi_2) \mapsto \tfrac{1}{2}\widehat{f}(\tfrac{\xi_1+\xi_2}{2})\widehat{g}(\tfrac{\xi_1-\xi_2}{2}).$$

(c) The Fourier transform of $(x_1, x_2, x_3) \mapsto f(x_1 - x_2 - x_3)g(x_2 + x_3)h(x_3)$ on $\mathbf{R}^3$ is

$$(\xi_1, \xi_2, \xi_3) \mapsto \widehat{f}(\xi_1)\widehat{g}(\xi_1 + \xi_2)\widehat{h}(\xi_3 - \xi_2).$$

2.1.2. (a) Compute the Fourier transform of the function $\chi_{[-\frac{1}{2}, \frac{1}{2}]}$ on the real line.
(b) Prove that for $\xi \in \mathbf{R}$ we have

$$\int_{-1}^{1}(1 - |t|)\cos(2\pi\xi t)\,dt = \frac{\sin^2(\pi\xi)}{\pi^2\xi^2},$$

with the proper interpretation when $\xi = 0$.
[*Hint:* Part (b): Use Example 1.6.3 and property (2.1.3) in Proposition 2.1.5.]

2.1.3. Prove that the convolution of two integrable radial functions on $\mathbf{R}^n$ is radial.

2.1.4. Recall that the function $\varphi(x) = e^{-\pi x^2}$ on the real line satisfies $\widehat{\varphi} = \varphi$.
(a) Prove that the Fourier transform of $u(x) = x\varphi(x)$ is $-iu(\xi)$.
(b) Construct a nonzero function v in $\mathscr{S}(\mathbf{R})$ such that $\widehat{v} = -v$.
(c) Construct a nonzero function w in $\mathscr{S}(\mathbf{R})$ such that $\widehat{w} = iw$.
[*Hint:* Parts (b) and (c): Use Proposition 2.1.8 (1) for the second and third derivatives.]

2.1.5. Let $\varphi_j, \varphi \in \mathscr{S}(\mathbf{R}^n)$ such that $\varphi_j \to \varphi$ in $\mathscr{S}$. Show that $\widehat{\varphi}_j \to \widehat{\varphi}$ in $\mathscr{S}$.
[*Hint:* Show that each $\rho_{\alpha,\beta}(\widehat{\varphi}_j - \widehat{\varphi})$ is bounded by a finite sum of $\rho_{\gamma,\delta}(\varphi_j - \varphi)$.]

2.1.6. Suppose that g is a nonzero integrable function on $\mathbf{R}$ with compact support. Prove that there is an entire function that coincides with the Fourier transform of g. Conclude that $\widehat{g}$ cannot vanish on a convergent sequence; in particular it cannot have compact support. Find formulas for the complex derivatives of $\widehat{g}$.

2.2 Fourier Inversion

Definition 2.2.1. Define the *inverse Fourier transform* of a function f in $L^1(\mathbf{R}^n)$ by
$$f^{\vee}(\xi) = \widehat{f}(-\xi), \quad \xi \in \mathbf{R}^n.$$

The inverse Fourier transform is the Fourier transform composed with the reflection $\xi \mapsto -\xi$ and has properties analogous to those listed in Proposition 2.1.6. Its name is justified by the following theorem.

Theorem 2.2.2. (1) (*Jumping hat identity*) For f, g in $L^1(\mathbf{R}^n)$ we have
$$\int_{\mathbf{R}^n} \widehat{f}(\xi) g(\xi) \, d\xi = \int_{\mathbf{R}^n} f(x) \widehat{g}(x) \, dx. \tag{2.2.1}$$

(2) (*Fourier inversion*) If both f and $\widehat{f}$ lie in $L^1(\mathbf{R}^n)$, then
$$(\widehat{f})^{\vee} = f = (f^{\vee})\widehat{} \qquad \text{a.e.} \tag{2.2.2}$$

Thus f is almost everywhere equal to a uniformly continuous function.

(3) (*Parseval's identity*) If $f, h, \widehat{h} \in L^1$, then
$$\int_{\mathbf{R}^n} f(x)\overline{h(x)} \, dx = \int_{\mathbf{R}^n} \widehat{f}(\xi) \overline{\widehat{h}(\xi)} \, d\xi \tag{2.2.3}$$

and
$$\int_{\mathbf{R}^n} f(x) h(x) \, dx = \int_{\mathbf{R}^n} \widehat{f}(x) h^{\vee}(x) \, dx. \tag{2.2.4}$$

Proof. (1) Identity (2.2.1) immediately follows from the definition of the Fourier transform and Fubini's theorem. The absolute convergence of the integrals is justified from the fact that $\widehat{f}$ and $\widehat{g}$ lie in L^{∞} [Proposition 2.1.5].

(2) For fixed $y \in \mathbf{R}^n$ and $f, \widehat{f}$ in $L^1(\mathbf{R}^n)$, we insert in (2.2.1) the function
$$g(\xi) = e^{2\pi i \xi \cdot y} e^{-\pi |\varepsilon \xi|^2}.$$

By Proposition 2.1.6 (6), (7) and Example 2.1.2, we have that

2.2 Fourier Inversion

$$\widehat{g}(x) = \frac{1}{\varepsilon^n} e^{-\pi \varepsilon^{-2}|x-y|^2},$$

and we note that $\frac{1}{\varepsilon^n} e^{-\pi \varepsilon^{-2}|\cdot|^2}$ is an approximate identity as $\varepsilon \to 0$. Now (2.2.1) gives

$$\int_{\mathbf{R}^n} f(x) \varepsilon^{-n} e^{-\pi \varepsilon^{-2}|x-y|^2} dx = \int_{\mathbf{R}^n} \widehat{f}(\xi) e^{2\pi i \xi \cdot y} e^{-\pi |\varepsilon \xi|^2} d\xi. \quad (2.2.5)$$

The left-hand side of (2.2.5) converges to $f(y)$ in L^1 by Theorem 1.9.4 as $\varepsilon \to 0$; thus by Theorem 1.1.9, for a subsequence $\varepsilon_j \to 0$, it converges to $f(y)$ a.e. The right-hand side of (2.2.5) converges to $(\widehat{f})^\vee(y)$ as $\varepsilon_j \to 0$ by the Lebesgue dominated convergence theorem. We conclude that $(\widehat{f})^\vee = f$ a.e. $\mathbf{R}^n$. But $(\widehat{f})^\vee$ is continuous, so f is a.e. equal to a continuous function. Replacing $f^\vee$ by $\widetilde{\widehat{f}}$ in $(f^\vee)^\wedge = (f^\vee)^{\widetilde{\vee}}$ and applying the left identity in (2.2.2) yields a.e. $(f^\vee)^\wedge = \widetilde{\widetilde{f}} = f$. Here $\widetilde{f}(\xi) = f(-\xi)$.

(3) Since both h and $\widehat{h}$ are in L^1, it follows that both functions are in L^∞. Thus all integrals in (2.2.3) and (2.2.4) converge absolutely. Define $g \in L^1$ by setting $\overline{g} = \widehat{h}$. We have that $\widehat{g} = \overline{h}$ by Fourier inversion. Then (2.2.3) is a consequence of (2.2.1) and (2.2.4) follows from (2.2.3) by replacing $\overline{h}$ by h. □

Next, we describe the behavior of the Fourier transform of integrable functions at infinity.

Proposition 2.2.3. *(Riemann–Lebesgue lemma)* *For a function f in $L^1(\mathbf{R}^n)$ we have that*

$$|\widehat{f}(\xi)| \to 0 \qquad \text{as} \qquad |\xi| \to \infty.$$

Proof. Given $\varepsilon > 0$ and f in $L^1(\mathbf{R}^n)$ there is $\varphi \in \mathscr{C}_0^\infty(\mathbf{R}^n)$ such that $\|f - \varphi\|_{L^1} < \varepsilon/2$. But $\widehat{\varphi}$ is a Schwartz function by Proposition 2.1.8. Thus there is an $M > 0$ such that for $|\xi| > M$ we have $|\widehat{\varphi}(\xi)| \le \varepsilon/2$. Then for $|\xi| > M$ we write

$$|\widehat{f}(\xi)| \le |\widehat{\varphi}(\xi)| + |\widehat{f}(\xi) - \widehat{\varphi}(\xi)| \le |\widehat{\varphi}(\xi)| + \|f - \varphi\|_{L^1} < \frac{\varepsilon}{2} + \frac{\varepsilon}{2} = \varepsilon,$$

which shows the claimed assertion. □

Next, we evaluate the Fourier transform of the integrable function $e^{-2\pi|x|}$ on $\mathbf{R}^n$. To achieve this we need the following lemma.

Lemma 2.2.4. *Let f be in $L^1(\mathbf{R})$. Then we have*

$$\int_{-\infty}^{+\infty} f(t - 1/t) dt = \int_{-\infty}^{+\infty} f(u) du. \quad (2.2.6)$$

Proof. Observe that the map $t \mapsto u = t - 1/t$ is a bijection from $(0, \infty)$ onto $(-\infty, \infty)$ and its inverse is $u \to t = \frac{1}{2}(u + \sqrt{u^2 + 4})$. Similarly the map $t \to t - 1/t$ is a bijection from $(-\infty, 0)$ onto $(-\infty, \infty)$ and its inverse is $u \mapsto t = \frac{1}{2}(u - \sqrt{u^2 + 4})$. Therefore

$$\int_0^\infty f(t-1/t)\,dt = \frac{1}{2}\int_{-\infty}^{+\infty} f(u)\left(du + d\sqrt{u^2+4}\right),$$

$$\int_{-\infty}^0 f(t-1/t)\,dt = \frac{1}{2}\int_{-\infty}^{+\infty} f(u)\left(du - d\sqrt{u^2+4}\right).$$

Summing the preceding two identities, we derive (2.2.6). □

Theorem 2.2.5. *The following identity concerning the Fourier transform of the function $e^{-2\pi|x|}$ on $\mathbf{R}^n$ is valid:*

$$\left(e^{-2\pi|\cdot|}\right)^\wedge(\xi) = \frac{\Gamma(\frac{n+1}{2})}{\pi^{\frac{n+1}{2}}} \frac{1}{(1+|\xi|^2)^{\frac{n+1}{2}}}, \qquad \xi \in \mathbf{R}^n. \qquad (2.2.7)$$

Proof. Applying Lemma 2.2.4 to $f(u) = e^{-\pi A u^2}$, for some $A > 0$, we get

$$2\int_0^{+\infty} e^{-\pi A t^2} e^{-\pi \frac{A}{t^2}} e^{2\pi A}\,dt = \int_{-\infty}^{+\infty} e^{-\pi A t^2} e^{-\pi \frac{A}{t^2}} e^{2\pi A}\,dt = \int_{-\infty}^{+\infty} e^{-\pi A u^2}\,du = \frac{1}{\sqrt{A}}$$

from which, by changing variables $s = \pi A t^2$, we obtain the *subordination* identity

$$e^{-2\pi A} = \frac{1}{\sqrt{\pi}} \int_0^\infty e^{-s} e^{-\frac{\pi^2 A^2}{s}} \frac{ds}{\sqrt{s}}, \qquad A > 0. \qquad (2.2.8)$$

Setting $A = |x|$, multiplying (2.2.8) by $e^{-2\pi i \xi \cdot x}$, and integrating with respect to dx we obtain

$$\left(e^{-2\pi|x|}\right)^\wedge(\xi) = \int_{\mathbf{R}^n} e^{-2\pi|x|} e^{-2\pi i \xi \cdot x}\,dx$$

$$= \int_{\mathbf{R}^n} \left(\frac{1}{\sqrt{\pi}} \int_0^\infty e^{-s - \frac{|\pi x|^2}{s}} \frac{ds}{\sqrt{s}}\right) e^{-2\pi i \xi \cdot x}\,dx$$

$$= \frac{1}{\sqrt{\pi}} \int_0^\infty e^{-s} \left(\int_{\mathbf{R}^n} e^{-\frac{|\pi x|^2}{s}} e^{-2\pi i \xi \cdot x}\,dx\right) \frac{ds}{\sqrt{s}}$$

$$= \frac{1}{\sqrt{\pi}} \int_0^\infty e^{-s} \left(\frac{\sqrt{s}}{\sqrt{\pi}}\right)^n e^{-\pi|\frac{\sqrt{s}}{\sqrt{\pi}}\xi|^2} \frac{ds}{\sqrt{s}},$$

where we used Tonelli's theorem and the fact that the Fourier transform of $e^{-\pi|\delta x|^2}$ is $\delta^{-n} e^{-\pi|x/\delta|^2}$ for $\delta > 0$; on this see Proposition 2.1.6 and Example 2.1.2 with the subsequent remark. To properly justify the application of Tonelli's theorem we needed that

$$\int_0^\infty \int_{\mathbf{R}^n} e^{-s} e^{-\frac{|\pi x|^2}{s}}\,dx \frac{ds}{\sqrt{s}} < \infty,$$

a fact that can be easily checked by first evaluating the x integral (which equals a constant multiple of $s^{n/2}$) and plugging this value into the s integral. Thus,

2.2 Fourier Inversion

$$(e^{-2\pi|\cdot|})^\widehat{}(\xi) = \frac{1}{\pi^{\frac{n+1}{2}}} \int_0^\infty e^{-s(1+|\xi|^2)} s^{\frac{n}{2}} \frac{ds}{\sqrt{s}}$$

$$= \frac{1}{\pi^{\frac{n+1}{2}}} \frac{1}{(1+|\xi|^2)^{\frac{n+1}{2}}} \int_0^\infty e^{-s} s^{\frac{n+1}{2}} \frac{ds}{s}$$

$$= \frac{\Gamma(\frac{n+1}{2})}{\pi^{\frac{n+1}{2}}} \frac{1}{(1+|\xi|^2)^{\frac{n+1}{2}}}$$

by the definition of the Γ function. This proves (2.2.7) and also proves that the Fourier transform of the function in (2.2.7) is $e^{-2\pi|x|}$, as for radial functions the Fourier transform and its inverse coincide. $\square$

The *Poisson kernel* on $\mathbf{R}^n$ is the following integrable function:

$$P(x) = \frac{\Gamma(\frac{n+1}{2})}{\pi^{\frac{n+1}{2}}} \frac{1}{(1+|x|^2)^{\frac{n+1}{2}}}.$$

This function plays a very important role in many areas of analysis and partial differential equations. It often appears as part of the family $\{P_t\}_{t>0}$, where $P_t(x) = t^{-n} P(x/t)$. The normalization in terms of the Gamma function ensures that the integral of P equals 1. Indeed, identity (2.2.7) gives $\int_{\mathbf{R}^n} P(x)\,dx = \widehat{P}(0) = e^{-2\pi \cdot 0} = 1$. Moreover, for $t, s > 0$ the Poisson kernel satisfies $P_t * P_s = P_{t+s}$ which can be easily seen by applying the Fourier transform.

Corollary 2.2.6. *The family of functions $P_\varepsilon(x) = \varepsilon^{-n} P(x/\varepsilon)$, $\varepsilon > 0$, is an approximate identity.*

Proof. Properties (i) and (ii) of approximate identities hold as $P \geq 0$ and $\|P\|_{L^1} = 1$. Property (iii) holds as for $\gamma > 0$, $\int_{|x| \geq \gamma} P_\varepsilon(x)\,dx = \int_{|x| \geq \gamma/\varepsilon} P(x)\,dx \to 0$ as $\varepsilon \to 0^+$. $\square$

Exercises

2.2.1. Suppose that $f, \widehat{f} \in L^1(\mathbf{R}^n)$. Prove that $f \in L^p(\mathbf{R}^n)$ for all $1 \leq p \leq \infty$.

2.2.2. Construct a nonnegative nonzero Schwartz function f on $\mathbf{R}^n$ whose Fourier transform is nonnegative and compactly supported. [*Hint:* Take $f = |\phi * \widetilde{\phi}|^2$, where ϕ is the inverse Fourier transforms of an odd, real-valued, and compactly supported function. Here $\widetilde{\phi}(x) = \phi(-x)$.]

2.2.3. Without computing derivatives, prove that the Poisson kernel P_t satisfies

$$\frac{\partial^2}{\partial t^2} P_t + \sum_{j=1}^n \partial_j^2 P_t = 0.$$

[*Hint:* Let $H(t,x)$ be the function on the left. As P_t is homogeneous of degree $-n$ in (x,t) and smooth on $\mathbf{S}^n$, we have $|H(t,x)| \leq C(|t|+|x|)^{-n-2}$. Thus $H(t,\cdot) \in L^1(\mathbf{R}^n)$. Show that the Fourier transform of $H(t,\cdot)$ is zero and then use Theorem 2.2.2 (2).]

2.2.4. For given $0 < a_1, \ldots, a_k < \infty$, prove the identity on $\mathbf{R}^n$
$$\left(e^{-\pi a_1 |\cdot|^2} * \cdots * e^{-\pi a_k |\cdot|^2}\right)(x) = \left(\frac{a}{a_1 \cdots a_k}\right)^{\frac{n}{2}} e^{-\pi a |x|^2},$$
where a is the harmonic mean of $a_1, \ldots, a_k$ (defined by $1/a = 1/a_1 + \cdots + 1/a_k$).

2.2.5. Let $P_y(x) = y^{-1} P(y^{-1} x)$, where P is the Poisson kernel and $y > 0$. Show that
$$\int_{\mathbf{R}} P_y(x-t) \frac{\sin(\pi t)}{\pi t} dt = \frac{y(1 - e^{-\pi y} \cos(\pi x)) + x e^{-\pi y} \sin(\pi x)}{\pi (x^2 + y^2)}.$$

[*Hint:* Use identity (2.2.3).]

2.2.6. Let f, g, h be Schwartz functions on $\mathbf{R}^n$. Derive the identity
$$\int_{\mathbf{R}^n} \int_{\mathbf{R}^n} f(x) g(y) h(x+y) dx dy = \int_{\mathbf{R}^n} \widehat{f}(\xi) \widehat{g}(\xi) \widehat{h}(-\xi) d\xi.$$

2.2.7. Fill in the gaps in the outlined procedure to prove that
$$\int_{-\infty}^{+\infty} \frac{\sin(\pi t)}{\pi t} dt = \lim_{\delta \to 0^+} \int_{|t| \leq \frac{1}{\delta}} \frac{\sin(\pi t)}{\pi t} dt = 1.$$

(a) Let φ be an even smooth function supported in $[-2,2]$ and equal to 1 on $[-1,1]$. Use the Riemann–Lebesgue lemma to show
$$\lim_{\delta \to 0} \int_{\frac{1}{\delta} \leq |t| \leq \frac{2}{\delta}} \frac{\sin(\pi t)}{\pi t} \varphi(\delta t) dt = 0.$$

(b) Use Fourier inversion to write
$$\int_{-\infty}^{+\infty} \frac{\sin(\pi t)}{\pi t} \varphi(\delta t) dt = \left(\chi_{[-\frac{1}{2},\frac{1}{2}]} * \frac{1}{\delta} \widehat{\varphi}\left(\frac{\cdot}{\delta}\right)\right)(0)$$

and let $\delta \to 0$.

2.2.8. Show that for $0 < \gamma < 1$ there are constants $0 < A_\gamma < B_\gamma < \infty$ such that
$$\frac{A_\gamma}{|\xi|^{1-\gamma}} \leq \left| \int_0^1 \frac{e^{-2\pi i \xi t}}{t^\gamma} dt \right| \leq \frac{B_\gamma}{|\xi|^{1-\gamma}}$$

for $|\xi|$ sufficiently large. Notice that one direction in the double inequality also holds when $\gamma = 0$. What do these inequalities say about the Riemann–Lebesgue lemma? [*Hint:* Change variables and integrate by parts. Use that $\int_0^\infty \frac{e^{-2\pi i t}}{t^\gamma} dt \neq 0$ in the $\geq$

inequality. To prove this, assuming $\int_0^\infty \frac{e^{-2\pi i t}}{t^\gamma} dt = 0$ implies $\int_0^\infty \frac{e^{-2\pi i t x}}{t^\gamma} dt = 0$ for all $x \in \mathbf{R} \setminus \{0\}$, and integrate against $xe^{-\pi x^2}$ for a contradiction; use Exercise 2.1.4 (a).]

2.2.9. Consider the change of variables $u = \frac{1}{2}(t - \frac{1}{t})$, related to that used in Lemma 2.2.4 for $f \in L^1(\mathbf{R})$, to derive the identity

$$\int_{-\infty}^{+\infty} f(t - 1/t) \frac{dt}{t^2 + 1} = \int_{-\infty}^{+\infty} f(2u) \frac{du}{u^2 + 1}.$$

[*Hint:* On the left integral change variables on each half-line using $\frac{dt}{t^2+1} = \frac{1}{2} \frac{du}{u^2+1}$.]

2.3 The Fourier Transform on L^2

The integral defining the Fourier transform does not converge absolutely for functions in $L^2(\mathbf{R}^n)$; however, the Fourier transform has a natural definition in this space accompanied by an elegant theory. We begin with the following proposition:

Proposition 2.3.1. *Suppose that f lies in $L^1(\mathbf{R}^n) \cap L^2(\mathbf{R}^n)$. Then $\widehat{f}$ lies in $L^2(\mathbf{R}^n)$ and*

$$\|\widehat{f}\|_{L^2} = \|f\|_{L^2}. \tag{2.3.1}$$

Proof. Given f in $L^1(\mathbf{R}^n) \cap L^2(\mathbf{R}^n)$ consider the function $h = f * \widetilde{\overline{f}}$ which lies in L^1 [as a convolution of two L^1 functions; see (1.6.3)] but also lies in L^∞ and is uniformly continuous (as a convolution of two L^2 functions; see Theorem 1.6.7). Using Propositions 2.1.5 and 2.1.6 we write

$$\widehat{h} = \widehat{f}\,\widehat{\widetilde{\overline{f}}} = \widehat{f}\,\overline{(\widehat{f})^\vee} = \widehat{f}\,\overline{\widehat{f}^\vee} = \widehat{f}\,\overline{\widehat{f}} = |\widehat{f}|^2.$$

Then Theorem 2.2.2(1) gives

$$\int_{\mathbf{R}^n} h(x) \frac{1}{\varepsilon^n} e^{-\pi |\frac{x}{\varepsilon}|^2} dx = \int_{\mathbf{R}^n} \widehat{h}(\xi) e^{-\pi |\varepsilon \xi|^2} d\xi = \int_{\mathbf{R}^n} |\widehat{f}(\xi)|^2 e^{-\pi |\varepsilon \xi|^2} d\xi.$$

If we let $\varepsilon \to 0$ the right-hand side tends to $\|\widehat{f}\|_{L^2}^2$ by the LMCT but the left-hand side tends to $h(0)$ by the continuity of h at 0 [Theorem 1.9.4 (b)]. We conclude that

$$\int_{\mathbf{R}^n} |\widehat{f}(\xi)|^2 d\xi = h(0) < \infty,$$

which implies that $\widehat{f}$ lies in L^2, and moreover

$$\|\widehat{f}\|_{L^2}^2 = h(0) = \int_{\mathbf{R}^n} f(y) \overline{\widetilde{f}(0-y)} dy = \int_{\mathbf{R}^n} f(y) \overline{f(y)} dy = \|f\|_{L^2}^2,$$

which yields identity (2.3.1). $\square$

Given a function f in $L^2(\mathbf{R}^n)$ we consider the sequence of $L^1 \cap L^2$ functions $f\chi_{B(0,N)}$ for $N \in \mathbf{Z}^+$. We observe that as a consequence of Proposition 2.3.1, the corresponding sequence of Fourier transforms is Cauchy in L^2. Indeed, for $N < N'$

$$\|(f\chi_{B(0,N)})\widehat{} - (f\chi_{B(0,N')})\widehat{}\|_{L^2}^2 = \|(f\chi_{B(0,N')\setminus B(0,N)})\widehat{}\|_{L^2}^2 = \|f\chi_{B(0,N')\setminus B(0,N)}\|_{L^2}^2,$$

which is smaller than $\int_{|x|\geq N} |f(x)|^2 dx$, which tends to zero as $N \to \infty$, by the LDCT.

Definition 2.3.2. Let $f \in L^2(\mathbf{R}^n)$. We define the Fourier transform $\widehat{f}$ of f as the L^2 limit of the sequence $\widehat{f_N}$, where

$$f_N = f\chi_{B(0,N)}. \tag{2.3.2}$$

Analogously define $f^\vee$ to be the L^2 limit of the sequence $(f_N)^\vee$.

Remark 2.3.3. If f lies in $L^1 \cap L^2$, then its Fourier transform as defined in (2.1.1) coincides with that given in Definition 2.3.2. Indeed in view of the LDCT, the L^1-Fourier transform of f is

$$\widehat{f}(\xi) = \int_{\mathbf{R}^n} f(x) e^{-2\pi i x \cdot \xi} dx = \lim_{N \to \infty} \int_{|x|<N} f(x) e^{-2\pi i x \cdot \xi} dx. \tag{2.3.3}$$

But as the sequence of integrals on the right in (2.3.3) converges in L^2 to the L^2-Fourier transform of f, a subsequence of it converges to it a.e. (Theorem 1.1.9), thus the L^1-Fourier transform and the L^2-Fourier transform coincide a.e. For this reason there is no ambiguity in using the same notation for the L^1-Fourier transform and L^2-Fourier transform of a function in $L^1 \cap L^2$.

Remark 2.3.4. If $f_N \in L^1(\mathbf{R}^n) \cap L^2(\mathbf{R}^n)$ is any sequence that tends to f in L^2, then $\widehat{f_N}$ tends to $\widehat{f}$ in L^2. To verify this, we notice that if f_N, g_N are two $L^1 \cap L^2$ sequences both converging to f in L^2 as $N \to \infty$, then

$$\|\widehat{f_N} - \widehat{g_N}\|_{L^2} = \|\widehat{f_N - g_N}\|_{L^2} = \|f_N - g_N\|_{L^2} \leq \|f_N - f\|_{L^2} + \|f - g_N\|_{L^2} \to 0$$

as $N \to \infty$. Picking $f_N = f\chi_{B(0,N)}$, we have $\widehat{f_N} \to \widehat{f}$ in L^2 and thus $\widehat{g_N} \to \widehat{f}$ in L^2.

In the next result, recall the operations on functions defined in (2.1.5).

Proposition 2.3.5. *Let $y \in \mathbf{R}^n$, $b \in \mathbf{C}$, $\lambda > 0$, and let A be an $n \times n$ matrix with nonzero determinant. Then for f, g in $L^2(\mathbf{R}^n)$ the following properties are valid:*

(1) $\widehat{f+g} = \widehat{f} + \widehat{g}$ a.e.

(2) $\widehat{bf} = b\widehat{f}$ a.e.

(3) $\widehat{\widetilde{f}} = \widetilde{\widehat{f}}$ a.e.

(4) $\overline{\widehat{f}} = \widetilde{\overline{f}}$ a.e.

2.3 The Fourier Transform on L^2 63

(5) $\widehat{\tau^y f}(\xi) = e^{-2\pi i y \cdot \xi} \widehat{f}(\xi)$ a.e.

(6) $(e^{2\pi i (\cdot) \cdot y} f)\widehat{} = \tau^y(\widehat{f})$ a.e.

(7) $[f(\lambda \cdot)]\widehat{} = \lambda^{-n} \widehat{f}(\lambda^{-1} \cdot)$ a.e.

(8) $\widehat{f \circ A^t} = \dfrac{1}{|\det A|} \widehat{f} \circ A^{-1}$ a.e., viewing elements of $\mathbf{R}^n$ as $n \times 1$ matrices.

Proof. The identities can be obtained from the corresponding assertions in Proposition 2.1.6 by taking limits. For instance, if $f_N = f \chi_{B(0,N)}$ and $g_N = g \chi_{B(0,N)}$, then $0 = \widehat{f_N + g_N} - \widehat{f_N} - \widehat{g_N}$ which converges in L^2 to $\widehat{f+g} - \widehat{f} - \widehat{g}$; thus $\widehat{f+g} - \widehat{f} - \widehat{g}$ must be zero a.e. Likewise we prove the remaining assertions (2)–(8). For instance, let us prove (7). The function $[f(\lambda \cdot)]\widehat{}$ is the L^2 limit of the sequence $[f_N(\lambda \cdot)]\widehat{}$, but this sequence is equal to $\lambda^{-n} \widehat{f_N}(\lambda^{-1} \cdot)$ whose L^2 limit is $\lambda^{-n} \widehat{f}(\lambda^{-1} \cdot)$. Thus these two functions are equal a.e. and (7) holds. □

In some situations the sequence of functions $\widehat{f_N}$ [defined in (2.3.2)] converges pointwise on the complement of a set of measure zero. In this case, we can identify $\widehat{f}$ with the pointwise limit of $\widehat{f_N}$. This is because an L^2 limit coincides a.e. with a pointwise limit. For the purposes of this section, let us denote by L^2_{ae} the space[1] of all functions f in $L^2(\mathbf{R}^n)$ with the property that $\widehat{f_N}$ converges pointwise except on a set of measure zero.

A shortfall of Proposition 2.3.5 is that the exceptional set that appear in the statements may depend on the auxiliary parameters b, y, λ, etc. However, on the space L^2_{ae} we can describe these exceptional sets.

Proposition 2.3.6. *Let $y \in \mathbf{R}^n$, $b \in \mathbf{C}$, $\lambda > 0$, and let A be an orthogonal matrix. Let f, g in $L^2_{ae}(\mathbf{R}^n)$ and define $f_N = f \chi_{B(0,N)}$ and $g_N = g \chi_{B(0,N)}$. Suppose that $\widehat{f_N} \to \widehat{f}$ pointwise on $\mathbf{R}^n \setminus E_f$ and $\widehat{g_N} \to \widehat{g}$ pointwise on $\mathbf{R}^n \setminus E_g$ as $N \to \infty$, where E_f and E_g are sets of measure zero. Then the following are valid:*

(1) $f+g$ lies in $L^2_{ae}(\mathbf{R}^n)$ and $\widehat{f+g} = \widehat{f} + \widehat{g}$ on $\mathbf{R}^n \setminus (E_f \cup E_g)$.

(2) bf lies in $L^2_{ae}(\mathbf{R}^n)$ and $\widehat{bf} = b\widehat{f}$ on $\mathbf{R}^n \setminus E_f$.

(3) $\widetilde{f}$ lies in $L^2_{ae}(\mathbf{R}^n)$ and $\widehat{\widetilde{f}} = \widetilde{\widehat{f}}$ on $\mathbf{R}^n \setminus (-E_f)$.

(4) $\overline{f}$ lies in $L^2_{ae}(\mathbf{R}^n)$ and $\widehat{\overline{f}} = \widetilde{\overline{\widehat{f}}}$ on $\mathbf{R}^n \setminus (-E_f)$.

(5) $e^{2\pi i (\cdot) \cdot y} f$ lies in $L^2_{ae}(\mathbf{R}^n)$ and $(e^{2\pi i (\cdot) \cdot y} f)\widehat{} = \tau^y(\widehat{f})$ on $\mathbf{R}^n \setminus (y + E_f)$.

(6) $f(\lambda \cdot)$ lies in L^2_{ae} and $[f(\lambda \cdot)]\widehat{} = \lambda^{-n} \widehat{f}(\lambda^{-1} \cdot)$ on $\mathbf{R}^n \setminus (\lambda E_f)$.

(7) $f \circ A$ lies in L^2_{ae} and $\widehat{f \circ A} = \widehat{f} \circ A$ on $\mathbf{R}^n \setminus (A^t E_f)$.

(8) if $f : \mathbf{R}^n \setminus \{0\} \to \mathbf{C}$ is radial and $E_f = \{0\}$, then $\widehat{f}$ is also radial.

---

[1] It is an important open question whether $L^2_{ae}(\mathbf{R}^n)$ coincides with $L^2(\mathbf{R}^n)$ in dimensions $n \geq 2$.

Here we used the notation $-E_f = \{-\xi : \xi \in E_f\}$, $y + E_f = \{y + \xi : \xi \in E_f\}$, $\lambda E_f = \{\lambda \xi : \xi \in E_f\}$, $A^t E_f = \{A^t \xi : \xi \in E_f\}$.

Proof. To prove (1), notice that $0 = \widehat{f_N + g_N} - \widehat{f_N} - \widehat{g_N}$ converges pointwise to $\widehat{f + g} - \widehat{f} - \widehat{g}$ on $\mathbf{R}^n \setminus (E_f \cup E_g)$, so $\widehat{f + g} = \widehat{f} + \widehat{g}$ on this set. To prove property (2) we note that $\widehat{(bf)_N} = \widehat{bf_N} = b\widehat{f_N}$ but $b\widehat{f_N}$ converges to $b\widehat{f}$ on $\mathbf{R}^n \setminus E_f$. Thus bf lies in L^2_{ae} and $\widehat{bf} = b\widehat{f}$. For (3) observe that $\widehat{(\overline{f})_N} = \widehat{\overline{f_N}} = \overline{\widetilde{f_N}}$ and that $\widetilde{f_N} \to \widetilde{f}$ on the complement of $-E_f$. Hence so does $\widehat{(\overline{f})_N}$, thus $\overline{f}$ lies in L^2_{ae} and $\widehat{\overline{f}} = \overline{\widetilde{f}}$ on $\mathbf{R}^n \setminus (-E_f)$. For (4) notice that conjugating $\widehat{f_N}(\xi) \to \widehat{f}(\xi)$ yields $\overline{(\widehat{f})_N}(-\xi) \to \overline{\widehat{f}}(-\xi)$ when $\xi \notin E_f$, thus $\widetilde{f}$ lies in L^2_{ae} and the identity in (4) holds on the complement of $-E_f$. For (5) we observe that $(e^{2\pi i(\cdot) \cdot y} f_N)\widehat{\ }(\xi) = \widehat{f_N}(\xi - y)$ which converges to $\widehat{f}(\xi - y)$ when ξ lies in $y + E_f$. Property (6) is proved similarly except that the translation is replaced by dilation. The proof of (7) relies on the identity $\widehat{f_N \circ A} = \widehat{f_N} \circ A$ [Proposition 2.1.6 (9)] and the fact that the balls $B(0,N)$ remain invariant under rotations. To prove (8), we note that as f is defined on $\mathbf{R}^n \setminus \{0\}$, then we have $f = f \circ A$ on $\mathbf{R}^n \setminus \{0\}$ for every orthogonal matrix A. The fact that $E_f = \{0\}$ yields that $A^t E_f = \{0\}$; on the complement of this set we have the identity $\widehat{f \circ A} = \widehat{f} \circ A$ by part (7). But $f = f \circ A$ implies $\widehat{f} = \widehat{f \circ A}$ and $\widehat{f \circ A} = \widehat{f} \circ A$ for every orthogonal matrix A, we deduce that $\widehat{f}$ is radial. $\square$

Proposition 2.3.7. *For f, g, h in L^2 we have*

(i) *(Plancherel's identity)* $\|f\|_{L^2(\mathbf{R}^n)} = \|\widehat{f}\|_{L^2(\mathbf{R}^n)}$

(ii) *(Parseval's identity)* $\int_{\mathbf{R}^n} f(x)\overline{g(x)}\,dx = \int_{\mathbf{R}^n} \widehat{f}(\xi)\overline{\widehat{g}(\xi)}\,d\xi$

(iii) *(Fourier inversion)* $(\widehat{f})^\vee = \widehat{f^\vee} = f$ a.e.

(iv) *(Jumping hat identity)* We have $\int_{\mathbf{R}^n} f(x)\widehat{h}(x)\,dx = \int_{\mathbf{R}^n} \widehat{f}(\xi)h(\xi)\,d\xi$.

Proof. Given $f \in L^2(\mathbf{R}^n)$, pick a sequence of Schwartz functions ϕ_N (which certainly lie in $L^1 \cap L^2$) such that $\phi_N \to f$ in $L^2(\mathbf{R}^n)$. Proposition 2.3.1 gives that $\|\phi_N\|_{L^2} = \|\widehat{\phi_N}\|_{L^2}$ for all N. As the definition of $\widehat{f}$ is independent of the sequence converging to the function, $\widehat{\phi_N} \to \widehat{f}$ in $L^2(\mathbf{R}^n)$. Thus $\|\phi_N\|_{L^2} \to \|f\|_{L^2}$ and $\|\widehat{\phi_N}\|_{L^2} \to \|\widehat{f}\|_{L^2}$ as $N \to \infty$ and thus assertion (i) follows.

To prove (ii) we use *polarization* as follows: We apply (i) to $f + g$, we expand both sides, and we use (i) for f and g to obtain

$$\text{Re}\int_{\mathbf{R}^n} f(x)\overline{g(x)}\,dx = \text{Re}\int_{\mathbf{R}^n} \widehat{f}(\xi)\overline{\widehat{g}(\xi)}\,d\xi.$$

We then apply (i) to $f + ig$ and likewise we obtain

$$\text{Re}\int_{\mathbf{R}^n} -if(x)\overline{g(x)}\,dx = \text{Re}\int_{\mathbf{R}^n} -i\widehat{f}(\xi)\overline{\widehat{g}(\xi)}\,d\xi$$

2.3 The Fourier Transform on L^2

which implies

$$\operatorname{Im} \int_{\mathbf{R}^n} f(x)\overline{g(x)}\,dx = \operatorname{Im} \int_{\mathbf{R}^n} \widehat{f}(\xi)\overline{\widehat{g}(\xi)}\,d\xi$$

as $\operatorname{Im} w = \operatorname{Re}(-iw)$. Thus we deduce (ii).

As $\widehat{\phi_N} \to \widehat{f}$ in L^2, it follows that $(\widehat{\phi_N})^\vee \to (\widehat{f})^\vee$ in L^2. Since $\phi_N \in \mathscr{S}(\mathbf{R}^n)$ we have $(\widehat{\phi_N})^\vee = \phi_N$, which converges to f in $L^2(\mathbf{R}^n)$. It follows that f and $(\widehat{f})^\vee$ are equal in $L^2(\mathbf{R}^n)$ and consequently equal almost everywhere. This proves (iii). To prove (iv) we simply take $\overline{g} = \widehat{h}$ (equivalently $h = \overline{\widehat{g}}$) in identity (ii). □

Example 2.3.8. We estimate the Fourier transform of the function $g(t) = t^{-\gamma}\chi_{t \geq 1}$ for $1/2 < \gamma < 1$. This function lies in $L^2(\mathbf{R})$ but does not lie in $L^1(\mathbf{R})$. If we can show that the limit

$$\lim_{N \to \infty} \int_1^N \frac{e^{-2\pi i t \xi}}{t^\gamma}\,dt = \int_1^\infty \frac{e^{-2\pi i t \xi}}{t^\gamma}\,dt$$

exists for all $\xi \neq 0$, then $\widehat{g}(\xi)$ can be identified with this limit. We make a few observations. When $\xi = 0$, this limit is infinite, so it is expected that $\widehat{g}(\xi)$ gets worse as $\xi \to 0$. We observe that $|\widehat{g}(\xi)| = |\widehat{g}(|\xi|)|$ as $\widehat{g}(\xi) = \overline{\widehat{g}(|\xi|)}$ for $\xi < 0$. Thus we may work with $|\xi|$ instead of $\xi \neq 0$. A change of variables gives

$$\widehat{g}(|\xi|) = \frac{1}{|\xi|^{1-\gamma}} \int_{|\xi|}^\infty \frac{e^{-2\pi i t}}{t^\gamma}\,dt = \frac{1}{|\xi|^{1-\gamma}} \left[\frac{e^{-2\pi i |\xi|}}{2\pi i |\xi|^\gamma} - \frac{\gamma}{2\pi i} \int_{|\xi|}^\infty \frac{e^{-2\pi i t}}{t^{\gamma+1}}\,dt \right],$$

where the second identity follows by an integration by parts. From this we obtain

$$|\widehat{g}(\xi)| \leq \frac{1}{\pi |\xi|} \qquad \text{for } |\xi| \geq 1, \tag{2.3.4}$$

in fact $|\widehat{g}(\xi)| \approx |\xi|^{-1}$ as $|\xi| \to \infty$. For $0 < |\xi| < 1$, writing

$$\widehat{g}(|\xi|) = \frac{1}{|\xi|^{1-\gamma}} \left[\int_{|\xi|}^1 \frac{e^{-2\pi i t}}{t^\gamma}\,dt + \int_1^\infty \frac{e^{-2\pi i t}}{t^\gamma}\,dt \right], \tag{2.3.5}$$

we deduce

$$|\widehat{g}(\xi)| \leq \frac{1}{|\xi|^{1-\gamma}} \left[\frac{1 - |\xi|^{1-\gamma}}{1-\gamma} + \frac{1}{\pi} \right] \qquad \text{for } 0 < |\xi| < 1.$$

We conclude

$$|\widehat{g}(\xi)| \leq \frac{(1-\gamma)^{-1} + \pi^{-1}}{|\xi|^{1-\gamma}} \qquad \text{for } 0 < |\xi| < 1. \tag{2.3.6}$$

Estimates (2.3.4) and (2.3.6) explain why $\widehat{g} \in L^2(\mathbf{R})$.

Having set down the basic facts concerning the action of the Fourier transform on L^1 and L^2, we extend its definition on $L^1 + L^2$, which in particular contains L^p for $1 < p < 2$.

Definition 2.3.9. The space $L^1(\mathbf{R}^n) + L^2(\mathbf{R}^n)$ consists of all functions of the form $f_1 + f_2$, where $f_1 \in L^1(\mathbf{R}^n)$ and $f_2 \in L^2(\mathbf{R}^n)$. Likewise the space $L^2(\mathbf{R}^n) + L^\infty(\mathbf{R}^n)$ consists of all functions of the form $f_2 + f_\infty$, where $f_2 \in L^2(\mathbf{R}^n)$ and $f_\infty \in L^\infty(\mathbf{R}^n)$. Given a function f in $L^1(\mathbf{R}^n) + L^2(\mathbf{R}^n)$ we define $\widehat{f} = \widehat{f_1} + \widehat{f_2}$, which is an element of $L^2(\mathbf{R}^n) + L^\infty(\mathbf{R}^n)$. Notice that $\widehat{f}$ is defined a.e. as $\widehat{f_2}$ does so.

This definition is independent of the choice of f_1 and f_2, for, if $f_1 + f_2 = h_1 + h_2$ a.e. for $f_1, h_1 \in L^1(\mathbf{R}^n)$ and $f_2, h_2 \in L^2(\mathbf{R}^n)$, we have $f_1 - h_1 = h_2 - f_2$ a.e. and belong to $L^1(\mathbf{R}^n)$, hence their Fourier transforms are equal. This gives

$$\widehat{f_1 - h_1}(\xi) = \widehat{h_2 - f_2}(\xi) \qquad \text{for all } \xi \in \mathbf{R}^n,$$

hence $\widehat{f_1} - \widehat{h_1} = \widehat{h_2} - \widehat{f_2}$ a.e., thus $\widehat{f_1} + \widehat{f_2} = \widehat{h_1} + \widehat{h_2}$ a.e. This definition also implies that if $f = f_1 + f_2 \in L^1 + L^2$ satisfies $\widehat{f} = 0$ a.e., then $f = 0$ a.e. Indeed, if $\widehat{f_1} + \widehat{f_2} = 0$ a.e., then $\widehat{f_2} = -\widehat{f_1}$ a.e., and as both functions lie in L^2, applying the inverse Fourier transform and using Proposition 2.3.7 (iii) yields $f_2 = -f_1$ a.e., that is, $f = 0$ a.e.

Also notice that $L^1(\mathbf{R}^n) + L^2(\mathbf{R}^n)$ contains $L^p(\mathbf{R}^n)$ for $1 < p < 2$, as given $f \in L^p$ we can express it as $f = f_1 + f_2$, where $f_1 = f\chi_{|f|>1} \in L^1$ and $f_2 = f\chi_{|f|\le 1} \in L^2$.

Next, we compute the Fourier transform of a function in $L^1(\mathbf{R}^n) + L^2(\mathbf{R}^n)$.

Example 2.3.10. We fix a complex number z satisfying $-n < \operatorname{Re} z < -\frac{n}{2}$ and we consider the function $F_z(x) = |x|^z$ on $\mathbf{R}^n \setminus \{0\}$. We observe that F_z can be written as the sum of an $L^1(\mathbf{R}^n)$ function F_z^1 and an $L^2(\mathbf{R}^n)$ function F_z^2 as follows:

$$F_z(x) = |x|^z = \underbrace{|x|^z \chi_{|x|<1}}_{F_z^1 \text{ in } L^1(\mathbf{R}^n)} + \underbrace{|x|^z \chi_{|x|\ge 1}}_{F_z^2 \text{ in } L^2(\mathbf{R}^n)}.$$

For a fixed $\xi \neq 0$ we claim that the limit

$$\lim_{N\to\infty} \int_{|x|\le N} F_z^2(x) e^{-2\pi i x \cdot \xi} dx = \lim_{N\to\infty} \int_{1\le |x|\le N} |x|^z e^{-2\pi i x \cdot \xi} dx \qquad (2.3.7)$$

exists. This will be shown using polar coordinates and an integration by parts. Note that in dimension $n = 1$, the existence of this limit was essentially shown in Example 2.3.8 (γ there could have been replaced by $\gamma + is$, s real). In dimensions $n \ge 2$, proving that the limit in (2.3.7) exists requires knowledge of the following asymptotic identity, which can found in Appendices B.4 and B.8 in [31]: for $r|\xi| \ge 1$

$$\int_{\mathbf{S}^{n-1}} e^{-2\pi i r \xi \cdot \theta} d\sigma(\theta) = \frac{2\pi}{|r\xi|^{\frac{n-2}{2}}} \left[\sqrt{\frac{2}{\pi 2\pi r |\xi|}} \cos\left(2\pi r|\xi| - \frac{\pi \frac{n-2}{2}}{2} - \frac{\pi}{4}\right) + R(r|\xi|) \right]$$

$$= \frac{e^{2\pi i r|\xi|} e^{-i\frac{\pi(n-1)}{4}} + e^{-2\pi i r|\xi|} e^{i\frac{\pi(n-1)}{4}}}{|r\xi|^{\frac{n-1}{2}}} + \frac{2\pi R(r|\xi|)}{|r\xi|^{\frac{n-2}{2}}}, \qquad (2.3.8)$$

where R is a function satisfying $|R(t)| \le C t^{-3/2}$ for all $t \ge 1$ and some $C > 0$.

2.3 The Fourier Transform on L^2

As the part of the integral in (2.3.7) over the region $1 \le |x| \le \max(|\xi|^{-1}, 1)$ produces a constant, we focus on the part over $\max(|\xi|^{-1}, 1) < |x| \le N$ [for $N > \max(|\xi|^{-1}, 1)$]. Expressing this part of the integral in (2.3.7) in terms of polar coordinates and inserting (2.3.8), we reduce to the existence of the limit

$$\lim_{N\to\infty} \int_{\max(|\xi|^{-1},1)}^{N} r^{z+n-1} \left[\frac{e^{2\pi i r |\xi|} e^{-i\frac{\pi(n-1)}{4}} + e^{-2\pi i r |\xi|} e^{i\frac{\pi(n-1)}{4}}}{r^{\frac{n-1}{2}} |\xi|^{\frac{n-1}{2}}} + \frac{2\pi R(r|\xi|)}{r^{\frac{n-2}{2}} |\xi|^{\frac{n-2}{2}}} \right] dr$$

as $N \to \infty$. In view of the bound $|R(r|\xi|)| \le C(r|\xi|)^{-3/2}$, the part of the integral containing $R(r|\xi|)$ converges absolutely as long as $\operatorname{Re} z + n - 1 - \frac{3}{2} - \frac{n-2}{2} < -1$, i.e., $\operatorname{Re} z < -\frac{n-1}{2}$, so for this part the limit exists. In the part of the integral containing the exponentials we write $e^{\pm 2\pi i r |\xi|} = (\pm 2\pi i |\xi|)^{-1} \frac{d}{dr} e^{\pm 2\pi i r |\xi|}$ and integrate by parts to deduce that the limit exists if $\operatorname{Re} z + n - 1 - 1 - \frac{n-1}{2} < -1$, i.e., $\operatorname{Re} z < -\frac{n-1}{2}$ as well. This argument shows that the Fourier transform of F_z^2 can be defined pointwise at every point $\xi \in \mathbf{R}^n \setminus \{0\}$, thus $E_{F_z^2} = \{0\}$ using the notation of Proposition 2.3.6.

As $F_z \in L^1 + L^2$, we wish to evaluate $\widehat{F}_z$. We begin with the observation that

$$F_z(\lambda x) = \lambda^z F_z(x) \qquad \text{for any } \lambda > 0 \text{ and } x \in \mathbf{R}^n \setminus \{0\}.$$

Applying the Fourier transform in this identity and using Proposition 2.1.6 (7) and Proposition 2.3.6 (7) (combined with the fact $E_{F_z^2} = \{0\}$) we obtain that

$$\lambda^{-n} \widehat{F}_z(\lambda^{-1}\xi) = \lambda^z \widehat{F}_z(\xi) \qquad \text{for any } \lambda > 0 \text{ and } \xi \in \mathbf{R}^n \setminus \{0\}.$$

This implies that $\widehat{F}_z$ is homogeneous of degree $-n - z$, i.e., $\widehat{F}_z(\lambda \xi) = \lambda^{-n-z} \widehat{F}_z(\xi)$ for all λ and $\xi \in \mathbf{R}^n \setminus \{0\}$.

In view of Corollary 2.1.7 for F_z^1 and Proposition 2.3.6 (8) for F_z^2 (which uses that $E_{F_z^2} = \{0\}$), we obtain that $\widehat{F}_z$ is a radial function, i.e., it has the form $g(|\xi|)$ for some function g on the line. Then for $|\xi| \ne 0$ we have

$$\widehat{F}_z(\xi) = |\xi|^{-n-z} \widehat{F}_z(\xi/|\xi|) = |\xi|^{-n-z} g(|\xi/|\xi||) = |\xi|^{-n-z} g(1).$$

It could be the case that $|g(1)| = \infty$, but in this case $\widehat{F}_z$ would equal infinity at every nonzero point, and thus it could not belong to $L^\infty(\mathbf{R}^n) + L^2(\mathbf{R}^n)$. We conclude that for some finite nonzero constant $c(z,n)$ we have

$$\widehat{F}_z(\xi) = c(z,n) |\xi|^{-n-z} \qquad \text{for } x \in \mathbf{R}^n \setminus \{0\}. \qquad (2.3.9)$$

Finally, notice that as $-n/2 < -\operatorname{Re} z - n < 0$, the part $c(z,n)|\xi|^{-n-z} \chi_{|\xi|<1}$ of $\widehat{F}_z(\xi)$ lies in $L^2(\mathbf{R}^n)$ and the part $c(z,n)|\xi|^{-n-z} \chi_{|\xi| \ge 1}$ is bounded.

Exercises

2.3.1. Let $f, g \in L^2(\mathbf{R}^n)$ and $h \in L^1(\mathbf{R}^n)$. Show that
(a) $\widehat{h * g} = \widehat{h}\,\widehat{g}$ a.e.
(b) $\widehat{fg} = \widehat{f} * \widehat{g}$.

2.3.2. Suppose that $f \in L^2_{\mathrm{ae}}(\mathbf{R})$ is associated with an exceptional set E_f and $y \in \mathbf{R}$. Show that $\widehat{\tau^y f} = e^{-2\pi i y(\cdot)} \widehat{f}$ on $\mathbf{R} \setminus E_f$.

2.3.3. Use Plancherel's identity to show that

$$\int_0^\infty \frac{\sin^2 t}{t^2}\,dt = \frac{\pi}{2}$$

and that

$$\int_0^\infty \frac{\sin^4 t}{t^4}\,dt = \frac{\pi}{3}.$$

2.3.4. Let $0 < \gamma < 1$ and consider the function g in Example 2.3.8. Find the range of indices $1 \leq p, q \leq \infty$ for which $g \in L^p(\mathbf{R}^n)$ and $\widehat{g} \in L^q(\mathbf{R})$.
[*Hint:* You may need that $\int_0^\infty t^{-\gamma} e^{-2\pi i t}\,dt \neq 0$; on this, see Exercise 2.2.8.]

2.3.5. Let $g(t) = t^{-1}\chi_{t \geq 1}$ defined on the real line. Prove that for $|\xi| \leq 1$ we have

$$|\widehat{g}(\xi)| \leq \frac{1}{\pi} + \log\left(\frac{1}{|\xi|}\right),$$

while $|\widehat{g}(\xi)| \leq (\pi |\xi|)^{-1}$ for $|\xi| \geq 1$.

2.3.6. Let k be a positive integer. Consider the function

$$F(z) = \int_0^\infty e^{-zt} t^k \frac{dt}{t} - \frac{(k-1)!}{z^k}$$

defined on the open half-space $\operatorname{Re} z > 0$ of the complex plane.
(a) Show that F is analytic on $\operatorname{Re} z > 0$ and notice that $F(x) = 0$ for all $x > 0$.
(b) Use the identity principle in complex analysis to prove that $F = 0$ identically, and then obtain that the Fourier transform of the function $t \mapsto e^{-2\pi t} t^{k-1} \chi_{[0,\infty)}(t)$ is

$$\frac{(k-1)!}{(2\pi)^k (1 + i\xi)^k}, \qquad \xi \in \mathbf{R}.$$

(c) Derive the identity

$$\frac{1}{\pi}\int_{-\infty}^{+\infty} \frac{dx}{(1+x^2)^k} = \frac{(2k-2)!}{2^{2k-2}((k-1)!)^2}.$$

2.3.7. For a function φ on $[0,\infty)$ with temperate growth (say $|\varphi(t)| \leq C(1+|t|)^M$ for all $t \geq 0$) define its *Laplace transform* by

$$\mathcal{L}\varphi(s) = \int_0^\infty e^{-st}\varphi(t)\,dt.$$

Assuming that $\mathcal{L}\varphi$ extends to an analytic function on $\operatorname{Re} z > 0$, prove that the Fourier transform of the function $t \mapsto e^{-2\pi t}\varphi(t)\chi_{[0,\infty)}(t)$ on the real line is $\mathcal{L}\varphi(2\pi(1+i\xi))$. Apply this to $\varphi_1(t) = \sin t$ and $\varphi_2(t) = \cos t$, as well as $\varphi_3(t) = t^\alpha$ for $\alpha > 0$.
[*Hint:* Use the idea of the preceding exercise.]

2.4 Complex Interpolation and the Hausdorff–Young Inequality

In this section we discuss interpolation between different Lebesgue spaces with an intermediate constant expressed as the geometric mean of the constants that appear on the given bounds. The techniques are based on elements of complex analysis, in particular on the maximum modulus principle.

Theorem 2.4.1. (Riesz–Thorin interpolation theorem) *Let $0 < p_0, p_1, q_0, q_1 \leq \infty$ and let (X,μ), (Y,ν) be two σ-finite measure spaces. Let T be a linear operator defined on the space of finitely simple functions of X and taking values in the space of measurable functions on Y with the property[2] $\int_B |T(\chi_A)|\,d\nu < \infty$ whenever $A \subseteq X$ and $B \subseteq Y$ satisfy $\mu(A) + \nu(B) < \infty$. For $0 < \theta < 1$ set*

$$\frac{1}{p} = \frac{1-\theta}{p_0} + \frac{\theta}{p_1} \quad \text{and} \quad \frac{1}{q} = \frac{1-\theta}{q_0} + \frac{\theta}{q_1}. \tag{2.4.1}$$

Suppose that for all finitely simple functions f on X we have

$$\|T(f)\|_{L^{q_0}} \leq M_0 \|f\|_{L^{p_0}}, \tag{2.4.2}$$
$$\|T(f)\|_{L^{q_1}} \leq M_1 \|f\|_{L^{p_1}}. \tag{2.4.3}$$

Then for all finitely simple functions f on X we have

$$\|T(f)\|_{L^q} \leq M_0^{1-\theta} M_1^\theta \|f\|_{L^p}. \tag{2.4.4}$$

Thus, T has a unique bounded extension from $L^p(X,\mu)$ to $L^q(Y,\nu)$ when $p < \infty$.

Proof. As X is a σ-finite measure space, we can work with a finitely simple function

$$f = \sum_{k=1}^K a_k e^{i\alpha_k} \chi_{A_k}$$

---

[2] The hypothesis $\int_B |T(\chi_A)|\,d\nu < \infty$ follows from (2.4.2) or (2.4.3) when $\max(q_0, q_1) \geq 1$ by Hölder's inequality, but is needed when $\max(q_0, q_1) < 1$.

defined on X, where $a_k > 0$, α_k are real, and A_k are pairwise disjoint subsets of X with finite strictly positive measure. Such functions are dense in $L^p(X)$ when $p < \infty$.
Case I: $\min(q_0, q_1) > 1$ and $p < \infty$. This forces $q > 1$, hence $q' < \infty$. We estimate

$$\|T(f)\|_{L^q(Y,v)} = \sup_g \left| \int_Y T(f)(y) g(y) \, dv(y) \right|,$$

where the supremum is taken over all finitely simple functions g on Y with $L^{q'}$ norm equal to 1. (Here we are using that Y is σ-finite in allowing g to be finitely simple, instead of simple, and also in the case $q = \infty$.) Write

$$g = \sum_{j=1}^{J} b_j e^{i\beta_j} \chi_{B_j},$$

where $b_j > 0$, β_j are real, and B_j are pairwise disjoint subsets of Y with finite v-measure. Let

$$P(z) = \frac{p}{p_0}(1-z) + \frac{p}{p_1} z \quad \text{and} \quad Q(z) = \frac{q'}{q_0'}(1-z) + \frac{q'}{q_1'} z. \qquad (2.4.5)$$

For z in the closed unit strip $\overline{S} = \{z \in \mathbf{C} : 0 \leq \operatorname{Re} z \leq 1\}$, define

$$f_z = \sum_{k=1}^{K} a_k^{P(z)} e^{i\alpha_k} \chi_{A_k}, \quad g_z = \sum_{j=1}^{J} b_j^{Q(z)} e^{i\beta_j} \chi_{B_j}, \qquad (2.4.6)$$

and

$$F(z) = \int_Y T(f_z)(y) g_z(y) \, dv(y).$$

Notice that $f_\theta = f$ and $g_\theta = g$. By linearity we have

$$F(z) = \sum_{k=1}^{K} \sum_{j=1}^{J} a_k^{P(z)} b_j^{Q(z)} e^{i\alpha_k} e^{i\beta_j} \int_Y T(\chi_{A_k})(y) \chi_{B_j}(y) \, dv(y).$$

Since $a_k, b_j > 0$, F is analytic in z, and the expression

$$\int_Y T(\chi_{A_k})(y) \chi_{B_j}(y) \, dv(y)$$

is a finite constant, as seen by Hölder's inequality with exponents q_0 and q_0'.
By the disjointness of the sets A_k we have (even when one of p_0, p_1 is infinity)

$$\|f_{it}\|_{L^{p_0}} = \|f\|_{L^p}^{\frac{p}{p_0}}, \quad \|f_{1+it}\|_{L^{p_1}} = \|f\|_{L^p}^{\frac{p}{p_1}},$$

since $|a_k^{P(it)}| = a_k^{\frac{p}{p_0}}$ and $|a_k^{P(1+it)}| = a_k^{\frac{p}{p_1}}$. By the disjointness of the sets B_j we have

2.4 Complex Interpolation and the Hausdorff–Young Inequality

$$\|g_{it}\|_{L^{q_0'}}^{q_0'} = \|g\|_{L^{q'}}^{q'} = 1, \qquad \|g_{1+it}\|_{L^{q_1'}}^{q_1'} = \|g\|_{L^{q'}}^{q'} = 1,$$

as $|b_j^{Q(it)}| = b_j^{\frac{q'}{q_0'}}$ and $|b_j^{Q(1+it)}| = b_j^{\frac{q'}{q_1'}}$. Hölder's inequality and the hypotheses give

$$|F(it)| \le \|T(f_{it})\|_{L^{q_0}} \|g_{it}\|_{L^{q_0'}} \le M_0 \|f_{it}\|_{L^{p_0}} = M_0 \|f\|_{L^p}^{\frac{p}{p_0}}, \tag{2.4.7}$$

$$|F(1+it)| \le \|T(f_{1+it})\|_{L^{q_1}} \|g_{1+it}\|_{L^{q_1'}} \le M_1 \|f_{1+it}\|_{L^{p_1}} \le M_1 \|f\|_{L^p}^{\frac{p}{p_1}}. \tag{2.4.8}$$

We observe that F is analytic in the unit strip $\mathbf{S}$ and continuous on its closure. Also, F is bounded on the closed unit strip (by some constant that depends on f and g). Therefore, (2.4.7), (2.4.8), and Corollary C.0.3 give

$$|F(\theta)| \le \left(M_0 \|f\|_{L^p}^{\frac{p}{p_0}}\right)^{1-\theta} \left(M_1 \|f\|_{L^p}^{\frac{p}{p_1}}\right)^{\theta} = M_0^{1-\theta} M_1^{\theta} \|f\|_{L^p}.$$

Observe that $P(\theta) = Q(\theta) = 1$ and hence

$$F(\theta) = \int_Y T(f) g \, dv.$$

Taking the supremum over all finitely simple functions g on Y with $L^{q'}$ norm equal to 1, we conclude the proof in the case where $\min(q_0, q_1) > 1$.

Case II: $\min(q_0, q_1) \le 1$ and $p < \infty$. In this case choose an $r > \max(1, q, q/q_0, q/q_1)$ such that $r < \infty$. We fix an arbitrary positive finitely simple function g with $\|g\|_{L^{r'}} = 1$ and we write

$$g = \sum_{j=1}^{J} c_j \chi_{E_j},$$

where $c_j > 0$, and E_j are pairwise disjoint measurable subsets of Y with $v(E_j) < \infty$. For $z \in \mathbf{S}$ set

$$g^z = \sum_{j=1}^{J} c_j^{R(z)} \chi_{E_j},$$

where

$$R(z) = r' \left[1 - \frac{q}{r} \left(\frac{1-z}{q_0} + \frac{z}{q_1} \right) \right].$$

Notice that

$$\|g^{it}\|_{L^{\left(\frac{rq_0}{q}\right)'}}^{\left(\frac{rq_0}{q}\right)'} = \|g\|_{L^{r'}}^{r'} = 1, \qquad \|g^{1+it}\|_{L^{\left(\frac{rq_1}{q}\right)'}}^{\left(\frac{rq_1}{q}\right)'} = \|g\|_{L^{r'}}^{r'} = 1. \tag{2.4.9}$$

Now consider the following function defined for $z \in \overline{\mathbf{S}}$:

$$G(z) = \int_Y |T(f_z)(y)|^{\frac{q}{r}} |g^z(y)| \, dv(y) = \sum_{j=1}^{J} \int_{E_j} \left| c_j^{\frac{r}{q} R(z)} \sum_{k=1}^{K} a_k^{P(z)} e^{i\alpha_k} T(\chi_{A_k})(y) \right|^{\frac{q}{r}} dv(y).$$

By assumption each $T(\chi_{A_k})$ is integrable over a set of finite measure. Thus for each fixed $j \in \{1,\ldots,J\}$, the mapping that takes $z \in \mathbf{S}$ to the function

$$y \mapsto c_j^{\frac{r}{q}R(z)} \sum_{k=1}^{K} a_k^{P(z)} e^{i\alpha_k} T(\chi_{A_k})(y), \qquad y \in E_j,$$

is well defined and analytic from $\mathbf{S}$ to the Banach space $L^1(E_j)$. Its analyticity can be checked by considering the integral of this function against bounded functions; see Theorem B.0.3 (Appendix B). As $q/r < 1$, it follows from Lemma B.0.5 that $\log G$ is subharmonic on S. Using Hölder's inequality with exponents $\frac{rq_0}{q} > 1$ and $\left(\frac{rq_0}{q}\right)'$, and (2.4.9), we obtain that

$$|G(it)| \le \left\{\int_Y |T(f_{it})(y)|^{q_0} d\nu(y)\right\}^{\frac{q}{rq_0}} \|g^{it}\|_{L^{(\frac{rq_0}{q})'}} \le \left(M_0 \|f\|_{L^p}^{\frac{p}{p_0}}\right)^{\frac{q}{r}}.$$

Likewise, we get

$$|G(1+it)| \le \left\{\int_Y |T(f_{1+it})(y)|^{q_1} d\nu(y)\right\}^{\frac{q}{rq_1}} \|g^{1+it}\|_{L^{(\frac{rq_1}{q})'}} \le \left(M_1 \|f\|_{L^p}^{\frac{p}{p_1}}\right)^{\frac{q}{r}}.$$

Notice that G is continuous and bounded on $\overline{\mathbf{S}}$. Applying Corollary C.0.3 in Appendix C, we obtain

$$\int_Y |T(f_\theta)(y)|^{\frac{q}{r}} g^\theta(y)\, d\nu(y) = G(\theta) \le \left(M_0 \|f\|_{L^p}^{\frac{p}{p_0}}\right)^{\frac{q}{r}(1-\theta)} \left(M_1 \|f\|_{L^p}^{\frac{p}{p_1}}\right)^{\frac{q}{r}\theta}.$$

Noticing that $g^\theta = g$ and recalling that $f_\theta = f$ we deduce that

$$\|T(f)\|_{L^q} = \left\| |T(f)|^{\frac{q}{r}} \right\|_{L^r}^{\frac{r}{q}}$$

$$= \sup\left\{\int |T(f)|^{\frac{q}{r}} g\, d\nu : g \ge 0,\ g \text{ is finitely simple},\ \|g\|_{L^{r'}} = 1\right\}^{\frac{r}{q}}$$

$$\le M_0^{1-\theta} M_1^\theta \|f\|_{L^p}.$$

This concludes the proof in the case $\min(q_0,q_1) \le 1$.

Case III: $p = \infty$. This forces $p_0 = p_1 = \infty$. Then Exercise 1.1.6 yields

$$\|T(f)\|_{L^q} \le \|T(f)\|_{L^{q_0}}^{1-\theta} \|T(f)\|_{L^{q_1}}^{\theta} \qquad (2.4.10)$$

and so inserting $\|T(f)\|_{L^{q_i}} \le M_i \|f\|_{L^\infty}$ ($i = 0,1$) in (2.4.10) yields (2.4.4). □

Proposition 2.4.2. (*Young's inequality*) Fix $1 < r < \infty$ and g in $L^r(\mathbf{R}^n)$. Let p, q be indices that satisfy $1 \le p \le r'$, $r \le q \le \infty$, and

$$\frac{1}{q} + 1 = \frac{1}{p} + \frac{1}{r}. \qquad (2.4.11)$$

2.4 Complex Interpolation and the Hausdorff–Young Inequality

Then for all $f \in L^p(\mathbf{R}^n)$ we have

$$\|f*g\|_{L^q(\mathbf{R}^n)} \leq \|f\|_{L^p(\mathbf{R}^n)}\|g\|_{L^r(\mathbf{R}^n)}. \tag{2.4.12}$$

Proof. On $L^1 \cup L^{r'}$ define the operator $T(f) = f*g$. By Minkowski's inequality $T: L^1 \to L^r$ with norm at most $\|g\|_{L^r}$. By Hölder's inequality $T: L^{r'} \to L^\infty$ with norm at most $\|g\|_{L^r}$. Theorem 2.4.1 gives that T has a bounded extension from L^p to L^q with norm at most $\|g\|_{L^r}^\theta \|g\|_{L^r}^{1-\theta} = \|g\|_{L^r}$, where

$$\frac{1}{p} = \frac{1-\theta}{1} + \frac{\theta}{r'} \quad \text{and} \quad \frac{1}{q} = \frac{1-\theta}{r} + \frac{\theta}{\infty}.$$

If we eliminate θ, these equations reduce to (2.4.11). This completes the proof of (2.4.12); see Figure 2.1. □

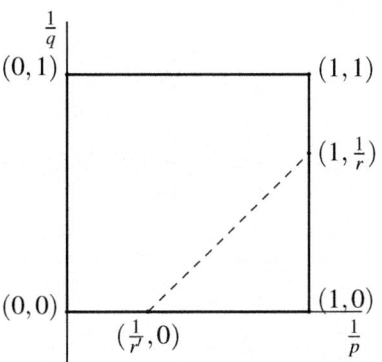

Fig. 2.1 Boundedness on the dotted line is obtained by interpolation between the endpoints $(1/r', 0)$ and $(1, 1/r)$.

Notice that Proposition 2.4.2 is also valid when $r = 1$ (in which case $p = q$) and when $r = \infty$ (in which case $p = 1$ and $q = \infty$) and both of these endpoint cases are just a restatement of Minkowski's convolution inequality (Theorem 1.6.6).

It turns out that for $g \in L^{r,\infty}$ we can obtain the stronger conclusion $\|f*g\|_{L^q(\mathbf{R}^n)} \leq \|f\|_{L^p(\mathbf{R}^n)}\|g\|_{L^{r,\infty}(\mathbf{R}^n)}$ interpolating between $L^1 \to L^{r,\infty}$ [Exercise 1.6.7 (b)] and $L^{r',1} \to L^\infty$ (duality between $L^{r',1}$ and $L^{r,\infty}$) using off-diagonal Marcinkiewicz interpolation ([31, Theorem 1.4.19]).

Example 2.4.3. On the real line consider the linear operator

$$L(g)(x) = \int_{2x}^{2x+1} g(x-t)\,dt, \qquad x \in \mathbf{R}.$$

We claim that L maps $L^p(\mathbf{R})$ to $L^{p/2}(\mathbf{R})$ for all $1 \leq p \leq \infty$. Obviously L maps L^∞ to itself. If we show that it also maps $L^1(\mathbf{R})$ to $L^{1/2}(\mathbf{R})$, then the conclusion for $p \in (1, \infty)$ will be a consequence of Theorem 2.4.1.

To achieve this we write a general $g \geq 0$ in $L^1(\mathbf{R})$ as $\sum_{k \in \mathbf{Z}} g_k$, where $g_k = g\chi_{[k,k+1)}$. Denoting by $[-x]$ the integer part of $-x$, we observe that

$$L(g)(x) = \sum_{k \in \mathbf{Z}} L(g_k)(x) = L(g_{[-x]})(x) + L(g_{[-x]-1})(x), \qquad x \in \mathbf{R},$$

as $L(g_k)$ is supported in $[-k-2, -k]$. Applying (1.1.9) we obtain

$$\|L(g)\|_{L^{1/2}} \leq 2 \sum_{i \in \{0,1\}} \left(\int_{\mathbf{R}} [L(g_{[-x]-i})(x)]^{\frac{1}{2}} dx \right)^2 \leq 2 \sum_{k \in \mathbf{Z}} \|L(g_k)\|_{L^{1/2}}.$$

Combining this inequality with the following application of the Cauchy–Schwarz inequality,

$$\|L(g_k)\|_{L^{1/2}} \leq 2 \|L(g_k)\|_{L^1} \leq 2 \|g_k\|_{L^1}, \tag{2.4.13}$$

we deduce $\|L(g)\|_{L^{1/2}} \leq 4 \|g\|_{L^1}$; that is, L maps $L^1(\mathbf{R})$ to $L^{1/2}(\mathbf{R})$.

Proposition 2.4.4. *(Hausdorff–Young inequality)* Let $1 \leq p \leq 2$. Then for every function f in $L^p(\mathbf{R}^n)$ we have the estimate

$$\|\widehat{f}\|_{L^{p'}} \leq \|f\|_{L^p}. \tag{2.4.14}$$

Proof. We apply Theorem 2.4.1 to interpolate between the estimates

$$\|\widehat{f}\|_{L^\infty} \leq \|f\|_{L^1}$$

[Estimate (2.1.2) in Proposition 2.1.5] and

$$\|\widehat{f}\|_{L^2} \leq \|f\|_{L^2}$$

to obtain (2.4.14). We conclude that, when $1 \leq p \leq 2$, the Fourier transform is a bounded operator from $L^p(\mathbf{R}^n)$ to $L^{p'}(\mathbf{R}^n)$ with norm at most 1. $\square$

Exercises

2.4.1. Let $1 \leq r \leq p \leq 2 \leq q \leq \infty$ be related as in $1/p + 1/q = 1/r$ and fix K in $L^r(\mathbf{R}^n)$. Prove that the linear operator $f \mapsto \widehat{f} * K$ maps $L^p(\mathbf{R}^n)$ to $L^q(\mathbf{R}^n)$.

2.4.2. Let $m \geq 3$ and $1 \leq p_1, p_2, \ldots, p_m, q \leq \infty$ be related as in

$$\frac{1}{p_1} + \cdots + \frac{1}{p_m} = \frac{1}{q} + m - 1.$$

Prove the inequality

$$\|f_1 * \cdots * f_m\|_{L^q(\mathbf{R}^n)} \leq \|f_1\|_{L^{p_1}(\mathbf{R}^n)} \cdots \|f_m\|_{L^{p_m}(\mathbf{R}^n)}.$$

2.4.3. *(Schur's test)* Let (X, μ), (Y, ν) be σ-finite measure spaces and let $K(x, y)$ be a nonnegative measurable function on $X \times Y$. Define

$$S(f)(x) = \int_Y K(x, y) f(y) \, d\nu(y)$$

for a nonnegative measurable function f on X. Assume that

$$A_0 = \operatorname*{ess.sup}_{y \in Y} \int_X K(x,y)\,d\mu(x) < \infty, \qquad A_1 = \operatorname*{ess.sup}_{x \in X} \int_Y K(x,y)\,d\nu(y) < \infty.$$

Show that S is well defined on $L^1(X) + L^\infty(X)$ and that it maps $L^p(X)$ to $L^p(Y)$ with bound at most $A_0^{\frac{1}{p}} A_1^{1-\frac{1}{p}}$ for $1 < p < \infty$.

2.5 Approximate Identities and Almost Everywhere Convergence

In this section we use the Hardy–Littlewood maximal operator to pointwise control averages with respect to approximate identities. As a result, we deduce almost everywhere convergence properties for approximate identities convolved with certain locally integrable functions.

For an integrable function K on $\mathbf{R}^n$ we define the L^1 dilations K_t of K by setting $K_t(x) = t^{-n} K(x/t)$ for $x \in \mathbf{R}^n$ and $t > 0$.

Theorem 2.5.1. *Let K in $L^1(\mathbf{R}^n)$ satisfy $|K(x)| \leq A |x|^{-n} \min(|x|^\gamma, |x|^{-\gamma})$, $A, \gamma > 0$, and let f be in $L^1_{\mathrm{loc}}(\mathbf{R}^n)$. Then for some constant $C_{n,\gamma} < \infty$ and all $x \in \mathbf{R}^n$ we have*

$$\sup_{t > 0} (|f| * |K_t|)(x) \leq C_{n,\gamma} A\, \mathcal{M}(f)(x). \tag{2.5.1}$$

Proof. We have

$$\begin{aligned}
(|f| * |K_t|)(x) &\leq \frac{A}{t^n} \int_{\mathbf{R}^n} |f(x-y)| \left|\frac{y}{t}\right|^{-n} \min\left(\left|\frac{y}{t}\right|^\gamma, \left|\frac{y}{t}\right|^{-\gamma}\right) dy \\
&= \frac{A}{t^n} \sum_{k=-\infty}^{\infty} \int_{2^k t < |y| \leq 2^{k+1} t} \min\left(\left|\frac{y}{t}\right|^\gamma, \left|\frac{y}{t}\right|^{-\gamma}\right) \left|\frac{y}{t}\right|^{-n} |f(x-y)|\, dy \\
&\leq \frac{A}{t^n} \sum_{k=-\infty}^{\infty} \min(2^{(k+1)\gamma}, 2^{-k\gamma}) \frac{1}{2^{kn}} \int_{2^k t < |y| \leq 2^{k+1} t} |f(x-y)|\, dy \\
&\leq A v_n 2^{n+\gamma} \sum_{k=-\infty}^{\infty} \min(2^{k\gamma}, 2^{-k\gamma}) \frac{1}{v_n (2^{(k+1)} t)^n} \int_{|y| \leq 2^{k+1} t} |f(x-y)|\, dy \\
&\leq C_{n,\gamma} A\, \mathcal{M}(f)(x),
\end{aligned}$$

where v_n is the volume of the unit ball in $\mathbf{R}^n$ and $C_{n,\gamma}$ is the finite constant $v_n 2^{n+\gamma} \sum_{k \in \mathbf{Z}} \min(2^{k\gamma}, 2^{-k\gamma}) = v_n 2^{n+\gamma} (1 + 2^{-\gamma})(1 - 2^{-\gamma})^{-1}$. $\square$

Corollary 2.5.2. *Let $A, \gamma > 0$ and G be a measurable function on $\mathbf{R}^n \times \mathbf{R}^+$ satisfying*

$$|G(x, \varepsilon)| \leq \frac{1}{\varepsilon^n} \frac{A}{(1 + |x|/\varepsilon)^{n+\gamma}}, \qquad x \in \mathbf{R}^n.$$

Let $f \in L^1_{\mathrm{loc}}(\mathbf{R}^n)$. Then for some constant $C_{n,\gamma} < \infty$ and all $x \in \mathbf{R}^n$ we have

$$\sup_{\varepsilon > 0} \big(|G(\cdot, \varepsilon)| * |f|\big)(x) \leq C_{n,\gamma} A\, \mathcal{M}(f)(x).$$

Proof. Use Theorem 2.5.1 with $K(x) = (1+|x|)^{-n-\gamma}$. □

The conditions on K in Theorem 2.5.1 are weakened in Exercise 2.5.3. Another proof of Theorem 2.5.1 can be given which explicitly relates the value of constant $C_{n,\gamma}$ in (2.5.1) to K.

Proposition 2.5.3. *Let $K(x)$ be a nonnegative integrable function on $\mathbf{R}^n$, which is radial and decreasing[3] on $[0,\infty)$ as a function of $|x|$. Then for $f \in L^1_{\mathrm{loc}}(\mathbf{R}^n)$ and any $x \in \mathbf{R}^n$ we have*

$$\sup_{t>0}(K_t * |f|)(x) \leq \|K\|_{L^1} \mathcal{M}(f)(x). \tag{2.5.2}$$

Proof. For a simple function of the form

$$L = \sum_{j=1}^{M} c_j \chi_{B(0,r_j)} = \sum_{i=0}^{M-1}(c_1 + \cdots + c_{M-i}) \chi_{B(0,r_{i+1}) \setminus B(0,r_i)} \tag{2.5.3}$$

with $c_j > 0$, $r_0 = 0 < r_1 < \cdots < r_M$, for $x \in \mathbf{R}^n$ and $t > 0$ we have

$$(L_t * |f|)(x) = \sum_{j=1}^{M} c_j |B(0,r_j)| \frac{(\chi_{B(0,tr_j)} * |f|)(x)}{|B(0,tr_j)|} \leq \|L\|_{L^1} \mathcal{M}(f)(x). \tag{2.5.4}$$

But an arbitrary nonnegative radially decreasing function on $\mathbf{R}^n$ can be pointwise approximated by an increasing sequence of functions L^k of the form (2.5.3). We then apply (2.5.4) to each L^k and take the limit as $k \to \infty$ applying the LMCT. Finally, taking the supremum over all $t > 0$, we deduce (2.5.2). □

The following is an application of Theorem 2.5.1.

Proposition 2.5.4. *The space*

$$\{\varphi \in \mathscr{S}(\mathbf{R}^n) : \widehat{\varphi} \in \mathscr{C}_0^\infty \text{ and } \widehat{\varphi} \text{ vanishes in a neighborhood of } 0\}$$

is dense in $L^p(\mathbf{R}^n)$ for $1 < p < \infty$.

Proof. Start with a $\mathscr{C}_0^\infty$ function $\widehat{\Phi}$ which is equal to 1 on the unit ball and vanishes outside the ball $B(0,2)$. Consider the family $1 - \widehat{\Phi}(\xi/\varepsilon)$ which converges pointwise to 1 for $\xi \neq 0$ and the family $\widehat{\Phi}(\varepsilon\xi) \to 1$ as $\varepsilon \to 0$. Then $(1 - \widehat{\Phi}(\xi/\varepsilon))\widehat{\Phi}(\varepsilon\xi)$ converges pointwise to $\chi_{\mathbf{R}^n \setminus \{0\}}(\xi)$ for all $\xi \in \mathbf{R}^n$ and vanishes for $|\xi| \geq 2/\varepsilon$ and for $|\xi| \leq \varepsilon$. Let $h \in \mathscr{S}(\mathbf{R}^n)$. Applying the LDCT we obtain

$$\lim_{\varepsilon \to 0} \int_{\mathbf{R}^n} \widehat{h}(\xi)\left(1 - \widehat{\Phi}\left(\tfrac{\xi}{\varepsilon}\right)\right)\widehat{\Phi}(\varepsilon\xi) e^{2\pi i x \cdot \xi} d\xi = \int_{\mathbf{R}^n} \widehat{h}(\xi) e^{2\pi i x \cdot \xi} d\xi = h(x)$$

for any $x \in \mathbf{R}^n$. In other words the sequence $h_\varepsilon = h * \Phi_\varepsilon - h * \Phi_{1/\varepsilon} * \Phi_\varepsilon$ converges pointwise everywhere to h and its Fourier transform has compact support and vanishes in a neighborhood of the origin. Moreover, for some constant C_Φ we have

---
[3] Such a function is said to be *radially decreasing*.

2.5 Approximate Identities and Almost Everywhere Convergence

$$|h_\varepsilon| \leq C_\Phi \left[M(h) + M(M(h)) \right] \in L^p(\mathbf{R}^n),$$

and so, applying the LDCT again we deduce $\lim_{\varepsilon \to 0} \|h_\varepsilon - h\|_{L^p} = 0$. □

We now derive the almost everywhere convergence of approximate identities using the maximal function. Recall that $K_t(x) = t^{-n} K(t^{-1} x)$ for $x \in \mathbf{R}^n$ and $t > 0$.

Theorem 2.5.5. Let K in $L^1(\mathbf{R}^n)$ satisfy $|K(x)| \leq A|x|^{-n} \min(|x|^\gamma, |x|^{-\gamma})$, where $A, \gamma > 0$ and let $c = \int_{\mathbf{R}^n} K(x) \, dx$. Then given $1 \leq p < \infty$ and $f \in L^p(\mathbf{R}^n)$ we have

$$\lim_{t \to 0} (K_t * f)(x) = c f(x) \qquad (2.5.5)$$

for almost all $x \in \mathbf{R}^n$.

Proof. We note that (2.5.5) holds pointwise everywhere for functions f in $\mathscr{C}_0^\infty$. We obtain the general case by approximation. Define the *oscillation* of a function g in $\cup_{1 \leq p < \infty} L^p(\mathbf{R}^n)$ by setting

$$\mathscr{O}_g = \limsup_{t \to 0} |K_t * g - c g|;$$

obviously $\mathscr{O}_g$ is well defined at the points where g is defined, in particular, it is defined on the Lebesgue set $\mathscr{L}_g$ of g. Notice that $\mathscr{O}_\varphi = 0$ everywhere if $\varphi \in \mathscr{C}_0^\infty(\mathbf{R}^n)$ [Theorem 1.9.7 (b)]. Fix a function $f \in L^p(\mathbf{R}^n)$ where $1 \leq p < \infty$. Then

$$\mathscr{O}_f \leq C_{n,\gamma} A \mathcal{M}(f) + |c| |f| \leq (C_{n,\gamma} A + |c|) \mathcal{M}(f) \qquad \text{on } \mathscr{L}_f,$$

by Theorem 2.5.1 and Corollary 1.5.5. Given $\varepsilon > 0$ there is a $\varphi \in \mathscr{C}_0^\infty$ such that $\|f - \varphi\|_{L^p} < \varepsilon$. Then

$$\mathscr{O}_f \leq \mathscr{O}_{f-\varphi} + \mathscr{O}_\varphi = \mathscr{O}_{f-\varphi} \leq \mathscr{O}_f + \mathscr{O}_\varphi = \mathscr{O}_f \qquad \text{on } \mathscr{L}_f = \mathscr{L}_{f-\varphi};$$

thus $\mathscr{O}_f = \mathscr{O}_{f-\varphi}$ on $\mathscr{L}_f$. Next we prove that for any $\delta > 0$ we have

$$\left| \{ x \in \mathbf{R}^n : \mathscr{O}_f(x) > \delta \} \right| = 0. \qquad (2.5.6)$$

Indeed, we have

$$\begin{aligned}
\left| \{ x \in \mathbf{R}^n : \mathscr{O}_f(x) > \delta \} \right| &= \left| \{ x \in \mathbf{R}^n : \mathscr{O}_{f-\varphi}(x) > \delta \} \right| \\
&\leq \left| \{ x \in \mathbf{R}^n : (C_{n,\gamma} A + |c|) \mathcal{M}(f - \varphi)(x) > \delta \} \right| \\
&\leq \left(\frac{3^n 2p}{p-1} \right)^p \frac{(C_{n,\gamma} A + |c|)^p}{\delta^p} \|f - \varphi\|_{L^p}^p \\
&\leq \left(\frac{3^n 2p}{p-1} \right)^p \frac{(C_{n,\gamma} A + |c|)^p}{\delta^p} \varepsilon^p,
\end{aligned}$$

having used Chebyshev's inequality and Corollary 1.4.7. Letting $\varepsilon \to 0$ we derive (2.5.6). This implies that $\mathscr{O}_f = 0$ a.e., and consequently (2.5.5) holds. □

Essentially the same proof yields the same more general result.

Theorem 2.5.6. *Fix $1 \leq p < \infty$. Let $\{T_t\}_{t>0}$ and T be linear operators defined on $L^p(\mathbf{R}^n)$ such that $T_t(\varphi)(x) \to T(\varphi)(x)$ as $t \to 0$ for all $\varphi \in \mathscr{C}_0^\infty(\mathbf{R}^n)$ and all $x \in \mathbf{R}^n$. Suppose that $T^{(*)}(f) = \sup_{t>0}|T_t(f)|$ is a bounded operator on $L^p(\mathbf{R}^n)$. Then for all $f \in L^p(\mathbf{R}^n)$, $T_t(f) \to T(f)$ a.e. as $t \to 0$.*

Proof. Adapt the proof of Theorem 2.5.5 replacing $K_t * f$ by $T_t(f)$ and $\mathcal{M}$ by $T^{(*)}$. □

Theorem 2.5.5 does not cover the case of $p = \infty$, in view of the lack of a nice dense subspace of L^∞. A different proof of Theorem 2.5.5 can be given that not only covers the case $p = \infty$, but also allows the function f to have moderate growth at infinity, or even be locally integrable, if K has compact support. But the most important ingredient of this proof is that it relates the set of almost everywhere convergence to the Lebesgue set $\mathscr{L}_f$ of f.

Theorem 2.5.7. *Let $K \in L^1(\mathbf{R}^n)$ satisfy $|K(x)| \leq A|x|^{-n}\min(|x|^\gamma, |x|^{-\gamma})$ when $x \neq 0$, where $A > 0$ and $0 < \gamma < n$. Let $f \in L^1_{\mathrm{loc}}(\mathbf{R}^n)$. Suppose that*

$$\lim_{t \to 0^+} \int_{|y| \geq \theta} |f(x-y)||K_t(y)|\,dy = 0 \qquad \text{for all } \theta > 0 \text{ and } x \in \mathbf{R}^n. \tag{2.5.7}$$

Then for every $x \in \mathscr{L}_f$ for which

$$\int_{|y| \leq 1} |f(x-y)||y|^{-n+\gamma}\,dy < \infty \tag{2.5.8}$$

we have

$$\lim_{t \to 0^+} (K_t * f)(x) = cf(x), \tag{2.5.9}$$

where $c = \int_{\mathbf{R}^n} K(y)\,dy$. Consequently, $K_t * f \to cf$ a.e. as $t \to 0^+$.

Proof. We fix f and K as in the statement of the theorem and $x_0 \in \mathscr{L}_f$ such that (2.5.8) is satisfied. We begin with the observation that (2.5.7) with $\theta = 1$ and (2.5.8), combined with the fact that $|K_t(y)| \leq t^{-n}|y/t|^{-n+\gamma}$, yield

$$(|f| * |K_t|)(x_0) < \infty \qquad \text{for } t \text{ sufficiently small depending on } x_0.$$

We will prove (2.5.9) for $x = x_0$.

Let $\varepsilon > 0$ be given. As $x_0 \in \mathscr{L}_f$ there is a $\delta_0 > 0$ (which we pick to satisfy $\delta_0 < 1$) such that

$$0 < r \leq \delta_0 \implies \frac{1}{v_n r^n}\int_{|y|<r}|f(x_0-y)-f(x_0)|\,dy < \frac{\gamma}{4\omega_{n-1}A}\varepsilon. \tag{2.5.10}$$

Here $v_n = |B(0,1)|$ and $\omega_{n-1} = |\mathbf{S}^{n-1}|$. Since $\int_{\mathbf{R}^n} K_t(y)\,dy = c$ for any $t > 0$, we write

$$(K_t * f)(x_0) - cf(x_0) = \int_{\mathbf{R}^n} K_t(y)\big(f(x_0-y) - f(x_0)\big)\,dy,$$

2.5 Approximate Identities and Almost Everywhere Convergence

and we bound the absolute value of the last integral by

$$\int_{|y|<\delta_0} |f(x_0-y)-f(x_0)||K_t(y)|\,dy + \int_{|y|\geq \delta_0} |f(x_0-y)-f(x_0)||K_t(y)|\,dy. \quad (2.5.11)$$

We first estimate the second integral in (2.5.11). For $t>0$ we have

$$\int_{|y|\geq \delta_0} |f(x_0-y)-f(x_0)||K_t(y)|\,dy$$
$$\leq \int_{|y|\geq \delta_0} |f(x_0-y)||K_t(y)|\,dy + |f(x_0)| A t^\gamma \int_{|y|\geq \delta_0} |y|^{-n-\gamma}\,dy.$$

We pick $\delta > 0$ such that the sum above is smaller than $\varepsilon/2$ when $0 < t < \delta$, in view of (2.5.7) and the appearance of t^γ. Note that δ depends on f, x_0, n, γ, and δ_0.

To handle the first integral in (2.5.11) we use polar coordinates to write

$$\frac{1}{r^n} \int_{|y|<r} |f(x_0-y)-f(x_0)|\,dy = \frac{1}{r^n} \int_0^r \rho^{n-1} \int_{S^{n-1}} |f(x_0-\rho\theta)-f(x_0)|\,d\theta\,d\rho$$
$$= \frac{1}{r^n} \int_0^r F(\rho)\,d\rho, \quad (2.5.12)$$

where

$$F(\rho) = \rho^{n-1} \int_{S^{n-1}} |f(x_0-\rho\theta)-f(x_0)|\,d\theta, \qquad \rho > 0.$$

Since $|f-f(x_0)|$ is integrable over any ball centered at x_0, it follows that $F(\rho)$ is defined for almost all $\rho > 0$. In view of (2.5.10), the expression in (2.5.12) is at most $\frac{\gamma v_n \varepsilon}{4\omega_{n-1} A}$ when $0 < r \leq \delta_0$. Now set $L(r) = A r^{-n} \min(r^\gamma, r^{-\gamma})$ defined for $r > 0$. This function is continuous on $(0,\infty)$ and continuously differentiable on $(0,1) \cup (1,\infty)$. Also the integration-by-parts identity

$$\int_0^b L\left(\frac{r}{t}\right)\phi'(r)\,dr = L(b)\phi(b) - \int_a^b \frac{1}{t} L'\left(\frac{r}{t}\right)\phi(r)\,dr \quad (2.5.13)$$

is valid for all $t > 0$, whenever ϕ is a differentiable function on $(0,b)$ satisfying

$$0 \leq \phi(r) \leq Cr^n \quad \text{and} \quad \int_0^b L\left(\frac{r}{t}\right)|\phi'(r)|\,dr < \infty. \quad (2.5.14)$$

If $b > 1$ this can be seen by splitting the interval of integration in $(0,1)$ and $(1,b)$ and summing the outputs using that $\lim_{\delta \to 0} L(\delta/t)\phi(\delta) = 0$. Since $\gamma < n$ we have $L' < 0$ on $(0,1) \cup (1,\infty)$ and L' is undefined at 1. Now for any $t > 0$ we write

$$\int_{|y|<\delta_0} |f(x_0-y)-f(x_0)||K_t(y)|\,dy$$
$$\leq \int_{|y|<\delta_0} |f(x_0-y)-f(x_0)| \frac{1}{t^n} L\left(\frac{|y|}{t}\right) dy$$

$$= \int_0^{\delta_0} \frac{d}{dr}\left[\int_0^r F(\rho)d\rho\right]\frac{1}{t^n}L\left(\frac{r}{t}\right)dr$$

$$= \left(\frac{1}{\delta_0^n}\int_0^{\delta_0} F(\rho)d\rho\right)\frac{\delta_0^n}{t^n}L\left(\frac{\delta_0}{t}\right) - \int_0^{\delta_0}\left(\frac{1}{r^n}\int_0^r F(\rho)d\rho\right)\frac{r^n}{t^n}\frac{1}{t}L'\left(\frac{r}{t}\right)dr$$

$$= Q_f(x_0),$$

having used (2.5.13) with $\phi(r) = \int_0^r F(\rho)d\rho$ and $b = \delta_0$. Since we picked $\delta_0 < 1$ it follows that for any $t > 0$

$$\int_0^{\delta_0} L\left(\frac{r}{t}\right)|\phi'(r)|dr = \int_{|y|<\delta_0} |f(x_0-y) - f(x_0)|L\left(\frac{|y|}{t}\right)dy < \infty$$

so (2.5.14) is valid and thus (2.5.13) is justified. Next we use (2.5.10) and the fact $-L' > 0$ to obtain the estimate

$$Q_f(x_0) \leq \frac{\gamma v_n \varepsilon}{4\omega_{n-1}A}\left[\frac{\delta_0^n}{t^n}L\left(\frac{\delta_0}{t}\right) - \int_0^{\delta_0}\frac{r^n}{t^n}\frac{1}{t}L'\left(\frac{r}{t}\right)dr\right]$$

$$= \frac{\gamma v_n \varepsilon}{4\omega_{n-1}A}\left[\frac{\delta_0^n}{t^n}L\left(\frac{\delta_0}{t}\right) - \int_0^{\delta_0/t} r^n L'(r)dr\right]$$

$$= \frac{\gamma v_n \varepsilon}{4\omega_{n-1}A} n\left[\int_0^{\delta_0/t} r^{n-1}L(r)dr\right]$$

$$\leq \frac{\gamma v_n \varepsilon}{4\omega_{n-1}A} n v_n A\left[\int_0^\infty \min(r^\gamma, r^{-\gamma})\frac{dr}{r}\right]$$

$$= \frac{\varepsilon}{2},$$

where the second equality is based on (2.5.13) with $\phi(r) = r^n$. Then for $0 < t < \delta$, combining the estimates derived for the two terms in (2.5.11), we deduce

$$|(K_t * f)(x_0) - cf(x_0)| < \frac{\varepsilon}{2} + \frac{\varepsilon}{2} = \varepsilon,$$

and this proves (2.5.9).

Finally we show that (2.5.8) is satisfied for almost all $x \in \mathscr{L}_f$, and thus the claimed almost convergence is valid. For every $N \in \mathbf{Z}^+$ we have

$$\int_{|x|<N}\left[\int_{|y|\leq 1} |f(x-y)|\frac{dy}{|y|^{n-\gamma}}\right]dx \leq \left(\int_{|y|\leq 1}\frac{dy}{|y|^{n-\gamma}}\right)\int_{|x'|\leq N+1} |f(x')|dx' < \infty.$$

Consequently the integral inside the square brackets is finite for almost all points x in the ball $B(0,N)$, so letting $N \to \infty$ through the positive integers we obtain (2.5.8) for almost all points x in $\mathbf{R}^n$. □

We note that there is no restriction in assuming that $\gamma < n$ as the size estimate on K deteriorates as γ decreases to 0.

2.5 Approximate Identities and Almost Everywhere Convergence

Remark 2.5.8. Theorem 2.5.7 could have been stated in the following form: (2.5.9) is valid whenever both (2.5.7) and (2.5.8) hold at a point $x \in \mathscr{L}_f$.

Remark 2.5.9. Let K be as in Theorem 2.5.7. If K has compact support, then condition (2.5.7) holds for any locally integrable function f. Indeed, if K is supported in a ball $B(0,M)$, then the integral in (2.5.7) is over the set $\theta \leq |y| \leq Mt$ and this set becomes empty when $t < \theta/M$, so the integral is zero for t sufficiently small.

We also observe that condition (2.5.7) can be derived from

$$\int_{\mathbf{R}^n} \frac{|f(z)|}{(1+|z|)^{n+\gamma}} dz < \infty. \tag{2.5.15}$$

Indeed, assuming (2.5.15), for any $x \in \mathbf{R}^n$, we obtain

$$\int_{\mathbf{R}^n} \frac{|f(x-y)|}{(1+|y|)^{n+\gamma}} dy = \int_{\mathbf{R}^n} \frac{|f(z)|}{(1+|x-z|)^{n+\gamma}} dz < \infty \tag{2.5.16}$$

by splitting the z integral in (2.5.16) in the regions $|z| \leq 2|x|$ and $|z| \geq 2|x|$; in the latter case $|z| \approx |z-x|$ so (2.5.15) applies. Also the integral over the region $|z| \leq 2|x|$ is finite as f is locally integrable. Then for $|y| \geq \theta$ and $t > 0$ we have

$$|K_t(y)| \leq \frac{A}{t^n} \left|\frac{y}{t}\right|^{-n} \left|\frac{y}{t}\right|^{-\gamma} = A \frac{t^\gamma}{|y|^{n+\gamma}} \leq At^\gamma \left(\frac{\theta+1}{\theta}\right)^{n+\gamma} \frac{1}{(1+|y|)^{n+\gamma}}.$$

Combining this estimate with (2.5.16), we deduce (2.5.7).

Example 2.5.10. Let $A > 0$, $0 < \gamma < n$ and $|K(x)| \leq A|x|^{-n}\min(|x|^\gamma, |x|^{-\gamma})$ when $x \neq 0$. Then Theorem 2.5.7 applies in the following situations:

(a) $f \in L^p(\mathbf{R}^n)$, $1 \leq p \leq \infty$.

(b) $|f(x)| \leq C(1+|x|)^\tau$ for $\tau < \gamma$.

(c) $|f(x)| \leq C(1+|x|)^\tau$ for $\tau < 1$ and K is he Poisson kernel P.

(d) $|f(x)| \leq Ce^{|x|^\delta}$ for $0 \leq \delta < 2$ and $K(x) = e^{-\pi|x|^2}$.

(e) $f \in L^1_{\text{loc}}(\mathbf{R}^n)$ and K has compact support.

Exercises

2.5.1. Verify that in the five cases of Example 2.5.10, condition (2.5.7) is satisfied.

2.5.2. Let $0 < \gamma < n$ and let x_0 be a Lebesgue point of a function f in $L^q(\mathbf{R}^n)$ where $\frac{n}{\gamma} < q \leq \infty$. Prove that

$$\lim_{\varepsilon \to 0} \frac{1}{\varepsilon^n} \int_{\mathbf{R}^n} \frac{f(x) - f(x_0)}{\left(1 + \frac{|x-x_0|}{\varepsilon}\right)^{n+\gamma}} dx = 0.$$

[*Hint:* Let $K(x) = (1+|x|)^{-n-\gamma}$. Show that condition (2.5.7) is valid for all $x \in \mathbf{R}^n$.]

2.5.3. Show that conditions on K in Theorem 2.5.1 can be relaxed as follows:
(a) $|K(x)| \leq L(|x|)$ for some decreasing function $L : (0,\infty) \to [0,\infty)$.
(b) $L(|x|)$ lies in $L^1(\mathbf{R}^n)$.
[*Hint:* Use that $(1-2^{-n})v_n \sum_{k\in\mathbf{Z}} 2^{(k+1)n} L(2^k) \leq 2^n \int_{\mathbf{R}^n} L(|x|)\,dx.$]

2.5.4. Under the hypotheses of Theorem 2.5.7, if additionally f lies in $L^\infty(\mathbf{R}^n)$ and is continuous on a closed ball $\overline{B(x_0,\delta_0)}$ on $\mathbf{R}^n$, prove that
$$(K_t * f)(x) \to cf(x_0) \qquad \text{as } (x,t) \to (x_0, 0^+).$$

2.5.5. (Borel–Cantelli lemma) Suppose that $\{f_t\}_{t>0}$ is a family of measurable functions on a compact subset K of $\mathbf{R}^n$ (or on any measure space with finite measure). Suppose that for any $\varepsilon > 0$ the sets $A_t(\varepsilon) = \{x \in K : |f_t(x)| \geq \varepsilon\}$ satisfy
$$\sum_{k=1}^\infty |A_{t_k}(\varepsilon)| < \infty$$
for any sequence $t_k > 0$ that tends to zero. Prove that $f_t \to 0$ a.e. as $t \to 0^+$.
[*Hint:* Show first that for any sequence $t_k \to 0^+$ we have $\left|\bigcap_{m=1}^\infty \bigcup_{k=m}^\infty A_{t_k}(\varepsilon)\right| = 0.$]

2.6 Tempered Distributions

An integrable function g is almost everywhere uniquely determined[4] by the integrals $\int_{\mathbf{R}^n} g\varphi\,dx$, where φ ranges over $\mathscr{C}_0^\infty(\mathbf{R}^n)$. For this reason we can identify g by the functional $L_g(\varphi) = \int_{\mathbf{R}^n} g\varphi\,dx$, acting on $\mathscr{C}_0^\infty(\mathbf{R}^n)$. Functionals acting on nice classes of functions are called *generalized functions* or *distributions*. Viewing functions as functionals allows us to perform operations to them that would normally not be possible. For instance, one can define the partial derivative of a function $g \in L^1(\mathbf{R}^n)$ to be the functional $\partial_1 L_g$ given by $\partial_1 L_g(\varphi) = -L_g(\partial_1\varphi)$ for all $\varphi \in \mathscr{C}_0^\infty(\mathbf{R}^n)$. For such reasons, the theory of distributions provides not only a mathematically sound but also a flexible framework to work with. The theory of distributions is vast and extensive, but here we focus only on some basic facts concerning tempered distributions.

A *linear functional* on u on the space of Schwartz functions $\mathscr{S}(\mathbf{R}^n)$ is a linear mapping from $\mathscr{S}(\mathbf{R}^n)$ to the complex numbers. The action $u(\varphi)$ of u on a Schwartz function φ is denoted by $\langle u, \varphi\rangle$. Recall that for $\varphi \in \mathscr{S}(\mathbf{R}^n)$ and multi-indices α, β the expressions
$$\rho_{\alpha,\beta}(\varphi) = \sup_{x \in \mathbf{R}^n} |x^\alpha \partial^\beta \varphi(x)| \qquad (2.6.1)$$
are called *Schwartz seminorms* of φ.

---

[4] Exercise 1.9.7.

2.6 Tempered Distributions

Definition 2.6.1. A linear functional u on $\mathscr{S}(\mathbf{R}^n)$ is a tempered distribution if and only if there exist $C > 0$ and M, K nonnegative integers such that

$$|\langle u, \varphi \rangle| \leq C \sum_{|\alpha| \leq M} \sum_{|\beta| \leq K} \rho_{\alpha,\beta}(\varphi) \quad \text{for all } \varphi \in \mathscr{S}(\mathbf{R}^n). \tag{2.6.2}$$

The class of all tempered distributions on $\mathbf{R}^n$ is denoted by $\mathscr{S}'(\mathbf{R}^n)$.

Definition 2.6.1 implies that tempered distributions are continuous functionals with respect to the Schwartz topology. This means that if $\varphi_j \to \varphi$ as $j \to \infty$ in $\mathscr{S}(\mathbf{R}^n)$ (i.e., in the Schwartz topology), then $\langle u, \varphi_j \rangle \to \langle u, \varphi \rangle$.

Examples 2.6.2. We discuss some important examples of tempered distributions.

1. The *Dirac mass* δ_{x_0} at a point $x_0 \in \mathbf{R}^n$. This is defined by

$$\langle \delta_{x_0}, \varphi \rangle = \varphi(x_0)$$

for $\varphi \in \mathscr{C}^\infty(\mathbf{R}^n)$. Then $\delta_{x_0} \in \mathscr{S}'(\mathbf{R}^n)$ since $|\varphi(x_0)| \leq \|\varphi\|_{L^\infty} = \rho_{0,0}(\varphi)$.

2. Any signed Borel measure μ with total variation $\|\mu\| < \infty$ is a tempered distribution via the action

$$\langle \mu, \varphi \rangle = \int_{\mathbf{R}^n} \varphi(x) \, d\mu. \tag{2.6.3}$$

As in the previous case we have $|\langle \mu, \varphi \rangle| \leq \|\mu\| \rho_{0,0}(\varphi)$.

3. A measurable function g that satisfies $|g(x)| \leq C(1+|x|)^M$ for all $x \in \mathbf{R}^n$ is called *tempered*. A function g on $\mathbf{R}^n$ that has controlled growth of the form

$$|g(x)| \leq C(1+|x|)^M \quad \text{for all } |x| \geq R,$$

for some $M, C, R > 0$, is called *tempered at infinity*. Tempered functions give rise to tempered distributions.[5] In fact, every locally integrable and tempered-at-infinity function g gives rise to a tempered distribution L_g via the correspondence

$$\langle L_g, \varphi \rangle = \int_{\mathbf{R}^n} g(x) \varphi(x) \, dx. \tag{2.6.4}$$

To verify that $L_g \in \mathscr{S}'$ we use (1.7.3) to write

$$|\langle L_g, \varphi \rangle| \leq \left(\int_{|x| \leq R} |g(x)| \, dx \right) \|\varphi\|_{L^\infty} + \left(\int_{|x| \geq R} \frac{CC' \, dx}{(1+|x|)^{n+1}} \right) \sum_{|\alpha| \leq [M]+n+1} \rho_{\alpha,0}(\varphi).$$

4. Let $1 \leq p \leq \infty$. Functions in L^p also give rise to tempered distributions in terms of (2.6.4). Indeed, given $g \in L^p(\mathbf{R}^n)$ ($1 \leq p \leq \infty$), Hölder's inequality gives

$$|\langle L_g, \varphi \rangle| \leq \int_{\mathbf{R}^n} |g(x)|(1+|x|)^{-n-1}|\varphi(x)|(1+|x|)^{n+1} dx$$

$$\leq \|g\|_{L^p} \|(1+|\cdot|)^{-n-1}\|_{L^{p'}} C' \sum_{|\alpha| \leq n+1} \rho_{\alpha,0}(\varphi),$$

---

[5] Hence the terminology *tempered distributions*.

and as the $L^{p'}$ integral produces a constant, the claim follows. If we replace g by $g_j - g$, this inequality yields that if $g_j \to g$ in L^p, then $L_{g_j} \to L_g$ in $\mathscr{S}'$.

5. Consider the following functional acting on functions $\varphi \in \mathscr{S}(\mathbf{R})$:

$$\langle u, \varphi \rangle = \lim_{\varepsilon \to 0} \int_{\varepsilon \leq |x| \leq 1} \varphi(x) \frac{dx}{x} = \lim_{\varepsilon \to 0} \int_{\varepsilon \leq |x| \leq 1} (\varphi(x) - \varphi(0)) \frac{dx}{x}.$$

We have that $|\langle u, \varphi \rangle| \leq 2\|\varphi'\|_{L^\infty} = 2\rho_{0,1}(\varphi)$ and this gives that $u \in \mathscr{S}'(\mathbf{R})$.

Motivated by Examples 3 and 4 above, it makes sense to ignore the distinction between g and L_g with a slight abuse of terminology, explained below.

Definition 2.6.3. We say that a locally integrable function g that is tempered at infinity *coincides* (or *agrees*) (or *can be identified*) with a tempered distribution u if (2.6.4) holds for all $\varphi \in \mathscr{S}(\mathbf{R}^n)$.

We introduce the notion of convergence in $\mathscr{S}'$ as follows:

Definition 2.6.4. Let $u_j, u \in \mathscr{S}'(\mathbf{R}^n)$. We say that $u_j \to u$ *in the sense of tempered distributions*, or simply *in* $\mathscr{S}'$, if $\langle u_j, \varphi \rangle \to \langle u, \varphi \rangle$ as $j \to \infty$ for all $\varphi \in \mathscr{S}$. The same definition can be given for families of the form $\{u_\varepsilon\}_{\varepsilon > 0}$ as $\varepsilon \to 0$.

Example 2.6.5. Let Φ be an integrable function on $\mathbf{R}^n$ with integral equal to 1. Let Φ_ε be the L^1 dilations of Φ. Then Φ_ε (or precisely L_{Φ_ε}) converge to the Dirac mass at the origin δ_0 in $\mathscr{S}'(\mathbf{R}^n)$ as $\varepsilon \to 0$. Indeed, for $\varphi \in \mathscr{S}(\mathbf{R}^n)$, then we have

$$\langle \Phi_\varepsilon, \varphi \rangle = \int_{\mathbf{R}^n} \Phi_\varepsilon(x) \varphi(x)\, dx = \int_{\mathbf{R}^n} \Phi_\varepsilon(x) \widetilde{\varphi}(0-x)\, dx = (\Phi_\varepsilon * \widetilde{\varphi})(0)$$

and this converges to $\widetilde{\varphi}(0) = \varphi(-0) = \langle \delta_0, \varphi \rangle$ by Theorem 1.9.4 (b).

Having discussed important examples of distributions, we turn to some of the operations we can perform on them. Suppose that φ and ψ are Schwartz functions and α a multi-index. Integrating by parts $|\alpha|$ times, we obtain

$$\int_{\mathbf{R}^n} (\partial^\alpha \varphi)(x) \psi(x)\, dx = (-1)^{|\alpha|} \int_{\mathbf{R}^n} \varphi(x) (\partial^\alpha \psi)(x)\, dx. \qquad (2.6.5)$$

If we wanted to define the derivative of a tempered distribution u, we would need to give a definition that extends the definition of the derivative of a function and satisfies the integration by parts property in (2.6.5). We just use Eq. (2.6.5) to define the derivative of a tempered distribution.

Definition 2.6.6. Let $u \in \mathscr{S}'$ and α be a multi-index. The αth derivative of u is the element of $\mathscr{S}'(\mathbf{R}^n)$ whose action on a Schwartz function φ is given by

$$\langle \partial^\alpha u, \varphi \rangle = (-1)^{|\alpha|} \langle u, \partial^\alpha \varphi \rangle. \qquad (2.6.6)$$

Note that $\rho_{\beta,\gamma}(\partial^\alpha \varphi) = \rho_{\beta,\alpha+\gamma}(\varphi)$, so the expression on the right in (2.6.6) is controlled by a finite sum of seminorms of φ. The tempered distribution $\partial^\alpha u$ is called the *distributional derivative of u* or the *derivative of u in the sense of distributions*.

2.6 Tempered Distributions

Example 2.6.7. Let $-\infty < a < b < \infty$. Then $(\chi_{[a,b]})' = \delta_a - \delta_b$. To see this, let φ be in $\mathscr{S}(\mathbf{R})$. Then

$$\langle \chi'_{[a,b]}, \varphi \rangle = -\langle \chi_{[a,b]}, \varphi' \rangle = -\int_a^b \varphi'(x) dx = \varphi(a) - \varphi(b) = \langle \delta_a - \delta_b, \varphi \rangle.$$

Motivated by identity (2.2.1) we give the following definition.

Definition 2.6.8. Let $u \in \mathscr{S}'$. We define the Fourier transform $\widehat{u}$ and the inverse Fourier transform $u^\vee$ of a tempered distribution u by the identities

$$\langle \widehat{u}, \varphi \rangle = \langle u, \widehat{\varphi} \rangle \quad \text{and} \quad \langle u^\vee, \varphi \rangle = \langle u, \varphi^\vee \rangle, \qquad (2.6.7)$$

for all functions φ in $\mathscr{S}(\mathbf{R}^n)$.

We explain why $\widehat{u}$ and $u^\vee$ indeed lie in $\mathscr{S}'(\mathbf{R}^n)$. Indeed, by Proposition 2.1.8 we have $\rho_{\beta,\gamma}(\widehat{\varphi}) = (2\pi)^{|\gamma|-|\beta|} \| [\partial^\beta((\cdot)^\gamma \varphi)]^\wedge \|_{L^\infty}$ and moreover

$$\|\widehat{h}\|_{L^\infty} \le \|(1+|\cdot|)^{n+1} h\|_{L^\infty} \int_{\mathbf{R}^n} (1+|y|)^{-n-1} dy \quad \text{for any } h \in \mathscr{S}(\mathbf{R}^n).$$

Finally, by the lower inequality in (1.7.3) we have that $\|(1+|\cdot|)^{n+1} \partial^\beta((\cdot)^\gamma \varphi)\|_{L^\infty}$ is bounded by a constant times a finite sum of seminorms of φ. Thus, so do the expressions on the right in (2.6.7) and this explains why the Fourier transform and the inverse Fourier transform of a tempered distribution lie in $\mathscr{S}'(\mathbf{R}^n)$.

Example 2.6.9. We have $\widehat{\delta_0} = 1$. More generally, for any multi-index α, $(\partial^\alpha \delta_0)^\wedge$ can be identified with function $\xi \mapsto (2\pi i \xi)^\alpha$. To see this, observe that for all φ in $\mathscr{S}(\mathbf{R}^n)$ we have

$$\begin{aligned}
\langle (\partial^\alpha \delta_0)^\wedge, \varphi \rangle &= \langle \partial^\alpha \delta_0, \widehat{\varphi} \rangle \\
&= (-1)^{|\alpha|} \langle \delta_0, \partial^\alpha \widehat{\varphi} \rangle \\
&= (-1)^{|\alpha|} \langle \delta_0, ((-2\pi i(\cdot))^\alpha \varphi)^\wedge \rangle \\
&= (-1)^{|\alpha|} ((-2\pi i(\cdot))^\alpha \varphi)^\wedge(0) \\
&= (-1)^{|\alpha|} \int_{\mathbf{R}^n} (-2\pi i x)^\alpha \varphi(x) dx \\
&= \int_{\mathbf{R}^n} (2\pi i x)^\alpha \varphi(x) dx.
\end{aligned}$$

This calculation indicates that $(\partial^\alpha \delta_0)^\wedge$ can be identified with the function $(2\pi i \xi)^\alpha$.

Example 2.6.10. We compute the Fourier transform of the Dirac mass at x_0.

$$\langle \widehat{\delta_{x_0}}, \varphi \rangle = \langle \delta_{x_0}, \widehat{\varphi} \rangle = \widehat{\varphi}(x_0) = \int_{\mathbf{R}^n} \varphi(x) e^{-2\pi i x \cdot x_0} dx, \qquad \varphi \in \mathscr{S}(\mathbf{R}^n),$$

that is, $\widehat{\delta_{x_0}}$ can be identified with the function $\xi \mapsto e^{-2\pi i \xi \cdot x_0}$.

Proposition 2.6.11. *(Fourier inversion for distributions)* For any $u \in \mathscr{S}'(\mathbf{R}^n)$ we have $\widehat{u}^\vee = \widehat{u^\vee} = u$.

Proof. For $\varphi \in \mathscr{S}(\mathbf{R}^n)$, using (2.2.2), we write
$$\langle \widehat{u}^\vee, \varphi \rangle = \langle \widehat{u}, \varphi^\vee \rangle = \langle u, \widehat{\varphi^\vee} \rangle = \langle u, \varphi \rangle,$$
and this shows $\widehat{u}^\vee = u$. Likewise we show $\widehat{u^\vee} = u$. $\square$

As a consequence of this result we obtain that Schwartz functions are exactly those tempered distributions whose Fourier transforms are also Schwartz functions.

Proposition 2.6.12. *If $\varphi \in \mathscr{S}(\mathbf{R}^n)$, then $\widehat{\varphi}$ and $\varphi^\vee$ lie in $\mathscr{S}(\mathbf{R}^n)$. Conversely, if the Fourier transform of a tempered distribution u on $\mathbf{R}^n$ coincides with a Schwartz function, then u also coincides with an element of $\mathscr{S}(\mathbf{R}^n)$.*

Proof. The fact that $\varphi \in \mathscr{S}(\mathbf{R}^n)$ implies $\widehat{\varphi} \in \mathscr{S}(\mathbf{R}^n)$ was proven in Proposition 2.1.8 (3); consequently, $\varphi^\vee = \widetilde{\widehat{\varphi}}$ also lies in $\mathscr{S}(\mathbf{R}^n)$. Conversely, if $\widehat{u} \in \mathscr{S}(\mathbf{R}^n)$ for some u in $\mathscr{S}'(\mathbf{R}^n)$, then the inverse Fourier transform of $\varphi = \widehat{u}$ lies in $\mathscr{S}(\mathbf{R}^n)$. By Proposition 2.6.11, $u = \varphi^\vee$ but $\varphi^\vee$ lies in $\mathscr{S}(\mathbf{R}^n)$, so $u \in \mathscr{S}(\mathbf{R}^n)$. $\square$

Now observe that the following are true for functions ϕ, ψ in $\mathscr{S}(\mathbf{R}^n)$:
$$\begin{aligned}
\int_{\mathbf{R}^n} \psi(x)\phi(x-y)\,dx &= \int_{\mathbf{R}^n} \psi(x+y)\phi(x)\,dx, \\
\int_{\mathbf{R}^n} \psi(tx)\phi(x)\,dx &= \int_{\mathbf{R}^n} \psi(x)t^{-n}\phi(t^{-1}x)\,dx, \\
\int_{\mathbf{R}^n} \widetilde{\psi}(x)\phi(x)\,dx &= \int_{\mathbf{R}^n} \psi(x)\widetilde{\phi}(x)\,dx,
\end{aligned} \quad (2.6.8)$$
for all $y \in \mathbf{R}^n$ and $t > 0$. Recall now the definitions of τ^y and $\sim$ given in (2.1.5). We also define the *dilation* f^t of a function f by setting $f^t(x) = f(tx)$ for $t > 0$. Also recall the L^1 dilation $f_t(x) = t^{-n}f(t^{-1}x)$ which is related to f^t by $f^t = t^{-n}f_{1/t}$. Motivated by (2.6.8), we give the following definition.

Definition 2.6.13. The *translation* $\tau^y u$, the *dilation* u^t, and the *reflection* $\widetilde{u}$ of a tempered distribution u are tempered distributions defined as follows:
$$\langle \tau^y u, \varphi \rangle = \langle u, \tau^{-y}\varphi \rangle, \qquad (2.6.9)$$
$$\langle u^t, \varphi \rangle = \langle u, \varphi_t \rangle, \qquad (2.6.10)$$
$$\langle \widetilde{u}, \varphi \rangle = \langle u, \widetilde{\varphi} \rangle, \qquad (2.6.11)$$
for all $y \in \mathbf{R}^n$, $t > 0$, and $\varphi \in \mathscr{S}(\mathbf{R}^n)$. Let A be an invertible matrix. The composition of $u \in \mathscr{S}'(\mathbf{R}^n)$ with an invertible matrix A is defined as the element of $\mathscr{S}'$
$$\langle u \circ A, \varphi \rangle = |\det A|^{-1} \langle u, \varphi \circ A^{-1} \rangle, \qquad (2.6.12)$$
where $\varphi \circ A^{-1}(x) = \varphi(A^{-1}x)$.

2.6 Tempered Distributions

One can check that the operations of translation, dilation, reflection, and differentiation are continuous on tempered distributions.

Example 2.6.14. Let $x_0 \in \mathbf{R}^n$. Then we have $\widetilde{\delta_{x_0}} = \delta_{-x_0}$ (in particular, $\widetilde{\delta_0} = \delta_0$), also $(\delta_0)^t = t^{-n}\delta_0$, and $\tau^{x_0}\delta_0 = \delta_{x_0}$.

We now define the product of a function and a distribution.

Definition 2.6.15. Let $u \in \mathscr{S}'$ and let h be a $\mathscr{C}^\infty$ tempered function whose derivatives are also tempered. This means that for all multi-indices γ there are $C_\gamma, k_\gamma > 0$ such that $|\partial^\gamma h(x)| \leq C_\gamma(1+|x|)^{k_\gamma}$. We define the product of h and u by setting

$$\langle hu, \varphi \rangle = \langle u, h\varphi \rangle, \qquad \varphi \in \mathscr{S}. \tag{2.6.13}$$

To verify that hu is a well-defined element of $\mathscr{S}'$, we first verify that $h\varphi$ lies in $\mathscr{S}$; indeed, for each pair of multi-indices α, β we have

$$\rho_{\alpha,\beta}(h\varphi) \leq \sum_{\gamma \leq \beta} C_\gamma C_{n,k_\gamma}^{-1} \binom{\beta_1}{\gamma_1} \cdots \binom{\beta_n}{\gamma_n} \sum_{|\delta| \leq k_\gamma} \rho_{\alpha+\delta,\beta-\gamma}(\varphi) < \infty,$$

in view of Leibniz's rule, where C_{n,k_γ} are the constants in (1.7.3). This implies that $|\langle hu, \varphi \rangle|$ is bounded by a finite sum of $\rho_{\gamma,\delta}(\varphi)$, thus hu lies in $\mathscr{S}'(\mathbf{R}^n)$.

To define the convolution of a function with a tempered distribution, we examine an identity for functions. Observe that for φ, ψ in $\mathscr{S}(\mathbf{R}^n)$ and any integrable function[6] g on $\mathbf{R}^n$ the identity holds:

$$\int_{\mathbf{R}^n} (\varphi * g)(x) \psi(x) \, dx = \int_{\mathbf{R}^n} g(x)(\widetilde{\varphi} * \psi)(x) \, dx. \tag{2.6.14}$$

Motivated by (2.6.14), we give the following definition:

Definition 2.6.16. Let $u \in \mathscr{S}'$ and $\varphi \in \mathscr{S}$. Define the *convolution* $\varphi * u$ as follows:

$$\langle \varphi * u, \psi \rangle = \langle u, \widetilde{\varphi} * \psi \rangle, \qquad \psi \in \mathscr{S}(\mathbf{R}^n). \tag{2.6.15}$$

We note that $\varphi * u$ lies in $\mathscr{S}'(\mathbf{R}^n)$, since for all multi-indices α, β we have

$$\rho_{\alpha,\beta}(\widetilde{\varphi} * \psi) \leq \sup_{x \in \mathbf{R}^n} \int_{\mathbf{R}^n} |x|^{|\alpha|} |\varphi(y-x)| |\partial^\beta \psi(y)| \, dy$$

$$\leq 2^{|\alpha|} \sup_{x \in \mathbf{R}^n} \int_{\mathbf{R}^n} (|y-x|^{|\alpha|} + |y|^{|\alpha|}) |\varphi(y-x)| |\partial^\beta \psi(y)| \, dy$$

$$\leq C_{\alpha,\beta,\varphi} \Big(\rho_{0,\beta}(\psi) + \sum_{|\gamma|=|\alpha|} \rho_{\gamma,\beta}(\psi) \Big),$$

using the inequality $|x|^{|\alpha|} \leq 2^{|\alpha|}|x-y|^{|\alpha|} + 2^{|\alpha|}|y|^{|\alpha|}$ and (1.7.3).

---

[6] In fact, any locally integrable function that is tempered at infinity.

Example 2.6.17. Let $u = \delta_{x_0}$ and $\varphi \in \mathscr{S}$. Then $\varphi * \delta_{x_0}$ coincides with the function $x \mapsto \varphi(x-x_0)$, since, for all $\psi \in \mathscr{S}$, we have

$$\langle \varphi * \delta_{x_0}, \psi \rangle = \langle \delta_{x_0}, \widetilde{\varphi} * \psi \rangle = (\widetilde{\varphi} * \psi)(x_0) = \int_{\mathbf{R}^n} \varphi(x-x_0)\psi(x)\,dx.$$

Thus, for $x_0 = 0$, $\varphi * \delta_0$ coincides with the function φ for all $\varphi \in \mathscr{S}(\mathbf{R}^n)$.

Let $u, v \in \mathscr{S}'(\mathbf{R}^n)$. Suppose that $\langle u-v, \psi \rangle = 0$ for all $\psi \in \mathscr{C}_0^\infty(\mathbf{R}^n)$. Given φ in $\mathscr{S}(\mathbf{R}^n)$ pick a sequence of $\mathscr{C}_0^\infty$ functions ψ_j such that $\rho_{\alpha,\beta}(\psi_j - \varphi) \to 0$ as $j \to \infty$, by Theorem 1.8.7. Then $\langle u-v, \phi \rangle = 0$, hence $u = v$. In other words, two elements of $\mathscr{S}'$ coincide if and only if their actions on $\mathscr{C}_0^\infty$ coincide, i.e.,

$$u = v \iff \langle u, \psi \rangle = \langle v, \psi \rangle \quad \text{for all } \psi \in \mathscr{C}_0^\infty(\mathbf{R}^n). \tag{2.6.16}$$

Now the integrable functions f, g coincide a.e. on an open set Ω if and only if

$$\int f(x)\varphi(x)\,dx = \int g(x)\varphi(x)\,dx \quad \text{for all } \varphi \text{ in } \mathscr{C}_0^\infty(\Omega). \tag{2.6.17}$$

(See Exercise 1.9.7.) Here $\mathscr{C}_0^\infty(\Omega)$ is the space of smooth functions whose support is compact and contained in Ω. Motivated by this, we give the following definition, which, with a slight abuse of terminology, treats distributions as functions.

Definition 2.6.18. We say that a tempered distribution u *coincides with a function h on an open set Ω*, or alternatively, we say *u agrees with h away from Ω^c* if

$$\langle u, \varphi \rangle = \int_{\mathbf{R}^n} h(x)\varphi(x)\,dx \quad \text{for all } \varphi \text{ in } \mathscr{C}_0^\infty(\Omega). \tag{2.6.18}$$

Example 2.6.19. The distribution $|x|^2 + \delta_{x_0}$, $x_0 \in \mathbf{R}^n$, coincides with the function $|x|^2$ on $\mathbf{R}^n \setminus \{x_0\}$. Also, the distribution in Example 2.6.2 (5) agrees with the function $x^{-1}\chi_{|x|\leq 1}$ away from the origin.

We observe that if a continuous function g is supported in a set K, then for all $f \in \mathscr{C}_0^\infty(K^c)$ we have

$$\int_{\mathbf{R}^n} f(x)g(x)\,dx = 0. \tag{2.6.19}$$

Moreover, the support of g is the intersection of all closed sets K such that (2.6.19) holds for all f in $\mathscr{C}_0^\infty(K^c)$. Based on this we give the following definition:

Definition 2.6.20. Let u be in $\mathscr{S}'(\mathbf{R}^n)$. The *support of u* (supp u) is the intersection of all closed sets K with the property

$$\varphi \in \mathscr{C}_0^\infty(\mathbf{R}^n), \quad \operatorname{supp}\varphi \subseteq \mathbf{R}^n \setminus K \implies \langle u, \varphi \rangle = 0. \tag{2.6.20}$$

Example 2.6.21. The support of $u = \delta_{x_0}$, the Dirac mass at x_0, is the set $\{x_0\}$. Indeed, $\{x_0\}$ is a closed set that satisfies (2.6.20) and the only proper subset of $\{x_0\}$ is the empty set, which obviously does not satisfy (2.6.20) if $u = \delta_{x_0}$.

Exercises

2.6.1. Show that the convolution of a tempered function with a Schwartz function is another tempered function.

2.6.2. Let a,b be real numbers.
(a) Prove that the distributional derivative of $\chi_{(a,\infty)}$ is δ_a.
(b) Prove that the distributional derivative of $\chi_{(-\infty,b)}$ is $-\delta_b$.
(c) Prove that $|\cdot|'' = 2\delta_0$ in the sense of $\mathscr{S}'(\mathbf{R})$.

2.6.3. Prove that the derivative of $\log|x| \in \mathscr{S}'(\mathbf{R})$ is the tempered distribution

$$\langle u, \varphi \rangle = \lim_{\varepsilon \to 0} \int_{\varepsilon \leq |x|} \varphi(x) \frac{dx}{x}.$$

2.6.4. Evaluate the $\partial_1 \partial_2 \cdots \partial_n$ distributional derivative of the function $\chi_{[0,\infty)^n}$ in $\mathscr{S}'(\mathbf{R}^n)$.

2.6.5. Show that for a given $f \in \mathscr{S}(\mathbf{R}^n)$ there is a unique $u \in \mathscr{S}(\mathbf{R}^n)$ such that

$$-\sum_{j=1}^{n} \partial_j^2 u + u = f.$$

2.6.6. Let f,g in $L^2(\mathbf{R}^n)$. Show that the distributional Fourier transform of $f * g$ coincides with the integrable function $\widehat{f}\,\widehat{g}$.

2.6.7. Let $z \in \mathbf{C}$. A distribution in $\mathscr{S}'(\mathbf{R}^n)$ is called *homogeneous of degree* z if for all $\lambda > 0$ and for all $\varphi \in \mathscr{S}(\mathbf{R}^n)$ we have

$$\langle u, \varphi^\lambda \rangle = \lambda^{-n-z} \langle u, \varphi \rangle.$$

(a) Prove that this definition agrees with the usual definition for functions.
(b) Show that δ_0 is homogeneous of degree $-n$.
(c) Prove that if u is homogeneous of degree z, then $\partial^\alpha u$ is homogeneous of degree $z - |\alpha|$.
(d) Show that u is homogeneous of degree z if and only if $\widehat{u}$ is homogeneous of degree $-n-z$. Verify this assertion for the distribution in Example 2.7.4.

2.7 Basic Operations with Tempered Distributions

Having completed the streak of required definitions concerning operations with distributions, we discuss properties of these operations.

We begin with the observation that for a given $\psi \in \mathscr{S}(\mathbf{R}^n)$ and $u \in \mathscr{S}'(\mathbf{R}^n)$ the function $x \mapsto \langle u, \tau^x \psi \rangle$ is tempered. Indeed, for any $x \in \mathbf{R}^n$ we write

$$\left|\langle u, \tau^x \psi \rangle\right| \leq C \sum_{\substack{|\alpha| \leq M \\ |\beta| \leq K}} \sup_{y \in \mathbf{R}^n} |y^\alpha \partial^\beta \psi(y-x)|$$

$$\leq C \sum_{\substack{|\alpha| \leq M \\ |\beta| \leq K}} 2^M \sup_{y \in \mathbf{R}^n} |y-x|^{|\alpha|} |\partial^\beta \psi(y-x)| + 2^M |x|^{|\alpha|} \sup_{y \in \mathbf{R}^n} |\partial^\beta \psi(y)|$$

$$\leq C_{M,K,\psi} (1+|x|)^M,$$

having used that $|y|^{|\alpha|} \leq 2^{|\alpha|} |y-x|^{|\alpha|} + 2^{|\alpha|} |x|^{|\alpha|}$ as well as (1.7.3).

Theorem 2.7.1. *Let $u \in \mathscr{S}'$ and $\varphi \in \mathscr{S}$. Then $\varphi * u$ coincides with the function $x \mapsto \langle u, \tau^x \widetilde{\varphi} \rangle$ for all $x \in \mathbf{R}^n$. Moreover, $\varphi * u$ is a $\mathscr{C}^\infty$ function that satisfies $\partial^\alpha(\varphi * u) = \partial^\alpha \varphi * u$ for any multi-index α. Also, there is a positive constant M such that for every multi-index α there is a constant $C_{\alpha,u,\varphi} > 0$ such that*

$$|\partial^\alpha(\varphi * u)(x)| \leq C_{\alpha,u,\varphi}(1+|x|)^M. \tag{2.7.1}$$

Proof. Let ψ be in $\mathscr{S}(\mathbf{R}^n)$. We have

$$\begin{aligned}
\langle \varphi * u, \psi \rangle &= \langle u, \widetilde{\varphi} * \psi \rangle \\
&= \left\langle u, \int_{\mathbf{R}^n} \widetilde{\varphi}(\cdot - y) \psi(y) \, dy \right\rangle \\
&= \left\langle u, \int_{\mathbf{R}^n} (\tau^y \widetilde{\varphi})(\cdot) \psi(y) \, dy \right\rangle \\
&= \int_{\mathbf{R}^n} \langle u, \tau^y \widetilde{\varphi} \rangle \psi(y) \, dy,
\end{aligned} \tag{2.7.2}$$

where the last step is justified by the continuity of u and by the fact that the Riemann sums of the inner integral in (2.7.2) converge to that integral in the topology of $\mathscr{S}$, a fact that will be justified in the subsequent Lemma 2.7.2. This calculation identifies $\varphi * u$ with the tempered function $x \mapsto (\varphi * u)(x) = \langle u, \tau^x \widetilde{\varphi} \rangle$, as claimed.

We now show that $\varphi * u$ is a $\mathscr{C}^\infty$ function. Let $e_j = (0, \ldots, 1, \ldots, 0)$ with 1 in the jth entry and zero elsewhere. Then

$$\frac{(\varphi * u)(x + te_j) - (\varphi * u)(x)}{t} = \left\langle u, \frac{\tau^{te_j} \tau^x \widetilde{\varphi} - \tau^x \widetilde{\varphi}}{t} \right\rangle \to \langle u, -\partial_j \tau^x \widetilde{\varphi} \rangle = \langle u, \tau^x \widetilde{(\partial_j \varphi)} \rangle$$

where the convergence is justified by the continuity of u and the fact that

$$\frac{\tau^{te_j}(\tau^x \widetilde{\varphi}) - \tau^x \widetilde{\varphi}}{t} \to -\partial_j \tau^x \widetilde{\varphi} = \tau^x \widetilde{(\partial_j \varphi)} \qquad \text{in } \mathscr{S}$$

as $t \to 0$; see Exercise 2.7.2. This gives that $\varphi * u$ has a jth partial derivative and precisely, $\partial_j(\varphi * u) = \partial_j \varphi * u$. Then we use induction to obtain $\varphi * u \in \mathscr{C}^\infty$ and that $\partial^\gamma(\varphi * u) = (\partial^\gamma \varphi) * u$ for all multi-indices γ.

Using that $\partial^\alpha(\varphi * u) = (\partial^\alpha \varphi) * u = \langle u, \tau^x \widetilde{\partial^\alpha \varphi} \rangle = (-1)^{|\alpha|} \langle u, \partial^\alpha \tau^x \widetilde{\varphi} \rangle$, it follows from (2.6.2) that for some C, M, and K we have

2.7 Basic Operations with Tempered Distributions

$$|\partial^\alpha(\varphi * u)(x)| \leq C \sum_{\substack{|\gamma| \leq M \\ |\beta| \leq K}} \sup_{y \in \mathbf{R}^n} |y^\gamma \partial^{\alpha+\beta}(\tau^x \widetilde{\varphi})(y)|$$

$$= C \sum_{\substack{|\gamma| \leq M \\ |\beta| \leq K}} \sup_{y \in \mathbf{R}^n} |(x+y)^\gamma (\partial^{\alpha+\beta} \widetilde{\varphi})(y)|$$

$$\leq (1+|x|)^M \Big[C M^n \sum_{|\beta| \leq K} \sup_{y \in \mathbf{R}^n} (1+|y|)^M |(\partial^{\alpha+\beta} \widetilde{\varphi})(y)| \Big],$$

and this yields (2.7.1), with $C_{\alpha,u,\varphi}$ being the expression in the square brackets. $\square$

Lemma 2.7.2. *The Riemann sums of the integral in (2.7.2) converge to this integral in the topology of $\mathscr{S}$.*

Proof. For each $N \in \mathbf{Z}^+$ we partition $[-N,N]^n$ into a union of $(2N^2)^n$ cubes Q_j of side length $1/N$ and we let y_j be the center of each Q_j. We will show that for multi-indices α, β the following Riemann sum minus the corresponding integral

$$D_N(x) = x^\alpha \Big[\sum_{j=1}^{(2N^2)^n} \psi(y_j) \partial_x^\beta \widetilde{\varphi}(x - y_j) |Q_j| - \int_{\mathbf{R}^n} \psi(y) \partial_x^\beta \widetilde{\varphi}(x - y) \, dy \Big] \quad (2.7.3)$$

converges to zero in $L^\infty(\mathbf{R}^n)$ as $N \to \infty$. We write

$$x^\alpha \Big[\sum_{j=1}^{(2N^2)^n} \partial_x^\beta \psi(y_j) \widetilde{\varphi}(x - y_j) |Q_j| - \sum_{j=1}^{(2N^2)^n} \int_{Q_j} \psi(y) \partial_x^\beta \widetilde{\varphi}(x - y) \, dy \Big]$$

$$= x^\alpha \sum_{j=1}^{(2N^2)^n} \int_{Q_j} \Big[\psi(y_j) \partial_x^\beta \widetilde{\varphi}(x - y_j) - \psi(y) \partial_x^\beta \widetilde{\varphi}(x - y) \Big] dy$$

$$= x^\alpha \sum_{j=1}^{(2N^2)^n} \int_{Q_j} \int_0^1 \nabla [\psi \partial_x^\beta \widetilde{\varphi}(x - \cdot)]((1-\theta)y + \theta y_j) \cdot (y_j - y) \, d\theta \, dy \quad (2.7.4)$$

by the mean value theorem. Using estimates for Schwartz functions and the simple inequality $|\nabla(FG)| \leq \sum_{k=1}^n (|\partial_k F| |G| + |F| |\partial_k G|)$, for $y \in Q_j$ we estimate

$$|x^\alpha \nabla[\partial_x^\beta \psi \widetilde{\varphi}(x - \cdot)](\xi) \cdot (y_j - y)| \leq \frac{C_M |x|^{|\alpha|}}{(1+|x-\xi|)^{M/2}} \frac{1}{(2+|\xi|)^M} \frac{\sqrt{n}}{2N}$$

when $M > 2|\alpha| + 2n$, where $\xi = (1-\theta)y + \theta y_j$. The last expression is bounded by

$$\frac{C_M |x|^{|\alpha|}}{(1+|x|)^{M/2}} \frac{1}{(2+|\xi|)^{M/2}} \frac{\sqrt{n}}{2N} \leq \frac{C_M |x|^{|\alpha|}}{(1+|x|)^{M/2}} \frac{1}{(1+|y|)^{M/2}} \frac{\sqrt{n}}{2N},$$

since $|\xi| \geq |y| - \theta|y - y_j| \geq |y| - \frac{\sqrt{n}}{2N} \geq |y| - 1$ for $N \geq \sqrt{n}$. Inserting this estimate in (2.7.4) and using (2.7.3) and the fact that $\mathbf{R}^n = \cup_j Q_j \cup ([-N,N]^n)^c$, we obtain

$$|D_N(x)| \le \frac{C'}{N} \frac{|x|^{|\alpha|}}{(1+|x|)^{M/2}} \int_{[-N,N]^n} \frac{dy}{(1+|y|)^{M/2}} + \int_{([-N,N]^n)^c} |x^\alpha \partial_x^\beta \widetilde{\varphi}(x-y) \psi(y)|\, dy$$

for $N \ge \sqrt{n}$. But the second integral in the preceding expression is bounded by

$$\int_{([-N,N]^n)^c} \frac{C''|x|^{|\alpha|}}{(1+|x-y|)^{M/2}(1+|y|)^M}\, dy \le \frac{C''|x|^{|\alpha|}}{(1+|x|)^{M/2}} \int_{([-N,N]^n)^c} \frac{dy}{(1+|y|)^{M/2}}.$$

Using these estimates we verify that $\lim_{N\to\infty} \sup_{x\in \mathbf{R}^n} |D_N(x)| = 0$. $\square$

We now extend the properties of the Fourier transform to tempered distributions.

Proposition 2.7.3. *Let $y \in \mathbf{R}^n$, $b \in \mathbf{C}$, $t > 0$, and α be a multi-index. Given u, v in $\mathscr{S}'(\mathbf{R}^n)$, $\varphi \in \mathscr{S}(\mathbf{R}^n)$, and h a $\mathscr{C}^\infty$ tempered function all of whose derivatives are also tempered functions, we have*

(1) $\widehat{u+v} = \widehat{u} + \widehat{v}$,

(2) $\widehat{bu} = b\widehat{u}$,

(3) *If $u_j \to u$ in $\mathscr{S}'$, then $\widehat{u}_j \to \widehat{u}$ in $\mathscr{S}'$,*

(4) $(\widetilde{u})\widehat{\ } = (\widehat{u})\widetilde{\ }$,

(5) $(\tau^y u)\widehat{\ } = e^{-2\pi i y \cdot \xi} \widehat{u}$,

(6) $(e^{2\pi i x \cdot y} u)\widehat{\ } = \tau^y \widehat{u}$,

(7) $(u^t)\widehat{\ } = (\widehat{u})_t = t^{-n}(\widehat{u})^{1/t}$,

(8) $(\partial^\alpha u)\widehat{\ } = (2\pi i \xi)^\alpha \widehat{u}$,

(9) $\partial^\alpha \widehat{u} = ((-2\pi i x)^\alpha u)\widehat{\ }$,

(10) $\widehat{\varphi * u} = \widehat{\varphi}\, \widehat{u}$,

(11) $\widehat{\varphi u} = \widehat{\varphi} * \widehat{u}$,

(12) **(Leibniz rule)** $\partial^\alpha(hu) = \sum_{\gamma \le \alpha} \binom{\alpha_1}{\gamma_1} \cdots \binom{\alpha_n}{\gamma_n} (\partial^\gamma h)(\partial^{\alpha-\gamma} u)$.

Proof. Properties (1) and (2) are straightforward while (3) is due to the identity $\langle \widehat{u}_j, \widehat{\varphi} \rangle = \langle u_j, \varphi \rangle$. Statements (4)–(11) can be obtained from related statements for Schwartz functions; indicatively, we prove (8): for $\varphi \in \mathscr{S}(\mathbf{R}^n)$ we have

$$\begin{aligned}
\langle (\partial^\alpha u)\widehat{\ }, \varphi \rangle &= \langle \partial^\alpha u, \widehat{\varphi} \rangle \\
&= (-1)^{|\alpha|} \langle u, \partial^\alpha \widehat{\varphi} \rangle \\
&= (-1)^{|\alpha|} \langle u, ((-2\pi i(\cdot))^\alpha \varphi)\widehat{\ } \rangle \\
&= (-1)^{|\alpha|} \langle \widehat{u}, (-2\pi i(\cdot))^\alpha \varphi \rangle \\
&= \langle (2\pi i(\cdot))^\alpha \widehat{u}, \varphi \rangle.
\end{aligned}$$

2.7 Basic Operations with Tempered Distributions

We now prove identity (12) when $\alpha = e_j$. In this case we have

$$\begin{aligned}\langle \partial_j(hu), \psi \rangle &= -\langle hu, \partial_j \psi \rangle \\ &= -\langle u, h\partial_j \psi \rangle \\ &= -\langle u, \partial_j(h\psi) \rangle + \langle u, \psi \partial_j h \rangle \\ &= \langle \partial_j u, h\psi \rangle + \langle (\partial_j h)u, \psi \rangle \\ &= \langle h\partial_j u + (\partial_j h)u, \psi \rangle\end{aligned}$$

for any function $\psi \in \mathscr{S}(\mathbf{R}^n)$. Thus $\partial_j(hu) = h\partial_j u + (\partial_j h)u$ and this establishes (12) when $\alpha = e_j$. The case of a general index α can be obtained by induction. $\square$

Example 2.7.4. For z a complex number with $-n < \operatorname{Re} z < 0$ we define the locally integrable function

$$u_z(x) = \frac{\pi^{\frac{z+n}{2}}}{\Gamma(\frac{z+n}{2})} |x|^z, \qquad x \in \mathbf{R}^n \setminus \{0\}.$$

We compute the distributional Fourier transform of u_z and we show that it coincides with the function u_{-n-z}, that is, we show that

$$\widehat{u_z} = u_{-n-z}, \qquad -n < \operatorname{Re} z < 0. \tag{2.7.5}$$

To prove this assertion, we temporarily fix z satisfying $-n < \operatorname{Re} z < -n/2$. Then $-n/2 < \operatorname{Re}(-n-z) < 0$, so both $|x|^z$ and $|x|^{-n-z}$ are locally integrable (and certainly tempered at infinity). For $\varphi \in \mathscr{S}(\mathbf{R}^n)$ we write

$$\langle \widehat{u_z}, \varphi \rangle = \langle u_z, \widehat{\varphi} \rangle, \tag{2.7.6}$$

but we choose $\varphi(x) = \widehat{\varphi}(x) = e^{-\pi|x|^2}$. Using the result in Example 2.3.10, which gives $\widehat{u_z}(x) = c(z,n)|x|^{-z-n}$ for every $x \neq 0$, we obtain

$$c(z,n) \int_{\mathbf{R}^n} |x|^{-n-z} e^{-\pi|x|^2} dx = \frac{\pi^{\frac{z+n}{2}}}{\Gamma(\frac{z+n}{2})} \int_{\mathbf{R}^n} |x|^z e^{-\pi|x|^2} dx.$$

Switching to polar coordinates yields

$$c(z,n) \int_0^\infty r^{-z} e^{-\pi r^2} \frac{dr}{r} = \frac{\pi^{\frac{z+n}{2}}}{\Gamma(\frac{z+n}{2})} \int_0^\infty r^{z+n} e^{-\pi r^2} \frac{dr}{r},$$

and this is equivalent to

$$c(z,n) \pi^{\frac{z}{2}} \int_0^\infty s^{-\frac{z}{2}} e^{-s} \frac{ds}{s} = \frac{\pi^{\frac{z+n}{2}}}{\Gamma(\frac{z+n}{2})} \pi^{-\frac{z+n}{2}} \int_0^\infty s^{\frac{z+n}{2}} e^{-s} \frac{ds}{s},$$

by the change of variables $s = \pi r^2$. From this we obtain the value

$$c(z,n) = \frac{\pi^{-\frac{z}{2}}}{\Gamma(-\frac{z}{2})}$$

and justify the validity of (2.7.5) for z satisfying $-n < \operatorname{Re} z < -n/2$. We now rewrite (2.7.6) for z satisfying $-n < \operatorname{Re} z < -n/2$ as

$$\frac{\pi^{-\frac{z}{2}}}{\Gamma(-\frac{z}{2})} \int_{\mathbf{R}^n} |x|^{-n-z} \varphi(x)\, dx = \frac{\pi^{\frac{z+n}{2}}}{\Gamma(\frac{z+n}{2})} \int_{\mathbf{R}^n} |x|^z \widehat{\varphi}(x)\, dx. \qquad (2.7.7)$$

If we knew that both functions in (2.7.7) are analytic on the region $-n < \operatorname{Re} z < 0$, then we appeal to the identity principle in complex analysis to deduce the validity of (2.7.7) for all such z. But this is a consequence of the following lemma.

Lemma 2.7.5. *Let B be a positive real number. Then for any $\delta > 0$ we have*

$$w \in \mathbf{C}, \quad 0 < |w| < \delta \implies \left| \frac{B^w - 1}{w} \right| \leq \frac{2}{\delta} \max\left(B^{2\delta}, \frac{1}{B^{2\delta}} \right). \qquad (2.7.8)$$

Proof. Let $w = x + iy$, where $x \neq 0$, $y \neq 0$ are real. Suppose $|w| < \delta$. Then

$$\left| \frac{B^w - 1}{w} \right| \leq B^x \left| \frac{B^{iy} - 1}{x + iy} \right| + \left| \frac{B^x - 1}{x + iy} \right|$$

$$\leq B^x \frac{|B^{iy} - 1|}{|y|} + \frac{|B^x - 1|}{|x|}$$

$$= B^x |\log B| \frac{|e^{iy \log B} - 1|}{|y \log B|} + \frac{|e^{x \log B} - 1|}{|x \log B|} |\log B|$$

$$\leq B^x |\log B| + \max\left(e^{x \log B}, 1 \right) |\log B|$$

$$\leq 2 \max\left(B^{|x|}, B^{-|x|} \right) |\log B|$$

$$\leq \frac{2}{\delta} \max\left(B^{\delta}, B^{-\delta} \right) |\log B^{\delta}|$$

$$\leq \frac{2}{\delta} \max\left(B^{2\delta}, \frac{1}{B^{2\delta}} \right),$$

having used that $\log t \leq t$ for $t \geq 1$. In the cases where one of x, y (but not both) is zero, simple modifications of the preceding argument yield (2.7.8) as well. $\square$

Lemma 2.7.6. *For $\psi \in \mathscr{S}(\mathbf{R}^n)$ the function*

$$w \mapsto \frac{\pi^{\frac{w+n}{2}}}{\Gamma(\frac{w+n}{2})} \int_{\mathbf{R}^n} |x|^w \psi(x)\, dx \qquad (2.7.9)$$

is analytic in the region $\operatorname{Re} w > -n$.

Proof. The analyticity of the Gamma function (and its reciprocal) in this region is a known fact and omitted here; in fact a simple modification of the subsequent

2.7 Basic Operations with Tempered Distributions

argument proves this assertion. So we prove that $w \mapsto \int_{\mathbf{R}^n} |x|^w \psi(x)\, dx$ is analytic. We fix w_0 with $\operatorname{Re} w_0 > -n$ and pick $\delta > 0$ such that $\operatorname{Re} w_0 - 2\delta > -n$. Then

$$\lim_{w \to 0} \frac{1}{w}\left[\int_{\mathbf{R}^n} |x|^{w+w_0} \psi(x)\, dx - \int_{\mathbf{R}^n} |x|^{w_0} \psi(x)\, dx\right] = \int_{\mathbf{R}^n} |x|^{w_0}(\log|x|)\psi(x)\, dx,$$

since $\lim_{w \to 0} \frac{|x|^w - 1}{w} = \log|x|$. The passing of the limit inside the integral is justified from the LDCT via the inequality in (2.7.8), which holds for $0 < |w| < \delta$, combined with the fact that $|x|^{w_0} \max(|x|^{2\delta}, |x|^{-2\delta})\psi(x) \log|x|$ is integrable over $\mathbf{R}^n$. □

It turns out that the function in (2.7.7) is entire and thus u_z extends to an entire-valued tempered distribution whose Fourier transform is u_{-z-n}. On this see [31].

Exercises

2.7.1. Let $\Phi \in \mathscr{S}(\mathbf{R}^n)$ with $\int_{\mathbf{R}^n} \Phi(x)\, dx = 1$ and for $\varepsilon > 0$ let $\Phi_\varepsilon(x) = \varepsilon^{-n}\Phi(\varepsilon^{-1}x)$. Show that $\Phi_\varepsilon \to \delta_0$ in $\mathscr{S}'(\mathbf{R}^n)$ and that $\Phi_\varepsilon * f \to f$ in $\mathscr{S}$ for every $f \in \mathscr{S}(\mathbf{R}^n)$. Conclude that $\Phi_\varepsilon * u \to u$ in $\mathscr{S}'$ for every $u \in \mathscr{S}'(\mathbf{R}^n)$.

2.7.2. For $\varphi \in \mathscr{S}(\mathbf{R}^n)$ prove that $(\tau^{-he_j}\varphi - \varphi)/h \to \partial_j \varphi$ in $\mathscr{S}$ as $h \to 0$.

2.7.3. On the real line consider the tempered distribution u_z of Example 2.7.4. Use the Taylor expansion at the origin of a function $\varphi \in \mathscr{S}(\mathbf{R})$

$$\varphi(x) = \sum_{k=0}^{N} \frac{\varphi^{(k)}(0)}{k!} x^k + \frac{x^{N+1}}{N!} \int_0^1 (1-t)^N \varphi^{(N+1)}(tx)\, dt$$

for an arbitrary even positive integer N, to write

$$\int_{|x|<1} |x|^z \varphi(x)\, dx = \sum_{\substack{k=0 \\ k \text{ even}}}^{N} \frac{\varphi^{(k)}(0)}{k!} \frac{\omega_{n-1}}{z+k+1} + \int_{|x|<1} \int_0^1 \frac{\varphi^{(N+1)}(tx)}{N!}(1-t)^N dt\, |x|^z x^{N+1} dx.$$

Deduce the analyticity of the function $z \mapsto \int_{|x|<1} |x|^z \varphi(x)\, dx$ on $\mathbf{C} \setminus E$, where $E = \{-1, -3, -5, \ldots\}$. Conclude from this that the function $z \mapsto \langle u_z, \varphi \rangle$ is entire. [Hint: The function $z \mapsto \Gamma(\frac{z+1}{2})^{-1}$ has zeros of order 1 at $-1, -3, -5, \ldots$.]

2.7.4. Suppose that f is a tempered distribution on $\mathbf{R}^n$ whose Fourier transform coincides with an integrable and compactly supported function. Prove that $f \in \mathscr{C}^\infty$. [Hint: Show first that f can be identified with the function $x \mapsto \int_{\mathbf{R}^n} \widehat{f}(\xi) 2^{2\pi i x \cdot \xi} d\xi$.]

2.7.5. Let $a > 0$. Using that the Fourier transform of $e^{-\pi|x|^2}$ is itself, show that
(a) The Fourier transform of $e^{-\pi a|x|^2}$ on $\mathbf{R}^n$ is $a^{-n/2} e^{-\pi|x|^2/a}$.
(b) The Fourier transform of $e^{-\pi(a+it)|x|^2}$ is $(a+it)^{-n/2} e^{-\pi|x|^2/(a+it)}$, $t \in \mathbf{R}$.
(c) The distributional Fourier transform of $e^{-i\pi t|x|^2}$ is $(it)^{-n/2} e^{i\pi|x|^2/t}$, $t \in \mathbf{R} \setminus \{0\}$.

[*Hint:* Note that $(it)^{1/2}$ is a well-defined complex number with argument $\pi/4$ if $t > 0$ and $-\pi/4$ if $t < 0$. For part (b) use part (a) and analytic continuation. Obtain part (c) by taking the limit in part (b) as $a \to 0^+$ using the LDCT.]

2.8 L^p Fourier Multipliers

We are interested in studying L^p boundedness properties of operators given by multiplication by a bounded function on $\mathbf{R}^n$ on the Fourier transform. Such operators are also expressed as convolution with certain tempered distributions.

Definition 2.8.1. Given $1 < p < \infty$, we denote by $\mathscr{M}_p(\mathbf{R}^n)$ the space of all L^∞ functions m on $\mathbf{R}^n$ such that the operator

$$T_m(\varphi) = (\widehat{\varphi} m)^\vee = \varphi * m^\vee, \qquad \varphi \in \mathscr{S}(\mathbf{R}^n), \tag{2.8.1}$$

admits a bounded extension from $L^p(\mathbf{R}^n)$ to $L^p(\mathbf{R}^n)$. Here $m^\vee$ is the distributional inverse Fourier transform of m. The norm of m in $\mathscr{M}_p(\mathbf{R}^n)$ is defined by

$$\|m\|_{\mathscr{M}_p} = \|T_m\|_{L^p \to L^p}.$$

Notice that $\|\cdot\|_{\mathscr{M}_p}$ is indeed a norm on $\mathscr{M}_p(\mathbf{R}^n)$. It certainly satisfies the triangle inequality and is homogeneous of degree 1; moreover if $\|m\|_{\mathscr{M}_p} = 0$, then $(me^{-\pi|\cdot|^2})^\vee = 0$, which implies that $m = 0$ a.e. by (2.2.2).

Example 2.8.2. (a) Let $b \in \mathbf{R}^n$. Then the function $m(\xi) = e^{2\pi i \xi \cdot b}$ lies in $\mathscr{M}_p$ with norm 1 as $T_m(f)(x) = f(x+b)$ is bounded on $L^p(\mathbf{R}^n)$ with operator norm 1.
(b) Given $K \in L^1(\mathbf{R}^n)$ we have that $\widehat{K}$ lies in $\mathscr{M}_p$ with norm at most $\|K\|_{L^1}$.

Proposition 2.8.3. *Let m_1, m_2 be in $\mathscr{M}_p$ and $c \in \mathbf{C}$. Then $m_1 + m_2$, cm_1, and $m_1 m_2$ lie also in $\mathscr{M}_p$.*

Proof. Observe that $m_1 m_2$ is the multiplier that corresponds to the operator $T_{m_1} T_{m_2} = T_{m_1 m_2}$ and thus

$$\|m_1 m_2\|_{\mathscr{M}_p} = \|T_{m_1} T_{m_2}\|_{L^p \to L^p} \leq \|m_1\|_{\mathscr{M}_p} \|m_2\|_{\mathscr{M}_p}.$$

The analogous conclusions for $m_1 + m_2$ and cm_1 are straightforward. □

Other properties of multipliers are summarized below:

Proposition 2.8.4. *Fix $1 < p < \infty$. For all $m \in \mathscr{M}_p$, $x_0 \in \mathbf{R}^n$, $\lambda > 0$, and all $n \times n$ matrices A with nonzero determinant we have*

$$\|\tau^{x_0} m\|_{\mathscr{M}_p} = \|m\|_{\mathscr{M}_p}, \tag{2.8.2}$$

$$\|m^\lambda\|_{\mathscr{M}_p} = \|m\|_{\mathscr{M}_p}, \tag{2.8.3}$$

2.8 L^p Fourier Multipliers

$$\left\|\widetilde{m}\right\|_{\mathcal{M}_p} = \|m\|_{\mathcal{M}_p},$$
$$\left\|e^{2\pi i(\cdot)\cdot x_0} m\right\|_{\mathcal{M}_p} = \|m\|_{\mathcal{M}_p},$$
$$\left\|m \circ A\right\|_{\mathcal{M}_p} = \|m\|_{\mathcal{M}_p}. \tag{2.8.4}$$

Proof. We first prove (2.8.3). For f a nonzero function in $\mathscr{S}(\mathbf{R}^n)$ we have

$$T_{m^\lambda}(f)(x) = \int_{\mathbf{R}^n} \widehat{f}(\xi) m(\lambda \xi) e^{2\pi i x \cdot \xi} d\xi = \int_{\mathbf{R}^n} \frac{1}{\lambda^n} \widehat{f}\left(\frac{\xi}{\lambda}\right) m(\xi) e^{2\pi i \frac{x}{\lambda} \cdot \xi} d\xi = T_m(f^\lambda)\left(\frac{x}{\lambda}\right),$$

where we used Proposition 2.1.6 (7). This implies that

$$\frac{\|T_{m^\lambda}(f)\|_{L^p}}{\|f\|_{L^p}} = \frac{\|T_m(f^\lambda)\|_{L^p} \lambda^{\frac{n}{p}}}{\|f\|_{L^p}} = \frac{\|T_m(f^\lambda)\|_{L^p} \lambda^{\frac{n}{p}}}{\lambda^{\frac{n}{p}} \|f^\lambda\|_{L^p}} = \frac{\|T_m(f^\lambda)\|_{L^p}}{\|f^\lambda\|_{L^p}},$$

so taking the supremum over all $f \in \mathscr{S}(\mathbf{R}^n)$, or equivalently over all $f^\lambda \in \mathscr{S}(\mathbf{R}^n)$, with $\|f\|_{L^p} \neq 0$ yields (2.8.3). We now prove (2.8.4). For f a nonzero function in $\mathscr{S}(\mathbf{R}^n)$, we use Proposition 2.1.6 (8) and $\|f \circ A\|_{L^p} = |\det A|^{-1/p} \|f\|_{L^p}$ to write

$$\frac{\|T_{m \circ A}(f)\|_{L^p}}{\|f\|_{L^p}} = \frac{\left\|\left[(m(\widehat{f} \circ A^{-1})) \circ A\right]^\vee\right\|_{L^p}}{\|f\|_{L^p}}$$

$$= \frac{\left\|(m(\widehat{f} \circ A^{-1}))^\vee \circ (A^t)^{-1}\right\|_{L^p}}{|\det A| \|f\|_{L^p}}$$

$$= \frac{|\det(A^t)^{-1}|^{-\frac{1}{p}} \left\|(m |\det A| \widehat{f \circ A^t})^\vee\right\|_{L^p}}{|\det A| \|f \circ A^t \circ (A^t)^{-1}\|_{L^p}}$$

$$= \frac{|\det(A^t)^{-1}|^{-\frac{1}{p}} \left\|(m \widehat{f \circ A^t})^\vee\right\|_{L^p}}{\|f \circ A^t\|_{L^p} |\det(A^t)^{-1}|^{-\frac{1}{p}}}$$

$$= \frac{\|T_m(f \circ A^t)\|_{L^p}}{\|f \circ A^t\|_{L^p}}.$$

The supremum over all $f \in \mathscr{S}(\mathbf{R}^n)$ with $\|f\|_{L^p} \neq 0$ is equal to the supremum over all $f \circ A^t \in \mathscr{S}(\mathbf{R}^n)$ with $\|f \circ A^t\|_{L^p} \neq 0$. This proves (2.8.4). We leave the remaining properties as exercises. $\square$

Theorem 2.8.5. *Let $u \in \mathscr{S}'$ and let $T(\varphi) = \varphi * u$ for $\varphi \in \mathscr{S}$. Then T admits a bounded extension from $L^2(\mathbf{R}^n)$ to $L^2(\mathbf{R}^n)$ if and only if the Fourier transform $\widehat{u}$ of u coincides with an L^∞ function. In this case we have*

$$\|T\|_{L^2 \to L^2} = \|\widehat{u}\|_{L^\infty}. \tag{2.8.5}$$

Proof. Suppose $\widehat{u} \in L^\infty$. Then for f in $\mathscr{S}(\mathbf{R}^n)$, we have $\widehat{f}\,\widehat{u} \in L^2$; thus $(\widehat{f}\,\widehat{u})^\vee = f * u$ also lies in L^2. Plancherel's theorem gives

$$\int_{\mathbf{R}^n} |f*u|^2 \, dx = \int_{\mathbf{R}^n} |\widehat{f*u}|^2 \, d\xi = \int_{\mathbf{R}^n} |\widehat{f}\,\widehat{u}|^2 \, d\xi \le \|\widehat{u}\|_{L^\infty}^2 \|f\|_{L^2}^2.$$

As $\mathscr{S}$ is a dense subspace of L^2, we obtain the inequality

$$\|T\|_{L^2 \to L^2} \le \|\widehat{u}\|_{L^\infty}. \tag{2.8.6}$$

Now suppose that T admits a bounded extension from $L^2(\mathbf{R}^n)$ to itself. We must show that the Fourier transform of u is a bounded function. First we prove that there is a function H in $L^2_{\text{loc}}(\mathbf{R}^n)$ which coincides with $\widehat{u}$ on every open ball. We begin with the observation [using Proposition 2.7.3 (10)] that for any $\varphi \in \mathscr{C}_0^\infty$ we have

$$\varphi \widehat{u} = (\varphi^\vee * u)\widehat{} = T(\varphi^\vee)\widehat{} \in L^2(\mathbf{R}^n), \tag{2.8.7}$$

as T maps L^2 to itself. We pick $\Phi \in \mathscr{C}^\infty(\mathbf{R}^n)$ equal to 1 on the closure of the unit ball and vanishing outside the ball of radius 2 centered at the origin. We define

$$H = \begin{cases} \Phi \widehat{u} & \text{on } \overline{B(0,1)}, \\ \Phi(\cdot/2^{m+1}) \widehat{u} & \text{on } \overline{B(0,2^{m+1})} \setminus B(0,2^m) \text{ for } m \in \mathbf{Z}^+ \cup \{0\}, \end{cases} \tag{2.8.8}$$

and we notice that $H \in L^2_{\text{loc}}(\mathbf{R}^n)$. Also observe that $H = \Phi(\cdot/2^{m+1}) \widehat{u}$ on the entire ball $\overline{B(0,2^{m+1})}$; indeed, $\Phi = \Phi(\cdot/2^{m+1})$ on $\overline{B(0,1)}$ and $\Phi(\cdot/2^{k+1}) = \Phi(\cdot/2^{m+1})$ on $\overline{B(0,2^{k+1})} \setminus B(0,2^k)$ for $1 \le k < m$.

Given $\psi \in \mathscr{C}_0^\infty(\mathbf{R}^n)$ supported in a ball $B(0,R)$, pick $m \in \mathbf{Z}^+$ such that $2^{m+1} > R$. Then

$$\langle \widehat{u}, \psi \rangle = \langle \widehat{u}, \Phi(\cdot/2^{m+1})\psi \rangle = \langle \Phi(\cdot/2^{m+1})\widehat{u}, \psi \rangle = \langle H, \psi \rangle$$

and

$$T(\psi^\vee) = (\psi \widehat{u})^\vee = (\psi \Phi(\cdot/2^{m+1}) \widehat{u})^\vee = (\psi H)^\vee. \tag{2.8.9}$$

These facts indicate that H coincides with $\widehat{u}$ on every open ball $B(0,R)$ and that (2.8.9) is valid for all $\psi \in \mathscr{C}_0^\infty(\mathbf{R}^n)$. We now show that (2.8.9) also holds for L^2 functions with compact support g. Indeed, let g be an L^2 function supported in the ball $B(0,K)$. Pick ψ_j a sequence of smooth functions supported in $B(0,2K)$ such that $\|g - \psi_j\|_{L^2} \to 0$ as $j \to \infty$. We have that $(\psi_j H)^\vee = \psi_j^\vee * u = T(\psi_j^\vee)$ which converges to $T(g^\vee)$ in L^2 as $j \to \infty$, since T is L^2-bounded. Then for a subsequence j_l we have $(\psi_{j_l} H)^\vee \to T(g^\vee)$ a.e. as $l \to \infty$. On the other hand,

$$\left\|(\psi_{j_l} H)^\vee - (gH)^\vee\right\|_{L^\infty} \le \left\|(\psi_{j_l} - g)H\right\|_{L^1} \le \|\psi_{j_l} - g\|_{L^2} \|H\|_{L^2(B(0,2K))} \to 0$$

as $l \to \infty$. We conclude that $T(g^\vee) = (gH)^\vee$ a.e. for all L^2 functions with compact support g. Plancherel's theorem and the boundedness of T give

$$\int_{\mathbf{R}^n} |g(y) H(y)|^2 \, dy = \int_{\mathbf{R}^n} |(gH)^\vee(y)|^2 \, dy = \int_{\mathbf{R}^n} |T(g^\vee)(y)|^2 \, dy \le \|T\|_{L^2 \to L^2}^2 \|g\|_{L^2}^2,$$

for g in L^2 with compact support. Thus for such g we have

2.8 L^p Fourier Multipliers

$$\int_{\mathbf{R}^n} \left(\|T\|^2_{L^2 \to L^2} - |H(y)|^2 \right) |g(y)|^2 \, dy \geq 0.$$

Picking $g(y) = |B(x,\varepsilon)|^{-\frac{1}{2}} \chi_{B(x,\varepsilon)}(y)$ we obtain

$$\frac{1}{|B(x,\varepsilon)|} \int_{B(x,\varepsilon)} \left(\|T\|^2_{L^2 \to L^2} - |H(y)|^2 \right) dy \geq 0. \tag{2.8.10}$$

Letting $\varepsilon \to 0$ and applying Theorem 1.5.1 we deduce that $\|T\|^2_{L^2 \to L^2} - |H|^2 \geq 0$ a.e. Hence H lies in fact in L^∞ and

$$\|H\|_{L^\infty} \leq \|T\|_{L^2 \to L^2}. \tag{2.8.11}$$

We conclude that $\widehat{u}$ coincides on any ball with the bounded function H; this implies that $\widehat{u}$ coincides with the bounded function H everywhere and hence (2.8.11) holds with $\widehat{u}$ in place of H. Combining (2.8.6) with (2.8.11), we derive (2.8.5). □

Proposition 2.8.6. *Suppose $1 < p < \infty$ and $m \in \mathscr{M}_p(\mathbf{R}^n)$. Then the operator T_m defined in (2.8.1) satisfies*

$$\|T_m\|_{L^{p'} \to L^{p'}} = \|T_m\|_{L^p \to L^p}. \tag{2.8.12}$$

Equivalently, we have $\mathscr{M}_{p'}(\mathbf{R}^n) = \mathscr{M}_p(\mathbf{R}^n)$ with $\|m\|_{\mathscr{M}_p} = \|m\|_{\mathscr{M}_{p'}}$.

Proof. Let u be the distributional inverse Fourier transform of m. We denote by T_m^t the transpose operator of T_m (see Appendix E). For $f, g \in \mathscr{S}(\mathbf{R}^n)$ we have

$$\int_{\mathbf{R}^n} f \, T_m^t(g) \, dx = \int_{\mathbf{R}^n} T_m(f) \, g \, dx$$

$$= \int_{\mathbf{R}^n} (f * u) \, g \, dx$$

$$= \int_{\mathbf{R}^n} f \, (g * \widetilde{u}) \, dx.$$

Therefore the transpose operator T_m^t of T_m is given by $T_m^t(\varphi) = \varphi * \widetilde{u}$ for $\varphi \in \mathscr{S}(\mathbf{R}^n)$. Next observe that for $\varphi \in \mathscr{S}(\mathbf{R}^n)$ we have $\widetilde{\varphi * u} = \widetilde{\varphi} * \widetilde{u}$, which yields

$$\varphi * \widetilde{u} = \widetilde{\widetilde{\varphi} * u}. \tag{2.8.13}$$

It follows from (2.8.13) that $\|\varphi * \widetilde{u}\|_{L^{p'}} = \|\widetilde{\varphi} * u\|_{L^{p'}}$ and thus

$$\|T_m^t\|_{L^{p'} \to L^{p'}} = \sup_{\substack{\varphi \in \mathscr{S}(\mathbf{R}^n) \\ \varphi \neq 0}} \frac{\|\varphi * \widetilde{u}\|_{L^{p'}}}{\|\varphi\|_{L^{p'}}} = \sup_{\substack{\varphi \in \mathscr{S}(\mathbf{R}^n) \\ \varphi \neq 0}} \frac{\|\widetilde{\varphi} * u\|_{L^{p'}}}{\|\widetilde{\varphi}\|_{L^{p'}}} = \|T_m\|_{L^{p'} \to L^{p'}}.$$

The fact that

$$\|T_m^t\|_{L^{p'} \to L^{p'}} = \|T_m\|_{L^p \to L^p}$$

yields (2.8.12). □

As a consequence of the preceding result, the normed spaces $\mathscr{M}_p$ are nested; that is, for $1 < p \le q \le 2$ we have

$$\mathscr{M}_p \subseteq \mathscr{M}_q \subseteq \mathscr{M}_2 = L^\infty. \tag{2.8.14}$$

To see this, take $m \in \mathscr{M}_p = \mathscr{M}_{p'}$ and $1 < p \le q \le 2 \le p'$, Theorem 2.4.1 yields

$$\big\|T_m\big\|_{L^q \to L^q} \le \big\|T_m\big\|_{L^p \to L^p}^{1-\theta} \big\|T_m\big\|_{L^{p'} \to L^{p'}}^{\theta} = \big\|T_m\big\|_{L^p \to L^p}, \tag{2.8.15}$$

where $1/q = (1-\theta)/p + \theta/p'$. Thus $\|m\|_{\mathscr{M}_q} \le \|m\|_{\mathscr{M}_p}$ whenever $1 < p \le q \le 2$, hence (2.8.14) holds; in particular, with $q = 2$ we obtain

$$\|m\|_{L^\infty} \le \|m\|_{\mathscr{M}_p} \tag{2.8.16}$$

for all $p \in (1, \infty)$ and all $m \in \mathscr{M}_p$.

Proposition 2.8.7. *Let $1 < p < \infty$. Suppose that $\sup_j \|m_j\|_{\mathscr{M}_p} < \infty$ and that $m_j \to m$ pointwise a.e. Then m lies in $\mathscr{M}_p$ and satisfies $\|m\|_{\mathscr{M}_p} \le \sup_j \|m_j\|_{\mathscr{M}_p}$.*

Proof. It follows from (2.8.16) that

$$\sup_j \|m_j\|_{L^\infty} \le \sup_j \|m_j\|_{\mathscr{M}_p} = C < \infty;$$

thus the m_j are uniformly bounded. Fix $\varphi \in \mathscr{S}$. For every $x \in \mathbf{R}^n$ we have

$$T_{m_j}(\varphi)(x) = \int_{\mathbf{R}^n} \widehat{\varphi}(\xi) m_j(\xi) e^{2\pi i x \cdot \xi}\, d\xi \to \int_{\mathbf{R}^n} \widehat{\varphi}(\xi) m(\xi) e^{2\pi i x \cdot \xi}\, d\xi = T_m(\varphi)(x)$$

by the Lebesgue dominated convergence theorem, since $C|\widehat{\varphi}|$ is an integrable upper bound of all integrands on the left in the preceding expression. An application of Fatou's lemma yields that

$$\begin{aligned}\int_{\mathbf{R}^n} |T_m(\varphi)|^p\, dx &= \int_{\mathbf{R}^n} \liminf_{j \to \infty} |T_{m_j}(\varphi)|^p\, dx \\ &\le \liminf_{j \to \infty} \int_{\mathbf{R}^n} |T_{m_j}(\varphi)|^p\, dx \\ &\le C^p \|\varphi\|_{L^p}^p,\end{aligned}$$

which implies that $m \in \mathscr{M}_p$. This shows that

$$\|m\|_{\mathscr{M}_p} \le \liminf_{j \to \infty} \|m_j\|_{\mathscr{M}_p} \le C = \sup_j \|m_j\|_{\mathscr{M}_p}$$

proving the claimed assertion. $\square$

Example 2.8.8. Suppose that $\chi_{[0,1]}$ lies in $\mathscr{M}_p(\mathbf{R})$, a fact that will be shown later. We show that $m = \chi_{[0,\infty)}$ also lies in $\mathscr{M}_p(\mathbf{R})$.

Dilations, in particular property (2.8.3) with $t = 1/j$, give that $m_j = \chi_{[0,j]}$ lie in $\mathcal{M}_p(\mathbf{R})$ and $\|m_j\|_{\mathcal{M}_p} = \|m_1\|_{\mathcal{M}_p}$ for $j \in \mathbf{Z}^+$. Proposition 2.8.7 then yields that $\|m\|_{\mathcal{M}_p} \leq \|m_1\|_{\mathcal{M}_p}$.

Exercises

2.8.1. Let $1 < p < \infty$.
  (a) Let $m \in L^\infty(\mathbf{R}^n)$ satisfy $m^\vee \in L^1(\mathbf{R}^n)$. Show that $\|m\|_{\mathcal{M}_p(\mathbf{R}^n)} \leq \|m^\vee\|_{L^1(\mathbf{R}^n)}$.
  (b) If m lies in $L^1(\mathbf{R}^n) \cap L^2(\mathbf{R}^n)$ prove that $\widehat{m}$ lies in $\mathcal{M}_p(\mathbf{R}^n)$.
  (c) Use part (a) to conclude that $\mathscr{S}(\mathbf{R}^n)$ is contained in $\mathcal{M}_p(\mathbf{R}^n)$.

2.8.2. Let $1 < p < \infty$.
  (a) If $m \in \mathcal{M}_p(\mathbf{R}^n)$ and $g \in L^1(\mathbf{R}^n)$, prove that $g * m$ lies in $\mathcal{M}_p(\mathbf{R}^n)$.
  (b) Let g be a function on $\mathbf{R}^n$ that satisfies $|g(x)| \leq A|x|^{-n} \min(|x|^\gamma, |x|^{-\gamma})$ for some $\gamma > 0$, $\int_{\mathbf{R}^n} g(x)\,dx \neq 0$, and suppose that $g_\varepsilon(x) = \varepsilon^{-n} g(\varepsilon^{-1} x)$ satisfies
$$\sup_{\varepsilon > 0} \|g_\varepsilon * m\|_{\mathcal{M}_p(\mathbf{R}^n)} < \infty,$$
where $m \in L^\infty(\mathbf{R}^n)$. Prove that $m \in \mathcal{M}_p(\mathbf{R}^n)$. [*Hint:* Use Theorem 2.5.7.]

2.8.3. Let $(m_1 \otimes \cdots \otimes m_n)(y_1, \ldots, y_n) = m_1(y_1) \cdots m_n(y_n)$, $m_j : \mathbf{R} \to \mathbf{C}$, $y_j \in \mathbf{R}$.
  (a) Verify that if $m_j \in \mathcal{M}_p(\mathbf{R})$ for $j = 1, \ldots, n$, then $m_1 \otimes \cdots \otimes m_n$ lies in $\mathcal{M}_p(\mathbf{R}^n)$.
  (b) Does the function $(\xi_1, \xi_2) \mapsto \sin^7(3\xi_1 + 4\xi_2) \cos^8(5\xi_1 + 6\xi_2)$ lie in $\mathcal{M}_p(\mathbf{R}^2)$?
[*Hint:* Write $T_{m_1 \otimes \cdots \otimes m_n} = T_{m_1}^{(1)} \circ \cdots \circ T_{m_n}^{(n)}$, where $T_{m_j}^{(j)}$ acts on the jth variable only.]

2.8.4. Assume that $\chi_{[0,1]}$ lies in $\mathcal{M}_p(\mathbf{R})$
  (a) Show that for all a, b satisfying $-\infty < a < b < \infty$ we have
$$\|\chi_{[a,b]}\|_{\mathcal{M}_p} = \|\chi_{[0,1]}\|_{\mathcal{M}_p}.$$
  (b) Prove that
$$\|\chi_{(a,b)}\|_{\mathcal{M}_p} \leq \|\chi_{[0,1]}\|_{\mathcal{M}_p}$$
when $-\infty \leq a < b \leq \infty$. Conclude that $\|\chi_{[0,1]}\|_{\mathcal{M}_p(\mathbf{R})} \geq 1$.

2.8.5. Let $1 < p < \infty$ and assume that $\chi_{[0,1]}$ lies in $\mathcal{M}_p(\mathbf{R})$. Suppose that m is a bounded and differentiable function on the line with the properties $\lim_{\xi \to \pm\infty} m(\xi) = 0$ and $m' \in L^1(\mathbf{R})$. Prove that $m \in \mathcal{M}_p(\mathbf{R})$.

2.8.6. Suppose that $\chi_{[0,\infty)}$ lies in $\mathcal{M}_p(\mathbf{R})$.
  (a) Show that the characteristic function of every half-plane lies in $\mathcal{M}_p(\mathbf{R}^2)$.
  (b) Show that the characteristic function of every triangle lies in $\mathcal{M}_p(\mathbf{R}^2)$.
  (c) Show that the characteristic function of every polygon lies in $\mathcal{M}_p(\mathbf{R}^2)$.
  (d) Show that the characteristic function of every angle lies in $\mathcal{M}_p(\mathbf{R}^2)$.

2.9 Van der Corput Lemma

We end this chapter by discussing how to obtain bounds for one-dimensional oscillatory integrals with phases that are not necessarily linear functions. For instance, we study the behavior of oscillatory integrals of the type $\int_{|t|\leq 1} e^{-2\pi i \xi t^2} dt$ as $\xi \to \infty$. If the function t^2 in the exponent is replaced by t, then the integral decays like ξ^{-1} as $\xi \to \infty$; however, the presence of the quadratic phase t^2 is responsible for the slower decay $\xi^{-1/2}$ as $\xi \to \infty$. Estimates in this sort are contained in the following lemma, which is the main result of this section.

Lemma 2.9.1. *(Van der Corput lemma)* Let $\lambda > 0$ and $-\infty < a < b < \infty$. Consider a real-valued function u defined on an open subset of $\mathbf{R}$ that contains $[a,b]$.
(a) Suppose that $u \in \mathscr{C}^2$, u' is monotonic, and satisfies $|u'| \geq 1$ on $[a,b]$. Then we have

$$\left| \int_a^b e^{i\lambda u(t)} dt \right| \leq \frac{3}{\lambda}. \tag{2.9.1}$$

(b) Suppose that $u \in \mathscr{C}^{k+1}$ for some $k \in \mathbf{Z}^+$, $k \geq 2$, and satisfies $|u^{(k)}| \geq 1$ on $[a,b]$. Then we have

$$\left| \int_a^b e^{i\lambda u(t)} dt \right| \leq \frac{5 \cdot 2^{k-1} - 2}{\lambda^{1/k}}. \tag{2.9.2}$$

Proof. (a) Without loss of generality we may assume that $u' \geq 1$ as we can always replace u by $-u$ noting that in this case the integral in (2.9.1) becomes the complex conjugate of the former. Notice that as u' is monotonic and $\mathscr{C}^1$, then so is $1/u'$ and consequently $1/u'$ is differentiable and its derivative is either strictly positive or strictly negative. We write

$$\int_a^b e^{i\lambda u(t)} dt = \int_a^b \frac{1}{i\lambda u'(t)} i\lambda u'(t) e^{i\lambda u(t)} dt$$
$$= \frac{1}{i\lambda u'(b)} e^{i\lambda u(b)} - \frac{1}{i\lambda u'(a)} e^{i\lambda u(a)} - \frac{1}{i\lambda} \int_a^b e^{i\lambda u(t)} \left(\frac{1}{u'}\right)'(t) dt$$

via an integration by parts. Taking absolute values and using that $u' \geq 1$ and that $|(1/u')'| = (1/u')'$ or $|(1/u')'| = -(1/u')'$ on $[a,b]$ we conclude that

$$\left| \int_a^b e^{i\lambda u(t)} dt \right| \leq \frac{2}{\lambda} + \frac{1}{\lambda} \int_a^b \left| \left(\frac{1}{u'}\right)'(t) \right| dt = \frac{2}{\lambda} + \frac{1}{\lambda} \left| \frac{1}{u'(b)} - \frac{1}{u'(a)} \right| \leq \frac{3}{\lambda}.$$

(b) We will derive the case $k = 2$ from the case $k = 1$ [Case (a)], the case $k = 3$ from the case $k = 2$, etc. To argue by induction, we may suppose that for some $k \geq 2$ there is a positive constant $C(k-1)$ such that the estimate

$$\left| \int_a^b e^{i\lambda v(t)} dt \right| \leq C(k-1) \lambda^{-\frac{1}{k-1}} \tag{2.9.3}$$

2.9 Van der Corput Lemma

holds for all $\lambda > 0$, for all intervals $[a,b]$, and all $\mathscr{C}^k$ functions v on an open set containing $[a,b]$ which satisfy $|v^{(k-1)}| \geq 1$ on $[a,b]$. Then for our fixed $\mathscr{C}^{k+1}$ function u which satisfies $|u^{(k)}| \geq 1$ on $[a,b]$ we will prove that

$$\left| \int_a^b e^{i\lambda u(t)} dt \right| \leq C(k) \lambda^{-\frac{1}{k}} \tag{2.9.4}$$

for some constant $C(k)$ explicitly related to $C(k-1)$. To prove (2.9.4) for $k=2$ we need to use (2.9.3) when $k=2$, i.e., the result proved in Case (a). But when $k=2$, the hypothesis $|u''| \geq 1$ implies that $u'' \geq 1$ or $u'' \leq -1$, thus u' is monotonic. Thus the additional assumption of the monotonicity of u' in Case (a) will automatically hold. This additional hypothesis is not needed for $k \geq 2$.

We now focus on establishing (2.9.4) assuming (2.9.3). As we may replace u by $-u$, there is no loss of generality to assume that $u^{(k)} \geq 1$. As $u^{(k-1)}$ is strictly increasing, the following cases are the only possibilities that appear:

(i) $u^{(k-1)}(a) \geq 0$. (See Figure 2.2.) Then for $0 < \delta < b-a$ and $a+\delta < x < b$ we have

$$u^{(k-1)}(x) = u^{(k-1)}(x) - u^{(k-1)}(a) + u^{(k-1)}(a) = u^{(k-1)}(a) + u^{(k)}(\xi)(x-a) > \delta,$$

for some $\xi \in (a,x)$. Therefore, when $0 < \delta < b-a$, applying (2.9.3) with $v = u/\delta$ we obtain

$$I = \left| \int_a^b e^{i\lambda u(t)} dt \right| \leq \left| \int_a^{a+\delta} e^{i\lambda u(t)} dt \right| + \left| \int_{a+\delta}^b e^{i\lambda u(t)} dt \right| \leq \delta + C(k-1)(\lambda\delta)^{-\frac{1}{k-1}}.$$

Note that the same estimate is valid when $\delta \geq b-a$, as in this case $I \leq b-a \leq \delta$.

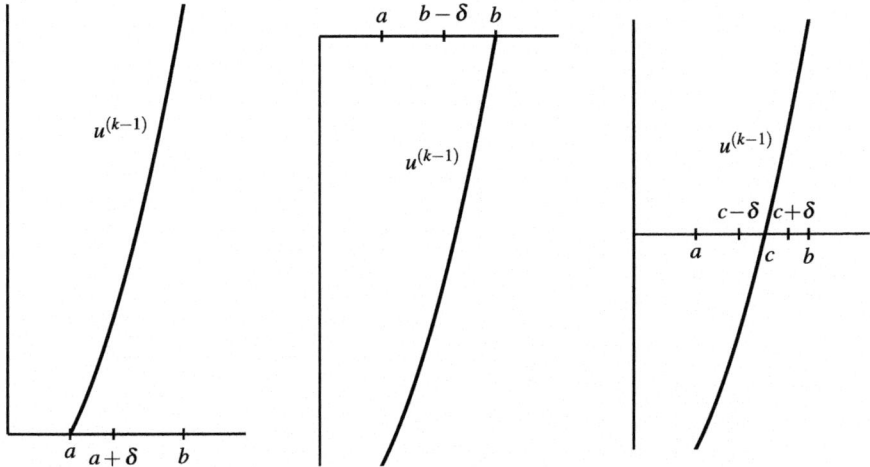

Fig. 2.2 Case (i)   **Fig. 2.3** Case (ii)   **Fig. 2.4** Case (iii)

(ii) $u^{(k-1)}(b) \leq 0$. (See Figure 2.3.) Then for $0 < \delta < b-a$ and $a < x < b - \delta$ we have

$$-u^{(k-1)}(x) = u^{(k-1)}(b) - u^{(k-1)}(x) - u^{(k-1)}(b) = -u^{(k-1)}(b) + u^{(k)}(\xi)(b-x) > \delta,$$

for some $\xi \in (x,b)$. Therefore, if $\delta < b-a$, applying (2.9.3) with $v = u/\delta$ we obtain

$$I = \left| \int_a^b e^{i\lambda u(t)} dt \right| \leq \left| \int_a^{b-\delta} e^{i\lambda u(t)} dt \right| + \left| \int_{b-\delta}^b e^{i\lambda u(t)} dt \right| \leq C(k-1)(\lambda\delta)^{-\frac{1}{k-1}} + \delta.$$

This estimate also holds if $\delta \geq b-a$, as $I \leq b-a \leq \delta \leq C(k-1)(\lambda\delta)^{-\frac{1}{k-1}} + \delta$.

(iii) $u^{(k-1)}(c) = 0$ for some $c \in (a,b)$. (See Figure 2.4.) For any $\delta > 0$ with $\delta < \min(b-c, c-a)$ we have

$$u^{(k-1)}(x) = u^{(k-1)}(x) - u^{(k-1)}(c) = u^{(k)}(\xi)(x-c) > \delta, \qquad \text{if } c + \delta < x < b$$

for some $\xi \in (c,x)$ or

$$-u^{(k-1)}(x) = u^{(k-1)}(c) - u^{(k-1)}(x) = u^{(k)}(\xi)(c-x) > \delta, \qquad \text{if } a < x < c - \delta,$$

for some $\xi \in (x,c)$. Applying the induction hypothesis, we write

$$I = \left| \int_a^b e^{i\lambda u(t)} dt \right| \leq \left| \int_a^{c-\delta} e^{i\lambda u(t)} dt \right| + \left| \int_{c-\delta}^{c+\delta} e^{i\lambda u(t)} dt \right| + \left| \int_{c+\delta}^b e^{i\lambda u(t)} dt \right|$$

$$\leq C(k-1)(\lambda\delta)^{-\frac{1}{k-1}} + 2\delta + C(k-1)(\lambda\delta)^{-\frac{1}{k-1}}$$

$$= 2C(k-1)(\lambda\delta)^{-\frac{1}{k-1}} + 2\delta.$$

This estimate is also valid when $\delta \geq \min(b-c, c-a)$. For instance, in the case $c - \delta \leq a$ and $c + \delta < b$ we estimate the integral over $[a, c+\delta]$ by $c + \delta - a \leq 2\delta$ and the integral over $[c+\delta, b]$ by $C(k-1)(\lambda\delta)^{-\frac{1}{k-1}}$. We argue similarly when $c - \delta > a$ and $c + \delta \geq b$. Finally, in the case $\delta \geq \max(b-c, c-a)$ we use $I \leq b-a \leq 2\delta$.

Thus, in all three cases (i), (ii), (iii), and for any $\delta > 0$ we have proved

$$\left| \int_a^b e^{i\lambda u(t)} dt \right| \leq 2C(k-1)(\lambda\delta)^{-\frac{1}{k-1}} + 2\delta.$$

Choosing $\delta = \lambda^{-\frac{1}{k}}$ we finally derive the estimate

$$\left| \int_a^b e^{i\lambda u(t)} dt \right| \leq (2C(k-1)+2)\lambda^{-\frac{1}{k}} = C(k)\lambda^{-\frac{1}{k}},$$

where $C(k) = 2C(k-1) + 2$. As $C(1) = 3$ we obtain $C(k) = 5 \cdot 2^{k-1} - 2$. □

Example 2.9.2. Let $k \in \mathbf{Z}^+$ and $k \geq 2$. We have the estimate

2.9 Van der Corput Lemma

$$\left| \int_0^b e^{-2\pi i \xi t^k} dt \right| \leq \frac{5 \cdot 2^{k-1} - 2}{(2\pi k!)^{1/k}} |\xi|^{-\frac{1}{k}} \tag{2.9.5}$$

for all $\xi \neq 0$ and any $b > 0$. Indeed, we apply Lemma 2.9.1 (b) with $u(t) = -(\operatorname{sgn} \xi) t^k / k!$, which satisfies $|u^{(k)}| \geq 1$ to deduce (2.9.5). In Exercise 2.9.2 estimate (2.9.5) is extended to noninteger values of k.

However, on an interval of the form $[a,b]$, where $0 < a < b < \infty$, applying Lemma 2.9.1 (a) to $-(\operatorname{sgn} \xi) t^k / k a^{k-1}$, we obtain the better estimate (in $|\xi| \neq 0$)

$$\left| \int_a^b e^{-2\pi i \xi t^k} dt \right| \leq \frac{3}{2\pi k a^{k-1}} |\xi|^{-1}. \tag{2.9.6}$$

Clearly estimate (2.9.6) is better than (2.9.5) in terms of the decay of $|\xi| \to \infty$, but obviously this gets worse as $a \downarrow 0^+$ if $k > 1$.

As the previous example indicates, in many cases one does not have $|u^{(k)}| \geq 1$, but $|u^{(k)}| \geq c_0$ for some constant c_0. Then we have to replace u by u/c_0 and λ by λc_0. We state this situation as a corollary.

Corollary 2.9.3. *Let $\lambda, c_0 > 0$ and $-\infty < a < b < \infty$. Let u be a continuously differentiable function on an open subset of $\mathbf{R}$ that contains $[a,b]$.*
(a) Suppose that u' is monotonic and satisfies $|u'| \geq c_0$ on $[a,b]$. Then

$$\left| \int_a^b e^{i\lambda u(t)} dt \right| \leq \frac{3}{\lambda c_0}.$$

(b) Let $k \geq 2$. Suppose that u is of class $\mathscr{C}^k$ and $|u^{(k)}| \geq c_0$ on $[a,b]$. Then

$$\left| \int_a^b e^{i\lambda u(t)} dt \right| \leq \frac{5 \cdot 2^{k-1} - 2}{(\lambda c_0)^{1/k}}.$$

The proof of this corollary has already been discussed.

Exercises

2.9.1. Show that for any real ξ satisfying $|\xi| \geq 1$ and any $k \in \mathbf{Z}$ with $k \geq 2$ we have

$$\left| \int_{|\xi|^{-\frac{1}{k-1}}}^1 e^{-i\xi t^k} dt \right| \leq \frac{1}{2k}.$$

[*Hint:* Use (2.9.6).]

2.9.2. Show that for any $\gamma > 1$ and $\gamma \notin \mathbf{Z}$ and $\lambda, b > 0$ we have

$$\left| \int_0^b e^{i\lambda t^\gamma} dt \right| \leq \frac{1 + \frac{3}{\gamma}}{\lambda^{1/\gamma}}.$$

[*Hint:* If $\lambda^{-1/\gamma} < b$, split the integral over the intervals $[0, \lambda^{-1/\gamma}]$ and $[\lambda^{-1/\gamma}, b]$ and apply Corollary 2.9.3 (a) to estimate the integral over the second interval.]

2.9.3. Let $\lambda > 0$, $d > 1$, and $0 < b \leq \frac{1}{e} - c$, where $0 < c < \frac{1}{e}$. Prove the following assertions:

$$\left| \int_0^1 e^{i\lambda t \log t} \, dt \right| \leq \frac{8}{\sqrt{\lambda}}, \quad \left| \int_1^d e^{i\lambda t \log t} \, dt \right| \leq \frac{3}{\lambda}, \quad \left| \int_0^b e^{i\lambda t \log t} \, dt \right| \leq \frac{3}{\lambda c e},$$

[*Hint:* For the last integral use that $|\log(1 - ce)| \geq ce$.]

2.9.4. Let a, b, k, λ and u be as in Lemma 2.9.1 and let φ be a $\mathscr{C}^k$ function on an open set containing the interval $[a, b]$. Show that

$$\left| \int_a^b \varphi(t) e^{i\lambda u(t)} \, dt \right| \leq \frac{5 \cdot 2^{k-1} - 2}{\lambda^{1/k}} \left[|\varphi(b)| + \int_a^b |\varphi'(t)| \, dt \right]$$

with the additional hypothesis that u' is monotonic when $k = 1$. (The expression in the brackets can be replaced by $\varphi(a)$ if φ is positive and decreasing.)

2.9.5. Let $k \in \mathbf{Z}^+$. Prove that there is a positive constant c_k such that

$$\sup_{\lambda \in \mathbf{R}} \sup_{0 < \varepsilon < 1} \left| \int_{\varepsilon \leq |t| \leq 1} e^{i\lambda t^k} \frac{dt}{t} \right| \leq c_k.$$

[*Hint:* For $|\lambda| \leq 1$ use the inequality $|e^{i\lambda t^k} - 1| \leq |\lambda t^k|$. If $|\lambda| \geq 1$ and $\varepsilon < |\lambda|^{-1/k}$ split the domains of integration into the regions $|t| \leq |\lambda|^{-1/k}$ and $|t| \geq |\lambda|^{-1/k}$ and use Exercise 2.9.4 in the second case. If $\varepsilon \geq |\lambda|^{-1/k}$ modify the previous argument.]

2.9.6. Show that

$$\lim_{\lambda \to 0^+} \int_1^\infty e^{it} \left(1 - e^{-it^3 \lambda}\right) \frac{dt}{t} = 0.$$

[*Hint:* For $\lambda > 10$ split the integral into the parts:

$$I = \int_{1/\sqrt{6\lambda}}^{1/\sqrt{\lambda}} e^{it} \left(1 - e^{-it^3 \lambda}\right) \frac{dt}{t},$$

$$II = \int_1^{1/\sqrt{6\lambda}} e^{it} \left(1 - e^{-it^3 \lambda}\right) \frac{dt}{t},$$

$$III = \int_{1/\sqrt{\lambda}}^\infty e^{it} \left(1 - e^{-it^3 \lambda}\right) \frac{dt}{t},$$

and apply the Riemann–Lebesgue lemma and Exercise 2.9.4 for part *I*. Parts *II* and *III* can be handled by integrating by parts.]

Chapter 3
Singular Integrals

3.1 The Hilbert Transform

Informally speaking, the Hilbert transform is given by convolution with $1/x$ on the real line. But as $1/x$ is not integrable, some care is needed to properly define this operation. This can be achieved by considering a tempered distribution that coincides with $1/x$ away from the origin.

We define a linear functional W_0 on $\mathscr{S}(\mathbf{R})$ as follows:

$$\langle W_0, \varphi \rangle = \lim_{\varepsilon \to 0} \int_{\varepsilon \leq |x| \leq 1} \frac{\varphi(x)}{x} \, dx + \int_{|x| \geq 1} \frac{\varphi(x)}{x} \, dx, \qquad \varphi \in \mathscr{S}(\mathbf{R}). \tag{3.1.1}$$

As the function $1/x$ has integral zero over $[-1, -\varepsilon] \cup [\varepsilon, 1]$, we replace $\varphi(x)$ by $\varphi(x) - \varphi(0)$ in the first integral in (3.1.1). Using the fact that $(\varphi(x) - \varphi(0))/x$ is bounded by $\|\varphi'\|_{L^\infty}$, we deduce that the limit in (3.1.1) exists. To verify that W_0 lies in $\mathscr{S}'(\mathbf{R})$ we notice that

$$|\langle W_0, \varphi \rangle| \leq 2\|\varphi'\|_{L^\infty} + 2\sup_{x \in \mathbf{R}} |x\varphi(x)| = 2[\rho_{0,1}(\varphi) + \rho_{1,0}(\varphi)]. \tag{3.1.2}$$

Definition 3.1.1. The *Hilbert transform* of $\varphi \in \mathscr{S}(\mathbf{R})$ is defined by

$$H(\varphi)(x) = \frac{1}{\pi}(\varphi * W_0)(x) = \frac{1}{\pi} \lim_{\varepsilon \to 0} \int_{|t| \geq \varepsilon} \frac{\varphi(x-t)}{t} \, dt. \tag{3.1.3}$$

Limits of integrals as the one in (3.1.3) can also be written without a limit. A *principal value integral* is an improper integral of the form

$$\text{p.v.} \int_{\mathbf{R}} \frac{\varphi(x-t)}{t} \, dt = \lim_{\varepsilon \to 0} \int_{|t| \geq \varepsilon} \frac{\varphi(x-t)}{t} \, dt.$$

Thus the Hilbert transform of a Schwartz function is a principal value integral.
According to Definition 3.1.1, the Hilbert transform of a Schwartz function φ is

$$H(\varphi)(x) = \frac{1}{\pi} \lim_{\varepsilon \to 0} \int_{|t| \geq \varepsilon} \frac{\varphi(x-t)}{t} \left[\chi_{|t|<1} + \chi_{|t|\geq 1} \right] dt \qquad (3.1.4)$$

$$= \frac{1}{\pi} \int_{|t|<1} \frac{\varphi(x-t) - \varphi(x)}{t} dt + \frac{1}{\pi} \int_{|t|\geq 1} \frac{\varphi(x-t)}{t} dt$$

$$= -\frac{1}{\pi} \int_{|t|<1} \int_0^1 \varphi'(x-t\theta) d\theta \, dt + \frac{1}{\pi} \int_{|t|\geq 1} \frac{\varphi(x-t)}{t} dt. \qquad (3.1.5)$$

Remark 3.1.2. $H(\varphi)$ is also well defined for other classes of functions φ.
(a) Let us assume that $|\varphi(t)| \leq C(1+|t|)^{-\delta}$ for some $C, \delta > 0$ so that the second integral in (3.1.5) converges. Then $H(\varphi)(x)$ is well defined if φ is a $\mathscr{C}^1$ function, since in this case the first integral in (3.1.5) also converges for all real x.
(b) More generally, if for a given $x \in \mathbf{R}$ there is an open interval $(x - \delta_x, x + \delta_x)$ on whose closure φ is continuously differentiable, then $H(\varphi)(x)$ is defined for this x. To make sense of this, we simply replace the intervals $|t| < 1$ and $|t| \geq 1$ in (3.1.4) and (3.1.5) by $|t| < \delta_x$ and $|t| \geq \delta_x$, respectively. An example of such a situation is $\varphi = \chi_{[a,b]}$ for some $a < b$, $x \in \mathbf{R} \setminus \{a,b\}$, and $\delta_x = \min(|x-a|, |x-b|)/2 > 0$.
(c) Given an integrable function g and $x \in \mathbf{R}$ such that the function $t \mapsto \frac{1}{|t|}|g(x-t)|$ is integrable on the line, then the limit in (3.1.4) exists by the LDCT and thus $H(g)$ is well defined at this given x. For instance, this happens when g has compact support and x lies outside the support of g. On this, see Example 3.1.4.

Remark 3.1.3. Let η be an even smooth function supported on $[-2,2]$ and equal to 1 on $[-1,1]$. Then the splitting $1 = \chi_{|t|<1} + \chi_{|t|\geq 1}$ in (3.1.4) can be replaced by $1 = \eta(t) + 1 - \eta(t)$. This yields the following equivalent identity for $H(\varphi)$:

$$H(\varphi)(x) = -\frac{1}{\pi} \int_{\mathbf{R}} \int_0^1 \varphi'(x-t\theta) d\theta \, \eta(t) dt + \frac{1}{\pi} \int_{\mathbf{R}} \varphi(x-t) \frac{1-\eta(t)}{t} dt. \qquad (3.1.6)$$

Example 3.1.4. Let φ be a $\mathscr{C}^1$ function on the line with compact support. Then for some constant C (depending on φ) one has

$$|H(\varphi)(x)| \leq \frac{C}{1+|x|} \qquad (3.1.7)$$

for all real x. Indeed, (3.1.5) implies that $H(\varphi)(x)$ is bounded everywhere, and this yields (3.1.7) for small x. If φ is supported in $[-K,K]$ (for some $K > 1$), then for $|x| > 2K$ the first integral in (3.1.5) is zero, while the second integral in (3.1.5) can be written as

$$\frac{1}{\pi} \int_{|x-t|\geq 1} \frac{\varphi(t)}{x-t} dt,$$

whose absolute value is bounded by $\frac{1}{\pi} \|\varphi\|_{L^1} \frac{2}{|x|}$, as

$$|x-t| \geq |x| - |t| \geq |x| - K \geq |x|/2$$

when $|x| \geq 2K$ and $|t| \leq K$. This implies (3.1.7) for large values of x.

3.1 The Hilbert Transform

Definition 3.1.5. The *truncated Hilbert transform* (at height ε) of a function f in $L^p(\mathbf{R})$, $1 \leq p < \infty$, is defined by

$$H^{(\varepsilon)}(f)(x) = \frac{1}{\pi} \int_{|y| \geq \varepsilon} \frac{f(x-y)}{y}\, dy. \tag{3.1.8}$$

Observe that $H^{(\varepsilon)}(f)$ is well defined for all $f \in L^p$, $1 \leq p < \infty$. This follows from Hölder's inequality, since $1/|x|$ is integrable to the power p' on the set $|x| \geq \varepsilon$.

We discuss the example $f = \chi_{[a,b]}$ which lies in all L^p spaces. As pointed out in Remark 3.1.2 (b), $H(\chi_{[a,b]})(x)$ is defined for all $x \notin \{a,b\}$. We calculate this value.

Example 3.1.6. We begin with the observation that if A, B are nonzero and satisfy $-\infty < A < B < \infty$, then we have

$$\text{p.v.} \int_A^B \frac{dt}{t} = \lim_{\varepsilon \to 0} \int_{|t| \geq \varepsilon} \frac{\chi_{[A,B]}(t)}{t}\, dt = \log \frac{|B|}{|A|}. \tag{3.1.9}$$

This can be shown by considering the cases $0 < A < B < \infty$, $-\infty < A < B < 0$, and $-\infty < A < 0 < B < \infty$. In the third case we consider the subcases $|A| < |B|$ and $|A| \geq |B|$. When $|A| < |B|$, taking $\varepsilon < |A|$ we split the interval of integration in (3.1.9) as $(A, -\varepsilon] \cup [\varepsilon, |A|) \cup [|A|, |B|)$ and we notice that $1/t$ has vanishing integral over $(A, -\varepsilon] \cup [\varepsilon, |A|)$. A similar argument is valid when $|A| \geq |B|$. Using (3.1.9) we obtain the following identity:

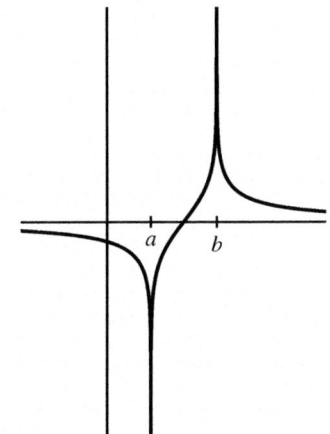

Fig. 3.1 The function $H(\chi_{[a,b]})(x)$.

$$H(\chi_{[a,b]})(x) = \text{p.v.} \frac{1}{\pi} \int_{x-b}^{x-a} \frac{dt}{t} = \frac{1}{\pi} \log \frac{|x-a|}{|x-b|}, \qquad x \notin \{a,b\} \tag{3.1.10}$$

and this function is plotted in Figure 3.1.

Note, in fact, that for $\varepsilon < \min(|x-a|, |x-b|)$ we actually have

$$H(\chi_{[a,b]})(x) = H^{(\varepsilon)}(\chi_{[a,b]})(x) = \frac{1}{\pi} \log \frac{|x-a|}{|x-b|}, \qquad x \notin \{a,b\}.$$

We now give an alternative characterization of the Hilbert transform using the Fourier transform. To achieve this we need to compute the Fourier transform of the distribution W_0 defined in (3.1.1). Recall the signum function

$$\mathrm{sgn}\, x = \begin{cases} +1 & \text{when } x > 0, \\ 0 & \text{when } x = 0, \\ -1 & \text{when } x < 0. \end{cases} \qquad (3.1.11)$$

Fix a Schwartz function φ on $\mathbf{R}$. Then

$$\left\langle \frac{1}{\pi}\widehat{W_0}, \varphi \right\rangle = \frac{1}{\pi}\langle W_0, \widehat{\varphi} \rangle \qquad (3.1.12)$$

$$= \frac{1}{\pi}\lim_{\varepsilon \to 0} \int_{|\xi| \geq \varepsilon} \widehat{\varphi}(\xi)\frac{d\xi}{\xi}$$

$$= \frac{1}{\pi}\lim_{\varepsilon \to 0} \int_{\frac{1}{\varepsilon} \geq |\xi| \geq \varepsilon} \int_{\mathbf{R}} \varphi(x) e^{-2\pi i x \xi}\, dx\, \frac{d\xi}{\xi}$$

$$= \lim_{\varepsilon \to 0} \int_{\mathbf{R}} \varphi(x) \left[\frac{1}{\pi} \int_{\frac{1}{\varepsilon} \geq |\xi| \geq \varepsilon} e^{-2\pi i x \xi}\, \frac{d\xi}{\xi} \right] dx$$

$$= \lim_{\varepsilon \to 0} \int_{\mathbf{R}} \varphi(x) \left[\frac{-i}{\pi} \int_{\frac{1}{\varepsilon} \geq |\xi| \geq \varepsilon} \sin(2\pi x \xi)\, \frac{d\xi}{\xi} \right] dx$$

$$= \lim_{\varepsilon \to 0} \int_{\mathbf{R}} \varphi(x) \left[\left(\frac{-i}{\pi}\mathrm{sgn}\, x\right) \int_{\frac{2\pi}{\varepsilon} \geq |\xi| \geq 2\pi\varepsilon} \sin(|x|\xi)\, \frac{d\xi}{\xi} \right] dx. \qquad (3.1.13)$$

Using the result of Exercise 2.2.7 we obtain that the integral inside the square brackets in (3.1.13) converges to π as $\varepsilon \to 0$, whenever $x \neq 0$ (and to 0 when $x = 0$). Moreover, we have (see Exercise 3.1.1)

$$\sup_{0 < a < b < \infty} \left| \int_{a \leq |t| \leq b} \frac{\sin t}{t}\, dt \right| \leq 8. \qquad (3.1.14)$$

These observations allow us to use the Lebesgue dominated convergence theorem to justify the passage of the limit inside the integral in (3.1.13). We obtain that

$$\left\langle \frac{1}{\pi}\widehat{W_0}, \varphi \right\rangle = \int_{\mathbf{R}} \varphi(x)(-i\,\mathrm{sgn}(x))\, dx. \qquad (3.1.15)$$

This implies that $\widehat{W_0}$ is indeed a function given by the formula

$$\frac{1}{\pi}\widehat{W_0}(\xi) = -i\,\mathrm{sgn}\,\xi. \qquad (3.1.16)$$

In view of identity (3.1.16) we have the following representation of H:

$$H(\varphi) = -i\left(\widehat{\varphi}\,\mathrm{sgn}(\cdot)\right)^{\vee}, \qquad \varphi \in \mathscr{S}(\mathbf{R}). \qquad (3.1.17)$$

This identity produces an alternative definition of the Hilbert transform in terms of the Fourier transform. An immediate consequence of (3.1.17) is that

$$\|H(\varphi)\|_{L^2} = \|\varphi\|_{L^2}, \qquad (3.1.18)$$

3.1 The Hilbert Transform

for all $\varphi \in \mathscr{S}(\mathbf{R})$. This implies that H has a unique extension on L^2 (still denoted by H), which is an isometry on $L^2(\mathbf{R})$. Moreover, as $(-i\operatorname{sgn}\xi)^2 = -1$ on $\mathbf{R}\setminus\{0\}$, we obtain that H satisfies on L^2

$$H^2 = H \circ H = -I,$$

where I is the identity operator.

The *adjoint operator* H^* of H is uniquely defined via the identity

$$\int_{\mathbf{R}} f \overline{H(g)}\, dx = \int_{\mathbf{R}} H^*(f)\,\overline{g}\, dx, \qquad f, g \in \mathscr{S}(\mathbf{R}),$$

and we can obtain that H^* corresponds to the multiplier $\overline{-i\operatorname{sgn}\xi} = i\operatorname{sgn}\xi$ by Parseval's identity. We conclude that $H^* = -H$, i.e., the Hilbert transform is an *anti-self-adjoint operator*.

Exercises

3.1.1. Show that

$$\sup_{a>0}\sup_{b>a}\left|\int_a^b \frac{\sin t}{t}\, dt\right| \le 4.$$

[*Hint:* Consider the cases $0 < a < b \le 1$, $a \le 1 < b$, and $1 \le a < b$. In the first case use $\sin t \le t$ (when $0 < t \le 1$), in the third case use integration by parts, and in the second case combine the other cases.]

3.1.2. Consider the function $g = -\chi_{[a,\frac{a+b}{2}]} + \chi_{[\frac{a+b}{2},b]}$. Show that for all $x \notin [a,b]$ we have

$$0 < H(g)(x) < \frac{1}{4\pi}\frac{(b-a)^2}{(x-a)(x-b)}.$$

Moreover, obtain that $H(g)(x)$ is proportional to $|x|^{-2}$ as $|x| \to \infty$. [*Hint:* Use identity (3.1.10) and that $\log(1+y) < y$ for $y > 0$.]

3.1.3. Let g be an integrable function supported in $[-K, K]$ for some $K > 0$. Prove that $H(g)(x)$ is well defined for all $x \notin [-K, K]$ and that there is a constant C (depending only on g) such that for $|x| \ge 2K$ we have

$$\left|H(g)(x) - \frac{1}{\pi x}\int_{-K}^{K} g(t)\, dt\right| \le C|x|^{-2}.$$

3.1.4. Let φ be in $\mathscr{S}(\mathbf{R})$. Prove that $H(\varphi)$ lies in $\mathscr{C}^\infty \cap L^\infty$ and that $H(\varphi)^{(m)} = H(\varphi^{(m)})$ for any $m \in \mathbf{Z}^+$. Conclude that $\varphi H(\varphi)$ lies in $\mathscr{S}$. [*Hint:* For $m = 1$ use identity (3.1.5) to differentiate in x. Then use induction on m.]

3.1.5. Let φ be in $\mathscr{S}(\mathbf{R})$. Prove that for any $m \in \mathbf{Z}^+ \cup \{0\}$ there is a constant $C = C(m,\varphi)$ such that

$$|H(\varphi^{(m)})(x)| \le \frac{C}{(1+|x|)^{m+1}}.$$

[*Hint:* Integrate by parts in the second integral in (3.1.6) and use Exercise 1.8.4.]

3.1.6. Let φ be a Schwartz function on the line and let $m \in \mathbf{Z}^+$. Prove that

$$H(\varphi)(x) = \frac{2}{\pi i}\sum_{k=1}^{m}(-1)^k\frac{\widehat{\varphi}^{(k-1)}(0)}{(2\pi i x)^k} + \frac{1}{(2\pi i x)^m}H\Big((2\pi i(\cdot))^m\varphi\Big)(x)$$

for all $x \in \mathbf{R} \setminus \{0\}$. Conclude that

$$\sup_{x \in \mathbf{R}}(1+|x|)^{m+1}|H(\varphi)(x)| < \infty \iff \int_{\mathbf{R}} t^k \varphi(t)\,dt = 0 \text{ for all } k \in \{0,\ldots,m-1\}.$$

[*Hint:* Factor $x^m - t^m$ in the identity

$$H(\varphi)(x) - \frac{1}{(2\pi i x)^m} H\Big((2\pi i(\cdot))^m\varphi\Big)(x) = \frac{1}{\pi}\int_{\mathbf{R}} \frac{x^m - t^m}{x^m}\frac{\varphi(t)}{x-t}\,dt.$$

Alternatively, integrate by parts m times in $\int_0^\infty \widehat{\varphi}(\xi)e^{2\pi i x\xi}d\xi$ and $\int_{-\infty}^0 \widehat{\varphi}(\xi)e^{2\pi i x\xi}d\xi$.]

3.2 Homogeneous Singular Integrals and Riesz Transforms

To extend the definition of the Hilbert transform to higher dimensions, we modify the function $1/|x|^n$ on $\mathbf{R}^n \setminus \{0\}$ in order to have integral zero over the unit sphere. We can draw inspiration from the one-dimensional case if we write

$$\frac{1}{x} = \frac{\operatorname{sgn} x}{|x|} = \frac{\operatorname{sgn}(x/|x|)}{|x|}, \qquad x \in \mathbf{R} \setminus \{0\}, \tag{3.2.1}$$

and we notice that only the values of sgn on $\mathbf{S}^0 = \{-1,1\}$ play a role. But on $\{-1,1\}$ the function sgn maps -1 to -1 and 1 to 1 and

$$\int_{\mathbf{S}^0} \operatorname{sgn} x\,dv = -1 \cdot v(\{-1\}) + 1 \cdot v(\{1\}) = -1 \cdot 1 + 1 \cdot 1 = 0,$$

where v here denotes counting measure on $\mathbf{S}^0$. This identity expresses the cancellation of $1/x$ on the line.

To extend this idea to higher dimensions, we suppose that Ω is a bounded function of the unit sphere $\mathbf{S}^{n-1}$ with mean value zero, i.e.,

$$\int_{\mathbf{S}^{n-1}} \Omega(\theta)\,d\theta = 0,$$

where $d\theta$ here denotes surface measure on $\mathbf{S}^{n-1}$. Then we consider the kernel

3.2 Homogeneous Singular Integrals and Riesz Transforms

$$K_\Omega(x) = \frac{\Omega(x/|x|)}{|x|^n}, \qquad x \in \mathbf{R}^n \setminus \{0\}, \tag{3.2.2}$$

which resembles the expression in (3.2.1) with Ω playing the role of sgn. The mean value zero property of Ω introduces cancellation to K_Ω and makes it have integral zero over all annuli centered at the origin.

To be able to define a singular integral operator related to K_Ω, we introduce a distribution W_Ω in $\mathscr{S}'(\mathbf{R}^n)$ by setting

$$\langle W_\Omega, \varphi \rangle = \lim_{\varepsilon \to 0} \int_{\varepsilon \leq |x| \leq 1} K_\Omega(x)\varphi(x)\,dx + \int_{|x|>1} K_\Omega(x)\varphi(x)\,dx \tag{3.2.3}$$

for $\varphi \in \mathscr{S}(\mathbf{R}^n)$, where in the first integral $\varphi(x)$ can be replaced by $\varphi(x) - \varphi(0)$ justifying its convergence. To see that W_Ω is a tempered distribution on $\mathbf{R}^n$ we write

$$|\langle W_\Omega, \varphi \rangle| = \left| \lim_{\varepsilon \to 0} \int_{\varepsilon \leq |x| \leq 1} \frac{\Omega(x/|x|)}{|x|^n}(\varphi(x) - \varphi(0))\,dx + \int_{|x|\geq 1} \frac{\Omega(x/|x|)}{|x|^n}\varphi(x)\,dx \right|$$

$$\leq \|\nabla\varphi\|_{L^\infty} \int_{|x|\leq 1} \frac{|\Omega(x/|x|)|}{|x|^{n-1}}\,dx + \sup_{y\in\mathbf{R}^n}|y|\,|\varphi(y)| \int_{|x|\geq 1} \frac{|\Omega(x/|x|)|}{|x|^{n+1}}\,dx$$

$$\leq C_1 \|\Omega\|_{L^\infty} \sum_{j=1}^n \rho_{0,e_j}(\varphi) + C_2 \|\Omega\|_{L^\infty} \sum_{|\alpha|\leq 1} \rho_{\alpha,0}(\varphi),$$

for suitable C_1 and C_2, where $\rho_{\alpha,\beta}$ is as in (2.6.1) and we used (1.7.2) in the last estimate. The distribution W_Ω coincides with the function K_Ω on $\mathbf{R}^n \setminus \{0\}$. [Here $e_j = (0,\ldots,1,\ldots,0)$ is the multi-index with 1 in the jth entry and 0 elsewhere.]

Example 3.2.1. For $1 \leq j \leq n$, the odd functions

$$\Omega_j(\theta_1,\ldots,\theta_n) = \theta_j \tag{3.2.4}$$

on $\mathbf{S}^{n-1}$ are bounded and have mean value zero. These give rise to kernels that are multiples of the *Riesz transforms*; see Definition 3.2.6.

Definition 3.2.2. Let Ω be a bounded function on the sphere $\mathbf{S}^{n-1}$ with mean value zero. We denote by T_Ω the singular integral operator whose kernel is the distribution W_Ω, that is,

$$T_\Omega(\varphi)(x) = (\varphi * W_\Omega)(x) = \lim_{\varepsilon \to 0} \int_{|y|\geq\varepsilon} \frac{\Omega(y/|y|)}{|y|^n} \varphi(x-y)\,dy, \tag{3.2.5}$$

defined for $\varphi \in \mathscr{S}(\mathbf{R}^n)$. For $\varepsilon > 0$ and $f \in \bigcup_{1\leq p<\infty} L^p(\mathbf{R}^n)$ we define the *truncated singular integral operator*

$$T_\Omega^{(\varepsilon)}(f)(x) = \int_{|y|\geq\varepsilon} f(x-y) \frac{\Omega(y/|y|)}{|y|^n}\,dy. \tag{3.2.6}$$

Then for $\varphi \in \mathscr{S}(\mathbf{R}^n)$ we have

$$T_\Omega(\varphi)(x) = \lim_{\varepsilon \to 0} T_\Omega^{(\varepsilon)}(\varphi)(x) \quad \text{pointwise for all } x \in \mathbf{R}^n.$$

We would like to compute the Fourier transform of W_Ω in order to determine the L^2 boundedness of T_Ω. We have the following result.

Proposition 3.2.3. *Let $n \geq 2$ and $\Omega \in L^\infty(\mathbf{S}^{n-1})$ have integral zero. Then the Fourier transform of W_Ω is the homogeneous-of-degree-zero function*

$$\widehat{W_\Omega}(\xi) = \int_{\mathbf{S}^{n-1}} \Omega(\theta) \left(\log \frac{1}{|\xi \cdot \theta|} - \frac{i\pi}{2} \operatorname{sgn}(\xi \cdot \theta) \right) d\theta \qquad (3.2.7)$$

for all $\xi \in \mathbf{R}^n \setminus \{0\}$.

We remark that for each $\xi \in \mathbf{R}^n \setminus \{0\}$, the function

$$\theta \mapsto \log \frac{1}{|\xi \cdot \theta|}$$

is integrable over the sphere (Appendix D3 in [31]). We also notice that in identity (3.2.7), the variable ξ could be replaced by $\xi' = \xi/|\xi|$. Before we return to the proof of Proposition 3.2.3, we discuss the following lemma:

Lemma 3.2.4. *Let a be a nonzero real number. Then for $0 < \varepsilon < N < \infty$ we have*

$$\lim_{\substack{\varepsilon \to 0 \\ N \to \infty}} \int_\varepsilon^N \frac{e^{-ira} - \cos(r)}{r} dr = \log \frac{1}{|a|} - i\frac{\pi}{2} \operatorname{sgn} a, \qquad (3.2.8)$$

$$\left| \int_\varepsilon^N \frac{e^{-ira} - \cos(r)}{r} dr \right| \leq 2 \left| \log \frac{1}{|a|} \right| + 4 \qquad \text{for all } N > \varepsilon > 0. \qquad (3.2.9)$$

Proof. We first prove the following assertions concerning the real parts of the expressions in (3.2.8):

$$\lim_{\substack{\varepsilon \to 0 \\ N \to \infty}} \int_\varepsilon^N \frac{\cos(ra) - \cos(r)}{r} dr = \log \frac{1}{|a|}, \qquad (3.2.10)$$

$$\left| \int_\varepsilon^N \frac{\cos(ra) - \cos(r)}{r} dr \right| \leq 2 \left| \log \frac{1}{|a|} \right| \qquad \text{for all } N > \varepsilon > 0. \qquad (3.2.11)$$

To verify these claims, by the fundamental theorem of calculus we write[1]

$$\int_\varepsilon^N \frac{\cos(ra) - \cos(r)}{r} dr = \int_\varepsilon^N \frac{\cos(r|a|) - \cos(r)}{r} dr$$

$$= -\int_\varepsilon^N \int_1^{|a|} \sin(tr) \, dt \, dr$$

$$= -\int_1^{|a|} \int_\varepsilon^N \sin(tr) \, dr \, dt$$

---

[1] The integrals $\int_1^{|a|}$ and $\int_N^{N|a|}$ are interpreted as $-\int_{|a|}^1$ and $-\int_{N|a|}^N$, respectively, if $0 < |a| < 1$.

3.2 Homogeneous Singular Integrals and Riesz Transforms

$$= -\int_1^{|a|} \frac{\cos(\varepsilon t)}{t}dt + \int_N^{N|a|} \frac{\cos(t)}{t}dt$$

$$= -\int_1^{|a|} \frac{\cos(\varepsilon t)}{t}dt + \frac{\sin(N|a|)}{N|a|} - \frac{\sin(N)}{N} + \int_N^{N|a|} \frac{\sin(t)}{t^2}dt,$$

where the last line follows from an integration by parts. The penultimate equality yields (3.2.11) by inserting absolute values inside, while the last equality yields (3.2.10), letting $\varepsilon \to 0$ and $N \to \infty$.

We finally derive (3.2.8) and (3.2.9) from (3.2.10) and (3.2.11). For this we need to know that the expressions

$$\left| \int_\varepsilon^N \frac{\sin(ra)}{r} dr \right| = \left| \int_{\varepsilon|a|}^{N|a|} \frac{\sin(r)}{r} dr \right| \tag{3.2.12}$$

tend to $\frac{\pi}{2}$ as $\varepsilon \to 0$ and $N \to \infty$ and are bounded by 4 uniformly in ε, N and a. These statements follow from Exercises 2.2.7 and 3.1.1. $\square$

Let us now prove Proposition 3.2.3.

Proof. Let us set $\xi' = \xi/|\xi|$ when $\xi \neq 0$. We write

$$\langle \widehat{W_\Omega}, \varphi \rangle = \langle W_\Omega, \widehat{\varphi} \rangle$$

$$= \lim_{\varepsilon \to 0} \int_{|x| \geq \varepsilon} \frac{\Omega(x/|x|)}{|x|^n} \widehat{\varphi}(x) dx$$

$$= \lim_{\substack{\varepsilon \to 0 \\ N \to \infty}} \int_{\varepsilon \leq |x| \leq N} \frac{\Omega(x/|x|)}{|x|^n} \widehat{\varphi}(x) dx$$

$$= \lim_{\substack{\varepsilon \to 0 \\ N \to \infty}} \int_{\mathbf{R}^n} \varphi(\xi) \int_{\varepsilon \leq |x| \leq N} \frac{\Omega(x/|x|)}{|x|^n} e^{-2\pi i x \cdot \xi} dx\, d\xi \qquad \text{(Fubini)}$$

$$= \lim_{\substack{\varepsilon \to 0 \\ N \to \infty}} \int_{\mathbf{R}^n} \varphi(\xi) \int_{S^{n-1}} \Omega(\theta) \int_\varepsilon^N e^{-2\pi i r \theta \cdot \xi} \frac{dr}{r} d\theta\, d\xi$$

$$= \lim_{\substack{\varepsilon \to 0 \\ N \to \infty}} \int_{\mathbf{R}^n} \varphi(\xi) \int_{S^{n-1}} \Omega(\theta) \int_\varepsilon^N \left(e^{-2\pi r |\xi| i \theta \cdot \xi'} - \cos(2\pi r |\xi|) \right) \frac{dr}{r} d\theta\, d\xi$$

$$= \lim_{\substack{\varepsilon \to 0 \\ N \to \infty}} \int_{\mathbf{R}^n \setminus \{0\}} \varphi(\xi) \int_{S^{n-1}} \Omega(\theta) \int_{2\pi |\xi| \varepsilon}^{2\pi |\xi| N} \frac{e^{-is\theta \cdot \xi'} - \cos(s)}{s} ds\, d\theta\, d\xi$$

$$= \int_{\mathbf{R}^n \setminus \{0\}} \varphi(\xi) \int_{S^{n-1}} \Omega(\theta) \left[\log \frac{1}{|\xi' \cdot \theta|} - \frac{i\pi}{2} \operatorname{sgn}(\xi' \cdot \theta) \right] d\theta\, d\xi,$$

where we subtracted $\cos(2\pi r|\xi|)$ from the r integral, as Ω has mean value zero over the sphere, and we used the LDCT to pass the limits inside. Lemma 3.2.4 provided the value of the limits; moreover, the use of the dominated convergence theorem is justified from (3.2.9) and the fact that

$$\int_{\mathbf{R}^n\setminus\{0\}} |\varphi(\xi)| \int_{\mathbf{S}^{n-1}} |\Omega(\theta)| \left(2\log \frac{1}{|\xi'\cdot\theta|} +4\right) d\theta d\xi$$
$$\leq \|\Omega\|_{L^\infty} \int_{\mathbf{R}^n\setminus\{0\}} |\varphi(\xi)| \int_{\mathbf{S}^{n-1}} \left(2\log \frac{1}{|\xi'\cdot\theta|} +4\right) d\theta d\xi$$
$$= \|\Omega\|_{L^\infty} \int_{\mathbf{R}^n\setminus\{0\}} |\varphi(\xi)| \int_{\mathbf{S}^{n-1}} \left(2\log \frac{1}{|\phi_1|} +4\right) d\phi d\xi < \infty,$$

using the spherical change of variables $\theta = A_\xi \phi$, where A_ξ is an orthogonal matrix with $(A_\xi)^t \xi' = e_1$, hence $\xi'\cdot\theta = \xi'\cdot A_\xi \phi = (A_\xi)^t \xi'\cdot\phi = e_1\cdot\phi = \phi_1$. For the convergence of the last spherical integral see Appendix D3 in [31]. □

We wonder if $\widehat{W_\Omega}$ is an essentially bounded function for every $\Omega \in L^\infty$. As the signum function sgn is bounded by 1, $\widehat{W_\Omega}$ is essentially bounded if and only if

$$\operatorname*{ess.sup}_{\xi'\in \mathbf{S}^{n-1}} \left| \int_{\mathbf{S}^{n-1}} \Omega(\theta) \log \frac{1}{|\xi'\cdot\theta|} d\theta \right| < \infty.$$

Inserting absolute values inside, and using that Ω is bounded, this assertion would be a consequence of

$$\operatorname*{ess.sup}_{\xi'\in \mathbf{S}^{n-1}} \int_{\mathbf{S}^{n-1}} \log \frac{1}{|\xi'\cdot\theta|} d\theta < \infty.$$

But a rotation yields that for all $\xi' \in \mathbf{S}^{n-1}$ we have

$$\int_{\mathbf{S}^{n-1}} \log \frac{1}{|\xi'\cdot\theta|} d\theta = \int_{\mathbf{S}^{n-1}} \log \frac{1}{|\theta_1|} d\theta,$$

which is a finite constant. These observations lead to the following result.

Corollary 3.2.5. *Let $\Omega \in L^\infty(\mathbf{S}^{n-1})$ have mean value zero. Then the associated T_Ω, initially defined on $\mathscr{S}(\mathbf{R}^n)$, admits a bounded extension from $L^2(\mathbf{R}^n)$ to $L^2(\mathbf{R}^n)$.*

Definition 3.2.6. For $1 \leq j \leq n$ let $\Omega_j(\theta) = \theta_j$ be as in (3.2.4). The *jth Riesz transform* R_j is a constant multiple of T_{Ω_j}; precisely, it is defined by

$$R_j(\varphi)(x) = \frac{\Gamma(\frac{n+1}{2})}{\pi^{\frac{n+1}{2}}} (\varphi * W_{\Omega_j})(x) = \frac{\Gamma(\frac{n+1}{2})}{\pi^{\frac{n+1}{2}}} \,\text{p.v.} \int_{\mathbf{R}^n} \frac{x_j - y_j}{|x-y|^{n+1}} \varphi(y)\, dy \quad (3.2.13)$$

for $\varphi \in \mathscr{S}(\mathbf{R}^n)$.

Proposition 3.2.7. *For every $\xi \in \mathbf{R}^n \setminus \{0\}$ we have*

$$\frac{\Gamma(\frac{n+1}{2})}{\pi^{\frac{n+1}{2}}} \widehat{W_{\Omega_j}}(\xi) = -i \frac{\xi_j}{|\xi|}. \quad (3.2.14)$$

Proof. We begin with the identity

3.2 Homogeneous Singular Integrals and Riesz Transforms

$$\int_{S^{n-1}} \operatorname{sgn}(\theta_k)\, \theta_j\, d\theta = \begin{cases} 0 & \text{if } k \neq j, \\ \int_{S^{n-1}} |\theta_j|\, d\theta & \text{if } k = j. \end{cases} \quad (3.2.15)$$

This is proved by noting that for $k \neq j$, $\operatorname{sgn}(\theta_k)$ has a constant sign on the hemispheres $\theta_k > 0$ and $\theta_k < 0$, on either of which the function $\theta \mapsto \theta_j$ has integral zero. It suffices to prove (3.2.14) for a unit vector ξ. Given $\xi \in S^{n-1}$, pick an orthogonal $n \times n$ matrix $A = (a_{kl})_{k,l}$ such that $Ae_j = \xi$. Then the jth column of the matrix A is the vector $(\xi_1, \xi_2, \ldots, \xi_n)^t$. We have

$$\int_{S^{n-1}} \operatorname{sgn}(\xi \cdot \theta)\, \theta_j\, d\theta = \int_{S^{n-1}} \operatorname{sgn}(Ae_j \cdot \theta)\, \theta_j\, d\theta$$

$$= \int_{S^{n-1}} \operatorname{sgn}(e_j \cdot A^t\theta)(AA^t\theta)_j\, d\theta$$

$$= \int_{S^{n-1}} \operatorname{sgn}(e_j \cdot \theta)(A\theta)_j\, d\theta$$

$$= \int_{S^{n-1}} \operatorname{sgn}(\theta_j)(a_{j1}\theta_1 + \cdots + \xi_j\theta_j + \cdots + a_{jn}\theta_n)\, d\theta$$

$$= \xi_j \int_{S^{n-1}} \operatorname{sgn}(\theta_j)\, \theta_j\, d\theta + \sum_{1 \leq m \neq j \leq n} a_{jm} \int_{S^{n-1}} \operatorname{sgn}(\theta_j)\, \theta_m\, d\theta$$

$$= \frac{\xi_j}{|\xi|} \int_{S^{n-1}} |\theta_j|\, d\theta + 0 = \frac{\xi_j}{|\xi|} \int_{S^{n-1}} |\theta_1|\, d\theta,$$

by rotational invariance. The identity in [31, Appendix D.2] gives

$$\int_{S^{n-1}} |\theta_1|\, d\theta = \int_{-1}^{1} |s| \int_{\sqrt{1-s^2}\, S^{n-2}} d\varphi\, \frac{ds}{\sqrt{1-s^2}}$$

$$= \omega_{n-2} \int_{-1}^{1} |s|(1-s^2)^{\frac{n-3}{2}}\, ds$$

$$= \omega_{n-2} \int_{0}^{1} u^{\frac{n-3}{2}}\, du = \frac{2\pi^{\frac{n-1}{2}}}{\Gamma\left(\frac{n+1}{2}\right)},$$

where $\omega_{n-2} = |S^{n-2}| = 2\pi^{\frac{n-1}{2}} \Gamma\left(\frac{n-1}{2}\right)^{-1}$; see [31, Appendix A.3]. Combining this fact with (3.2.7) yields (3.2.14). $\square$

Proposition 3.2.8. *The Riesz transforms satisfy the identity*

$$-I = \sum_{j=1}^{n} R_j^2 \quad (3.2.16)$$

as operators acting on L^2. Here I is the identity operator and R_j^2 denotes $R_j \circ R_j$.

Proof. Use the Fourier transform and the identity

$$\sum_{j=1}^{n}(-i\xi_j/|\xi|)^2 = -1$$

to deduce (3.2.16) for any f in $L^2(\mathbf{R}^n)$. $\square$

Exercises

3.2.1. Show that the nonlinear operator

$$\varphi \mapsto Q(\varphi) = \Big(\sum_{j=1}^{n}|R_j(\varphi)|^2\Big)^{1/2}$$

satisfies $\|Q(\varphi)\|_{L^2(\mathbf{R}^n)} = \|\varphi\|_{L^2(\mathbf{R}^n)}$ for all $\varphi \in \mathscr{S}(\mathbf{R}^n)$.

3.2.2. Prove that for φ in $\mathscr{S}(\mathbf{R}^n)$ and $1 \leq j,k \leq n$ and all $x \in \mathbf{R}^n$ we have

$$\partial_j \partial_k \varphi(x) = -R_j R_k \Delta \varphi(x).$$

Here $\Delta\varphi = \sum_{j=1}^{n} \partial_j^2 \varphi$ is the *Laplacian* of φ, given by multiplication by $-4\pi^2|\xi|^2$ on the Fourier transform. Conclude that

$$\|\partial_{j_1}\cdots\partial_{j_{2m}}\varphi\|_{L^2(\mathbf{R}^n)} \leq \|\Delta^m\varphi\|_{L^2(\mathbf{R}^n)}$$

for all $m \in \mathbf{Z}^+$, $j_1,\ldots,j_{2m} \in \{1,\ldots,n\}$ and all $\varphi \in \mathscr{S}(\mathbf{R}^n)$.

3.2.3. Let $c_n = \Gamma(\frac{n+1}{2})\pi^{-\frac{n+1}{2}}$ and $\varphi \in \mathscr{C}^\infty(\mathbf{R}^n)$ be supported in $\overline{B(0,K)}$ for some $K > 0$. Prove that for any $\delta > 0$ and any $|x| \geq (1+\delta)K$ we have

$$\left|R_j(\varphi)(x) - c_n \frac{x_j}{|x|^{n+1}} \int_{\mathbf{R}^n} \varphi(y)\,dy\right| \leq \frac{nc_n(\frac{1+\delta}{\delta})^{n+1}}{|x|^{n+1}} \int_{\mathbf{R}^n} |\varphi(y)||y|dy.$$

3.2.4. Use identity (2.7.5) with $z = -n+1$ to obtain another proof of Proposition 3.2.7. [*Hint:* First show that $\partial_j |x|^{-n+1} = (1-n)W_{\Omega_j}$ by comparing the action of both distributions on a Schwartz function $\varphi = \varphi_e + \varphi_o$, where φ_e is even and φ_o is odd.]

3.2.5. For any φ in $\mathscr{S}(\mathbf{R}^n)$ and any multi-index α prove that there is a constant $C_{n,\alpha,\varphi} > 0$ such that for all $x \in \mathbf{R}^n$ one has

$$|R_j(\partial^\alpha \varphi)(x)| \leq C_{n,\alpha,\varphi}(1+|x|)^{-n-|\alpha|}.$$

[*Hint:* Use a smooth splitting as in (3.1.6) and integrate by parts to handle one term.]

3.3 Calderón–Zygmund Singular Integrals

In the previous two sections of this chapter we discussed singular integrals given by convolution with distributions on $\mathbf{R}^n$ that are homogeneous of degree $-n$. In this section we consider a class of singular integrals whose kernels are not necessarily homogeneous distributions.

We make some remarks about the smoothness of the kernels of the Hilbert transform and Riesz transforms. Note that

$$\left|\left(\frac{1}{x}\right)'\right| = \left|-\frac{1}{x^2}\right| = \frac{1}{|x|^{1+1}}, \qquad x \in \mathbf{R} \setminus \{0\}.$$

Moreover,

$$\left|\nabla \frac{x_j}{|x|^{n+1}}\right| = \frac{\sqrt{|x|^2 + x_j^2(n^2-1)}}{|x|^{n+2}} \leq \frac{n}{|x|^{n+1}} \qquad (3.3.1)$$

and so the kernels of the Hilbert and the Riesz transforms have the common smoothness property

$$\left|\nabla K(x)\right| \leq \frac{A_2'}{|x|^{n+1}}, \qquad x \in \mathbf{R}^n \setminus \{0\} \qquad (3.3.2)$$

for some positive constant A_2'. This smoothness is captured by the weaker condition

$$\sup_{y \neq 0} \int_{|x| \geq 2|y|} |K(x-y) - K(x)| \, dx = A_2 < \infty,$$

for some other constant A_2. Indeed, assuming (3.3.2), for $y \in \mathbf{R}^n \setminus \{0\}$ we write

$$\int_{|x| \geq 2|y|} |K(x-y) - K(x)| \, dx = \int_{|x| \geq 2|y|} \left|\int_0^1 \nabla K(x - \theta y) \cdot y \, d\theta\right| dx$$

$$\leq \int_{|x| \geq 2|y|} \int_0^1 \frac{A_2'|y|}{|x - \theta y|^{n+1}} \, d\theta \, dx$$

$$\leq \int_{|x| \geq 2|y|} \int_0^1 \frac{A_2'|y|}{(|x|/2)^{n+1}} \, d\theta \, dx$$

$$\leq 2^{n+1} |y| A_2' \omega_{n-1} \int_{r=2|y|}^\infty \frac{r^{n-1}}{r^{n+1}} \, dr$$

$$= 2^n A_2' \omega_{n-1} = A_2,$$

as $|x - \theta y| \geq |x| - |y| \geq |x|/2$ when $|x| \geq 2|y|$. Here $\omega_{n-1} = |\mathbf{S}^{n-1}|$.

Based on this discussion, we consider a measurable function K on $\mathbf{R}^n \setminus \{0\}$ which satisfies the *size condition*

$$|K(x)| \leq \frac{A_1}{|x|^n} < \infty, \qquad (3.3.3)$$

the *smoothness condition*

$$\sup_{y\neq 0}\int_{|x|\geq 2|y|}|K(x-y)-K(x)|\,dx=A_2<\infty, \qquad (3.3.4)$$

and the *cancellation condition*

$$\sup_{\delta>0}\sup_{N>\delta}\left|\int_{\delta<|x|<N}K(x)\,dx\right|=A_3<\infty, \qquad (3.3.5)$$

for some $A_1, A_2, A_3 > 0$. Such functions will give rise to kernels of general singular integrals that are not necessarily of homogeneous type. Condition (3.3.4) is often referred to as *Hörmander's integral smoothness condition*.

Taking $N=1$, condition (3.3.5) implies that there are δ_k in $(0,1)$, $k=1,2,\ldots,$ such that $\delta_k \to 0$ and that

$$\lim_{k\to\infty}\int_{1\geq|x|\geq\delta_k}K(x)\,dx=L \qquad (3.3.6)$$

exists. Obviously, $|L|\leq A_3$. We define a tempered distribution W associated with K and the choice of sequence δ_k as follows:

$$\langle W,\varphi\rangle=\int_{|x|\leq 1}K(x)(\varphi(x)-\varphi(0))\,dx+\varphi(0)L+\int_{|x|\geq 1}K(x)\varphi(x)\,dx \qquad (3.3.7)$$

for $\varphi \in \mathscr{S}$. Equivalently, we write the identity in (3.3.7) as

$$\langle W,\varphi\rangle=\lim_{k\to\infty}\int_{|x|\geq\delta_k}K(x)\varphi(x)\,dx. \qquad (3.3.8)$$

Let us now show that the functional W defined in this way is an element of $\mathscr{S}'(\mathbf{R}^n)$.

Using (3.3.7) we obtain the estimate

$$|\langle W,\varphi\rangle|\leq \|\nabla\varphi\|_{L^\infty}\int_{|x|\leq 1}|x||K(x)|\,dx+|L|\|\varphi\|_{L^\infty}+\int_{|x|\geq 1}|K(x)\varphi(x)|\,dx$$

$$\leq \left(\int_{|x|\leq 1}\frac{A_1\,dx}{|x|^{n-1}}\right)\|\nabla\varphi\|_{L^\infty}+A_3\|\varphi\|_{L^\infty}+\left(\int_{|x|\geq 1}\frac{A_1\,dx}{|x|^{n+1}}\right)\sup_{x\in\mathbf{R}^n}|x||\varphi(x)|$$

$$\leq A_1\omega_{n-1}\sum_{j=1}^n \rho_{0,e_j}(\varphi)+A_3\rho_{0,0}(\varphi)+A_1\omega_{n-1}\sum_{j=1}^n \rho_{e_j,0}(\varphi),$$

recalling the definition of $\rho_{\alpha,\beta}$ given in (2.6.1). This shows that W lies in $\mathscr{S}'(\mathbf{R}^n)$. Finally, we notice that if φ is a Schwartz function supported in $\mathbf{R}^n\setminus\{0\}$, then

$$\langle W,\varphi\rangle=\int_{\mathbf{R}^n}K(x)\varphi(x)\,dx.$$

Thus W coincides with K on $\mathbf{R}^n\setminus\{0\}$.

Definition 3.3.1. A *Calderón–Zygmund singular integral* is an operator given by convolution with a distribution W, associated as in (3.3.8) with a sequence δ_k in

3.3 Calderón–Zygmund Singular Integrals

$(0,1)$ that tends to zero and with a kernel K that satisfies (3.3.3), (3.3.4), and (3.3.5). Precisely it is an operator of the form

$$T^W(\varphi) = \varphi * W, \qquad \varphi \in \mathscr{S}(\mathbf{R}^n).$$

For $\varphi \in \mathscr{S}(\mathbf{R}^n)$ and $x \in \mathbf{R}^n$, we can explicitly write

$$T^W(\varphi)(x) = \lim_{k \to \infty} \int_{|y| \geq \delta_k} K(y)\varphi(x-y)dy \qquad (3.3.9)$$

$$= L\varphi(x) + \int_{|y| \leq 1} K(y)(\varphi(x-y) - \varphi(x))dy + \int_{|y| \geq 1} K(y)\varphi(x-y)dy.$$

Example 3.3.2. Let τ be a nonzero real number and let $K(x) = \frac{1}{|x|^{n+i\tau}}$ be defined for $x \neq 0$. Notice that (3.3.3) is clearly satisfied for K and also (3.3.4) is valid, as $|\nabla K(x)| \leq |n+i\tau| |x|^{-n-1}$. Finally, (3.3.5) is also satisfied, as for $0 < \varepsilon < N < \infty$,

$$\left| \int_{\varepsilon < |x| < N} \frac{1}{|x|^{n+i\tau}} dx \right| = \omega_{n-1} \left| \frac{N^{-i\tau} - \varepsilon^{-i\tau}}{-i\tau} \right| \leq \frac{2\omega_{n-1}}{|\tau|} = A_3.$$

Consider the following two sequences $\delta_k^1 = e^{-(2k+1)\pi/\tau}$ and $\delta_k^2 = e^{-2k\pi/\tau}$ indexed by $k = 1, 2, \dots$ if $\tau > 0$ and by $k = -1, -2, -3, \dots$ if $\tau < 0$. Both sequences lie in $(0,1)$ and tend to zero. For a Schwartz function φ on $\mathbf{R}^n$ define distributions

$$\langle W^1, \varphi \rangle = \lim_{k \to \infty} \int_{|x| \geq \delta_k^1} \varphi(x) \frac{dx}{|x|^{n+i\tau}} \qquad (3.3.10)$$

and

$$\langle W^2, \varphi \rangle = \lim_{k \to \infty} \int_{|x| \geq \delta_k^2} \varphi(x) \frac{dx}{|x|^{n+i\tau}}. \qquad (3.3.11)$$

We have that

$$\int_{\delta_k^1 < |x| < 1} \frac{1}{|x|^{n+i\tau}} dx = \omega_{n-1} \frac{1^{-i\tau} - (\delta_k^1)^{-i\tau}}{-i\tau} = \omega_{n-1} \frac{1 - e^{-(2k+1)i\pi}}{-i\tau} = \frac{2i\omega_{n-1}}{\tau},$$

$$\int_{\delta_k^2 < |x| < 1} \frac{1}{|x|^{n+i\tau}} dx = \omega_{n-1} \frac{1^{-i\tau} - (\delta_k^2)^{-i\tau}}{-i\tau} = \omega_{n-1} \frac{1 - e^{-2ki\pi}}{-i\tau} = 0,$$

and as these expressions are constant for all integers k, they have limits $L_1 = 2i\omega_{n-1}/\tau$ and $L_2 = 0$ as $|k| \to \infty$, respectively. In general, note that the subsequence $\varepsilon_k = e^{-(2k+\alpha)\pi/\tau}$ yields the limit $\omega_{n-1} \frac{1-e^{-i\pi\alpha}}{-i\tau}$ which varies with $\alpha \in [0,1]$.

Notice that both W^1 and W^2 agree on $\mathbf{R}^n \setminus \{0\}$ but are not the same tempered distribution. In fact, it is not hard to see that $W^1 - W^2 = c\,\delta_0$, where $c = \frac{2i\omega_{n-1}}{\tau}$. Thus the associated Calderón–Zygmund singular integral operators are T^{W^1} and T^{W^2}, which differ by a constant multiple of the identity operator.

In general, the difference of two Calderón–Zygmund operators (Definition 3.3.1) associated with the same function K on $\mathbf{R}^n \setminus \{0\}$ and two sequences δ_k^1 and δ_k^2 is cI, where c is a constant satisfying $|c| \le 2A_3$.

Exercises

3.3.1. Prove the equality in (3.3.1).

3.3.2. Let F be a bounded $\mathscr{C}^1$ function on the real line with F' in $L^\infty(\mathbf{R})$ that has the property
$$\sup_{-\infty < A < B < \infty} \left| \int_A^B F(t)\,dt \right| < \infty.$$
Prove that the kernel
$$K(x) = \frac{F(\log|x|)}{|x|^n}, \qquad x \in \mathbf{R}^n \setminus \{0\},$$
satisfies (3.3.3), (3.3.4), and (3.3.5). An example of such a function is $F(t) = \sin t/t$. A multitude of examples arise by taking $F = G'$, where G, G', G'' are bounded.

3.3.3. Let $\delta > 0$ and η be a smooth function on the real line supported in $[-2,2]$ and equal to 1 on the $[-1,1]$. Show that the kernels
$$K_1(x) = \frac{\sin(|x|^{-\delta})}{|x|^n}(1-\eta(|x|)), \qquad K_2(x) = \frac{\sin(|x|^\delta)}{|x|^n}\eta(|x|),$$
defined on $\mathbf{R}^n \setminus \{0\}$, satisfy (3.3.3), (3.3.2), and (3.3.5).

3.3.4. Suppose that a function K on $\mathbf{R}^n \setminus \{0\}$ satisfies condition (3.3.3) with constant A_1 and condition (3.3.4) with constant A_2. Let $A_1' = A_1 \omega_{n-1} \log 2$.
(a) Show that the functions $K(x)\chi_{|x|\ge \varepsilon}$ also satisfy condition (3.3.4) uniformly in $\varepsilon > 0$ with constant $\max(A_1', A_2)$ in place of A_2.
(b) Use part (a) to obtain that the truncations $K(x)\chi_{|x|<N}$ satisfy (3.3.4) uniformly in $N > 0$, with constant $2\max(A_1', A_2)$.
(c) Deduce from parts (a), (b) that the double truncations $K^{(\varepsilon,N)}(x) = K(x)\chi_{\varepsilon \le |x| < N}$ also satisfy condition (3.3.4) uniformly in $N, \varepsilon > 0$ with constant $2\max(A_1', A_2)$.
$\bigl[$*Hint:* Part (b): Write $K(x)\chi_{|x|<N} = K(x) - K(x)\chi_{|x|\ge N}$. Part (c): Use $K(x)\chi_{\varepsilon \le |x| < N} = K(x)\chi_{|x|\ge \varepsilon} - K(x)\chi_{|x|\ge N}.\bigr]$

3.3.5. (a) Prove that for all $x, y \in \mathbf{R}^n$ that satisfy $0 \ne x \ne y$ we have
$$\left| \frac{x-y}{|x-y|} - \frac{x}{|x|} \right| \le 2\frac{|y|}{|x|}.$$

(b) Let Ω be a bounded function with mean value zero on the sphere $\mathbf{S}^{n-1}$. Suppose that for some $\alpha \in (0,1)$, Ω satisfies the *Lipschitz condition*

$$|\Omega(\theta_1) - \Omega(\theta_2)| \le B_0 |\theta_1 - \theta_2|^\alpha$$

for all $\theta_1, \theta_2 \in \mathbf{S}^{n-1}$. Prove that $K(x) = \Omega(x/|x|)/|x|^n$ satisfies condition (3.3.4) with constant at most a multiple of $B_0 + \|\Omega\|_{L^\infty}$.
[*Hint:* Part (a): Add and subtract $\frac{x-y}{|x|}$. Part (b): Use part (a).]

3.4 L^2 Boundedness of Calderón–Zygmund Operators

We now turn to the issue of the $L^2(\mathbf{R}^n)$ boundedness of operators given by convolution with distributions W associated with sequences $\delta_k \in (0,1)$ that tend to zero and with kernels K that satisfy (3.3.3), (3.3.4), and (3.3.5). We would like to compute the Fourier transform of the distribution W. To do this, for $\varphi \in \mathscr{S}(\mathbf{R}^n)$, we write

$$\begin{aligned}
\langle \widehat{W}, \varphi \rangle &= \langle W, \widehat{\varphi} \rangle \\
&= \lim_{k \to \infty} \int_{|x| \ge \delta_k} K(x) \widehat{\varphi}(x) \, dx \\
&= \lim_{k \to \infty} \int_{\delta_k \le |x| \le k} K(x) \int_{\mathbf{R}^n} \varphi(\xi) e^{-2\pi i x \cdot \xi} \, d\xi \, dx \\
&= \lim_{k \to \infty} \int_{\mathbf{R}^n} \varphi(\xi) \left[\int_{\delta_k \le |x| \le k} K(x) e^{-2\pi i x \cdot \xi} \, dx \right] d\xi. \quad (3.4.1)
\end{aligned}$$

To be able to insert the limit inside the ξ integral, we need to show that the expressions inside the square brackets (a) have a limit as $k \to \infty$ and (b) they are bounded (uniformly in ξ and k), so that we can justify using the LDCT.

Before achieving this goal we prove a lemma.

Lemma 3.4.1. *Given A, B measurable subsets of $\mathbf{R}^n$ and F, G integrable functions on $A \cup B$, we have*

$$\int_A F \, dx - \int_B G \, dx = \int_B (F - G) \, dx + \int_{A \setminus B} F \, dx - \int_{B \setminus A} F \, dx. \quad (3.4.2)$$

Proof. We write

$$\int_A F \, dx = \int_{A \setminus B} F \, dx + \int_{A \cap B} F \, dx = \int_{A \setminus B} F \, dx + \int_B F \, dx - \int_{B \setminus A} F \, dx,$$

and subtracting $\int_B G \, dx$ from both sides we deduce (3.4.2). □

We also make the observation that (3.3.3) implies

$$\int_{R \le |x| \le 2R} |K(x)| \, dx \le A_1 \int_{R \le |x| \le 2R} |x|^{-n} \, dx = \omega_{n-1} A_1 \log 2 < \omega_{n-1} A_1 \quad (3.4.3)$$

uniformly in all $R > 0$. Here $\omega_{n-1} = |\mathbf{S}^{n-1}|$.

Theorem 3.4.2. *Assume that K satisfies (3.3.3), (3.3.4), and (3.3.5). Then*

$$\sup_{\varepsilon>0}\sup_{N>\varepsilon}\sup_{\xi\in\mathbf{R}^n}\left|\int_{\varepsilon<|x|<N}K(x)e^{-2\pi ix\cdot\xi}\,dx\right|\leq 9\omega_{n-1}A_1+A_2+A_3. \quad (3.4.4)$$

Moreover, let W in $\mathscr{S}'(\mathbf{R}^n)$ be associated with K via a sequence $\delta_k\to 0$. Then $\widehat{W}$ coincides with a function on $\mathbf{R}^n\setminus\{0\}$ which is bounded by $A=9\omega_{n-1}A_1+A_2+A_3$.

Proof. The integrals in (3.4.4) are bounded by A_3 when $\xi=0$, so in proving (3.4.4) we can restrict the third supremum to $\mathbf{R}^n\setminus\{0\}$. Thus we fix $\xi\neq 0$, $0<\varepsilon<N<\infty$, and we set

$$K^{(\varepsilon,N)}(x)=K(x)\chi_{\varepsilon<|x|<N}.$$

We consider the following three cases.

Case 1: $\varepsilon<|\xi|^{-1}<N$. In this case we write $\widehat{K^{(\varepsilon,N)}}(\xi)=I_1^\varepsilon(\xi)+I_2^N(\xi)$, where

$$I_1^\varepsilon(\xi)=\int_{\varepsilon<|x|<|\xi|^{-1}}K(x)e^{-2\pi ix\cdot\xi}\,dx,\qquad I_2^N(\xi)=\int_{|\xi|^{-1}<|x|<N}K(x)e^{-2\pi ix\cdot\xi}\,dx.$$

We split $I_1^\varepsilon(\xi)$ as

$$I_1^\varepsilon(\xi)=\int_{\varepsilon<|x|<|\xi|^{-1}}K(x)\,dx+\int_{\varepsilon<|x|<|\xi|^{-1}}K(x)\left(e^{-2\pi ix\cdot\xi}-1\right)dx. \quad (3.4.5)$$

It follows that

$$|I_1^\varepsilon(\xi)|\leq A_3+2\pi|\xi|\int_{|x|<|\xi|^{-1}}|x|\,|K(x)|\,dx<7\omega_{n-1}A_1+A_3 \quad (3.4.6)$$

uniformly in ε. Let us now examine $I_2^N(\xi)$. Let $z=\frac{\xi}{2|\xi|^2}$ so that $e^{2\pi iz\cdot\xi}=-1$ and $2|z|=|\xi|^{-1}$. Via the change of variables $x=x'-z$, we rewrite I_2^N as

$$I_2^N(\xi)=-\int_{|\xi|^{-1}<|x'-z|<N}K(x'-z)e^{-2\pi ix'\cdot\xi}\,dx';$$

hence averaging gives

$$I_2^N(\xi)=\frac{1}{2}\int_{|\xi|^{-1}<|x|<N}K(x)e^{-2\pi ix\cdot\xi}\,dx-\frac{1}{2}\int_{|\xi|^{-1}<|x-z|<N}K(x-z)e^{-2\pi ix\cdot\xi}\,dx.$$

Now use that (3.4.2) to write

$$I_2^N(\xi)=J_1^N(\xi)+J_2^N(\xi)+J_3^N(\xi)+J_4^N(\xi)+J_5^N(\xi),$$

where

$$J_1^N(\xi)=+\frac{1}{2}\int_{|\xi|^{-1}<|x-z|<N}(K(x)-K(x-z))e^{-2\pi ix\cdot\xi}\,dx,$$

3.4 L^2 Boundedness of Calderón–Zygmund Operators

$$J_2^N(\xi) = +\frac{1}{2} \int_{\substack{|\xi|^{-1} < |x| < N \\ |x-z| \leq |\xi|^{-1}}} K(x) e^{-2\pi i x \cdot \xi} dx,$$

$$J_3^N(\xi) = +\frac{1}{2} \int_{\substack{|\xi|^{-1} < |x| < N \\ |x-z| \geq N}} K(x) e^{-2\pi i x \cdot \xi} dx,$$

$$J_4^N(\xi) = -\frac{1}{2} \int_{\substack{|\xi|^{-1} < |x-z| < N \\ |x| \leq |\xi|^{-1}}} K(x) e^{-2\pi i x \cdot \xi} dx,$$

$$J_5^N(\xi) = -\frac{1}{2} \int_{\substack{|\xi|^{-1} < |x-z| < N \\ |x| \geq N}} K(x) e^{-2\pi i x \cdot \xi} dx.$$

Since $2|z| = |\xi|^{-1}$, we have

$$|J_1^N(\xi)| \leq \frac{1}{2} \int_{|\xi|^{-1} < |x-z|} |K(x) - K(x-z)| dx = \frac{1}{2} \int_{2|y| < |x'|} |K(x'-y) - K(x')| dx'$$

(with $y = -z$ and $x' = x - z$), which is bounded by $\frac{1}{2} A_2$, in view of (3.3.4).

Next observe that $|\xi|^{-1} \leq |x| \leq \frac{3}{2}|\xi|^{-1}$ in $J_2^N(\xi)$, while $\frac{1}{2}|\xi|^{-1} \leq |x| \leq |\xi|^{-1}$ in $J_4^N(\xi)$; hence either of $J_2^N(\xi)$, $J_4^N(\xi)$ is bounded by $\frac{1}{2}\omega_{n-1}A_1$ by (3.4.3). Also we have

$$\frac{1}{2}N < N - \frac{1}{2}|\xi|^{-1} < |x| < N \quad \text{in } J_3^N(\xi) \tag{3.4.7}$$

and

$$N \leq |x| < N + \frac{1}{2}|\xi|^{-1} < \frac{3}{2}N \quad \text{in } J_5^N(\xi). \tag{3.4.8}$$

Thus both $J_3^N(\xi)$ and $J_5^N(\xi)$ are bounded above by $\frac{1}{2}\omega_{n-1}A_1$ uniformly in N. Combining these facts, we deduce

$$|I_2^N(\xi)| < \frac{1}{2}A_2 + 4\left(\frac{1}{2}\omega_{n-1}A_1\right) = 2\omega_{n-1}A_1 + \frac{1}{2}A_2. \tag{3.4.9}$$

Adding estimates (3.4.6) and (3.4.9), we obtain the claimed bound in Case 1.

Case 2: $\varepsilon < N \leq |\xi|^{-1}$. Here we write

$$\int_{\varepsilon < |x| < N} K(x) e^{-2\pi i x \cdot \xi} dx = \int_{\varepsilon < |x| < N} K(x) dx + \int_{\varepsilon < |x| < N} K(x)(e^{-2\pi i x \cdot \xi} - 1) dx$$

which is bounded in absolute value by

$$A_3 + 2\pi|\xi| \int_{|x| \leq |\xi|^{-1}} |K(x)| |x| dx \leq 2\pi \omega_{n-1} A_1 + A_3 < A.$$

Case 3: $|\xi|^{-1} \leq \varepsilon < N$. In this case we write

$$\int_{\varepsilon < |x| < N} K(x) e^{-2\pi i x \cdot \xi} dx = \int_{|\xi|^{-1} < |x| < N} K(x) e^{-2\pi i x \cdot \xi} dx - \int_{|\xi|^{-1} < |x| < \varepsilon} K(x) e^{-2\pi i x \cdot \xi} dx,$$

and the expression on the right is equal to $I_2^N(\xi) - I_2^\varepsilon(\xi)$ and each one of these terms was shown to be bounded by $2\omega_{n-1} A_1 + \frac{1}{2} A_2$ in (3.4.9). Thus we obtain the bound $4\omega_{n-1} A_1 + A_2 < A$ in Case 3. Hence (3.4.4) holds in all cases.

Having proven (3.4.4), we now turn to the fact that for any $\xi \in \mathbf{R}^n \setminus \{0\}$, the expression inside the square brackets in (3.4.1) has a limit, which is of course bounded by A. This fact, combined with the identity leading to (3.4.1), would imply that $\widehat{W}$ coincides with a bounded function on $\mathbf{R}^n \setminus \{0\}$.

Fix $\xi \in \mathbf{R}^n \setminus \{0\}$ and pick k_0 such that for all $k \geq k_0$ we have $\delta_k < |\xi|^{-1} < k$. We use the decomposition in Case 1 with $\varepsilon = \delta_k$ and $N = k$ to show that the expressions $I_1^{\delta_k}$ and I_2^k have limits as $k \to \infty$. Using (3.4.5), we see that $I_1^{\delta_k}$ converges to

$$\lim_{k \to \infty} \int_{\delta_k < |x| < |\xi|^{-1}} K(x) \, dx + \int_{|x| < |\xi|^{-1}} K(x) \left(e^{-2\pi i x \cdot \xi} - 1 \right) dx. \quad (3.4.10)$$

Additionally, (3.4.7) and (3.4.8) give that

$$|J_3^k(\xi)| \leq \frac{1}{2} A_1 \omega_{n-1} \log \frac{k}{k - \frac{1}{2}|\xi|^{-1}}, \qquad |J_5^k(\xi)| \leq \frac{1}{2} A_1 \omega_{n-1} \log \frac{k + \frac{1}{2}|\xi|^{-1}}{k},$$

and hence these expressions tend to zero as $k \to \infty$. Moreover

$$J_1^k(\xi) \to \frac{1}{2} \int_{|\xi|^{-1} < |x-z|} \left(K(x) - K(x-z) \right) e^{-2\pi i x \cdot \xi} dx \quad (3.4.11)$$

as $k \to \infty$, while

$$J_2^k(\xi) \to \frac{1}{2} \int_{\substack{|\xi|^{-1} < |x| \\ |x-z| \leq |\xi|^{-1}}} K(x) e^{-2\pi i x \cdot \xi} dx \quad (3.4.12)$$

and

$$J_4^k(\xi) \to -\frac{1}{2} \int_{\substack{|\xi|^{-1} < |x-z| \\ |x| \leq |\xi|^{-1}}} K(x) e^{-2\pi i x \cdot \xi} dx \quad (3.4.13)$$

as $k \to \infty$, and all of the above are absolutely convergent integrals. The bounded function $\widehat{W}(\xi)$ is equal to the sum of the expressions in (3.4.10), (3.4.11), (3.4.12), and (3.4.13) when $\xi \neq 0$. $\square$

We finally observe that

$$\int_{1 \leq |x| \leq k} K(x) \, dx$$

3.4 L^2 Boundedness of Calderón–Zygmund Operators

may not have a limit as $k \to \infty$, and thus $\widehat{W}(0)$ is not defined; see Example 3.3.2.

Theorem 3.4.2 yields that Calderón–Zygmund singular integral operators are bounded on L^2.

Corollary 3.4.3. *Let W be as in Theorem 3.4.2. Then the associated operator T^W given by convolution with W has a bounded extension on $L^2(\mathbf{R}^n)$ with bound $9\omega_{n-1}A_1 + A_2 + A_3$.*

Proof. This is a consequence of Theorems 2.8.5 and 3.4.2. $\square$

Exercises

3.4.1. Let $f \in L^1(\mathbf{R}^n)$. Use the averaging idea in Theorem 3.4.2 to prove that
$$|\widehat{f}(\xi)| \leq \frac{1}{2} \int_{\mathbf{R}^n} |f(x) - f(x - \xi/(2|\xi|^2))|\,dx, \quad \xi \in \mathbf{R}^n \setminus \{0\}.$$
Conclude that
$$|\widehat{f}(\xi)| \leq \||\nabla f|\|_{L^1} (4|\xi|)^{-1}$$
when $f \in \mathscr{C}^1(\mathbf{R}^n) \cap L^1(\mathbf{R}^n)$.

3.4.2. Let W be a tempered distribution on $\mathbf{R}^n$. Suppose that the operator $T^W(\varphi) = \varphi * W$, initially defined on $\varphi \in \mathscr{S}(\mathbf{R}^n)$, admits a bounded extension from $L^2(\mathbf{R}^n)$ to itself. Prove that for any $\varphi \in \mathscr{S}(\mathbf{R}^n)$ we have
$$\sup_{x \in \mathbf{R}^n} |(\varphi * W)(x)| \leq \|T^W\|_{L^2 \to L^2} \|\widehat{\varphi}\|_{L^1}.$$

3.4.3. Let K satisfy (3.3.3), (3.3.4), and (3.3.5) and let $W \in \mathscr{S}'$ be an extension of K on $\mathbf{R}^n$ associated with a sequence $\delta_j \in (0,1)$ that tends to zero. Let φ be a compactly supported $\mathscr{C}^1$ function on $\mathbf{R}^n$ with mean value zero. Prove that the function $T^W(\varphi) = \varphi * W$ lies in $L^1(\mathbf{R}^n)$.

3.4.4. Let K and W be as in the preceding exercise and suppose that $\widehat{W}$ is nonconstant and homogeneous of degree zero. Suppose that $f \in L^1(\mathbf{R}^n) \cap L^2(\mathbf{R}^n)$. Assuming that $T^W(f)$ is integrable over $\mathbf{R}^n$, prove that f must have integral zero.

3.4.5. Suppose that $\Omega \in S^1$ is defined by $\Omega(\theta) = \chi_I - \chi_J$, where I is an interval[2] of length less than $1/2$ in S^1 and $J = \{\theta \in S^1 : -\theta \in I\}$. Prove that $\Omega(x/|x|)|x|^{-2}$ defined on $\mathbf{R}^2 \setminus \{0\}$ satisfies Hörmander's integral smoothness condition (3.3.4).

3.4.6. Let K be a function on $\mathbf{R}^n \setminus \{0\}$ that satisfies (3.3.3). Let W be a tempered distribution on $\mathbf{R}^n$ that coincides with K on $\mathbf{R}^n \setminus \{0\}$. Suppose that operator T^W

---

[2] An interval on the unit circle is the intersection of the circle with a cone centered at the origin.

given by convolution with W maps $L^2(\mathbf{R}^n)$ to itself. Prove that K must satisfy (3.3.5). [*Hint:* For $0 < \varepsilon < N < \infty$ use that

$$\int_{\varepsilon < |x| < N} K \, dx = \int_{\mathbf{R}^n} K(\varphi^N - \varphi^\varepsilon) \, dx + \int_{\varepsilon < |x| < 2\varepsilon} K \varphi^\varepsilon \, dx - \int_{N < |x| < 2N} K \varphi^N \, dx$$

where $\varphi^t(x) = \varphi(x/t)$ and φ is a $\mathscr{C}_0^\infty$ function with $0 \leq \varphi \leq 1$ that equals 1 on $\overline{B(0,1)}$ and vanishes outside $\overline{B(0,2)}$. Use Exercise 3.4.2 to estimate one term.]

3.5 The Calderón–Zygmund Decomposition

We will obtain L^p bounds for singular integrals via the Calderón–Zygmund[3] decomposition. We describe this decomposition for dyadic cubes.

Definition 3.5.1. A *dyadic cube* in $\mathbf{R}^n$ is a set of the form

$$[2^k m_1, 2^k(m_1+1)) \times \cdots \times [2^k m_n, 2^k(m_n+1)),$$

where $k, m_1, \ldots, m_n \in \mathbf{Z}$.

We observe that two dyadic cubes are either disjoint or are related by inclusion. Each dyadic cube has a unique *ancestor*, i.e., a dyadic cube of twice its length that contains it. Dyadic cubes have 2^n *descendants*, i.e., dyadic cubes of half their length.

The decomposition of a function $f = g + b$ described below is called its *Calderón–Zygmund decomposition* at height α. The function g is called the *good function* of the decomposition, since it is both integrable and bounded. The function b is called the *bad function*, since it contains the singular part of f, but it is carefully chosen to have mean value zero.

Theorem 3.5.2. *Let $f \in L^1(\mathbf{R}^n)$ and $\alpha > 0$. Then there exist functions g and b on $\mathbf{R}^n$ and a collection $\{Q_j\}_j$ of disjoint dyadic cubes such that*

(i) $f = g + b$,

(ii) $b = \sum_j b_j$, *where each b_j is supported in Q_j,*

(iii) $\|b_j\|_{L^1} \leq 2^{n+1} \alpha |Q_j|$,

(iv) $\int_{Q_j} b_j(x) \, dx = 0$,

(v) $\sum_j |Q_j| \leq \alpha^{-1} \|f\|_{L^1}$,

---
[3] Occasionally abbreviated as CZ.

3.5 The Calderón–Zygmund Decomposition

(vi) $\|g\|_{L^1} \leq \|f\|_{L^1}$, $\|g\|_{L^\infty} \leq 2^n \alpha$, and

$$\|g\|_{L^r} \leq (2^n \alpha)^{\frac{1}{r'}} \|f\|_{L^1}^{\frac{1}{r}}$$

for any $1 \leq r \leq \infty$.

Proof. Pick a positive integer N such that

$$2^{-nN} \|f\|_{L^1} \leq \alpha. \tag{3.5.1}$$

We consider all dyadic cubes of the form $2^N m + [0, 2^N)^n$, where m varies over $\mathbf{Z}^n$. We call these cubes *of generation zero* and we note that their union is $\mathbf{R}^n$. Subdivide each cube of generation zero into 2^n congruent cubes by bisecting each of its sides. This way we obtain a collection of dyadic cubes, which we call *of generation one*. Select a cube Q of generation one if

$$\frac{1}{|Q|} \int_Q |f| \, dx > \alpha. \tag{3.5.2}$$

Let $S^{(1)}$ be the set of all selected cubes of generation one. Now subdivide each unselected cube of generation one into 2^n congruent subcubes by bisecting each of its sides; call these cubes of generation two. Then select all cubes Q of generation two for which (3.5.2) holds. Let $S^{(2)}$ be the set of all selected cubes *of generation two*. Repeat this procedure indefinitely or until it is terminated. See Figure 3.2. We obtain a collection of selected cubes $\bigcup_{m=1}^\infty S^{(m)}$. Note that $S^{(m)}$ may be empty for some m and that (3.5.1) forces all selected cubes to be of generation at least one, i.e., $m \geq 1$.

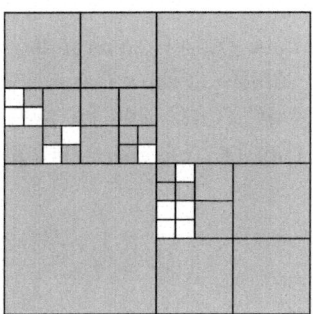

Fig. 3.2 The selected cubes are shown in dark. The unselected ones, shown in white, keep being subdivided.

If the set of all selected cubes $\bigcup_{m=1}^\infty S^{(m)}$ is empty, then we set $b = 0$ and $g = f$. Otherwise, $\bigcup_{m=1}^\infty S^{(m)}$ consists of countably many cubes which we denote by $\{Q_j\}_j$. Let us observe that the selected cubes are disjoint, for otherwise some Q_k would be a proper subset of some Q_j, which is impossible since the selected cube Q_j was never subdivided. Now define

$$b_j = \left(f - \frac{1}{|Q_j|} \int_{Q_j} f \, dx\right) \chi_{Q_j}. \tag{3.5.3}$$

For a selected cube Q_j there exists a unique unselected cube Q' with twice its side length that contains Q_j. Let us call this cube the parent of Q_j. Since the parent Q' of Q_j was not selected, we have $|Q'|^{-1} \int_{Q'} |f| \, dx \leq \alpha$. Then

$$\frac{1}{|Q_j|}\int_{Q_j}|f|\,dx \le \frac{1}{|Q_j|}\int_{Q'}|f|\,dx = \frac{2^n}{|Q'|}\int_{Q'}|f|\,dx \le 2^n\alpha.$$

Consequently,

$$\int_{Q_j}|b_j|\,dx \le \int_{Q_j}|f|\,dx + |Q_j|\left|\frac{1}{|Q_j|}\int_{Q_j}f\,dx\right| \le 2\int_{Q_j}|f|\,dx \le 2^{n+1}\alpha|Q_j|,$$

which proves (iii). In particular b_j is integrable over Q_j and in view of (3.5.3) it has integral zero; thus (iv) holds. To prove (v), simply observe that

$$\sum_j |Q_j| \le \frac{1}{\alpha}\sum_j \int_{Q_j}|f|\,dx = \frac{1}{\alpha}\int_{\bigcup_j Q_j}|f|\,dx \le \frac{1}{\alpha}\|f\|_{L^1}.$$

We define $b = \sum_j b_j$ so that (ii) holds and also define $g = f - b$; finally, we turn our attention to (vi). We obviously have

$$g = \begin{cases} f & \text{on } \mathbf{R}^n \setminus \bigcup_j Q_j, \\ \frac{1}{|Q_j|}\int_{Q_j}f\,dx & \text{on } Q_j. \end{cases} \qquad (3.5.4)$$

On the cube Q_j, g is equal to the constant $|Q_j|^{-1}\int_{Q_j}f\,dx$, and this is bounded by $2^n\alpha$. It suffices to show that g is bounded outside the union of the Q_j. Indeed, for each $x \in \mathbf{R}^n \setminus \bigcup_j Q_j$ and for each $k = 0, 1, 2, \ldots$ there exists a unique unselected dyadic cube $Q_x^{(k)}$ of generation k that contains x. Then for each $k \ge 0$, we have

$$\left|\frac{1}{|Q_x^{(k)}|}\int_{Q_x^{(k)}}f(y)\,dy\right| \le \frac{1}{|Q_x^{(k)}|}\int_{Q_x^{(k)}}|f(y)|\,dy \le \alpha.$$

The intersection of the closures of the cubes $Q_x^{(k)}$ is the singleton $\{x\}$. Using Corollary 1.5.6 we deduce that for almost all $x \in \mathbf{R}^n \setminus \bigcup_j Q_j$ we have

$$f(x) = \lim_{k\to\infty}\frac{1}{|Q_x^{(k)}|}\int_{Q_x^{(k)}}f(y)\,dy.$$

Since these averages are at most α, we conclude that $|f| \le \alpha$ a.e. on $\mathbf{R}^n \setminus \bigcup_j Q_j$, hence $|g| \le \alpha$ a.e. on this set. Finally, (3.5.4) gives

$$\|g\|_{L^1} = \int_{\mathbf{R}^n\setminus\bigcup_j Q_j}|f|\,dx + \sum_j \int_{Q_j}\left|\frac{1}{|Q_j|}\int_{Q_j}f\,dy\right|dx = \int_{\mathbf{R}^n\setminus\bigcup_j Q_j}|f|\,dx + \sum_j\left|\int_{Q_j}f\,dy\right|$$

and this implies that $\|g\|_{L^1} \le \|f\|_{L^1}$. This completes the proof. $\square$

Remark 3.5.3. The cubes $\{Q_j\}_j$ selected in the proof of Theorem 3.5.2 are exactly the maximal (with respect to inclusion) dyadic cubes such that (3.5.2) holds. This is

3.5 The Calderón–Zygmund Decomposition

because if a cube is selected, then all of its ancestors were not selected, so this cube is maximal (with respect to inclusion) satisfying (3.5.2).

Example 3.5.4. Let D be a dyadic cube in $\mathbf{R}^n$ and let $\alpha > 0$. We find the CZ decomposition of χ_D. Let Q_m be the largest dyadic cube containing D such that

$$\frac{|D|}{|Q_m|} = \frac{1}{|Q_m|} \int_{Q_m} \chi_D \, dy > \alpha. \tag{3.5.5}$$

If $\alpha \geq 1$, no such dyadic cube Q_m exists, so in this case $b = 0$ and $g = \chi_D$. Suppose now that $2^{-n} \leq \alpha < 1$. Then the largest dyadic cube Q_m satisfying (3.5.5) is D itself and in this case $b = \chi_D - \chi_D = 0$ and $g = \chi_D$. Now for $\alpha < 2^{-n}$ there is a largest dyadic cube Q_m that contains D and satisfies (3.5.5). In this case the CZ decomposition of χ_D is

$$b = \chi_D - \frac{|D|}{|Q_m|}\chi_{Q_m}, \qquad g = \frac{|D|}{|Q_m|}\chi_{Q_m}.$$

Example 3.5.5. Let f be a finite linear combination of characteristic functions of disjoint dyadic cubes. We claim that the CZ decomposition of f at height $\alpha > 0$ contains only finitely many cubes Q_j (cf. Theorem 3.5.2).

To prove this, we pick $M \in \mathbf{Z}$ such that the smallest cube appearing in the definition of f has side length 2^{-M}. We also pick the least integers N' and N'' such that $[-2^{N'}, 2^{N'}]^n$ contains the support of f and that $2^{-nN''}\|f\|_{L^1} \leq \alpha$. Set $N = \max(N', N'')$ and let $\mathscr{G}_{N,M}$ be the set of all dyadic cubes of side length 2^{-M} contained in $[-2^N, 2^N)^n$. There is a subset $\mathscr{G}'_{N,M}$ of $\mathscr{G}_{N,M}$ and $\lambda_Q \in \mathbf{C} \setminus \{0\}$ for any $Q \in \mathscr{G}'_{N,M}$ such that

$$f = \sum_{Q \in \mathscr{G}'_{N,M}} \lambda_Q \chi_Q. \tag{3.5.6}$$

We consider the following list of numbers

$$0 = \alpha_0 < \alpha_1 < \alpha_2 < \cdots < \alpha_m < \alpha_{m+1} = \infty$$

so that $\{\alpha_1, \ldots, \alpha_m\} = \{|\lambda_Q| : Q \in \mathscr{G}'_{N,M}\}$. For our given $\alpha > 0$, pick $s \in \{0, 1, \ldots, m\}$ such that $\alpha_s \leq \alpha < \alpha_{s+1}$. No dyadic subcube R of a given $Q \in \mathscr{G}'_{N,M}$ with $|\lambda_Q| \leq \alpha_s$ is selected as $|R|^{-1} \int_R |f| \, dy = |\lambda_Q| \leq \alpha_s \leq \alpha$. Moreover, no dyadic subcube of a cube in $\mathscr{G}_{N,M} \setminus \mathscr{G}'_{N,M}$ is selected as f vanishes there. Now every Q in $\mathscr{G}'_{N,M}$ with $|\lambda_Q| \geq \alpha_{s+1} > \alpha$ satisfies (3.5.2), so it is either selected, or one of its ancestors was previously selected. So all selected cubes have side length at least 2^{-M} and are contained in $[-2^N, 2^N]^n$; thus there are finitely many selected cubes $\{Q_j\}_j$.

Exercises

3.5.1. Let $\alpha > 0$ and $f \in L^1(\mathbf{R}^n)$ which satisfies $|f| \leq \alpha$ a.e. What are the functions b and g constructed in Theorem 3.5.2 in this case?

3.5.2. For a given $f \in L^1(\mathbf{R}^n)$, let $\{Q_j^\alpha\}_j$ be the cubes obtained in the CZ decomposition at height $\alpha > 0$. For given $0 < \alpha < \beta < \infty$ prove that

$$\bigcup_j Q_j^\beta \subseteq \bigcup_i Q_i^\alpha.$$

3.5.3. Show that the number of CZ cubes $\{Q_j\}$ in Example 3.5.5 is at most

$$2^n 2^{nM} \max\left[\frac{2^n \|f\|_{L^1}}{\alpha}, 2^{nN'}\right],$$

where $[-2^{N'}, 2^{N'}]$ contains the support of f and 2^{-M} is the side length of the smallest dyadic cube appearing in the definition of f.

3.5.4. Prove that finite linear combinations of characteristic functions of dyadic cubes are dense in $L^p(\mathbf{R}^n)$ for any $0 < p < \infty$.

3.5.5. On $\mathbf{R}^2$, let $Q_1 = [0, \frac{1}{2})^2$, and $Q_j = [\frac{1}{2} + \cdots + \frac{1}{2^{j-1}}, \frac{1}{2} + \cdots + \frac{1}{2^j}) \times [0, \frac{1}{2^j})$ for $j \geq 2$. Prove that $(1,0)$ lies in $\overline{\bigcup_{j=1}^\infty Q_j}$ but not in $\bigcup_{j=1}^\infty Q_j^*$. Here Q_j^* is a cube with the same center as Q_j but $\ell(Q_j^*) = 2\ell(Q_j)$. Construct a similar example when $\ell(Q_j^*) = 2^N \ell(Q_j)$ for a given $N \in \mathbf{Z}^+$.

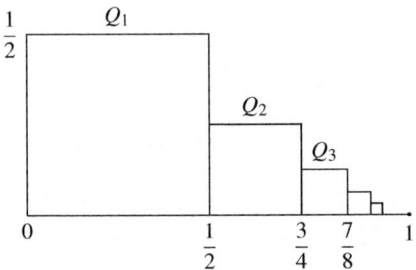

3.5.6. (**Calderón–Zygmund decomposition on L^q**) Fix a function $f \in L^q(\mathbf{R}^n)$ for some $1 \leq q < \infty$ and let $\alpha > 0$. Then there exist functions g and b on $\mathbf{R}^n$ such that

(1) $f = g + b$.

(2) $\|g\|_{L^q} \leq \|f\|_{L^q}$ and $\|g\|_{L^\infty} \leq 2^{\frac{n}{q}} \alpha$.

(3) $b = \sum_j b_j$, where each b_j is supported in a cube Q_j. Furthermore, the cubes Q_k and Q_j have disjoint interiors when $j \neq k$.

(4) $\|b_j\|_{L^q}^q \leq 2^{n+q} \alpha^q |Q_j|$.

(5) $\int_{Q_j} b_j(x)\,dx = 0.$

(6) $\sum_j |Q_j| \leq \alpha^{-q}\|f\|_{L^q}^q.$

(7) $\|b\|_{L^q} \leq 2^{\frac{n+q}{q}}\|f\|_{L^q}$ and $\|b\|_{L^1} \leq 2\alpha^{1-q}\|f\|_{L^q}^q.$

[*Hint:* Imitate the basic idea of the proof of Theorem 3.5.2, but select a cube Q if $\left(\frac{1}{|Q|}\int_Q |f(x)|^q\,dx\right)^{1/q} > \alpha$. Define g and b as in the proof of Theorem 3.5.2.]

3.5.7. (**Calderón–Zygmund decomposition with bounded overlap**) Let f be in $L^1(\mathbf{R}^n)$ and let $\alpha > 0$. Prove that there exist functions g and b on $\mathbf{R}^n$ such that

(1) $f = g + b.$

(2) $\|g\|_{L^1} \leq \|f\|_{L^1}$, $\|g\|_{L^\infty} \leq (10\sqrt{n})^n \alpha.$

(3) $b = \sum_j b_j$, where each b_j is supported in a dyadic cube Q_j. Furthermore, the interiors of Q_k and Q_j are disjoint when $j \neq k.$

(4) $\int_{Q_j} b_j(x)\,dx = 0.$

(5) $\|b_j\|_{L^1} \leq 2(10\sqrt{n})^n \alpha |Q_j|.$

(6) $\sum_j |Q_j| \leq \alpha^{-1}\|f\|_{L^1}.$

(7) $\sum_j \chi_{Q_j^*} \leq 2^n$, where Q_j^* has the same center as Q_j and $\ell(Q_j^*) = (1+\varepsilon)\ell(Q_j)$, for any ε with $0 < \varepsilon < 1/4.$

[*Hint:* Let $\Omega = \{M_c(f) > \alpha\}$, where M_c is the uncentered maximal operator with respect to cubes in $\mathbf{R}^n$. Write $\Omega = \{Q_j\}_j$ in terms of the Whitney decomposition of Theorem 7.5.2. Define

$$b_j = \left(f - \frac{1}{|Q_j|}\int_{Q_j} f\,dx\right)\chi_{Q_j},$$

$b = \sum_j b_j$, and $g = f - b$.]

3.6 L^2 Boundedness Implies L^p Boundedness

This next theorem provides the most classical application of the CZ decomposition.

Theorem 3.6.1. *Let K be a function on $\mathbf{R}^n \setminus \{0\}$ that satisfies (3.3.3) and (3.3.4) for some $A_1, A_2 < \infty$. Let W be the distribution defined as in (3.3.8) with respect to a sequence $\delta_k \downarrow 0$, and suppose that the operator T given by convolution with W has a bounded extension that maps $L^2(\mathbf{R}^n)$ to itself with norm B. Then T has a unique bounded extension from $L^1(\mathbf{R}^n)$ to $L^{1,\infty}(\mathbf{R}^n)$ with norm*

$$\|T\|_{L^1 \to L^{1,\infty}} \leq C'_n (A_2 + B), \qquad (3.6.1)$$

and T also extends to a bounded operator from $L^p(\mathbf{R}^n)$ to itself for $1 < p < \infty$ with norm

$$\|T\|_{L^p \to L^p} \leq C_n \max\left(p, (p-1)^{-1}\right)(A_2 + B), \qquad (3.6.2)$$

where C_n, C'_n are constants that depend on the dimension but not on p.

Proof. It will suffice to prove (3.6.1) for functions in $L^1 \cap L^2$, which is dense in L^1. If this is known, for a given $f \in L^1$, pick $f_k \in L^2 \cap L^1$ that converge to f in L^1. Then $\{f_k\}_k$ is a Cauchy sequence in L^1, and by (3.6.1) it follows that $\{T(f_k)\}_k$ is a Cauchy sequence in $L^{1,\infty}$. Proposition 1.2.12 gives that this sequence is convergent in $L^{1,\infty}$; thus $T(f)$ can be defined as the $L^{1,\infty}$ limit of $\{T(f_k)\}_k$. Note that this definition does not depend on the choice of the sequence $\{f_k\}_k$ as for another sequence $\{f_k^\#\}_k$ that also converges to f in L^1, it follows from (3.6.1) that $T(f_k) - T(f_k^\#)$ tends to 0 in $L^{1,\infty}$. This proves that T has a unique bounded extension from L^1 to $L^{1,\infty}$.

So we fix $f \in L^1 \cap L^2$ and let $\alpha > 0$ be given. We apply the Calderón–Zygmund decomposition to f at height $\gamma \alpha$, where γ is a positive constant to be chosen later. That is, write the function f as the sum

$$f = g + b = g + \sum_j b_j,$$

where conditions (i)–(vi) of Theorem 3.5.2 are satisfied with $\gamma \alpha$ in place of α. We denote by $\ell(Q)$ the side length of a cube Q. Let Q_j^* be the unique cube with sides parallel to the axes having the same center as Q_j with side length $\ell(Q_j^*) = 2\sqrt{n}\,\ell(Q_j)$. As f lies in L^2 and so does g, it follows that b lies in L^2. Hence each b_j lies in L^2 and so $T(b_j)$ is a well-defined L^2 function.

We observe that for all j and all $x \notin Q_j^*$ the LDCT and (3.3.3) give

$$T(b_j)(x) = \lim_{k \to \infty} \int_{k \geq |x-y| \geq \delta_k} K(x-y) b_j(y)\, dy = \int_{Q_j} K(x-y) b_j(y)\, dy,$$

as the last integral converges absolutely. Moreover, we note that, since $f \in L^1 \cap L^2$ and $g \in L^2$, we must have $b \in L^2$, and thus $T(b)$ is a well-defined L^2 function. We claim that for any $\alpha > 0$

$$\left|\left\{x \notin \bigcup_i Q_i^* : |T(b)(x)| > \tfrac{\alpha}{2}\right\}\right| \leq \left|\left\{x \notin \bigcup_i Q_i^* : \sum_{j=1}^\infty |T(b_j)(x)| > \tfrac{\alpha}{2}\right\}\right|. \qquad (3.6.3)$$

To see this,[4] consider an arbitrary enumeration of the b_j. Then for $N \in \mathbf{Z}^+$ we have

$$\left|T\left(\sum_{j=1}^\infty b_j\right)\right| = \left|\sum_{j=1}^N T(b_j) + T\left(\sum_{j=N+1}^\infty b_j\right)\right| \leq \sum_{j=1}^\infty |T(b_j)| + E_N,$$

---

[4] One could work with functions f that are finite linear combinations of characteristic functions of dyadic cubes. By Example 3.5.5, the CZ decompositions of such functions f contain only finitely many cubes; so (3.6.3) would trivially hold in this case by linearity.

3.6 L^2 Boundedness Implies L^p Boundedness

where $E_N = |T(\sum_{j=N+1}^{\infty} b_j)|$. But

$$\left|\sum_{j=N+1}^{\infty} b_j\right| \to 0$$

pointwise and is bounded by $|b| = |f - g| \in L^2$, so it converges to zero in L^2 as $N \to \infty$ by the LDCT. Then $\|E_N\|_{L^2} \to 0$ by the boundedness of T on L^2. Thus $|\{|E_N| > \alpha(1-\delta)/2\}| \to 0$ as $N \to \infty$ for any $\delta < 1$ and yields (3.6.3) letting $\delta \uparrow 1$.

Let y_j be the center of Q_j. We use the cancellation of b_j in the following way:

$$\int_{(\cup_i Q_i^*)^c} \sum_j |T(b_j)(x)| \, dx$$

$$= \int_{(\cup_i Q_i^*)^c} \sum_j \left| \int_{Q_j} b_j(y)(K(x-y) - K(x-y_j)) \, dy \right| dx$$

$$\leq \sum_j \int_{(Q_j^*)^c} \int_{Q_j} |b_j(y)| |K(x-y) - K(x-y_j)| \, dy \, dx$$

$$= \sum_j \int_{Q_j} |b_j(y)| \int_{(Q_j^*)^c} |K(x-y) - K(x-y_j)| \, dx \, dy$$

$$= \sum_j \int_{Q_j} |b_j(y)| \int_{-y_j + (Q_j^*)^c} |K(x-(y-y_j)) - K(x)| \, dx \, dy$$

$$\leq \sum_j \int_{Q_j} |b_j(y)| \int_{|x| \geq 2|y-y_j|} |K(x-(y-y_j)) - K(x)| \, dx \, dy$$

$$\leq A_2 \sum_j \|b_j\|_{L^1}$$

$$\leq A_2 2^{n+1} \|f\|_{L^1},$$

having used (3.3.4) and that $-y_j + (Q_j^*)^c \subseteq \{x : |x| \geq 2|y-y_j|\}$.

To verify the last assertion we argue as follows: If $x \in -y_j + (Q_j^*)^c$, then it holds that

$$|x| \geq \frac{1}{2}\ell(Q_j^*) = \sqrt{n}\ell(Q_j).$$

As $y - y_j$ lies in $-y_j + Q_j$ we must have

$$|y - y_j| \leq \frac{\sqrt{n}}{2}\ell(Q_j).$$

Thus $|x| \geq 2|y-y_j|$. These inequalities have geometric interpretations related to distances; see Figure 3.3.

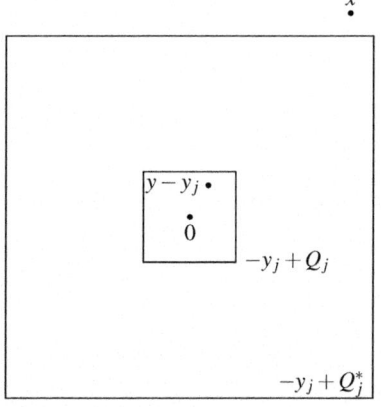

Fig. 3.3 The cubes $-y_j + Q_j$ and $-y_j + Q_j^*$.

Thus we proved that

$$\int_{(\cup_i Q_i^*)^c} \sum_j |T(b_j)(x)|\,dx \le 2^{n+1} A_2 \|f\|_{L^1},$$

an inequality we use below. Appealing to (3.6.3), we write

$$\begin{aligned}
|\{x \in \mathbf{R}^n : |T(f)(x)| > \alpha\}| \\
&\le \left|\left\{x \in \mathbf{R}^n : |T(g)(x)| > \frac{\alpha}{2}\right\}\right| + \left|\left\{x \in \mathbf{R}^n : |T(b)(x)| > \frac{\alpha}{2}\right\}\right| \\
&\le \frac{4}{\alpha^2} \|T(g)\|_{L^2}^2 + \left|\bigcup_i Q_i^*\right| + \left|\left\{x \notin \bigcup_i Q_i^* : |T(\sum_j b_j)(x)| > \frac{\alpha}{2}\right\}\right| \\
&= \frac{4}{\alpha^2} \|T(g)\|_{L^2}^2 + \left|\bigcup_i Q_i^*\right| + \left|\left\{x \notin \bigcup_i Q_i^* : \sum_j |T(b_j)(x)| > \frac{\alpha}{2}\right\}\right| \\
&\le \frac{4}{\alpha^2} B^2 \|g\|_{L^2}^2 + \sum_i |Q_i^*| + \frac{2}{\alpha} \int_{(\cup_i Q_i^*)^c} \sum_j |T(b_j)(x)|\,dx \\
&\le \frac{4}{\alpha^2} 2^n B^2 (\gamma \alpha) \|f\|_{L^1} + (2\sqrt{n})^n \frac{\|f\|_{L^1}}{\gamma \alpha} + \frac{2}{\alpha} 2^{n+1} A_2 \|f\|_{L^1} \\
&\le \left(\frac{(2^{n+1} B\gamma)^2}{2^n \gamma} + \frac{(2\sqrt{n})^n}{\gamma} + 2^{n+2} A_2 \right) \frac{\|f\|_{L^1}}{\alpha}.
\end{aligned}$$

Choosing $\gamma = B^{-1}$, we deduce estimate (3.6.1) with $C_n' = (2\sqrt{n})^n + 2^{n+2}$.

By the density argument discussed at the beginning of the proof, T is well defined on L^1, and thus on L^p which is contained in $L^1 + L^2$ for $1 < p < 2$. Using Theorem 1.3.3 (Marcinkiewicz's interpolation theorem) we obtain that

$$\|T\|_{L^p \to L^p} \le 2 \left(\frac{p}{p-1} + \frac{p}{2-p} \right) C_n'(A_2 + B), \qquad 1 < p < 2. \tag{3.6.4}$$

We now observe that the transpose operator T^t of T has kernel $K^t(x) = K(-x)$ which also satisfies (3.3.3) and (3.3.4) for some $A_1, A_2 < \infty$ and moreover T^t maps $L^2(\mathbf{R}^n)$ to itself with the same norm B. Then T^t satisfies (3.6.4), which implies

$$\|T\|_{L^p \to L^p} \le 2 \left(\frac{p'}{p'-1} + \frac{p'}{2-p'} \right) C_n'(A_2 + B), \qquad 2 < p < \infty. \tag{3.6.5}$$

Now we employ Theorem 1.3.3 to interpolate between $L^{5/4}$ and L^5. We obtain

$$\|T\|_{L^p \to L^p} \le 2 \left(\frac{p}{p-\frac{5}{4}} + \frac{p}{5-p} \right) C_n''(A_2 + B), \qquad \frac{5}{4} < p < 5. \tag{3.6.6}$$

For $1 < p \le \frac{3}{2}$, we use (3.6.4) to obtain $\|T\|_{L^p \to L^p} \le \frac{4p}{p-1} C_n'(A_2+B)$. For $3 \le p < \infty$, (3.6.5) yields the bound $\|T\|_{L^p \to L^p} \le 4p C_n'(A_2+B)$. Finally, for $\frac{3}{2} < p < 3$ we use (3.6.6) to obtain the bound $\|T\|_{L^p \to L^p} \le \frac{8p}{p-1} C_n''(A_2+B)$. Combining these cases, we deduce (3.6.2). $\square$

3.6 L^2 Boundedness Implies L^p Boundedness

Corollary 3.6.2. *Let K be a function on $\mathbf{R}^n \setminus \{0\}$ that satisfies (3.3.3), (3.3.4), and (3.3.5) for some $A_1, A_2, A_3 < \infty$. Let W be the distribution associated with K as in (3.3.8). Then T has an extension on L^p for all $p \in [1, \infty)$ that satisfies (3.6.1) and (3.6.2) with $B = 9\omega_{n-1}A_1 + A_2 + A_3$.*

Proof. Conditions (3.3.3), (3.3.4), and (3.3.5) imply that T is L^2 bounded with bound $B \leq 9\omega_{n-1}A_1 + A_2 + A_3$ in view of Theorem 3.4.2. Then the L^2 hypothesis in Theorem 3.6.1 also holds and the conclusion follows. □

Corollary 3.6.3. *The Hilbert transform and the Riesz transforms are bounded from L^1 to $L^{1,\infty}$ and from L^p for all $1 < p < \infty$ with bounds $C(n) \max((p-1)^{-1}, p)$.*

Proof. This is a direct consequence of Corollary 3.6.2. □

Having established the L^p boundedness of the Hilbert transform and of the Riesz transforms for $1 < p < \infty$, we turn to general odd singular integrals of homogeneous type. We begin with the observation that the Hilbert transform acting on the first variable on $\mathbf{R}^n$ composed with the identity operator in the remaining variables

$$\mathscr{H}_{e_1}(f)(x_1, x_2, \ldots, x_n) = \lim_{\varepsilon \to 0^+} \frac{1}{\pi} \int_{|t| \geq \varepsilon} f(x_1 - t, x_2, \ldots, x_n) \frac{dt}{t}, \quad (3.6.7)$$

defined for $f \in \mathscr{S}(\mathbf{R}^n)$, is bounded on $L^p(\mathbf{R}^n)$ with bound $C_p = \|H\|_{L^p \to L^p}$. To verify this, we raise the absolute values of both sides in (3.6.7) to the power p and then integrate in x_1. Using the boundedness of the Hilbert transform we estimate the right-hand side by

$$C_p^p \int_{\mathbf{R}} |f(x_1, x_2, \ldots, x_n)|^p dx_1.$$

Integrating over the remaining variables implies the conclusion. Next, for a unit vector $\theta \in \mathbf{S}^{n-1}$ we define

$$\mathscr{H}_\theta(f)(x) = \lim_{\varepsilon \to 0^+} \frac{1}{\pi} \int_{|t| \geq \varepsilon} f(x - t\theta) \frac{dt}{t}, \quad f \in \mathscr{S}(\mathbf{R}^n), \quad (3.6.8)$$

called the *directional Hilbert transform* in the direction θ. We observe that the following identity is valid for all matrices $A \in O(n)$:

$$\mathscr{H}_{Ae_1}(f)(x) = \mathscr{H}_{e_1}(f \circ A)(A^{-1}x). \quad (3.6.9)$$

Now given $\theta \in \mathbf{S}^{n-1}$ pick $A \in O(n)$ such that $Ae_1 = \theta$. This implies that

$$\|\mathscr{H}_\theta(f)\|_{L^p} = \|\mathscr{H}_{e_1}(f \circ A) \circ A^{-1}\|_{L^p} = \|\mathscr{H}_{e_1}(f \circ A)\|_{L^p} \leq C_p \|f \circ A\|_{L^p} = C_p \|f\|_{L^p},$$

and this yields that $\mathscr{H}_\theta$ maps $L^p(\mathbf{R}^n)$ to itself $L^p(\mathbf{R}^n)$ uniformly in θ. We use this to obtain boundedness for T_Ω (Definition 3.2.2) when Ω is an odd bounded function. Such functions clearly have vanishing integral.

Corollary 3.6.4. *Let $\Omega \in L^\infty(\mathbf{S}^{n-1})$ be odd and let T_Ω be as in Definition 3.2.2. Then for $1 < p < \infty$, T_Ω admits a bounded extension from $L^p(\mathbf{R}^n)$ to itself with norm at most $\frac{\pi \omega_{n-1}}{2} \|\Omega\|_{L^\infty} \|H\|_{L^p \to L^p}$.*

Proof. Let $f \in \mathscr{S}(\mathbf{R}^n)$. We express a general singular integral T_Ω with Ω odd and bounded on the sphere as follows. For given $x \in \mathbf{R}^n$ and $\varepsilon > 0$ we write

$$\int_{|y|\geq \varepsilon} \frac{\Omega(y/|y|)}{|y|^n} f(x-y)\,dy = +\int_{S^{n-1}} \Omega(\theta) \int_{r=\varepsilon}^{\infty} f(x-r\theta)\frac{dr}{r}\,d\theta$$

$$= \int_{S^{n-1}} \Omega(-\theta) \int_{r=\varepsilon}^{\infty} f(x+r\theta)\frac{dr}{r}\,d\theta,$$

$$= -\int_{S^{n-1}} \Omega(\theta) \int_{r=\varepsilon}^{\infty} f(x+r\theta)\frac{dr}{r}\,d\theta,$$

where the first identity follows by switching to polar coordinates, the second one uses the change of variables $\theta \mapsto -\theta$, and the third one expresses that Ω is an odd function on the sphere. Averaging the first and third identities on the right, we obtain

$$\int_{|y|\geq \varepsilon} \frac{\Omega(y/|y|)}{|y|^n} f(x-y)\,dy$$

$$= \frac{\pi}{2}\int_{S^{n-1}} \Omega(\theta)\frac{1}{\pi}\int_{r=\varepsilon}^{\infty} \frac{f(x-r\theta)-f(x+r\theta)}{r}\,dr\,d\theta. \qquad (3.6.10)$$

We write $\frac{1}{r}\big(f(x-r\theta)-f(x+r\theta)\big) = -\int_{-1}^{1} \nabla f(x-sr\theta)\cdot\theta\,ds$ and we note that the first expression has rapid decay (when $r > 2|x|$) and that the second expression is bounded (when $r \leq 2|x|$). Then by the Lebesgue dominated convergence theorem, we can pass the limit as $\varepsilon \to 0^+$ inside the integral in (3.6.10). This yields

$$T_\Omega(f)(x) = \frac{\pi}{2}\int_{S^{n-1}} \Omega(\theta)\lim_{\varepsilon \to 0^+}\frac{1}{\pi}\int_{r=\varepsilon}^{\infty} \frac{f(x-r\theta)-f(x+r\theta)}{r}\,dr\,d\theta$$

$$= \frac{\pi}{2}\int_{S^{n-1}} \Omega(\theta)\lim_{\varepsilon \to 0^+}\frac{1}{\pi}\int_{|r|\geq \varepsilon} f(x-r\theta)\frac{dr}{r}\,d\theta$$

$$= \frac{\pi}{2}\int_{S^{n-1}} \Omega(\theta)\,\mathscr{H}_\theta(f)(x)\,d\theta \qquad (3.6.11)$$

for $x \in \mathbf{R}^n$ and $f \in \mathscr{S}(\mathbf{R}^n)$. The boundedness of T_Ω on $L^p(\mathbf{R}^n)$ to itself is then a straightforward consequence of (3.6.11) via Minkowski's integral inequality. $\square$

Exercises

3.6.1. Assume that T is a linear operator acting on measurable functions on $\mathbf{R}^n$ such that whenever a function f is supported in a cube Q, then $T(f)$ is supported in $Q^* = \rho Q$ for some $\rho > 1$. Suppose that T maps L^2 to L^2 with norm B. Prove that T extends to a bounded operator from L^1 to $L^{1,\infty}$ with norm a constant multiple of B.

3.6.2. Let K satisfy (3.3.3) and (3.3.4) for some $A_1, A_2 > 0$. Let $W \in \mathscr{S}'(\mathbf{R}^n)$ be associated as in (3.3.8) with a sequence $\delta_j \in (0,1)$ that tends to zero and let T be the operator given by convolution with W. Suppose that T maps $L^\infty(\mathbf{R}^n)$ to itself with constant B. Prove that T has an extension on $L^1 + L^\infty$ that satisfies (3.6.1) and for $1 < p < \infty$

$$\|T\|_{L^p \to L^p} \leq 2C_n' \left(\frac{p}{p-1}\right)^{1/p} (A_2 + B).$$

[*Hint:* Apply the Calderón–Zygmund decomposition $f = g + b$ at height $\alpha\gamma$, where $\gamma = (2^{n+1}B)^{-1}$. Since $|g| \leq 2^n \alpha\gamma$, observe that $|\{|T(f)| > \alpha\}| \leq |\{|T(b)| > \alpha/2\}|$.]

3.6.3. Let $-\infty \leq a_j < b_j \leq +\infty$ for $j \in \{1,\ldots,n\}$. Prove that $\chi_{[a_1,b_1] \times \cdots \times [a_n,b_n]}$ lies in $\mathscr{M}_p(\mathbf{R}^n)$ with bound independent of a_j and b_j. [*Hint:* When $n = 1$ express $\frac{I + iH}{2}$ as a multiplier operator. Use Exercise 2.8.3 (a).]

3.6.4. Let $1 < p < \infty$. Prove that the norm of the operator $\mathscr{H}_\theta$ defined (3.6.8) from $L^p(\mathbf{R}^n)$ to itself is the same as that of the Hilbert transform from $L^p(\mathbf{R})$ to itself.

3.6.5. Let $\Delta^m = \Delta \circ \cdots \circ \Delta$ be the m-fold composition of the Laplacian with itself. Show that for any $1 < p < \infty$ there exists an $A_{p,n} > 0$ such that for all $f \in \mathscr{S}(\mathbf{R}^n)$ and all $j_1, \ldots, j_{2m} \in \{1,\ldots,n\}$ we have

$$\|\partial_{j_1} \cdots \partial_{j_{2m}} f\|_{L^p(\mathbf{R}^n)} \leq A_{p,n}^m \|\Delta^m f\|_{L^p(\mathbf{R}^n)}.$$

[*Hint:* Prove first the inequality in the case $m = 1$ using the Riesz transforms.]

3.6.6. Prove that for all $1 < p < \infty$ there exists a constant $C_p > 0$ such that for every $f \in \mathscr{S}(\mathbf{R}^2)$ we have

$$\|\partial_1 f\|_{L^p(\mathbf{R}^2)} + \|\partial_2 f\|_{L^p(\mathbf{R}^2)} \leq C_p \|(\partial_1 + i\partial_2) f\|_{L^p(\mathbf{R}^2)}.$$

Likewise, show that there exists a constant $C_{p,m} > 0$ such that for all $f \in \mathscr{S}(\mathbf{R}^n)$ and all $j_1, \ldots, j_m \in \{1,2\}$ we have

$$\|\partial_{j_1} \cdots \partial_{j_m} f\|_{L^p(\mathbf{R}^2)} \leq C_{p,m} \|(\partial_1 + i\partial_2)^m f\|_{L^p(\mathbf{R}^2)}.$$

3.7 The Hilbert Transform and the Poisson Kernel

We investigate connections between the Hilbert transform and the Poisson kernel. Recall the Poisson kernel $P(x) = \frac{1}{\pi}\frac{1}{x^2+1}$ given in Example 1.9.2. Consider the family of kernels $P_y(x) = \frac{1}{y}P(\frac{x}{y}) = \frac{1}{\pi}\frac{y}{x^2+y^2}$ for $y > 0$. Then for a real-valued function g in $L^p(\mathbf{R})$, $1 \leq p < \infty$, we have

$$(P_y * g)(x) = \frac{1}{\pi}\int_{-\infty}^{+\infty} \frac{y}{(x-t)^2 + y^2} g(t)\,dt, \qquad (3.7.1)$$

and the integral in (3.7.1) converges absolutely by Hölder's inequality, since the function $t \mapsto ((x-t)^2 + y^2)^{-1}$ belongs to $L^{p'}(\mathbf{R})$ whenever $y > 0$.

When $z = x + iy$ for $x \in \mathbf{R}$ and $y > 0$ we can write
$$\frac{i}{\pi}\frac{1}{z} = \frac{1}{\pi}\frac{y}{x^2+y^2} + i\frac{1}{\pi}\frac{x}{x^2+y^2},$$
which implies
$$(P_y * g)(x) = \operatorname{Re}\left(\frac{i}{\pi}\int_{-\infty}^{+\infty}\frac{g(t)}{x-t+iy}dt\right) = \operatorname{Re}\left(\frac{i}{\pi}\int_{-\infty}^{+\infty}\frac{g(t)}{z-t}dt\right).$$

On the upper half-space $\mathbf{R}^2_+ = \{z = x + iy : y > 0\}$ the function
$$F_g(z) = \frac{i}{\pi}\int_{-\infty}^{+\infty}\frac{g(t)}{z-t}dt$$
is analytic. To verify this, for $z = x + iy$ with $y > 0$, notice that
$$\frac{F_g(z+h) - F_g(z)}{h} = \frac{i}{\pi}\int_{\mathbf{R}}\frac{-g(t)\,dt}{(z+h-t)(z-t)} \to -\frac{i}{\pi}\int_{\mathbf{R}}\frac{g(t)}{(z-t)^2}dt \quad \text{as } h \to 0$$
by the LDCT; the use of this theorem is based on the fact that for $|h| \leq \frac{1}{2}|y|$,
$$\left|\frac{g(t)}{(z+h-t)(z-t)}\right| \leq \frac{2}{|y|}\frac{|g(t)|}{|z-t|} \in L^1(dt)$$
by Hölder's inequality ($g \in L^p(\mathbf{R})$ and $|z - \cdot|^{-1} \in L^{p'}(\mathbf{R})$).

The real part of $F_g(x+iy)$ is $(P_y * g)(x)$. The imaginary part of $F_g(x+iy)$ is
$$\operatorname{Im}\left(\frac{i}{\pi}\int_{-\infty}^{+\infty}\frac{g(t)}{x-t+iy}dt\right) = \frac{1}{\pi}\int_{-\infty}^{+\infty}\frac{g(t)(x-t)}{(x-t)^2+y^2}dt = (g * Q_y)(x),$$
where $Q_y(x) = \frac{1}{\pi}\frac{x}{x^2+y^2} = \frac{1}{y}Q(\frac{x}{y})$ is the L^1 dilation of the *conjugate Poisson kernel*
$$Q(x) = \frac{1}{\pi}\frac{x}{x^2+1}. \tag{3.7.2}$$

Thus, if g is real-valued, we have
$$F_g(x+iy) = \frac{i}{\pi}\int_{-\infty}^{+\infty}\frac{g(t)}{x+iy-t}dt = (g * P_y)(x) + i(g * Q_y)(x) \tag{3.7.3}$$
and this is analytic on the upper half-space. Hence the functions $x + iy \mapsto (g * P_y)(x)$ and $x + iy \mapsto (g * Q_y)(x)$ are *conjugate harmonic functions*. This explains the choice of name given to the conjugate Poisson kernel.

Up until this point, the function g was real-valued. We now define F_f for any complex-valued function f in $\cup_{1 \leq p < \infty} L^p(\mathbf{R})$ by simply replacing g by f in (3.7.3).

The following lemma reveals an intimate relationship between the Poisson kernel and its conjugate.

3.7 The Hilbert Transform and the Poisson Kernel

Lemma 3.7.1. *The identity $H(P) = Q$ holds. More generally, one has $H(P_\varepsilon) = Q_\varepsilon$ for any $\varepsilon > 0$. We also have $H(Q) = -P$ and $H(Q_\varepsilon) = -P_\varepsilon$ for any $\varepsilon > 0$.*

Proof. Recalling that $\widehat{P}(\xi) = e^{-2\pi|\xi|}$ [Theorem 2.2.5] and identity (3.1.16), we write

$$H(P)(x) = \int_{-\infty}^{+\infty} e^{-2\pi|\xi|}(-i\operatorname{sgn}\xi)e^{2\pi i x\xi}\,d\xi$$

$$= 2\int_0^\infty e^{-2\pi\xi}\sin(2\pi x\xi)\,d\xi$$

$$= \frac{1}{\pi}\int_0^\infty e^{-\xi}\sin(x\xi)\,d\xi$$

$$= \frac{1}{\pi}\operatorname{Im}\int_0^\infty e^{-\xi(1-ix)}\,d\xi$$

$$= \frac{1}{\pi}\operatorname{Im}\frac{1}{1-ix}$$

$$= Q(x).$$

This proves that $H(P) = Q$. The fact $H(Q) = -P$ follows from the identity $H^2 = -I$. The assertions about P_ε and Q_ε are obtained via dilations. $\square$

Given any $\varepsilon > 0$ and $f \in L^p(\mathbf{R})$, $1 < p < \infty$, the function $H(f)$ also lies in $L^p(\mathbf{R})$, and as P_ε, Q_ε lie in $L^{p'}(\mathbf{R})$, it follows from Theorem 1.6.7 that both $f * P_\varepsilon$ and $H(f) * P_\varepsilon$ are uniformly continuous functions.

Lemma 3.7.2. *Let $f \in L^p(\mathbf{R})$ for some $1 < p < \infty$. Then for any $\varepsilon > 0$ we have*

$$f * Q_\varepsilon = H(f) * P_\varepsilon \tag{3.7.4}$$

pointwise everywhere on the real line.

Proof. We first prove (3.7.4) for a function $f \in \mathscr{S}(\mathbf{R})$. Applying the Fourier transform, we see that (3.7.4) is equivalent to the identity $H(P_\varepsilon) = Q_\varepsilon$, proved in the Lemma 3.7.1. So (3.7.4) holds when $f \in \mathscr{S}(\mathbf{R})$. Now given $f \in L^p(\mathbf{R})$, $1 < p < \infty$, there is a sequence $\phi_j \in \mathscr{S}(\mathbf{R})$ such that $\|f - \phi_j\|_{L^p} \to 0$ as $j \to \infty$. The boundedness of the Hilbert transform on L^p yields $\|H(f) - H(\phi_j)\|_{L^p} \to 0$ as well as $j \to \infty$. Also, P_ε, Q_ε lie in $L^{p'}(\mathbf{R})$ ($p' > 1$ since $p < \infty$), so Hölder's inequality yields (3.7.4) for general $f \in L^p(\mathbf{R})$. $\square$

As the family $\{P_\varepsilon\}_{\varepsilon > 0}$ is an approximate identity, it follows from Theorem 1.9.4 that $P_\varepsilon * f \to f$ in $L^p(\mathbf{R})$ as $\varepsilon \to 0$. The following question therefore arises: What is the L^p limit of $f * Q_\varepsilon$ as $\varepsilon \to 0$? As $Q \notin L^1(\mathbf{R})$, the family $\{Q_\varepsilon\}_{\varepsilon > 0}$ is not an approximate identity, so this question cannot be addressed in terms of Theorem 1.9.4. It can be answered, however, via the use of the preceding lemma which relates $\{Q_\varepsilon\}_{\varepsilon > 0}$ to the approximate identity $\{P_\varepsilon\}_{\varepsilon > 0}$.

Theorem 3.7.3. *Let $1 < p < \infty$ and $f \in L^p(\mathbf{R})$. Then we have $f * Q_\varepsilon \to H(f)$ in L^p and almost everywhere as $\varepsilon \to 0$. Consequently,*

$$F_f(\cdot + i\varepsilon) = \frac{i}{\pi} \int_{-\infty}^{+\infty} \frac{f(t)}{(\cdot) + i\varepsilon - t} dt \to f + iH(f) \qquad \text{in } L^p \text{ and a.e.} \qquad (3.7.5)$$

as $\varepsilon \downarrow 0$. Moreover, if f and $H(f)$ are uniformly continuous in a neighborhood of a subset B of the real line, then the convergence in (3.7.5) also holds uniformly on B.

Proof. Let us fix $f \in L^p(\mathbf{R})$ for some $1 < p < \infty$. We have $H(f) * P_\varepsilon \to H(f)$ in L^p and a.e. in view of Theorems 1.9.7 (with $A = 1$) and Theorem 2.5.5 (with $c = 1$). But $f * Q_\varepsilon = H(f) * P_\varepsilon$ by (3.7.4), so $f * Q_\varepsilon \to H(f)$ in L^p and a.e. as $\varepsilon \to 0$. We conclude that

$$F_f(\cdot + i\varepsilon) = f * (P_\varepsilon + iQ_\varepsilon) = (f + iH(f)) * P_\varepsilon \to f + iH(f) \qquad \text{in } L^p \text{ and a.e.}$$

Moreover, this convergence is uniform on B by Theorem 1.9.4 (b), provided f and $H(f)$ are uniformly continuous in a neighborhood of B in the sense of (1.9.1). □

The set of ideas we have discussed allows us to find the limits of the truncated Hilbert transforms for general functions in L^p.

Theorem 3.7.4. *Let $1 < p < \infty$ and $f \in L^p(\mathbf{R})$. Consider the truncated Hilbert transforms $H^{(\varepsilon)}(f)$ (Definition 3.1.5). Then $H^{(\varepsilon)}(f) \to H(f)$ in L^p and a.e. as $\varepsilon \to 0$.*

Proof. Let us fix $f \in L^p(\mathbf{R})$ for some $1 < p < \infty$. Using (3.7.4) we write

$$H^{(\varepsilon)}(f) = H^{(\varepsilon)}(f) - f * Q_\varepsilon + H(f) * P_\varepsilon. \qquad (3.7.6)$$

Next we observe that

$$H^{(\varepsilon)}(f)(x) - f * Q_\varepsilon(x) = \frac{1}{\pi} \int_{|t| \geq \varepsilon} \frac{f(x-t)}{t} dt - (Q_\varepsilon * f)(x) = (f * \psi_\varepsilon)(x),$$

where $\psi_\varepsilon(x) = \varepsilon^{-1}\psi(\varepsilon^{-1}x)$ and

$$\psi(t) = \frac{1}{\pi} \begin{cases} \dfrac{1}{t} - \dfrac{t}{t^2+1} & \text{when } |t| \geq 1, \\ -\dfrac{t}{t^2+1} & \text{when } |t| < 1. \end{cases} \qquad (3.7.7)$$

Notice that

$$|\psi(t)| \leq \frac{1}{\pi}\frac{1}{t^2+1} \qquad (3.7.8)$$

so ψ is integrable over the line and has integral zero, as it is odd. Thus $\{\psi_\varepsilon\}_{\varepsilon > 0}$ is an approximate zero family. It follows from Theorem 1.9.7 (with $A = 0$) that $f * \psi_\varepsilon \to 0$ in L^p. Also Theorem 2.5.5 (with $c = 0$) implies that $f * \psi_\varepsilon \to 0$ a.e. as $\varepsilon \to 0$. □

3.7 The Hilbert Transform and the Poisson Kernel

Returning to (3.7.6), we observe that the identity $H^{(\varepsilon)}(f) = f * \psi_\varepsilon + H(f) * P_\varepsilon$, combined with Proposition 2.5.3, gives

$$\sup_{\varepsilon>0} |H^{(\varepsilon)}(f)| \leq \mathcal{M}(f) + \mathcal{M}(H(f)), \tag{3.7.9}$$

noting that the integral of the Poisson kernel, i.e., the function on the right in (3.7.8), equals 1. For $f \in \bigcup_{1 \leq p < \infty} L^p(\mathbf{R})$ we define the operator

$$H^{(*)}(f) = \sup_{\varepsilon>0} |H^{(\varepsilon)}(f)|,$$

called the *maximal Hilbert transform*. As a consequence of (3.7.9) and of Theorem 3.6.3 and Corollary 1.4.7 we obtain

Corollary 3.7.5. *There is a constant C such that*

$$\|H^{(*)}(f)\|_{L^p(\mathbf{R})} \leq C \max(p, (p-1)^{-2}) \|f\|_{L^p(\mathbf{R})}$$

for $1 < p < \infty$ and all $f \in L^p(\mathbf{R})$.

This bound will be improved to $\max(p, (p-1)^{-1})$ in Corollary 3.8.2.

Exercises

3.7.1. Prove that

$$\lim_{y \downarrow 0} \int_{-\infty}^{+\infty} \frac{dt}{\pi(1+t^2)(x+iy-t)} = \frac{1}{x+i}, \quad \lim_{y \downarrow 0} \int_{-\infty}^{+\infty} \frac{t\,dt}{\pi(1+t^2)(x+iy-t)} = \frac{-i}{x+i}$$

for every real number x.

3.7.2. Prove that

$$\lim_{y \downarrow 0} \int_{-\infty}^{+\infty} \frac{\sin(\pi t)}{\pi t(x+iy-t)}\,dt = \frac{1-e^{i\pi x}}{x}$$

for every real number x, where the function on the right is 1 when $x = 0$.

3.7.3. Let f, g be real-valued Schwartz functions on the real line. Show that

$$H(f)H(g) - fg = H(fH(g) + gH(f)),$$

where H is the Hilbert transform. Then prove that this identity also holds a.e. for real-valued square-integrable functions f, g on the line.
[*Hint:* Consider the boundary values of the analytic function $F_f F_g$.]

3.7.4. Let $P(x) = \frac{1}{\pi}\frac{1}{x^2+1}$ and $Q(x) = \frac{1}{\pi}\frac{x}{x^2+1}$. Prove that $H(PQ) = \frac{1}{2}Q^2 - \frac{1}{2}P^2$.
[*Hint:* Use the preceding exercise.]

3.7.5. For $1 < p < \infty$ let $A_p = \|H\|_{L^p \to L^p}$ and $B_p = \|\mathcal{M}\|_{L^p \to L^p}$, where H is the Hilbert transform and $\mathcal{M}$ is the (centered) Hardy–Littlewood maximal function on $\mathbf{R}$. Let Q be the conjugate Poisson kernel. Prove that for any $f \in L^p(\mathbf{R})$ we have

$$\sup_{\varepsilon > 0} \|Q_\varepsilon * f\|_{L^p(\mathbf{R})} \le A_p \|f\|_{L^p(\mathbf{R})},$$

$$\Big\| \sup_{\varepsilon > 0} |Q_\varepsilon * f| \Big\|_{L^p(\mathbf{R})} \le A_p B_p \|f\|_{L^p(\mathbf{R})}.$$

[*Hint:* Use Lemma 3.7.2 and Proposition 2.5.3.]

3.7.6. Let $2 \le p < \infty$. Prove that for any real-valued function $f \in L^p(\mathbf{R})$ we have

$$\sup_{\varepsilon > 0} \|(P_\varepsilon * f) + i(Q_\varepsilon * f)\|_{L^p(\mathbf{R})} \le (1 + A_p^2)^{\frac{1}{2}} \|f\|_{L^p(\mathbf{R})},$$

where $A_p = \|H\|_{L^p \to L^p}$ and H is the Hilbert transform.
[*Hint:* Use the preceding exercise and the subadditivity of the $L^{p/2}$ norm.]

3.7.7. On $\mathbf{R}^n$ define the *j*th *conjugate Poisson kernel* $Q^{(j)}$ by

$$Q^{(j)}(x) = \frac{\Gamma(\frac{n+1}{2})}{\pi^{\frac{n+1}{2}}} \frac{x_j}{(|x|^2 + 1)^{\frac{n+1}{2}}}, \qquad 1 \le j \le n.$$

Let $Q_y^{(j)}$ be the L^1 dilation of $Q^{(j)}$ for $y > 0$. Prove that

$$(Q_y^{(j)})^\wedge(\xi) = -i \frac{\xi_j}{|\xi|} e^{-2\pi y |\xi|}.$$

Conclude that $R_j(P_y) = Q_y^{(j)}$ and that for all f in $L^p(\mathbf{R}^n)$, $1 < p < \infty$, we have $R_j(f) * P_y = f * Q_y^{(j)}$ for $y > 0$.

3.7.8. Let $f \in L^p(\mathbf{R}^n)$ where $1 < p < \infty$. Prove that the truncated Riesz transforms $R_j^{(\varepsilon)}(f)$ converge to $R_j(f)$ in L^p and a.e. as $\varepsilon \to 0$.
[*Hint:* Using Exercise 3.7.7, write $R_j^{(\varepsilon)}(f) = R_j^{(\varepsilon)}(f) - f * Q_\varepsilon^{(j)} + R_j(f) * P_\varepsilon$ and then apply the idea in Theorem 3.7.4.]

3.7.9. Let η be an even smooth function on the real line such that $\eta(t) = 1$ for $|t| \ge 1$ and η vanishes for $|t| \le \frac{1}{2}$. Define the *smoothly truncated Hilbert transform* (associated with η) acting on a function $f \in L^p(\mathbf{R})$ ($1 < p < \infty$) by

$$H_\eta^{(\varepsilon)}(f)(x) = \int_{\mathbf{R}} f(x-t) \frac{\eta(t/\varepsilon)}{t} dt.$$

Given $1 < p < \infty$ and $f \in L^p(\mathbf{R})$, prove that $H_\eta^{(\varepsilon)}(f) \to H(f)$ in L^p and a.e. as $\varepsilon \to 0$.

3.8 Maximal Singular Integrals

We introduce maximal singular integrals and we derive their boundedness under the general conditions of size, smoothness and cancellations. For a function K that satisfies (3.3.3) we define the associated *truncated singular integral operator*

$$T^{(\varepsilon)}(f)(x) = \int_{|x-y|\geq \varepsilon} K(x-y)f(y)\,dy, \qquad x \in \mathbf{R}^n, \tag{3.8.1}$$

whenever $f \in \bigcup_{1\leq p<\infty} L^p(\mathbf{R}^n)$. Note that as $y \mapsto |x-y|^{-n}\chi_{|y-x|\geq \varepsilon}$ lies in $L^{p'}(\mathbf{R}^n)$, $p' > 1$, Hölder's inequality yields that the integral in (3.8.1) converges absolutely for all $x \in \mathbf{R}^n$. We also set

$$T^{(*)}(f) = \sup_{\varepsilon>0} |T^{(\varepsilon)}(f)|, \qquad f \in \bigcup_{1\leq p<\infty} L^p(\mathbf{R}^n),$$

and we call this operator the *maximal singular integral operator* associated with K.

Theorem 3.8.1. *(Cotlar's inequality)* Let $\delta > 0$, $0 < A_1, A_2, A_3 < \infty$, and suppose that K is defined on $\mathbf{R}^n \setminus \{0\}$ and satisfies the size condition (3.3.3), the smoothness condition

$$|K(x-y) - K(x)| \leq A_2 |y|^\delta |x|^{-n-\delta}, \qquad |x| \geq 2|y|, \tag{3.8.2}$$

and the cancellation condition (3.3.5). Suppose that W is a tempered distribution on $\mathbf{R}^n$ defined in terms of (3.3.8) (associated with a sequence $\delta_k \to 0$) and let T be the operator given by convolution with W. Then for $0 < r < 1$ there is a constant $C_{n,\delta,r}$ such that the following inequality is valid:

$$T^{(*)}(f) \leq 3^{\frac{1-r}{r}} \left[M(|T(f)|^r) \right]^{\frac{1}{r}} + C_{n,\delta,r}(A_1 + A_2 + A_3) M(f), \tag{3.8.3}$$

for all $f \in \bigcup_{1\leq p<\infty} L^p(\mathbf{R}^n)$, where M is the Hardy–Littlewood maximal operator.

Note that (3.8.2) implies the Hörmander integral smoothness condition (3.3.4).

Proof. We fix r satisfying $0 < r < 1$, $\varepsilon > 0$, $f \in \bigcup_{1\leq p<\infty} L^p(\mathbf{R}^n)$, and $x_0 \in \mathbf{R}^n$. We pick a smooth radial function η with values in $[0,1]$ which equals 1 on the unit ball and vanishes outside $B(0,2)$. Then we define

$$f_0^{\varepsilon,x_0}(x) = f(x)\eta(\tfrac{x-x_0}{\varepsilon}) \qquad f_\infty^{\varepsilon,x_0}(x) = f(x)(1-\eta(\tfrac{x-x_0}{\varepsilon}))$$

at the points $x \in \mathbf{R}^n$ for which f is defined. We write

$$T^{(\varepsilon)}(f)(x_0) - T(f_\infty^{\varepsilon,x_0})(x_0) = \int_{|x_0-y|\geq \varepsilon} K(x_0-y)f(y)\,dy - \int_{\mathbf{R}^n} K(x_0-y) f_\infty^{\varepsilon,x_0}(y)\,dy$$

$$= \int_{2\varepsilon \geq |x_0-y| \geq \varepsilon} K(x_0-y)\,\eta(\tfrac{y-x_0}{\varepsilon})f(y)\,dy,$$

in view of the support properties of η. It follows from this and (3.3.3) that

$$|T^{(\varepsilon)}(f)(x_0) - T(f_\infty^{\varepsilon,x_0})(x_0)| \le C_n A_1 M(f)(x_0). \tag{3.8.4}$$

For $z \in B(x_0, \frac{\varepsilon}{2})$ we have $|x_0 - z| \le \frac{1}{2}|x_0 - y|$ whenever $|y - x_0| \ge \varepsilon$ and thus

$$\begin{aligned}|T(f_\infty^{\varepsilon,x_0})(x_0) - T(f_\infty^{\varepsilon,x_0})(z)| &= \left| \int_{|y-x_0|\ge\varepsilon} (K(x_0 - y) - K(z - y)) f(y)(1 - \eta(\tfrac{y-x_0}{\varepsilon})) dy \right| \\ &\le \int_{|y-x_0|\ge\varepsilon} |K(x_0 - y - (x_0 - z)) - K(x_0 - y)| |f(y)| dy \\ &\le |x_0 - z|^\delta \int_{|y-x_0|\ge\varepsilon} \frac{A_2 |f(y)|}{|x_0 - y|^{n+\delta}} dy \\ &\le 2^{n+\delta} \left(\frac{\varepsilon}{2}\right)^\delta \int_{|y-x_0|\ge\varepsilon} \frac{A_2 |f(y)|}{(|y - x_0| + \varepsilon)^{n+\delta}} dy \\ &\le C_{n,\delta} A_2 M(f)(x_0),\end{aligned}$$

where the last estimate is a consequence of Theorem 2.5.1. We conclude that for all $z \in B(x_0, \frac{\varepsilon}{2})$, we have

$$\begin{aligned}|T^{(\varepsilon)}(f)(x_0)| &\le |T(f_\infty^{\varepsilon,x_0})(x_0)| + C_n A_1 M(f)(x_0) \\ &\le |T(f_\infty^{\varepsilon,x_0})(x_0) - T(f_\infty^{\varepsilon,x_0})(z)| + |T(f_\infty^{\varepsilon,x_0})(z)| + C_n A_1 M(f)(x_0) \\ &\le C_{n,\delta}(A_1 + A_2) M(f)(x_0) + |T(f - f_0^{\varepsilon,x_0})(z)| \\ &\le C_{n,\delta}(A_1 + A_2) M(f)(x_0) + |T(f)(z)| + |T(f_0^{\varepsilon,x_0})(z)|.\end{aligned}$$

Raising to the power $r < 1$ we obtain

$$|T^{(\varepsilon)}(f)(x_0)|^r \le C_{n,\delta}^r (A_1 + A_2)^r M(f)(x_0)^r + |T(f)(z)|^r + |T(f_0^{\varepsilon,x_0})(z)|^r \tag{3.8.5}$$

for all $z \in B(x_0, \frac{\varepsilon}{2})$. First we average over $z \in B(x_0, \frac{\varepsilon}{2})$, then we take the supremum over $\varepsilon > 0$ (only) in the second term, and then we raise to the power $\frac{1}{r}$ making use of the inequality $(a + b + c)^{1/r} \le 3^{(1-r)/r}(a^{1/r} + b^{1/r} + c^{1/r})$ for $a, b, c \ge 0$. We deduce

$$|T^{(\varepsilon)}(f)(x_0)| \le 3^{\frac{1-r}{r}} \left[C_{n,\delta}(A_1 + A_2) M(f)(x_0) + [M(|T(f)|^r)(x_0)]^{\frac{1}{r}} + \left(\frac{1}{|B(x_0, \frac{\varepsilon}{2})|} \int_{B(x_0, \frac{\varepsilon}{2})} |T(f_0^{\varepsilon,x_0})(z)|^r dz \right)^{\frac{1}{r}} \right].$$

The third term in the square brackets is estimated, via Exercise 1.2.6 (a), by

$$\left(\frac{1}{|B(x_0, \frac{\varepsilon}{2})|} \frac{\|T\|_{L^1 \to L^{1,\infty}}^r}{1-r} |B(x_0, \tfrac{\varepsilon}{2})|^{1-r} \|f_0^{\varepsilon,x_0}\|_{L^1}^r \right)^{\frac{1}{r}} \le C_{n,r}(A_1 + A_2 + A_3) M(f)(x_0).$$

Inserting this estimate in the inequality bounding $|T^{(\varepsilon)}(f)(x_0)|$ yields (3.8.3). $\square$

3.8 Maximal Singular Integrals

Corollary 3.8.2. *Let K, A_1, A_2, A_3, W, and T be as in the preceding theorem. Then $T^{(*)}$ admits a bounded extension on L^p for all $1 < p < \infty$ and also maps L^1 to $L^{1,\infty}$. Moreover, there are constants C_n, C_n' such that*

$$\|T^{(*)}\|_{L^1 \to L^{1,\infty}} \leq C_n'(A_1 + A_2 + A_3),$$
$$\|T^{(*)}\|_{L^p \to L^p} \leq C_n \max(p, (p-1)^{-1})(A_1 + A_2 + A_3).$$

Thus, there are constants c, c' such that the maximal Hilbert transform $H^{(*)}$ satisfies

$$\|H^{(*)}\|_{L^1(\mathbf{R}) \to L^{1,\infty}(\mathbf{R})} \leq c',$$
$$\|H^{(*)}\|_{L^p(\mathbf{R}) \to L^p(\mathbf{R})} \leq c \max(p, (p-1)^{-1}),$$

and there are constants c_n, c_n' such that the maximal Riesz transforms $R_j^{(*)}$ satisfy

$$\|R_j^{(*)}\|_{L^1(\mathbf{R}^n) \to L^{1,\infty}(\mathbf{R}^n)} \leq c_n',$$
$$\|R_j^{(*)}\|_{L^p(\mathbf{R}^n) \to L^p(\mathbf{R}^n)} \leq c_n \max(p, (p-1)^{-1}).$$

Proof. To show that $T^{(*)}$ maps L^1 to $L^{1,\infty}$ we need to use that the Hardy–Littlewood maximal operator maps $L^{p,\infty}$ to $L^{p,\infty}$ for all $1 < p < \infty$; on this see Exercise 1.4.8. For all $0 < p, q < \infty$ note the identity

$$\||f|^q\|_{L^{p,\infty}} = \|f\|_{L^{pq,\infty}}^q$$

which can easily be deduced from Definition 1.2.5. We use (3.8.3) with $r = 1/2$. The difficult term is the one involving $T(f)$. We estimate this as follows:

$$\begin{aligned}\|M(|T(f)|^{\frac{1}{2}})^2\|_{L^{1,\infty}} &= \|M(|T(f)|^{\frac{1}{2}})\|_{L^{2,\infty}}^2 \\ &\leq \widetilde{C}_n \||T(f)|^{\frac{1}{2}}\|_{L^{2,\infty}}^2 \\ &= \widetilde{C}_n \|T(f)\|_{L^{1,\infty}} \\ &\leq C_n'(A_1 + A_2 + A_3)\|f\|_{L^1},\end{aligned}$$

where we made use that M maps $L^{2,\infty}$ to itself (Exercise 1.4.8) and that T maps L^1 to $L^{1,\infty}$ with bound a multiple of $A_1 + A_2 + A_3$ in the last estimate; this is a consequence of Corollary 3.6.2.

We essentially repeat the preceding argument to obtain the L^p boundedness of $T^{(*)}$ for $1 < p < \infty$. Recall that the maximal function is bounded on $L^{2p}(\mathbf{R}^n)$ with norm at most

$$3^{\frac{n}{2p}} \frac{2 \cdot 2p}{2p - 1} \leq 4 \cdot 3^{\frac{n}{2}}$$

by Corollary 1.4.7. We have

$$\begin{aligned}
\|M(|T(f)|^{\frac{1}{2}})^2\|_{L^p} &= \|M(|T(f)|^{\frac{1}{2}})\|_{L^{2p}}^2 \\
&\le (4\cdot 3^{\frac{n}{2}})^2 \||T(f)|^{\frac{1}{2}}\|_{L^{2p}}^2 \\
&= 16\cdot 3^n \|T(f)\|_{L^p} \\
&\le C_n \max(\tfrac{1}{p-1},p)(A_1+A_2+A_3)\|f\|_{L^p},
\end{aligned}$$

where we used the L^p boundedness of T in the last estimate. $\square$

Corollary 3.8.3. *Let $1 \le p < \infty$. Assume that K, W (associated with a sequence $\delta_k \to 0$), and T are as in Theorem 3.8.1. Then for a given function $f \in L^p(\mathbf{R}^n)$ we have $T^{(\delta_k)}(f) \to T(f)$ a.e. as $k \to \infty$. Moreover, for $p > 1$, $T^{(\delta_k)}(f) \to T(f)$ in L^p.*

Proof. We begin with the observation that $T^{(\delta_k)}(\varphi)(x) \to T(\varphi)(x)$ as $k \to \infty$ for all $x \in \mathbf{R}^n$ whenever $\varphi \in \mathscr{S}(\mathbf{R}^n)$; this is a consequence of (3.3.7) (applied to $\tau^x\widetilde{\varphi}$). Combining Theorems 3.8.1 and 2.5.6 yields the asserted a.e. convergence for functions f in $L^p(\mathbf{R}^n)$. The claimed convergence in L^p for $p > 1$ is a consequence of the LDCT as $T^{(\delta_k)}(f) \to T(f)$ a.e. and $|T^{(\delta_k)}(f)| \le T^{(*)}(f) \in L^p(\mathbf{R}^n)$ for all k. $\square$

We conclude this section with an analog of Corollary 3.6.4.

Corollary 3.8.4. *Let Ω be an odd bounded function on $\mathbf{S}^{n-1}$. Then $T_\Omega^{(*)}$ is bounded on $L^p(\mathbf{R}^n)$ with norm at most $\frac{\pi \omega_{n-1}}{2}\|\Omega\|_{L^\infty}\|H^{(*)}\|_{L^p \to L^p}$.*

Proof. Let $f \in \mathscr{S}(\mathbf{R}^n)$. We begin with

$$\int_{|y|\ge \varepsilon} \frac{\Omega(y/|y|)}{|y|^n} f(x-y)\,dy = \frac{\pi}{2}\int_{\mathbf{S}^{n-1}} \Omega(\theta)\,\mathscr{H}_\theta^{(\varepsilon)}(f)(x)\,d\theta$$

derived in (3.6.10). Inserting absolute values and taking the supremum over all $\varepsilon > 0$ yields

$$T_\Omega^{(*)}(f)(x) \le \frac{\pi}{2}\int_{\mathbf{S}^{n-1}} |\Omega(\theta)|\,\mathscr{H}_\theta^{(*)}(f)(x)\,d\theta, \qquad x \in \mathbf{R}^n,$$

for $f \in \mathscr{S}(\mathbf{R}^n)$. From this we obtain the claimed L^p bound for $T_\Omega^{(*)}$ by Minkowski's integral inequality (Theorem 1.1.12). Using the result in Appendix D, we extend this operator to $L^p(\mathbf{R}^n)$. $\square$

Exercises

3.8.1. Let Ω be a bounded function on $\mathbf{S}^{n-1}$ with vanishing integral and let T_Ω be as in Definition 3.2.5. Assume that either Ω is odd or Ω satisfies the *Lipschitz condition* $|\Omega(\theta_1) - \Omega(\theta_2)| \le B_0|\theta_1 - \theta_2|^\alpha$ for some $\alpha \in (0,1)$, $B_0 > 0$, and all $\theta_1, \theta_2 \in \mathbf{S}^{n-1}$. Prove that for any $1 < p < \infty$ and all $f \in L^p(\mathbf{R}^n)$, $T_\Omega^{(\varepsilon)}(f) \to T_\Omega(f)$ a.e. and in L^p. [*Hint:* Use Exercise 3.3.5, Theorem 2.5.6, and Theorem 3.8.1.]

3.8 Maximal Singular Integrals

3.8.2. Let K be a function on $\mathbf{R}^n \setminus \{0\}$ that satisfies $|K(x)| \leq A_1 |x|^{-n}$. Let η be a smooth function that equals 1 when $|x| \geq 2$ and vanishes for $|x| \leq 1$. For $f \in L^p(\mathbf{R}^n)$, $1 \leq p < \infty$, define the *smoothly truncated singular integral* by

$$T_\eta^{(\varepsilon)}(f)(x) = \int_{\mathbf{R}^n} \eta(y/\varepsilon) K(y) f(x-y)\, dy, \qquad x \in \mathbf{R}^n.$$

Show that the maximal singular integral

$$T^{(*)} = \sup_{\varepsilon > 0} |T^{(\varepsilon)}|$$

is bounded from $L^p(\mathbf{R}^n)$ to $L^p(\mathbf{R}^n)$ for $1 < p < \infty$ if and only if the *smoothly truncated maximal singular integral*

$$T_\eta^{(*)} = \sup_{\varepsilon > 0} |T_\eta^{(\varepsilon)}|$$

is bounded from $L^p(\mathbf{R}^n)$ to $L^p(\mathbf{R}^n)$. [*Hint:* Use Corollary 2.5.2.]

3.8.3. (**Simpler form of Cotlar's inequality**) Suppose that K is a function defined on $\mathbf{R}^n \setminus \{0\}$ that satisfies (3.3.3), (3.8.2), and (3.3.5). Let W be a tempered distribution on $\mathbf{R}^n$ associated with K as in (3.3.8) and let T be the operator given by convolution with W. Follow the steps below to prove the inequality

$$T^{(*)}(f) \leq c M(T(f)) + C_{n,\delta}(A_1 + A_2 + A_3) M(f), \qquad f \in L^p(\mathbf{R}^n).$$

(a) Notice that the operator $T^{(\varepsilon)}$ is given by convolution with $K\chi_{B(0,\varepsilon)^c}$. (b) Pick a smooth function ϕ supported in the ball $\overline{B(0,1/2)}$ with integral 1. Write

$$K\chi_{B(0,\varepsilon)^c} = \phi_\varepsilon * W + \left(K\chi_{B(0,\varepsilon)^c} - \phi_\varepsilon * W \right).$$

(c) Show that

$$\sup_{\varepsilon > 0} |f * \phi_\varepsilon * W| \leq c M(T(f))$$

for $f \in L^p(\mathbf{R}^n)$. In (d), (e) prove that

$$\left| \left(K\chi_{B(0,\varepsilon)^c} - \phi_\varepsilon * W \right)(x) \right| \leq \frac{1}{\varepsilon^n} \frac{(A_1 + A_2 + A_3) C_{\phi,\delta,n}}{(1 + |x/\varepsilon|)^{n+\delta}},$$

where $C_{\phi,\delta,n}$ depends on the indicated parameters. Then apply Corollary 2.5.2. (d) In the case $|x| \geq \varepsilon$ write

$$K(x) = \int_{\mathbf{R}^n} K(x) \phi_\varepsilon(y)\, dy$$

and use (3.8.2) to obtain the inequality in (c) with $C_{\phi,\delta,n} = 2^{n+\delta} \int_{\mathbf{R}^n} |y|^\delta |\phi(y)|\, dy$. (e) To prove the inequality in (c) in the case $|x| < \varepsilon$ begin with

$$(\phi_\varepsilon * W)(x) = \lim_{\delta_j \to 0} \int_{|x-y| \geq \delta_j} K(x-y)\phi_\varepsilon(y)\,dy$$

for some sequence $\delta_j \in (0,1)$ tending to zero which defines W. Then write $(\phi_\varepsilon * W)(x)$ as the sum of the following expressions:

$$\int_{|x-y| > \frac{\varepsilon}{8}} K(x-y)\phi_\varepsilon(y)\,dy,$$

$$\int_{|x-y| \leq \frac{\varepsilon}{8}} K(x-y)\left[\phi_\varepsilon(y) - \phi_\varepsilon(x)\right] dy,$$

$$\phi_\varepsilon(x) \lim_{\delta_j \to 0} \int_{\delta_j \leq |x-y| \leq \frac{\varepsilon}{8}} K(x-y)\,dy.$$

Finally, estimate all these expressions by $(A_1 + A_3)\varepsilon^{-n} C(\phi, n)$.

Chapter 4
Vector-Valued Singular Integrals and Littlewood–Paley Theory

4.1 The Vector-Valued Calderón–Zygmund Theorem

Let T be a bounded operator from $L^p(\mathbf{R}^n)$ to itself for some $p \in (1,\infty)$. One may wonder if a stronger estimate of the form

$$\left\| \Big(\sum_{j=1}^{N} |T(f_j)|^s \Big)^{\frac{1}{s}} \right\|_{L^p} \leq C \left\| \Big(\sum_{j=1}^{N} |f_j|^q \Big)^{\frac{1}{q}} \right\|_{L^p} \qquad (4.1.1)$$

might hold, where $f_j \in L^p(\mathbf{R}^n)$ and $1 \leq q, s \leq \infty$ (with the obvious modifications when q or s is infinite). Naturally, we would like this estimate to hold with a constant C independent of N, so that we can let $N \to \infty$. We will derive estimates of the form (4.1.1) by introducing operators acting on finite sequences of functions.

We fix positive integers M, N and $1 \leq q, s \leq \infty$. We denote by ℓ^q_N the Banach space of all finite sequences $(a_1, \ldots, a_N)$ of complex numbers equipped with the norm

$$\|(a_1, \ldots, a_N)\|_{\ell^q} = \begin{cases} \big(\sum_{j=1}^{N} |a_j|^q \big)^{\frac{1}{q}} & \text{when } q < \infty, \\ \sup\{|a_j| : j = 1, \ldots, N\} & \text{when } q = \infty. \end{cases}$$

Bounded linear operators from ℓ^q_N to ℓ^s_M can be identified with $M \times N$ matrices $Y = (y_{ij})_{1 \leq i \leq M, 1 \leq j \leq N}$. The norm of a such a matrix Y is denoted by $\|Y\|_{\ell^q_N \to \ell^s_M}$. The precise form of $\|Y\|_{\ell^q_N \to \ell^s_M}$ is not needed in general and can be calculated in some instances; upper and lower estimates for $\|Y\|_{\ell^q_N \to \ell^s_M}$ are given in Exercise 4.3.1.

Let $0 < p \leq \infty$ and $1 \leq q \leq \infty$. We will be working with spaces of finite sequences $\{f_j\}_{j=1}^{N}$ of measurable functions on $\mathbf{R}^n$ that satisfy

$$\big\| \|(f_1, \ldots, f_N)\|_{\ell^q} \big\|_{L^p(\mathbf{R}^n)} = \left(\int_{\mathbf{R}^n} \Big(\sum_{j=1}^{N} |f_j(x)|^q \Big)^{\frac{p}{q}} dx \right)^{\frac{1}{p}} < \infty,$$

with the obvious modification when p or q is infinity. We define

$$L^p(\mathbf{R}^n, \ell_N^q) = \left\{ \{f_j\}_{j=1}^N : \big\| \|(f_1, \ldots, f_N)\|_{\ell^q} \big\|_{L^p(\mathbf{R}^n)} < \infty \right\}$$

and we note that this is a normed space when $p \geq 1$ and a quasi-normed space when $p < 1$. In fact these spaces are also complete; see Exercise 4.1.2. When $p \geq 1$, we define the integral of an element $\{f_j\}_{j=1}^N$ of $L^p(\mathbf{R}^n, \ell_N^q)$ over a compact subset B of $\mathbf{R}^n$ by setting

$$\int_B \{f_j\}_{j=1}^N(x)\,dx = \left\{ \int_B f_j(x)\,dx \right\}_{j=1}^N.$$

We consider the following situation. Suppose that for all $x \in \mathbf{R}^n \setminus \{0\}$ there is an $M \times N$ matrix

$$\vec{K}(x) = \begin{pmatrix} K_{11}(x) & K_{12}(x) & \cdots & K_{1N}(x) \\ K_{21}(x) & K_{22}(x) & \cdots & K_{2N}(x) \\ \vdots & \vdots & \vdots & \vdots \\ K_{M1}(x) & K_{M2}(x) & \cdots & K_{MN}(x) \end{pmatrix} = (K_{ij}(x))_{1 \leq i \leq M, 1 \leq j \leq N},$$

where each K_{ij} lies in $L^1_{\mathrm{loc}}(\mathbf{R}^n \setminus \{0\})$, i.e., it is an integrable function on every compact subset of $\mathbf{R}^n \setminus \{0\}$. Also suppose that K_{ij} satisfy the size condition

$$|K_{ij}(x)| \leq A_1^{ij} |x|^{-n}, \qquad x \neq 0, \tag{4.1.2}$$

for some $A_1^{ij} < \infty$ and, for some $1 \leq q, s \leq \infty$, the regularity condition

$$\sup_{y \in \mathbf{R}^n \setminus \{0\}} \int_{|x| \geq 2|y|} \|\vec{K}(x-y) - \vec{K}(x)\|_{\ell_N^q \to \ell_M^s}\,dx \leq A_2 < \infty. \tag{4.1.3}$$

Finally, we assume that there is a sequence δ_k in $(0,1)$ with $\delta_k \to 0$ as $k \to \infty$ and an $M \times N$ complex matrix $\vec{K}_0 = (K_{ij}^0)_{1 \leq i \leq M, 1 \leq j \leq N}$, such that

$$\lim_{k \to \infty} \int_{\delta_k \leq |x| \leq 1} K_{ij}(x)\,dx = K_{ij}^0. \tag{4.1.4}$$

Note that (4.1.4) would be a consequence of the assumption

$$\sup_{1 \leq i \leq M} \sup_{1 \leq j \leq N} \sup_{\varepsilon > 0} \sup_{R > \varepsilon} \left| \int_{\varepsilon \leq |x| \leq R} K_{ij}(x)\,dx \right| \leq A_3 < \infty. \tag{4.1.5}$$

For a compact set B that does not contain the origin, we define the integral of $\vec{K}$ over B to be the matrix of the integrals of the coordinates over B, i.e.,

$$\int_B \vec{K}(x)\,dx = \left(\int_B K_{ij}(x)\,dx \right)_{1 \leq i \leq M, 1 \leq j \leq N}.$$

Using this notation we write

4.1 The Vector-Valued Calderón–Zygmund Theorem

$$\lim_{k\to\infty}\int_{\delta_k\leq|y|\leq 1}\vec{K}(y)\,dy=\vec{K}^0. \tag{4.1.6}$$

Given $\vec{K}$ satisfying assumptions (4.1.2), (4.1.3), and (4.1.5), we define an operator $\vec{T}$ acting on N-tuples of smooth functions with compact support $\{f_j\}_{j=1}^N$ as follows:

$$\vec{T}(\{f_j\}_{j=1}^N)(x)=\lim_{k\to\infty}\int_{|y|\geq\delta_k}\vec{K}(y)(\{f_j(x-y)\}_{j=1}^N)\,dy. \tag{4.1.7}$$

This equals:

$$\begin{pmatrix} \lim_{k\to\infty}\int_{|y|\geq\delta_k}K_{11}(y)f_1(x-y)\,dy+\cdots+\lim_{k\to\infty}\int_{|y|\geq\delta_k}K_{1N}(y)f_N(x-y)\,dy \\ \lim_{k\to\infty}\int_{|y|\geq\delta_k}K_{21}(y)f_1(x-y)\,dy+\cdots+\lim_{k\to\infty}\int_{|y|\geq\delta_k}K_{2N}(y)f_N(x-y)\,dy \\ \vdots \\ \lim_{k\to\infty}\int_{|y|\geq\delta_k}K_{M1}(y)f_1(x-y)\,dy+\cdots+\lim_{k\to\infty}\int_{|y|\geq\delta_k}K_{MN}(y)f_N(x-y)\,dy \end{pmatrix},$$

and each one of these limits exists and is similar to the limit in (3.3.9).

Assume $r=q=s$ in (4.1.1) and suppose $1<r<\infty$. Then (4.1.1) holds trivially when $p=r$ with $C=\|T\|_{L^p\to L^p}$. Thus, it is natural to consider $p=r$ (instead of $p=2$) as the initial estimate in the following vector-valued adaptation of Theorem 3.6.1. As above, we fix below $M,N\in\mathbf{Z}^+$ and $1\leq q,s\leq\infty$.

Theorem 4.1.1. *Suppose that for each $x\in\mathbf{R}^n\setminus\{0\}$, $\vec{K}(x)$ is a matrix that satisfies (4.1.2) and (4.1.3) for some $A_1^{ij},A_2>0$ and (4.1.6) for some sequence $\delta_k\downarrow 0$ and some $M\times N$ complex matrix $\vec{K}_0$. Let $\vec{T}$ be the operator associated with $\vec{K}$ as defined in (4.1.7). Assume that $\vec{T}$ is a bounded linear operator from $L^r(\mathbf{R}^n,\ell_N^q)$ to $L^r(\mathbf{R}^n,\ell_M^s)$ with norm $B_\star$ for some $1<r\leq\infty$. Then $\vec{T}$ has well-defined extensions on $L^p(\mathbf{R}^n,\ell_N^q)$ for all $1\leq p<\infty$ and there exist constants C_n, C_n' such that these extensions satisfy*

$$\|\vec{T}(F)\|_{L^{1,\infty}(\mathbf{R}^n,\ell_M^s)}\leq C_n'(A_2+B_\star)\|F\|_{L^1(\mathbf{R}^n,\ell_N^q)} \tag{4.1.8}$$

for all F in $L^1(\mathbf{R}^n,\ell_N^q)$ and

$$\|\vec{T}(F)\|_{L^p(\mathbf{R}^n,\ell_M^s)}\leq C_n C(p)(A_2+B_\star)\|F\|_{L^p(\mathbf{R}^n,\ell_N^q)} \tag{4.1.9}$$

for all F in $L^p(\mathbf{R}^n,\ell_N^q)$ when $1<p<\infty$. Here $C(p)=\max\bigl(p,(p-1)^{-1}\bigr)$ if $r<\infty$ and $C(p)=p(p-1)^{-1}$ if $r=\infty$.

It is remarkable that the constant C_n' in (4.1.8) and C_n in (4.1.9) are in fact independent of A_1, q, N, s, M, r.

Proof. We fix $F=\{f_k\}_{k=1}^N$, where each f_k is a finite linear combination of characteristic functions of dyadic cubes. Such functions F are dense in $L^1(\mathbf{R}^n,\ell_N^q)$ by Exercise 4.1.3. Notice that $\|F\|_{\ell_N^q}$ is also a finite linear combination of characteristic functions of dyadic cubes, hence its Calderón–Zygmund decomposition contains only finitely many cubes, cf. Example 3.5.5. We apply the Calderón–Zygmund decomposition to $\|F\|_{\ell_N^q}$ at height $\gamma\alpha$, where $\gamma = 2^{-n-1}(A_2 + B_\star)^{-1}$ extracting a finite collection of closed dyadic cubes $\{Q_j\}_j$ satisfying $\sum_j |Q_j| \leq (\gamma\alpha)^{-1}\|F\|_{L^1(\mathbf{R}^n,\ell_N^q)}$. We define the good function G of the decomposition by

$$G(x) = \begin{cases} F(x) & \text{for } x \notin \cup_i Q_i, \\ \frac{1}{|Q_j|}\int_{Q_j} F(x)\,dx & \text{for } x \in Q_j. \end{cases}$$

Also define the bad function $B = F - G$. Then $B = \sum_j B_j$, where each B_j is supported in the cube Q_j and has mean value zero over Q_j. Moreover,

$$\|G\|_{L^1(\mathbf{R}^n,\ell_N^q)} \leq \|F\|_{L^1(\mathbf{R}^n,\ell_N^q)}, \tag{4.1.10}$$

$$\|G\|_{L^\infty(\mathbf{R}^n,\ell_N^q)} \leq 2^n\gamma\alpha, \tag{4.1.11}$$

and $\|B_j\|_{L^1(\mathbf{R}^n,\ell_N^q)} \leq 2^{n+1}\gamma\alpha|Q_j|$, by an argument similar to that given in the proof of Theorem 3.5.2. We only verify (4.1.11). On the cube Q_j, G is equal to the constant $|Q_j|^{-1}\int_{Q_j} F(x)\,dx$, and this is bounded by $2^n\gamma\alpha$. For each $x \in \mathbf{R}^n \setminus \cup_j Q_j$ and for each $k = 0, 1, 2, \ldots$ there exists a unique nonselected dyadic cube $Q_x^{(k)}$ of generation k that contains x. Then for each $k \geq 0$, we have by Theorem 1.1.12

$$\left\|\frac{1}{|Q_x^{(k)}|}\int_{Q_x^{(k)}} F(y)\,dy\right\|_{\ell_N^q} \leq \frac{1}{|Q_x^{(k)}|}\int_{Q_x^{(k)}} \|F(y)\|_{\ell_N^q}\,dy \leq \gamma\alpha.$$

The intersection of the closures of the cubes $Q_x^{(k)}$ is the singleton $\{x\}$. Using Corollary 1.5.6, we deduce that for almost all $x \in \mathbf{R}^n \setminus \cup_i Q_i$ we have

$$F(x) = \lim_{k\to\infty} \frac{1}{|Q_x^{(k)}|}\int_{Q_x^{(k)}} F(y)\,dy.$$

Since these averages are at most $\gamma\alpha$, we conclude that $\|F\|_{\ell_N^q} \leq \gamma\alpha$ almost everywhere on $\mathbf{R}^n \setminus \cup_j Q_j$; hence $\|G\|_{\ell_N^q} \leq \gamma\alpha$ a.e. on this set. This proves (4.1.11).

We begin with the estimate concerning the good function G. Suppose first that $r < \infty$. Then

$$\left|\{x \in \mathbf{R}^n : \|\vec{T}(G)(x)\|_{\ell_M^s} > \alpha/2\}\right| \leq \left(\frac{2}{\alpha}\right)^r \|\vec{T}(G)\|_{L^r(\mathbf{R}^n,\ell_M^s)}^r$$

$$\leq \left(\frac{2B_\star}{\alpha}\right)^r \|G\|_{L^r(\mathbf{R}^n,\ell_N^q)}^r$$

$$\leq \left(\frac{2B_\star}{\alpha}(2^n\alpha\gamma)^{\frac{1}{r'}}\|F\|_{L^1(\mathbf{R}^n,\ell_N^q)}^{\frac{1}{r}}\right)^r$$

$$\leq \frac{2(A_2+B_\star)}{\alpha}\|F\|_{L^1(\mathbf{R}^n,\ell_N^q)}, \tag{4.1.12}$$

4.1 The Vector-Valued Calderón–Zygmund Theorem

where the third inequality follows from (4.1.10) and (4.1.11) (Exercise 1.1.6). Now if $r = \infty$, we have

$$\|\vec{T}(G)\|_{L^\infty(\mathbf{R}^n, \ell^s_M)} \leq B_\star \|G\|_{L^\infty(\mathbf{R}^n, \ell^q_N)} \leq 2^n \gamma \alpha B_\star < \frac{\alpha}{2}$$

which implies that

$$\left|\{x \in \mathbf{R}^n : \|\vec{T}(G)(x)\|_{\ell^s_M} > \alpha/2\}\right| = 0.$$

Thus estimate (4.1.12) also holds when $r = \infty$.

We now turn our attention to the bad function B. As $B = \sum_j B_j$ we have that $\vec{T}(B) = \sum_j \vec{T}(B_j)$, since the Calderón–Zygmund decomposition of F contains finitely many cubes Q_j. Let $Q_j^* = 2\sqrt{n} Q_j$. We write

$$\left|\{x \in \mathbf{R}^n : \|\vec{T}(B)(x)\|_{\ell^s_M} > \alpha/2\}\right|$$

$$\leq \left|\bigcup_j Q_j^*\right| + \left|\left\{x \notin \bigcup_j Q_j^* : \|\vec{T}(B)(x)\|_{\ell^s_M} > \alpha/2\right\}\right|$$

$$\leq \frac{(2\sqrt{n})^n}{\gamma} \frac{\|F\|_{L^1(\mathbf{R}^n, \ell^q_N)}}{\alpha} + \frac{2}{\alpha} \int_{(\cup_j Q_j^*)^c} \|\vec{T}(B)(x)\|_{\ell^s_M} dx$$

$$\leq \frac{(2\sqrt{n})^n}{\gamma} \frac{\|F\|_{L^1(\mathbf{R}^n, \ell^q_N)}}{\alpha} + \frac{2}{\alpha} \sum_j \int_{(Q_j^*)^c} \|\vec{T}(B_j)(x)\|_{\ell^s_M} dx. \quad (4.1.13)$$

It suffices to estimate the last sum. Denoting by y_j the center of the cube Q_j and using the fact that B_j has mean value zero over Q_j, for $x \notin Q_j^*$, using the LDCT and (4.1.2), we write

$$\vec{T}(B_j)(x) = \int_{Q_j} \left(\vec{K}(x-y) - \vec{K}(x-y_j)\right)(B_j(y)) dy. \quad (4.1.14)$$

For the argument below refer to Figure 3.3. Using (4.1.14) we write

$$\sum_j \int_{(Q_j^*)^c} \|\vec{T}(B_j)(x)\|_{\ell^s_M} dx$$

$$= \sum_j \int_{(Q_j^*)^c} \left\|\int_{Q_j} \left(\vec{K}(x-y) - \vec{K}(x-y_j)\right)(B_j(y)) dy\right\|_{\ell^s_M} dx$$

$$\leq \sum_j \int_{Q_j} \|B_j(y)\|_{\ell^q_N} \int_{(Q_j^*)^c} \|\vec{K}(x-y) - \vec{K}(x-y_j)\|_{\ell^q_N \to \ell^s_M} dx\, dy$$

$$\leq \sum_j \int_{Q_j} \|B_j(y)\|_{\ell^q_N} \int_{|x-y_j| \geq 2|y-y_j|} \|\vec{K}(x-y) - \vec{K}(x-y_j)\|_{\ell^q_N \to \ell^s_M} dx\, dy$$

$$\leq A_2 \sum_j \|B_j\|_{L^1(\mathbf{R}^n, \ell^q_N)}$$

$$\leq 2^{n+1} A_2 \|F\|_{L^1(\mathbf{R}^n, \ell^q_N)}, \quad (4.1.15)$$

where we used the fact that $|x-y_j| \geq 2|y-y_j|$ for all $x \notin Q_j^*$ and $y \in Q_j$ and (4.1.3). We also used Minkowski's integral inequality in the first inequality above. Combining (4.1.12), (4.1.13), (4.1.15), the estimates for the good and bad functions, we deduce

$$\left|\{x \in \mathbf{R}^n : \|\vec{T}(F)(x)\|_{\ell_M^s} > \alpha\}\right|$$
$$\leq \frac{2(A_2+B_\star)}{\alpha}\|F\|_{L^1(\mathbf{R}^n,\ell_N^q)} + \frac{(2\sqrt{n})^n}{\gamma}\frac{\|F\|_{L^1(\mathbf{R}^n,\ell_N^q)}}{\alpha} + \frac{2}{\alpha}2^{n+1}A_2\|F\|_{L^1(\mathbf{R}^n,\ell_N^q)}$$
$$\leq \left(2+(2\sqrt{n})^n 2^{n+1}+2^{n+2}\right)(A_2+B_\star)\frac{\|F\|_{L^1(\mathbf{R}^n,\ell_N^q)}}{\alpha}.$$

Combining the estimates leading to (4.1.12), (4.1.13), and (4.1.15) yields (4.1.8) with $C_n' = 2+(2\sqrt{n})^n 2^{n+1} + 2^{n+2}$. Thus $\vec{T}$ has an extension that maps $L^1(\mathbf{R}^n, \ell_N^q)$ to $L^{1,\infty}(\mathbf{R}^n, \ell_M^s)$ with constant at most $C_n'(A_2+B_\star)$. In obtaining the extension from $L^1(\mathbf{R}^n, \ell_N^q)$ to $L^{1,\infty}(\mathbf{R}^n, \ell_M^s)$ we use the completeness of weak L^1 (Proposition 1.2.12).

Next we interpolate between (4.1.8) and $\vec{T}: L^r(\mathbf{R}^n, \ell_N^q) \to L^r(\mathbf{R}^n, \ell_M^s)$. Using Exercise 1.3.7 we obtain

$$\|\vec{T}\|_{L^p(\mathbf{R}^n,\ell_N^q) \to L^p(\mathbf{R}^n,\ell_M^s)} \leq 2\left(\frac{p}{p-1} + \frac{p}{r-p}\right)^{\frac{1}{p}} C_n'(A_2+B_\star), \quad (4.1.16)$$

when $1 < p < r$. Taking $C(p) = 2p(p-1)^{-1}$ completes the argument when $r = \infty$.

We now consider the case $r < \infty$. Notice that $\vec{T}^t$ maps $L^{r'}(\mathbf{R}^n, \ell_M^{s'})$ to $L^{r'}(\mathbf{R}^n, \ell_N^{q'})$ with constant $B_\star$ (see Exercise 4.1.5). As the kernel of $\vec{T}^t$ satisfies the same estimates as that of $\vec{T}$, it follows that $\vec{T}^t$ also admits a bounded extension from $L^1(\mathbf{R}^n, \ell_M^{s'})$ to $L^{1,\infty}(\mathbf{R}^n, \ell_N^{q'})$ with bound at most $C_n'(A_2+B_\star)$. By interpolation (Exercise 1.3.7) we obtain

$$\|\vec{T}^t\|_{L^{p'}(\mathbf{R}^n,\ell_M^{s'}) \to L^{p'}(\mathbf{R}^n,\ell_N^{q'})} \leq 2\left(\frac{p'}{p'-1} + \frac{p'}{r'-p'}\right)^{\frac{1}{p'}} C_n'(A_2+B_\star), \quad (4.1.17)$$

when $1 < p' < r'$. In view of Exercise 4.1.5, the dual of estimate (4.1.17) implies

$$\|\vec{T}\|_{L^p(\mathbf{R}^n,\ell_N^q) \to L^p(\mathbf{R}^n,\ell_M^s)} \leq 2\left(\frac{p'}{p'-1} + \frac{p'}{r'-p'}\right)^{\frac{1}{p'}} C_n'(A_2+B_\star) \quad (4.1.18)$$

when $r < p < \infty$. Estimates (4.1.16) and (4.1.18) cover the entire region $1 < p < \infty$, but in order to obtain a better constant that does not depend on r we use interpolation.

Restricting (4.1.16) and (4.1.18) to smaller regions we obtain

$$\|\vec{T}\|_{L^p(\mathbf{R}^n,\ell_N^q) \to L^p(\mathbf{R}^n,\ell_M^s)} \leq 4C_n'(A_2+B_\star)(p')^{\frac{1}{p}}, \quad 1 < p < \frac{r+1}{2}, \quad (4.1.19)$$

$$\|\vec{T}\|_{L^p(\mathbf{R}^n,\ell_N^q) \to L^p(\mathbf{R}^n,\ell_M^s)} \leq 4C_n'(A_2+B_\star)p^{\frac{1}{p'}}, \quad 2r-1 < p < \infty. \quad (4.1.20)$$

4.1 The Vector-Valued Calderón–Zygmund Theorem

These estimates prove (4.1.9) for $p < \frac{r+1}{2}$ and for $p > 2r-1$. For the remaining values of p we interpolate between $p=1$ with bound $C'_n(A_2+B_\star)$ and $p=4r$ with bound $4C'_n(A_2+B_\star)(4r)^{1/(4r)'}$ coming from (4.1.20). Using again Exercise 1.3.7, for $1 < p < 4r$ we obtain

$$\|\vec{T}\|_{L^p(\mathbf{R}^n,\ell^q_N) \to L^p(\mathbf{R}^n,\ell^s_M)} \leq 8C'_n(A_2+B_\star)\left(\frac{p}{p-1}+\frac{p}{4r-p}\right)^{\frac{1}{p}}\left((4r)^{\frac{1}{(4r)'}}\right)^{\frac{1-\frac{1}{p}}{1-\frac{1}{4r}}}.$$

For $p \in [\frac{r+1}{2}, 2r-1]$, $(4r-p)^{-1} \leq (p-1)^{-1}$ and $4r \leq 8p$, so the quantity on the right is bounded by $16C'_n(A_2+B_\star)(p/(p-1))^{1/p}(8p)^{1-1/p}$. This yields

$$\|\vec{T}\|_{L^p(\mathbf{R}^n,\ell^q_N) \to L^p(\mathbf{R}^n,\ell^s_M)} \leq \frac{128C'_n(A_2+B_\star)p}{(p-1)^{1/p}}, \quad \frac{r+1}{2} \leq p \leq 2r-1. \quad (4.1.21)$$

Combining (4.1.19), (4.1.20), and (4.1.21) we deduce (4.1.9). □

We remark that instead of assuming that each coordinate of F is a finite linear combination of characteristic functions of dyadic cubes, we could have assumed that F lies in $L^1(\mathbf{R}^n, \ell^q_N) \cap L^r(\mathbf{R}^n, \ell^q_N)$. In this case, one would have to show a property analogous to (3.6.3).

Exercises

4.1.1. Let $Y = (y_{ij})_{1 \leq i \leq M, 1 \leq j \leq N}$ be a complex $M \times N$ matrix and let $1 \leq q, s \leq \infty$. Prove the following:

$$\sup_{1 \leq j \leq N}\left(|y_{1j}|^s + \cdots + |y_{Mj}|^s\right)^{\frac{1}{s}} \leq \|Y\|_{\ell^q_N \to \ell^s_M} \leq \left(\sum_{i=1}^M \left(|y_{i1}|^{q'} + \cdots + |y_{iN}|^{q'}\right)^{\frac{s}{q'}}\right)^{\frac{1}{s}},$$

with the obvious modification when $s = \infty$ or $q = 1$. Notice that the (worse) estimates

$$\sup_{1 \leq i \leq M}\sup_{1 \leq j \leq N}|y_{ij}| \leq \|Y\|_{\ell^q_N \to \ell^s_M} \leq \sum_{1 \leq i \leq M}\sum_{1 \leq j \leq N}|y_{ij}|$$

are valid for all $1 \leq q, s \leq \infty$.

4.1.2. Let $0 < p \leq \infty$, $N \in \mathbf{Z}^+$, and $1 \leq r \leq \infty$. Prove that $L^p(\mathbf{R}^n, \ell^r_N)$ is a Banach space if $p \geq 1$ and a quasi-Banach space when $p < 1$. Also show that $L^{1,\infty}(\mathbf{R}^n, \ell^r_N)$ is a quasi-Banach space.
[*Hint:* $\sup_{1 \leq j \leq N}\|f_j\|_{L^p}^{\min(1,p)} \leq \|\|\{f_j\}_{j=1}^N\|_{\ell^q_N}\|_{L^p}^{\min(1,p)} \leq \sum_{j=1}^N \|f_j\|_{L^p}^{\min(1,p)}.$]

4.1.3. Let $\mathscr{D}$ be a dense subspace of $L^p(\mathbf{R}^n)$, where $0 < p < \infty$. Let $1 \leq q \leq \infty$. Show that the space $\underbrace{\mathscr{D} \times \cdots \times \mathscr{D}}_{N \text{ times}}$ is dense in $L^p(\mathbf{R}^n, \ell^q_N)$.

4.1.4. Let $\{a_j\}_{j=1}^N$ be a finite sequence of complex numbers. Prove that for any q satisfying $1 \leq q \leq \infty$ we have

$$\left\|\{a_j\}_{j=1}^N\right\|_{\ell_N^q} = \max_{\|\{b_j\}_{j=1}^N\|_{\ell_N^{q'}}=1} \left|\sum_{j=1}^N a_j b_j\right|.$$

4.1.5. Prove that for any $1 \leq q \leq \infty$, $1 < p < \infty$, and $f_j \in L^p(\mathbf{R}^n)$ we have

$$\left\|\{f_j\}_{j=1}^N\right\|_{L^p(\mathbf{R}^n,\ell_N^q)} = \max_{\|\{g_j\}_{j=1}^N\|_{L^{p'}(\mathbf{R}^n,\ell_N^{q'})}=1} \left|\int_{\mathbf{R}^n} \sum_{j=1}^N f_j g_j \, dx\right|.$$

4.2 Applications of Vector-Valued Inequalities

An important consequence of Theorem 4.1.1 is the following:

Corollary 4.2.1. *Fix $A, B > 0$, $1 < r < \infty$, and let K_j be a sequence of functions on $\mathbf{R}^n \setminus \{0\}$ that satisfy, for some $A_1^j < \infty$,*

$$|K_j(x)| \leq A_1^j |x|^{-n}, \qquad x \neq 0, \tag{4.2.1}$$

$$\lim_{\delta_k \to 0} \int_{\delta_k \leq |x| \leq 1} K_j(x) \, dx = L_j, \tag{4.2.2}$$

for certain complex constants L_j and a sequence $\delta_k \in (0,1)$ that tends to zero, and

$$\sup_{y \in \mathbf{R}^n \setminus \{0\}} \int_{|x| \geq 2|y|} \sup_{j \in \mathbf{Z}} |K_j(x-y) - K_j(x)| \, dx \leq A_2. \tag{4.2.3}$$

Let $W_j \in \mathscr{S}'(\mathbf{R}^n)$ be associated with K_j as in (3.3.8) and let T_j be the operator given by convolution with W_j. Assume that $\widehat{W_j}$ coincide with bounded functions satisfying $\sup_j \|\widehat{W_j}\|_{L^\infty} \leq B$. Then for all $1 < p < \infty$, T_j admit bounded extensions from $L^p(\mathbf{R}^n)$ to itself and from $L^1(\mathbf{R}^n)$ to $L^{1,\infty}(\mathbf{R}^n)$, and there exist $C_n, C_n' > 0$ such that for all $f_j \in L^p(\mathbf{R}^n)$ we have

$$\left\|\left(\sum_j |T_j(f_j)|^r\right)^{\frac{1}{r}}\right\|_{L^{1,\infty}} \leq C_n' \max\left(r, \frac{1}{r-1}\right)(A_2+B)\left\|\left(\sum_j |f_j|^r\right)^{\frac{1}{r}}\right\|_{L^1},$$

$$\left\|\left(\sum_j |T_j(f_j)|^r\right)^{\frac{1}{r}}\right\|_{L^p} \leq C_n \max\left(r, \frac{1}{r-1}\right)\max\left(p, \frac{1}{p-1}\right)(A_2+B)\left\|\left(\sum_j |f_j|^r\right)^{\frac{1}{r}}\right\|_{L^p}.$$

Proof. The assumption $\sup_j \|\widehat{W_j}\|_{L^\infty} \leq B$ implies that all T_j are L^2 bounded with norms at most B. It follows from Theorem 3.6.1 that all T_j are of weak type $(1,1)$

4.2 Applications of Vector-Valued Inequalities

with bounds at most $C_n(A_2+B)$ and also bounded on L^r with bounds at most $C_n \max(r,(r-1)^{-1})(A_2+B)$, uniformly in j.

It suffices to prove the claimed inequalities for $|j| \leq N$, where $N \in \mathbf{Z}^+$. Then the LMCT implies the conclusion for all $j \in \mathbf{Z}$ (in the case of $L^{1,\infty}$ refer to Exercise 1.2.2). We fix $N \in \mathbf{Z}^+$ and we define

$$\vec{T}(\{f_j\}_{|j|\leq N}) = \{f_j * W_j\}_{|j|\leq N}$$

for $\{f_j\}_j$ in $\mathscr{S}(\mathbf{R}^n)^{2N+1}$ which is a dense subspace of $L^r(\mathbf{R}^n, \ell^r_{2N+1})$. It is immediate to verify the second claimed inequality with $p = r$, i.e., that $\vec{T}$ maps $L^r(\mathbf{R}^n, \ell^r_{2N+1})$ to itself with norm

$$B_\star \leq C_n \max(r,(r-1)^{-1})(A_2+B).$$

The kernel of $\vec{T}$ is $\vec{K}$ in $L(\ell^r_{2N+1}, \ell^r_{2N+1})$ is a diagonal matrix defined by

$$\vec{K}(x)(\{t_j\}_{|j|\leq N}) = \{K_j(x)t_j\}_{|j|\leq N}, \qquad \{t_j\}_{|j|\leq N} \in \ell^r_{2N+1}.$$

Obviously, we have

$$\big\|\vec{K}(x)\big\|_{\ell^r_{2N+1} \to \ell^r_{2N+1}} \leq \sup_j |K_j(x)|.$$

This implies that

$$\big\|\vec{K}(x-y) - \vec{K}(x)\big\|_{\ell^r_{2N+1} \to \ell^r_{2N+1}} \leq \sup_j |K_j(x-y) - K_j(x)|,$$

and therefore condition (4.1.3) holds for $\vec{K}$ as a consequence of (4.2.3). Moreover, (4.1.2) and (4.1.6) with $\vec{K}_0 = \{L_j\}_j$ are also valid for this $\vec{K}$, in view of assumptions (4.2.1) and (4.2.2). The desired conclusion follows from Theorem 4.1.1. □

Remark 4.2.2. Note that in Corollary 4.2.1, if $A_1 = \sup_j A_1^j < \infty$, the hypothesis that $\sup_{j \in \mathbf{Z}} \|\widehat{W_j}\|_{L^\infty} \leq B < \infty$ could have been replaced by

$$\sup_{j \in \mathbf{Z}} \sup_{\delta > 0} \sup_{N > \delta} \left| \int_{\delta \leq |x| \leq N} K_j(x)\,dx \right| \leq A_3. \tag{4.2.4}$$

In that case the constant $A_2 + B$ in the conclusion of Corollary 4.2.1 should be replaced by $A_1 + A_2 + A_3$. This is because under assumptions (4.2.1), (4.2.3), and (4.2.4) it follows from Theorem 3.4.2 that the operators T_j are bounded on L^2 by some constant $B \leq c_n(A_1+A_2+A_3)$ uniformly in j.

Corollary 4.2.3. *Let K be a function on $\mathbf{R}^n \setminus \{0\}$ that satisfies*

$$|K(x)| \leq A_1 |x|^{-n}, \qquad x \neq 0,$$

$$\lim_{\delta_k \to 0} \int_{\delta_k \leq |x| \leq 1} K(x)\,dx = L,$$

for some $A_1 < \infty$, $L \in \mathbf{C}$, and $\delta_k \in (0,1)$ that tend to zero as $k \to \infty$, and

$$\sup_{y\in \mathbf{R}^n\setminus\{0\}} \int_{|x|\geq 2|y|} |K(x-y)-K(x)|\,dx \leq A_2. \qquad (4.2.5)$$

Let $W \in \mathscr{S}'(\mathbf{R}^n)$ be associated with K as in (3.3.8) and let T be the operator given by convolution with W. Let $1 < r < \infty$ and $B > 0$. Assume that $\widehat{W}$ coincides with a bounded function satisfying $\|\widehat{W}\|_{L^\infty} \leq B$. Then for all $1 < p < \infty$, T admits a bounded extension from $L^p(\mathbf{R}^n)$ to itself and from $L^1(\mathbf{R}^n)$ to $L^{1,\infty}(\mathbf{R}^n)$, and there exist positive constants C_n, C_n' such that for all $f_j \in L^p(\mathbf{R}^n)$ we have

$$\Big\|\Big(\sum_j |T(f_j)|^r\Big)^{\frac{1}{r}}\Big\|_{L^{1,\infty}} \leq C_n' \max\Big(r, \frac{1}{r-1}\Big)(A_2+B)\Big\|\Big(\sum_j |f_j|^r\Big)^{\frac{1}{r}}\Big\|_{L^1},$$

$$\Big\|\Big(\sum_j |T(f_j)|^r\Big)^{\frac{1}{r}}\Big\|_{L^p} \leq C_n \max\Big(r, \frac{1}{r-1}\Big)\max\Big(p, \frac{1}{p-1}\Big)(A_2+B)\Big\|\Big(\sum_j |f_j|^r\Big)^{\frac{1}{r}}\Big\|_{L^p}.$$

In particular, these inequalities are valid for the Hilbert transform and the Riesz transforms.

Proof. Apply Corollary 4.2.1 with $K_j = K$, $W_j = W$, and $T_j = T$ for all j. $\square$

We now discuss an application of Theorem 4.1.1 when $r = \infty$. This provides another proof of the boundedness of the maximal operator appearing in Corollary 2.5.2 and of the Hardy–Littlewood maximal operator.

Example 4.2.4. Let $A, \gamma > 0$ and G be a measurable function on $\mathbf{R}^n \times \mathbf{R}^+$ satisfying

$$|G(x,t)| \leq \frac{1}{t^n}\frac{A}{(1+|x|/t)^{n+\gamma}} = A\frac{\Phi(x/t)}{t^n} = A\Phi_t(x),$$

for all $x \in \mathbf{R}^n$ and $t > 0$, where

$$\Phi(x) = (1+|x|)^{-n-\gamma}.$$

Consider the maximal operator

$$\mathcal{N}(f) = \sup_{t>0} |G(\cdot,t) * f|$$

defined for f in $\cup_{1\leq q\leq \infty} L^q(\mathbf{R}^n)$. By Corollary 2.5.2, the operator $\mathcal{N}$ is pointwise controlled by a constant multiple of the Hardy–Littlewood maximal operator $\mathcal{M}$, and, for a certain function G, it essentially coincides with it. Indeed, for the choice $G(x,t) = v_n^{-1} t^{-n} \chi_{|x|\leq t}$ we have $\mathcal{N}(|f|) = \mathcal{M}(f)$. Next we note that

$$A^{-1}\mathcal{N}(|f|) \leq \sup_{t>0} \Phi_t * |f| \leq 2^n \sup_{j\in \mathbf{Z}} \Phi_{2^j} * |f|, \qquad (4.2.6)$$

since for $2^{j-1} \leq t < 2^j$ ($j \in \mathbf{Z}$) and $x \in \mathbf{R}^n$ we have

$$\Phi_t(x) = \frac{1}{t^n}\frac{1}{(1+|x|/t)^{n+\gamma}} \leq \frac{1}{(2^{j-1})^n}\frac{1}{(1+|x|/2^j)^{n+\gamma}} = 2^n \Phi_{2^j}(x).$$

4.2 Applications of Vector-Valued Inequalities

For an odd positive integer M consider the vector-valued operator

$$f \mapsto (\Phi_{2^j} * f)_{|j| \leq [\frac{M}{2}]}$$

which we think of as a mapping from $L^p(\mathbf{R}^n) = L^p(\mathbf{R}^n, \mathbf{C})$ to $L^p(\mathbf{R}^n, \ell_M^\infty)$. The kernel of this operator is the $M \times 1$ matrix

$$\vec{K}(x) = (\Phi_{2^{-[M/2]}}(x), \ldots, \Phi_{2^{[M/2]}}(x))^t, \quad x \in \mathbf{R}^n,$$

which acts on complex numbers a as follows:

$$a \mapsto (\Phi_{2^{-[M/2]}}(x), \ldots, \Phi_{2^{[M/2]}}(x))^t a = (\Phi_{2^{-[M/2]}}(x)a, \ldots, \Phi_{2^{[M/2]}}(x)a)^t.$$

This operator maps $\mathbf{C}$ to ℓ_M^∞ with norm

$$\|\vec{K}(x)\|_{\mathbf{C} \to \ell_M^\infty} = \sup_{|j| \leq [\frac{M}{2}]} |\Phi_{2^j}(x)|.$$

Now (4.1.2) is valid as

$$\sup_{|j| \leq [\frac{M}{2}]} |\Phi_{2^j}(x)| = \sup_{|j| \leq [\frac{M}{2}]} \frac{1}{2^{jn}} \frac{1}{(1+|x|/2^j)^{n+\gamma}} \leq \frac{1}{|x|^n}, \quad (4.2.7)$$

and likewise, for $x \neq 0$, we have

$$\sup_{|j| \leq [\frac{M}{2}]} |\nabla \Phi_{2^j}(x)| = \sup_{|j| \leq [\frac{M}{2}]} \frac{1}{2^{j(n+1)}} \frac{n+\gamma}{(1+\frac{|x|}{2^j})^{n+\gamma+1}} \left|\left(\frac{x_1}{|x|}, \ldots, \frac{x_n}{|x|}\right)\right| \leq \frac{n+\gamma}{|x|^{n+1}}.$$

Now given $x, y \in \mathbf{R}^n$ such that $|x| \geq 2|y|$, by the mean value theorem, for each j there is a ξ_j on the line segment joining y to the origin, such that

$$\|\vec{K}(x-y) - \vec{K}(x)\|_{\mathbf{C} \to \ell_M^\infty} = \sup_{|j| \leq [\frac{M}{2}]} |\Phi_{2^j}(x-y) - \Phi_{2^j}(x)|$$

$$= \sup_{|j| \leq [\frac{M}{2}]} |\nabla \Phi_{2^j}(x - \xi_j) \cdot y|$$

$$\leq \sup_{|j| \leq [\frac{M}{2}]} \frac{(n+\gamma)|y|}{|x - \xi_j|^{n+1}}$$

$$\leq (n+\gamma) 2^{n+1} \frac{|y|}{|x|^{n+1}} \quad (4.2.8)$$

as $|x - \xi_j| \geq |x| - |\xi_j| \geq |x| - |y| \geq |x|/2$. This estimate implies the validity of (4.1.3) for some constant A_2 that depends only on n and γ.

Also notice that (4.1.5) is valid with $A_3 = \|\Phi\|_{L^1}$. Thus (4.1.6) holds for any sequence $\delta_k \downarrow 0$. Finally, the estimate

$$\Big\| \sup_{|j|\leq [\frac{M}{2}]} |\Phi_{2^j} * f| \Big\|_{L^\infty} \leq B_\star \|f\|_{L^\infty}$$

is valid with $B_\star = \|\Phi\|_{L^1}$ and is obtained by inserting the L^∞ norm inside the supremum and applying Theorem 1.6.6 (with $p = \infty$). An application of Theorem 4.1.1 (with $r = \infty$) yields the inequalities

$$\Big\| \sup_{|j|\leq [\frac{M}{2}]} |\Phi_{2^j} * f| \Big\|_{L^{1,\infty}} \leq C'_{n,\gamma} \|f\|_{L^1} \tag{4.2.9}$$

and

$$\Big\| \sup_{|j|\leq [\frac{M}{2}]} |\Phi_{2^j} * f| \Big\|_{L^p} \leq C_{n,\gamma} \frac{p}{p-1} \|f\|_{L^p}. \tag{4.2.10}$$

Letting $M \to \infty$ in (4.2.9) and (4.2.10) (through the odd integers) and using (4.2.6) we deduce the boundedness of $\mathcal{N}$ from $L^1(\mathbf{R}^n)$ to $L^{1,\infty}(\mathbf{R}^n)$ and from $L^p(\mathbf{R}^n)$ to itself when $1 < p < \infty$ with bounds proportional to those in (4.2.9) and (4.2.10) times the constant A. These estimates provide a proof of Corollary 1.4.7, i.e., of the boundedness of the Hardy–Littlewood maximal operator on $L^p(\mathbf{R}^n)$ that is not based on a covering lemma but on Calderón–Zygmund theory.

Exercises

4.2.1. Assume that in Corollary 4.2.1, the hypothesis $\sup_j \|\widehat{W_j}\|_{L^\infty} \leq B$ is replaced by the assumption that $\sup_j \|T_j\|_{L^r \to L^r} \leq B$. Show that the conclusion of this corollary can be strengthened as follows: For any $1 < p < \infty$, T_j admit bounded extensions from $L^p(\mathbf{R}^n)$ to itself and from $L^1(\mathbf{R}^n)$ to $L^{1,\infty}(\mathbf{R}^n)$, and there exist $C_n, C'_n > 0$ such that for all $f_j \in L^p(\mathbf{R}^n)$ we have

$$\Big\| \Big(\sum_j |T_j(f_j)|^r \Big)^{\frac{1}{r}} \Big\|_{L^{1,\infty}} \leq C'_n (A_2 + B) \Big\| \Big(\sum_j |f_j|^r \Big)^{\frac{1}{r}} \Big\|_{L^1},$$

$$\Big\| \Big(\sum_j |T_j(f_j)|^r \Big)^{\frac{1}{r}} \Big\|_{L^p} \leq C_n \max\Big(p, \frac{1}{p-1}\Big)(A_2 + B) \Big\| \Big(\sum_j |f_j|^r \Big)^{\frac{1}{r}} \Big\|_{L^p}.$$

4.2.2. For each $j \in \mathbf{Z}$ let I_j be an open interval in $\mathbf{R}$ (which could be half infinite or the entire line) and let T_j be the operator given by convolution with $(\chi_{I_j})^\vee$. Prove that there exists a constant $C > 0$ such that for all $1 < p, r < \infty$ and for all $f_j \in L^p(\mathbf{R})$ we have

$$\Big\| \Big(\sum_j |T_j(f_j)|^r \Big)^{\frac{1}{r}} \Big\|_{L^{1,\infty}(\mathbf{R})} \leq C \max\Big(r, \frac{1}{r-1}\Big) \Big\| \Big(\sum_j |f_j|^r \Big)^{\frac{1}{r}} \Big\|_{L^1(\mathbf{R})},$$

$$\Big\| \Big(\sum_j |T_j(f_j)|^r \Big)^{\frac{1}{r}} \Big\|_{L^p(\mathbf{R})} \leq C \max\Big(r, \frac{1}{r-1}\Big) \max\Big(p, \frac{1}{p-1}\Big) \Big\| \Big(\sum_j |f_j|^r \Big)^{\frac{1}{r}} \Big\|_{L^p(\mathbf{R})}.$$

4.2 Applications of Vector-Valued Inequalities

[*Hint:* Let I be the identity operator, H be the Hilbert transform, $M^a(f)(x) = f(x)e^{2\pi i a x}$, and $I_j = \chi_{(a_j,b_j)}$. Show that: (i) if $-\infty < a_j < b_j = +\infty$, then $T_j = \frac{1}{2}(I + iM^{a_j}HM^{-a_j})$; (ii) if $-\infty = a_j < b_j < +\infty$, then $T_j = \frac{1}{2}(I - iM^{b_j}HM^{-b_j})$; (iii) if $-\infty < a_j < b_j < +\infty$, then $T_j = \frac{i}{2}(M^{a_j}HM^{-a_j} - M^{b_j}HM^{-b_j})$; (iv) if $a_j = -\infty$ and $b_j = +\infty$, then $T_j = I$. Split the intervals in four groups and use Corollary 4.2.3.]

4.2.3. Let $R_j = (a_j^1, b_j^1) \times \cdots \times (a_j^n, b_j^n)$, where $-\infty \le a_j^k < b_j^k \le +\infty$, and define an operator S_j given by convolution by $(\chi_{R_j})^\vee$. Prove that there exists a constant $C_n < \infty$ such that for $1 < p, r < \infty$ and all functions f_j in $L^p(\mathbf{R}^n)$ we have

$$\left\| \left(\sum_j |S_j(f_j)|^r \right)^{\frac{1}{r}} \right\|_{L^p(\mathbf{R}^n)} \le C_n \max\left(r, \frac{1}{r-1}\right)^n \max\left(p, \frac{1}{p-1}\right)^n \left\| \left(\sum_j |f_j|^r \right)^{\frac{1}{r}} \right\|_{L^p(\mathbf{R}^n)}.$$

[*Hint:* Write S_j as the composition of n one-dimensional operators and use the preceding exercise.]

4.2.4. For fixed θ in $\mathbf{S}^{n-1}$ let $H_\theta = \{x \in \mathbf{R}^n : x \cdot \theta > 0\}$ be a half space of $\mathbf{R}^n$. Let T_{H_θ} be an operator given by convolution with $(\chi_{H_\theta})^\vee$. Prove that there exists a $C < \infty$ (independent of n) such that for $1 < p, r < \infty$ and all f_j in $L^p(\mathbf{R}^n)$ we have

$$\left\| \left(\sum_j |T_{H_\theta}(f_j)|^r \right)^{\frac{1}{r}} \right\|_{L^p(\mathbf{R}^n)} \le C \max\left(r, \frac{1}{r-1}\right) \max\left(p, \frac{1}{p-1}\right) \left\| \left(\sum_j |f_j|^r \right)^{\frac{1}{r}} \right\|_{L^p(\mathbf{R}^n)}.$$

Conclude the validity of this inequality when H_θ is replaced by $z + H_\theta$ for any $z \in \mathbf{R}^n$ with the same constant on the right. [*Hint:* Apply a rotation.]

4.2.5. Let A be the set of points inside an open angle whose vertex is at zero and let a_j be a sequence of points in $\mathbf{R}^2$. Consider the operator T_{a_j+A} given by convolution with $(\chi_{a_j+A})^\vee$. (Here $a_j + A$ is the set obtained by translating the set A by a_j.) Prove that there exists a $C < \infty$ such that for $1 < p, r < \infty$ and all f_j in $L^p(\mathbf{R}^2)$ we have

$$\left\| \left(\sum_j |T_{a_j+A}(f_j)|^r \right)^{\frac{1}{r}} \right\|_{L^p(\mathbf{R}^2)} \le \left[C \max\left(r, \tfrac{1}{r-1}\right) \max\left(p, \tfrac{1}{p-1}\right) \right]^2 \left\| \left(\sum_j |f_j|^r \right)^{\frac{1}{r}} \right\|_{L^p(\mathbf{R}^2)}.$$

[*Hint:* Use Exercise 4.2.4.]

4.2.6. An n-simplex is the convex hull of $n+1$ points $z_0, z_1, \ldots, z_n$ in $\mathbf{R}^n$ that are *affinely independent*, which means that $z_1 - z_0, \ldots, z_n - z_0$ are linearly independent. Prove that the characteristic function of an n-simplex lies in $\mathcal{M}_p(\mathbf{R}^n)$ for $1 < p < \infty$. Use this information to show that the characteristic function of any polyhedron also lies in $\mathcal{M}_p(\mathbf{R}^n)$ for $1 < p < \infty$. (Compare with Exercise 2.8.6.) [*Hint:* An n-simplex is the intersection of the characteristic functions of $n+1$ half-spaces. A polyhedron is a finite union of n-simplices.]

4.2.7. Fix an n-simplex Q and a sequence of points a_j in $\mathbf{R}^n$. Consider the operator T_{a_j+Q} given by convolution with $(\chi_{a_j+Q})^\vee$. Prove that there is a constant C (independent of the dimension n) such that for $1 < p, r < \infty$ and f_j in $L^p(\mathbf{R}^n)$ we have

$$\Big\|\Big(\sum_j |T_{a_j+Q}(f_j)|^r\Big)^{\frac{1}{r}}\Big\|_{L^p(\mathbf{R}^n)} \leq \Big[C\max(r,\tfrac{1}{r-1})\max(p,\tfrac{1}{p-1})\Big]^{n+1}\Big\|\Big(\sum_j |f_j|^r\Big)^{\frac{1}{r}}\Big\|_{L^p(\mathbf{R}^n)}.$$

Deduce an analogous estimate if Q is a polyhedron with an additional factor on the right of the number of n-simplices that comprise it. [*Hint:* Use Exercises 4.2.4, 4.2.6 and that $T_{a_j+Q} = M^{a_j} T_Q M^{-a_j}$; here $M^a(f)(x) = e^{2\pi i x \cdot a} f(x)$.]

4.3 A Matrix-Valued Calderón–Zygmund Theorem and Its Applications

We saw in Example 4.2.4 how to obtain L^p bounds for maximal averages via the vector-valued Calderón–Zygmund theorem. In this section we derive a matrix-valued version of Theorem 4.1.1 and from this we deduce $L^p(\ell^r)$ bounds for vectors of maximal functions. To formulate this extension in a general setting, we consider an $L \times M$ matrix

$$\vec{K} = \begin{pmatrix} K_{11} & K_{12} & \cdots & K_{1M} \\ K_{21} & K_{22} & \cdots & K_{2M} \\ \vdots & \vdots & \cdots & \vdots \\ K_{L1} & K_{L2} & \cdots & K_{LM} \end{pmatrix}$$

of integrable functions defined on $\mathbf{R}^n$. Let $\vec{F} = (f_{ij})$ be an $M \times N$ matrix of L^p functions. We would like to study linear operators of the form

$$\begin{pmatrix} f_{11} & f_{12} & \cdots & f_{1N} \\ f_{21} & f_{22} & \cdots & f_{2N} \\ \vdots & \vdots & \cdots & \vdots \\ f_{M1} & f_{M2} & \cdots & f_{MN} \end{pmatrix} \mapsto \begin{pmatrix} K_{11} & K_{12} & \cdots & K_{1M} \\ K_{21} & K_{22} & \cdots & K_{2M} \\ \vdots & \vdots & \cdots & \vdots \\ K_{L1} & K_{L2} & \cdots & K_{LM} \end{pmatrix} * \begin{pmatrix} f_{11} & f_{12} & \cdots & f_{1N} \\ f_{21} & f_{22} & \cdots & f_{2N} \\ \vdots & \vdots & \cdots & \vdots \\ f_{M1} & f_{M2} & \cdots & f_{MN} \end{pmatrix},$$

where the preceding convolution of matrices is defined as follows:

$$\vec{K} * \vec{F} = \begin{pmatrix} \sum_{m=1}^{M} K_{1m} * f_{m1} & \sum_{m=1}^{M} K_{1m} * f_{m2} & \cdots & \sum_{m=1}^{M} K_{1m} * f_{mN} \\ \sum_{m=1}^{M} K_{2m} * f_{m1} & \sum_{m=1}^{M} K_{2m} * f_{m2} & \cdots & \sum_{m=1}^{M} K_{2m} * f_{mN} \\ \vdots & \vdots & \cdots & \vdots \\ \sum_{m=1}^{M} K_{Lm} * f_{m1} & \sum_{m=1}^{M} K_{Lm} * f_{m2} & \cdots & \sum_{m=1}^{M} K_{Lm} * f_{mN} \end{pmatrix}.$$

We apply a norm on the matrices $\vec{F}$ and $\vec{K} * \vec{F}$ that produces scalar-valued functions whose L^p norms can be evaluated. For $1 \leq q, r \leq \infty$ we introduce the $\ell_N^q(\ell_M^r)$ of an $M \times N$ matrix, to be the ℓ^q norm of the vector formed by the ℓ^r norms of its columns. For $1 \leq q, r \leq \infty$ and $1 \leq p < \infty$, we introduce the space $L^p(\mathbf{R}^n, \ell_N^q(\ell_M^r))$

4.3 A Matrix-Valued Calderón–Zygmund Theorem and Its Applications

of all $M \times N$ matrices of measurable functions $\vec{F} = (f_{mj})_{1\le m\le M, 1\le j\le N}$ on $\mathbf{R}^n$ such that

$$\|\vec{F}\|_{L^p(\mathbf{R}^n,\ell^q_N(\ell^r_M))} = \Big\|\|\vec{F}\|_{\ell^q_N(\ell^r_M)}\Big\|_{L^p(\mathbf{R}^n)} =: \bigg\|\bigg(\sum_{j=1}^N \Big[\sum_{m=1}^M |f_{mj}|^r\Big]^{\frac{q}{r}}\bigg)^{\frac{1}{q}}\bigg\|_{L^p(\mathbf{R}^n)} < \infty,$$

with the obvious modification when $q = \infty$ or $r = \infty$. Likewise we define the space $L^{1,\infty}(\mathbf{R}^n, \ell^q_N(\ell^r_M))$.

In this section, for $1 < p < \infty$, we are interested in estimates of the form

$$\|\vec{K}*\vec{F}\|_{L^p(\ell^q_N(\ell^\infty_L))} \le C \|\vec{F}\|_{L^p(\ell^q_N(\ell^\infty_M))} \tag{4.3.1}$$

with C independent of $\vec{F}, L, M, N$. When $1 < p < \infty$, the dual space of $L^p(\mathbf{R}^n, \ell^q_N(\ell^\infty_M))$ is $L^{p'}(\mathbf{R}^n, \ell^{q'}_N(\ell^1_M))$, where

$$\|\vec{F}\|_{L^{p'}(\mathbf{R}^n, \ell^{q'}_N(\ell^1_M))} = \bigg\|\bigg(\sum_{j=1}^N \Big[\sum_{m=1}^M |f_{mj}|\Big]^{q'}\bigg)^{\frac{1}{q'}}\bigg\|_{L^{p'}(\mathbf{R}^n)} < \infty$$

with the obvious modification when $q = 1$. The dual estimate to (4.3.1) is

$$\|\vec{\widetilde{K}}^t * \vec{G}\|_{L^{p'}(\ell^{q'}_N(\ell^1_M))} \le C \|\vec{G}\|_{L^{p'}(\ell^{q'}_N(\ell^1_L))}, \tag{4.3.2}$$

where $\vec{\widetilde{K}}^t$ is the transpose of the matrix obtained by replacing each K_{lm} by $\widetilde{K_{lm}}$. Note that estimate (4.3.2) is completely equivalent to (4.3.1) (Exercise 4.3.4), so we can focus on either one.

We discuss two important examples of the situation just mentioned: (4.3.2) and (4.3.1), both with $M = 1$.

Example 4.3.1. (a) Let $\vec{K} = (K_1, \ldots, K_L)$ be a $1 \times L$ matrix of integrable functions defined on $\mathbf{R}^n$. Consider an operator of the form

$$\begin{pmatrix} f_{11} & f_{12} & \cdots & f_{1N} \\ f_{21} & f_{22} & \cdots & f_{2N} \\ \vdots & \vdots & \cdots & \vdots \\ f_{L1} & f_{L2} & \cdots & f_{LN} \end{pmatrix} \mapsto \vec{K} * \begin{pmatrix} f_{11} & f_{12} & \cdots & f_{1N} \\ f_{21} & f_{22} & \cdots & f_{2N} \\ \vdots & \vdots & \cdots & \vdots \\ f_{L1} & f_{L2} & \cdots & f_{LN} \end{pmatrix} = \bigg(\sum_{l=1}^L K_l * f_{l1}, \ldots, \sum_{l=1}^L K_l * f_{lN}\bigg)$$

acting on functions $\{\{f_{ij}\}_{i=1}^L\}_{j=1}^N$ in $L^p(\mathbf{R}^n, \ell^q_N(\ell^1_L))$ for $1 < p < \infty$. We are interested in obtaining bounds for this operator from $L^p(\mathbf{R}^n, \ell^q_N(\ell^1_L))$ to $L^p(\mathbf{R}^n, \ell^q_N)$. Whenever $x \in \mathbf{R}^n$ is such that all $K_l(x)$ are defined, $\vec{K}(x) = (K_1(x), \ldots, K_L(x))$ is the linear mapping acting on $L \times N$ matrices of complex numbers as follows:

$$\bigl(K_1(x),\ldots,K_L(x)\bigr)\begin{pmatrix} a_{11} & a_{12} & \cdots & a_{1N} \\ a_{21} & a_{22} & \cdots & a_{2N} \\ \vdots & \vdots & \cdots & \vdots \\ a_{L1} & a_{L2} & \cdots & a_{LN} \end{pmatrix} = \left(\sum_{l=1}^{L} K_l(x)a_{l1},\ldots,\sum_{l=1}^{L} K_l(x)a_{lN}\right).$$

For every $x \in \mathbf{R}^n$ for which all $K_l(x)$ are defined, $\vec{K}(x)$ maps $\ell_N^q(\ell_L^1)$ to ℓ_N^q and satisfies

$$\|\vec{K}(x)\|_{\ell_N^q(\ell_L^1) \to \ell_N^q} \leq \sup_{1 \leq l \leq L} |K_l(x)|. \tag{4.3.3}$$

(b) Now let $\vec{K} = (K_1,\ldots,K_L)^t$ be an $L \times 1$ matrix of integrable functions defined on $\mathbf{R}^n$. For $\{f_j\}_{j=1}^N$ in $L^p(\mathbf{R}^n, \ell_N^q)$ consider the linear operator

$$\bigl(f_1, f_2, \ldots, f_N\bigr) \mapsto \begin{pmatrix} K_1 \\ K_2 \\ \vdots \\ K_L \end{pmatrix} * \bigl(f_1 \; f_2 \; \cdots \; f_N\bigr) = \begin{pmatrix} K_1 * f_1 & K_1 * f_2 & \cdots & K_1 * f_N \\ K_2 * f_1 & K_2 * f_2 & \cdots & K_2 * f_N \\ \vdots & \vdots & \cdots & \vdots \\ K_L * f_1 & K_L * f_2 & \cdots & K_L * f_N \end{pmatrix}$$

for which we are interested in obtaining bounds from $L^p(\mathbf{R}^n, \ell_N^q)$ to $L^p(\mathbf{R}^n, \ell_N^q(\ell_L^\infty))$. For all $x \in \mathbf{R}^n$ for which all $K_j(x)$ are defined, $\vec{K}(x)$ is the linear mapping

$$\begin{pmatrix} K_1(x) \\ K_2(x) \\ \vdots \\ K_L(x) \end{pmatrix} \bigl(a_1 \; a_2 \; \cdots \; a_N\bigr) = \begin{pmatrix} K_1(x)a_1 & K_1(x)a_2 & \cdots & K_1(x)a_N \\ K_2(x)a_1 & K_2(x)a_2 & \cdots & K_2(x)a_N \\ \vdots & \vdots & \cdots & \vdots \\ K_L(x)a_1 & K_L(x)a_2 & \cdots & K_L(x)a_N \end{pmatrix}$$

acting on sequences of complex numbers. This maps ℓ_N^q to $\ell_N^q(\ell_L^\infty)$ with norm

$$\|\vec{K}(x)\|_{\ell_N^q \to \ell_N^q(\ell_L^\infty)} = \sup_{1 \leq l \leq L} |K_l(x)|. \tag{4.3.4}$$

Motivated by the discussion in the preceding example, and in particular estimates (4.3.3) and (4.3.4), it seems reasonable that the smoothness conditions of a vector kernel $(K_1,\ldots,K_L)$ concern the supremum $\sup_{1 \leq l \leq L} |K_l|$.

The following two conditions on the vector $(K_1,\ldots,K_L)$ are analogous to (3.3.3) and (3.3.4): suppose there are finite constants $A_1^1,\ldots,A_1^L$ and A_2 such that

$$|K_l(x)| \leq \frac{A_1^l}{|x|^n} \qquad \text{for almost all } x \in \mathbf{R}^n \setminus \{0\} \tag{4.3.5}$$

for all $l \in \{1,\ldots,L\}$ and

$$\sup_{y \neq 0} \int_{|x| \geq 2|y|} \sup_{1 \leq l \leq L} |K_l(x-y) - K_l(x)|\, dx \leq A_2. \tag{4.3.6}$$

4.3 A Matrix-Valued Calderón–Zygmund Theorem and Its Applications

Notice that the analog of (3.3.5) [or (4.1.5)] in this case is valid with $A_3 = \sup_{1 \leq l \leq L} \|K_l\|_{L^1} < \infty$.

The next result provides an extension of Theorem 4.1.1 in the two situations of Example 4.3.1.

Theorem 4.3.2. Let $1 \leq q \leq \infty$, $1 < r \leq \infty$, $N, L \in \mathbf{Z}^+$, and $A_1^1, \ldots, A_1^L, A_2 > 0$. Suppose that $K_1, \ldots, K_L$ are L^1 functions defined on $\mathbf{R}^n$ that satisfy (4.3.5) and (4.3.6).
(a) Define a linear operator by

$$\vec{T}(\vec{F})(x) = (K_1(x), \ldots, K_L(x)) * \vec{F}, \qquad x \in \mathbf{R}^n,$$

where $\vec{F}$ is an $L \times N$ matrix whose entries are functions in $\cup_{1 \leq p < \infty} L^p(\mathbf{R}^n)$. Assume that $\vec{T}$ is bounded from $L^r(\mathbf{R}^n, \ell_N^q(\ell_L^1))$ to $L^r(\mathbf{R}^n, \ell_N^q)$ with norm $B_\star$. Let $C(p) = \max(p, (p-1)^{-1})$ if $r < \infty$ and $C(p) = p(p-1)^{-1}$ if $r = \infty$. Then there exist dimensional constants C_n, C_n' such that

$$\|\vec{T}(\vec{F})\|_{L^{1,\infty}(\mathbf{R}^n, \ell_N^q)} \leq C_n'(A_2 + B_\star)\|\vec{F}\|_{L^1(\mathbf{R}^n, \ell_N^q(\ell_L^1))} \qquad (4.3.7)$$

for all $\vec{F}$ in $L^1(\mathbf{R}^n, \ell_N^q(\ell_L^1))$ and

$$\|\vec{T}(\vec{F})\|_{L^p(\mathbf{R}^n, \ell_N^q)} \leq C_n C(p)(A_2 + B_\star)\|\vec{F}\|_{L^p(\mathbf{R}^n, \ell_N^q(\ell_L^1))} \qquad (4.3.8)$$

for all $\vec{F}$ in $L^p(\mathbf{R}^n, \ell_N^q(\ell_L^1))$, whenever $1 < p < \infty$.
(b) Define

$$\vec{S}(\vec{G})(x) = (K_1(x), \ldots, K_L(x))^t * \vec{G}, \qquad x \in \mathbf{R}^n,$$

where $\vec{G}$ is a $1 \times N$ vector whose entries are functions in $\cup_{1 \leq p < \infty} L^p(\mathbf{R}^n)$. Assume that $\vec{S}$ is a bounded linear operator from $L^r(\mathbf{R}^n, \ell_N^q)$ to $L^r(\mathbf{R}^n, \ell_N^q(\ell_L^\infty))$ with norm $B_\star$. Then there exist constants C_n, C_n' such that

$$\|\vec{S}(\vec{G})\|_{L^{1,\infty}(\mathbf{R}^n, \ell_N^q(\ell_L^\infty))} \leq C_n'(A_2 + B_\star)\|\vec{G}\|_{L^1(\mathbf{R}^n, \ell_N^q)} \qquad (4.3.9)$$

for all $\vec{G}$ in $L^1(\mathbf{R}^n, \ell_N^q)$ and

$$\|\vec{S}(\vec{G})\|_{L^p(\mathbf{R}^n, \ell_N^q(\ell_L^\infty))} \leq C_n C(p)(A_2 + B_\star)\|\vec{G}\|_{L^p(\mathbf{R}^n, \ell_N^q)} \qquad (4.3.10)$$

for all $\vec{G}$ in $L^p(\mathbf{R}^n, \ell_N^q)$, whenever $1 < p < \infty$.

Proof. We first derive (4.3.7) and (4.3.9) and use these to deduce (4.3.8) and (4.3.10) via duality and interpolation. We discuss (4.3.7) and (4.3.9) in cases (a) and (b) below, respectively.

(a) In this case $\vec{F} = \{f_j\}_{j=1}^N$ is thought of as a row vector consisting of columns $f_j = (f_{ij})_{i=1}^L$. Then $\vec{T}(\vec{F})$ is the row vector $(\sum_{i=1}^L K_i * f_{i1}, \ldots, \sum_{i=1}^L K_i * f_{iN})$. To obtain (4.3.7) we repeat the proof of the corresponding estimate (4.1.8) in Theorem 4.1.1 by simply changing ℓ_N^q in the domain by $\ell_N^q(\ell_L^1)$ and ℓ_M^s in the range by ℓ_N^q.

(b) Here $\vec{G} = \{g_j\}_{j=1}^N$ is a row vector but $\vec{S}(\vec{G})$ is now a row vector consisting of columns of length L. The jth column of $\vec{S}(\vec{G})$ is $(K_i * g_j)_{i=1}^L$. In this case we obtain (4.3.9) by repeating the proof of estimate (4.1.8) and replacing any appearance of ℓ_M^s in the range by $\ell_N^q(\ell_L^\infty)$. Note that the domain $\ell_N^q(\ell_1^1) = \ell_N^q$ remains unchanged.

The transpose operator $\vec{T}^t$ of $\vec{T}$ has kernel $(\widetilde{K_1}, \ldots, \widetilde{K_L})^t$ and the transpose operator $\vec{S}^t$ of $\vec{S}$ has kernel $(\widetilde{K_1}, \ldots, \widetilde{K_L})$, and these kernels obviously satisfy (4.3.5) and (4.3.6). So, modulo the reflection of the K_j, the operators $\vec{T}$ and $\vec{S}$ are transposes of one another. Next we interpolate between $\vec{T}: L^r(\mathbf{R}^n, \ell_N^q(\ell_L^1)) \to L^r(\mathbf{R}^n, \ell_N^q)$ and estimate (4.3.7). Using Exercise 1.3.7, we obtain for $1 < p < r$

$$\|\vec{T}\|_{L^p(\mathbf{R}^n, \ell_N^q(\ell_L^1)) \to L^p(\mathbf{R}^n, \ell_N^q)} \leq 2C_n'\left(\frac{p}{p-1} + \frac{p}{r-p}\right)^{\frac{1}{p}}(A_2 + B_\star). \quad (4.3.11)$$

If $r = \infty$, the constant in (4.3.11) raised to the power $1/p$ is bounded by $C(p) = p(p-1)^{-1}$. Now if $r < \infty$, notice that $\vec{T}^t$ maps $L^{r'}(\mathbf{R}^n, \ell_N^{q'})$ to $L^{r'}(\mathbf{R}^n, \ell_N^{q'}(\ell_L^\infty))$ with bound $B_\star$. As the kernel of $\vec{T}^t$ satisfies the same estimates as that of $\vec{S}$, it follows that $\vec{T}^t$ also admits a bounded extension from $L^1(\mathbf{R}^n, \ell_N^{q'})$ to $L^{1,\infty}(\mathbf{R}^n, \ell_N^{q'}(\ell_L^\infty))$ with bound at most $C_n'(A_2 + B_\star)$. By interpolation (Exercise 1.3.7) we obtain for $1 < p' < r'$

$$\|\vec{T}^t\|_{L^{p'}(\mathbf{R}^n, \ell_N^{q'}) \to L^{p'}(\mathbf{R}^n, \ell_N^{q'}(\ell_L^\infty))} \leq 2C_n'\left(\frac{p'}{p'-1} + \frac{p'}{r'-p'}\right)^{\frac{1}{p'}}(A_2 + B_\star). \quad (4.3.12)$$

Estimates (4.3.11) and (4.3.12) imply statements analogous to (4.1.19) and (4.1.20) with ℓ_N^q replaced by $\ell_N^q(\ell_L^1)$ and ℓ_M^s replaced by ℓ_N^q. The rest of the argument proceeds as that in the proof of Theorem 4.1.1 A completely analogous argument is also valid for $\vec{S}$. Combining these ingredients completes the proof. $\square$

We now pass to an application which extends the discussion in Example 4.2.4. Let $\Phi(x) = (1 + |x|)^{-n-\gamma}$ be as in that example. For a fixed odd positive integer L let $\{t_1, \ldots, t_L\} = \{2^{-[L/2]}, 2^{-[L/2]+1}, \ldots, 2^{[L/2]}\}$. We consider the $L \times 1$ matrix $\vec{K}(x) = (\Phi_{t_1}(x), \ldots, \Phi_{t_L}(x))^t$ defined on $\mathbf{R}^n$. This matrix can be viewed as the operator

$$(a_1, a_2, \ldots, a_N) \mapsto \begin{pmatrix} \Phi_{t_1}(x) \\ \Phi_{t_2}(x) \\ \vdots \\ \Phi_{t_L}(x) \end{pmatrix} (a_1 \; a_2 \; \cdots \; a_N) = \begin{pmatrix} \Phi_{t_1}(x)a_1 & \Phi_{t_1}(x)a_2 & \cdots & \Phi_{t_1}(x)a_N \\ \Phi_{t_2}(x)a_1 & \Phi_{t_2}(x)a_2 & \cdots & \Phi_{t_2}(x)a_N \\ \vdots & \vdots & \vdots & \vdots \\ \Phi_{t_L}(x)a_1 & \Phi_{t_L}(x)a_2 & \cdots & \Phi_{t_L}(x)a_N \end{pmatrix},$$

which maps ℓ_N^r to $\ell_N^r(\ell_L^\infty)$ with norm

$$\|\vec{K}(x)\|_{\ell_N^r \to \ell_N^r(\ell_L^\infty)} = \sup_{1 \leq i \leq L} |\Phi_{t_i}(x)|.$$

Properties (4.1.2) and (4.1.3) are proved via the arguments yielding (4.2.7) and (4.2.8), respectively. Additionally, for $1 < r < \infty$, the estimate

4.3 A Matrix-Valued Calderón–Zygmund Theorem and Its Applications

$$\left\|\left(\sum_{j=1}^{N}[\sup_{1\le l\le L}\Phi_{t_l}*|f_j|]^r\right)^{\frac{1}{r}}\right\|_{L^r} \le B_\star\left\|\left(\sum_{j=1}^{N}|f_j|^r\right)^{\frac{1}{r}}\right\|_{L^r} \qquad (4.3.13)$$

is valid with $B_\star = c_n(1+\|\Phi\|_{L^1})\frac{r}{r-1}$ and can be obtained from

$$\left\|\sup_{t>0}|\Phi_t*f|\right\|_{L^r} \le B_\star\|f\|_{L^r},$$

which is a consequence of (4.2.10) (letting $M\to\infty$) and (4.2.6); i.e., the L^r boundedness of the operator $\mathcal{N}$ in Example 4.2.4. Applying Theorem 4.3.2 (b) we obtain

$$\left\|\left(\sum_{j=1}^{N}[\sup_{1\le l\le L}\Phi_{t_l}*|f_j|]^r\right)^{\frac{1}{r}}\right\|_{L^{1,\infty}} \le C'_{n,\gamma}\frac{r}{r-1}\left\|\left(\sum_{j=1}^{N}|f_j|^r\right)^{\frac{1}{r}}\right\|_{L^1} \qquad (4.3.14)$$

and

$$\left\|\left(\sum_{j=1}^{N}[\sup_{1\le l\le L}\Phi_{t_l}*|f_j|]^r\right)^{\frac{1}{r}}\right\|_{L^p} \le \frac{C_{n,\gamma}r}{r-1}\max\left(p,\frac{1}{p-1}\right)\left\|\left(\sum_{j=1}^{N}|f_j|^r\right)^{\frac{1}{r}}\right\|_{L^p}. \qquad (4.3.15)$$

These imply analogous estimates for the Hardy–Littlewood maximal operator $\mathcal{M}$.

Theorem 4.3.3. *(Fefferman–Stein vector-valued maximal function inequality)*
For $1 < p < \infty$ and $1 < r < \infty$ we have

$$\left\|\left(\sum_{k\in\mathbf{Z}}|\mathcal{M}(f_k)|^r\right)^{\frac{1}{r}}\right\|_{L^{1,\infty}} \le C'_n\frac{r}{r-1}\left\|\left(\sum_{k\in\mathbf{Z}}|f_k|^r\right)^{\frac{1}{r}}\right\|_{L^1}, \qquad (4.3.16)$$

$$\left\|\left(\sum_{k\in\mathbf{Z}}|\mathcal{M}(f_k)|^r\right)^{\frac{1}{r}}\right\|_{L^p} \le C_n\frac{r}{r-1}\max\left(p,\frac{1}{p-1}\right)\left\|\left(\sum_{k\in\mathbf{Z}}|f_k|^r\right)^{\frac{1}{r}}\right\|_{L^p}. \qquad (4.3.17)$$

Proof. We notice that $2^{-n-1}\chi_{|x|\le 1}\le (1+|x|)^{-n-1}=\Phi(x)$; thus for this choice of Φ we have

$$\mathcal{M}(f)\le \frac{2^{n+1}}{v_n}\sup_{t>0}\Phi_t*|f|\le \frac{2^{n+1}}{v_n}2^n\sup_{j\in\mathbf{Z}}\Phi_{2^j}*|f|, \qquad (4.3.18)$$

where the second inequality is a consequence of the second inequality in (4.2.6). Letting $L\uparrow\infty$ first (through the odd integers) and then $N\uparrow\infty$ in (4.3.14) and (4.3.15) and applying the LMCT, we deduce (4.3.16) and (4.3.17), respectively, with the aid of (4.3.18). $\square$

Note that (4.3.16) and (4.3.17) are also valid when $r=\infty$ (with $\frac{r}{r-1}=1$). Also, (4.3.17) is valid when $p=r=\infty$ with the understanding that $\frac{r}{r-1}=1$ and that $\max(p,\frac{1}{p-1})$ is replaced by $\frac{p}{p-1}$. These statements are left as exercises.

Exercises

4.3.1. Let $1 \leq q \leq \infty$, $1 \leq s \leq t \leq \infty$, and let $Y = (y_{ij})_{1\leq i \leq L, 1 \leq j \leq M}$ be a complex $L \times M$ matrix viewed as a linear operator acting on complex $M \times N$ matrices by multiplication. The $\ell_N^q(\ell_M^s)$ norm of an $M \times N$ matrix is the ℓ^q norm of the vector of the ℓ^s norms of its columns. Prove that

$$\sup_{1 \leq j \leq M} \left(|y_{1j}|^t + \cdots + |y_{Lj}|^t \right)^{\frac{1}{t}} \leq \|Y\|_{\ell_N^q(\ell_M^s) \to \ell_N^q(\ell_L^t)} \leq \left(\sum_{i=1}^{L} \left(|y_{i1}|^{s'} + \cdots + |y_{iM}|^{s'} \right)^{\frac{s}{s'}} \right)^{\frac{1}{s}},$$

with the obvious modification when $s, t \in \{1, \infty\}$. Deduce the simpler estimate

$$\sup_{1 \leq i \leq L} \sup_{1 \leq j \leq M} |y_{ij}| \leq \|Y\|_{\ell_N^q(\ell_M^s) \to \ell_N^q(\ell_L^t)} \leq \sum_{i=1}^{L} \sum_{j=1}^{M} |y_{ij}|.$$

4.3.2. Let $\{\{a_{ij}\}_{i=1}^L\}_{j=1}^N$ be a doubly indexed finite sequence of complex numbers. Prove that for any $1 \leq q, s \leq \infty$ we have

$$\left\| \left\{ \|\{a_{ij}\}_{i=1}^L \|_{\ell^s} \right\}_{j=1}^N \right\|_{\ell^q} = \max_{\|\{\|\{b_{ij}\}_{i=1}^L\|_{\ell^{s'}}\}_{j=1}^N\|_{\ell^{q'}} = 1} \left| \sum_{i=1}^{L} \sum_{j=1}^{N} a_{ij} b_{ij} \right|.$$

[*Hint:* When $q, s < \infty$, set $b_{ij} = 0$ if $a_{ij} = 0$ and

$$b_{ij} = \frac{\overline{a_{ij}}}{|a_{ij}|^{2-s}} \frac{1}{\left(\sum_{k=1}^{L} |a_{kj}|^s \right)^{1-\frac{q}{s}}} \frac{1}{\left\| \{ \| \{a_{ij}\}_{i=1}^L \|_{\ell^s} \}_{j=1}^N \right\|_{\ell^q}^{q-1}}$$

if $a_{ij} \neq 0$. Use Exercise 4.1.4 when q or s is infinite.]

4.3.3. Show that $L^p(\mathbf{R}^n, \ell_N^q(\ell_L^s))$ is complete when $1 \leq q, s \leq \infty$ and $0 < p \leq \infty$. Also show that $L^{1,\infty}(\mathbf{R}^n, \ell_N^q(\ell_L^s))$ is complete. [*Hint:* Use the inequalities

$$\sup_{\substack{1 \leq j \leq N \\ 1 \leq i \leq L}} \|f_{ij}\|_{L^p}^{\min(1,p)} \leq \left\| \left\| \{ \|\{f_{ij}\}_{i=1}^L\|_{\ell^s} \}_{j=1}^N \right\|_{\ell^q} \right\|_{L^p}^{\min(1,p)} \leq \sum_{j=1}^{N} \sum_{i=1}^{L} \|f_{ij}\|_{L^p}^{\min(1,p)}$$

and an analogous one with $L^{1,\infty}$ in place of L^p with an extra factor NL on the right.]

4.3.4. Prove that for any $1 \leq q, s \leq \infty$ and $1 < p < \infty$ we have

$$\left\| \left\| \{ \|\{f_{ij}\}_{i=1}^L\|_{\ell^s} \}_{j=1}^N \right\|_{\ell^q} \right\|_{L^p} = \max_{\|\|\{\|\{g_{ij}\}_{i=1}^L\|_{\ell^{s'}}\}_{j=1}^N\|_{\ell^{q'}}\|_{L^{p'}} = 1} \left| \int_{\mathbf{R}^n} \sum_{i=1}^{L} \sum_{j=1}^{N} f_{ij} g_{ij} \, dx \right|.$$

[*Hint:* When $q, s < \infty$, set $g_{ij}(x) = 0$ if $f_{ij}(x) = 0$ and

$$g_{ij}(x) = \frac{\overline{f_{ij}(x)}}{|f_{ij}(x)|^{2-s}} \frac{1}{\left(\sum_{k=1}^{L}|f_{kj}(x)|^s\right)^{1-\frac{q}{s}}} \frac{1}{\left(\sum_{l=1}^{N}\left(\sum_{k=1}^{L}|f_{kl}(x)|^s\right)^{\frac{q}{s}}\right)^{1-\frac{p}{q}}} \frac{1}{Q^{p-1}}$$

if $f_{ij}(x) \neq 0$, where $Q = \|\|\{\|\{f_{ij}\}_{i=1}^{L}\|_{\ell^s}\}_{j=1}^{N}\|_{\ell^q}\|_{L^p}.]$

4.3.5. On the real line show that the following endpoint cases of estimate (4.3.17) fail: (a) $p = \infty$ and $1 < r < \infty$. (b) $1 < p < \infty$ and $r = 1$. [*Hint:* Part (a): Take $f_j = \chi_{[2^{j-1}, 2^j]}$. Part (b): Take $f_j = \chi_{[\frac{j-1}{N}, \frac{j}{N}]}$, $j = 1, 2, \ldots, N$.]

4.3.6. Let $\{Q_j\}_j$ be a countable collection of cubes in $\mathbf{R}^n$ with disjoint interiors, with centers c_j and side lengths d_j. For $\varepsilon > 0$, define the *Marcinkiewicz function* associated with the family $\{Q_j\}_j$ as follows:

$$M_\varepsilon(x) = \sum_j \frac{d_j^{n+\varepsilon}}{|x - c_j|^{n+\varepsilon} + d_j^{n+\varepsilon}}.$$

Prove that there are constants $C_{n,\varepsilon,p}$ and $C_{n,\varepsilon}$ such that

$$\|M_\varepsilon\|_{L^p} \leq C_{n,\varepsilon,p}\left(\sum_j |Q_j|\right)^{\frac{1}{p}}, \qquad p > \frac{n}{n+\varepsilon},$$

$$\|M_\varepsilon\|_{L^{\frac{n}{n+\varepsilon},\infty}} \leq C_{n,\varepsilon}\left(\sum_j |Q_j|\right)^{\frac{n+\varepsilon}{n}},$$

and consequently $\int_{\mathbf{R}^n} M_\varepsilon(x)\,dx \leq C_{n,\varepsilon,1}\sum_j |Q_j|$.
[*Hint*: Verify that

$$\frac{d_j^{n+\varepsilon}}{|x - c_j|^{n+\varepsilon} + d_j^{n+\varepsilon}} \leq CM(\chi_{Q_j})(x)^{\frac{n+\varepsilon}{n}}$$

and use Theorem 4.3.3.]

4.3.7. Let $M^{(j)}$ denote the Hardy–Littlewood maximal function on $\mathbf{R}^n$ acting only on the jth variable and define $\mathbf{M} = M^{(1)} \circ \cdots \circ M^{(n)}$. Prove that there is constant A_n such that for all $1 < p, r < \infty$ and for all functions $f_j \in L^p(\mathbf{R}^n)$ we have

$$\left\|\left(\sum_j |\mathbf{M}(f_j)|^r\right)^{\frac{1}{r}}\right\|_{L^p(\mathbf{R}^n)} \leq A_n \left(\frac{r}{r-1}\max\left(p, \frac{1}{p-1}\right)\right)^n \left\|\left(\sum_j |f_j|^r\right)^{\frac{1}{r}}\right\|_{L^p(\mathbf{R}^n)}.$$

The *strong maximal function* $\mathscr{M}(f)(x)$ is defined as the supremum of the averages of a measurable function $|f|$ over all rectangles with sides parallel to the axes that contain a given $x \in \mathbf{R}^n$. Derive the same estimate for $\mathscr{M}$.

4.4 Littlewood–Paley Theory

In this section we obtain a characterization of the L^p norm of a function in terms of the restrictions of its Fourier transform on dyadic annuli.

Definition 4.4.1. Let Ψ be an integrable function on $\mathbf{R}^n$ and $j \in \mathbf{Z}$. We define the *Littlewood–Paley operator* Δ_j^Ψ (associated with Ψ) as the operator given by convolution with $\Psi_{2^{-j}}$. Here $\Psi_{2^{-j}}(x) = 2^{jn}\Psi(2^j x)$ for all x in $\mathbf{R}^n$, equivalently $\widehat{\Psi_{2^{-j}}}(\xi) = \widehat{\Psi}(2^{-j}\xi)$ for all ξ in $\mathbf{R}^n$. The operator Δ_j^Ψ is well defined on $\cup_{1 \le p \le \infty} L^p(\mathbf{R}^n)$.

In most applications we choose Ψ to be a smooth function whose Fourier transform is supported in an annulus $0 < c_1 < |\xi| < c_2 < \infty$. Then the Fourier transform of $\Delta_j^\Psi(f)$ is supported in the annulus $c_1 2^j < |\xi| < c_2 2^j$; in other words, it is localized near the frequency $|\xi| \approx 2^j$. Thus Δ_j^Ψ isolates frequencies near $|\xi| \approx 2^j$. The *Littlewood–Paley square function* associated with Ψ is the function

$$f \mapsto \Big(\sum_{j \in \mathbf{Z}} |\Delta_j^\Psi(f)|^2 \Big)^{\frac{1}{2}}.$$

The next theorem concerns the Littlewood–Paley square function.

Theorem 4.4.2. *(Littlewood–Paley theorem)* Let $B, \delta > 0$. Suppose that Ψ is a $\mathscr{C}^1$ function on $\mathbf{R}^n$ with mean value zero that satisfies for all $x \in \mathbf{R}^n$

$$|\Psi(x)| + |\nabla \Psi(x)| \le B(1+|x|)^{-n-\delta}. \tag{4.4.1}$$

Then there exists a constant $C_{n,\delta} < \infty$ such that for all $1 < p < \infty$ and all f in $L^p(\mathbf{R}^n)$ we have

$$\Big\| \Big(\sum_{j \in \mathbf{Z}} |\Delta_j^\Psi(f)|^2 \Big)^{\frac{1}{2}} \Big\|_{L^p(\mathbf{R}^n)} \le C_{n,\delta} B \max\Big(p, \frac{1}{p-1}\Big) \|f\|_{L^p(\mathbf{R}^n)}. \tag{4.4.2}$$

There also exists a $C'_{n,\delta} < \infty$ such that for all f in $L^1(\mathbf{R}^n)$ we have

$$\Big\| \Big(\sum_{j \in \mathbf{Z}} |\Delta_j^\Psi(f)|^2 \Big)^{\frac{1}{2}} \Big\|_{L^{1,\infty}(\mathbf{R}^n)} \le C'_{n,\delta} B \|f\|_{L^1(\mathbf{R}^n)}. \tag{4.4.3}$$

Proof. We first prove (4.4.2) when $p = 2$. Using Plancherel's theorem, we rewrite (4.4.2) when $p = 2$ as

$$\sum_{j \in \mathbf{Z}} \int_{\mathbf{R}^n} |\widehat{\Psi}(2^{-j}\xi)|^2 |\widehat{f}(\xi)|^2 \, d\xi \le C_{n,\delta}^2 4 B^2 \int_{\mathbf{R}^n} |\widehat{f}(\xi)|^2 \, d\xi.$$

So (4.4.2) when $p = 2$ will be a consequence of the inequality

$$\sum_{j \in \mathbf{Z}} |\widehat{\Psi}(2^{-j}\xi)|^2 \le c_{n,\delta}^2 B^2 \tag{4.4.4}$$

4.4 Littlewood–Paley Theory

for some $c_{n,\delta} < \infty$. So we prove (4.4.4). As Ψ has mean value zero we write

$$\widehat{\Psi}(\xi) = \int_{\mathbf{R}^n} e^{-2\pi i x \cdot \xi} \Psi(x)\, dx = \int_{\mathbf{R}^n} (e^{-2\pi i x \cdot \xi} - 1)\Psi(x)\, dx, \qquad (4.4.5)$$

from which, with the aid of (4.4.1), we obtain the estimate

$$|\widehat{\Psi}(\xi)| \leq \int_{\mathbf{R}^n} 2^{1-\gamma}(2\pi|\xi||x|)^\gamma |\Psi(x)|\, dx \leq c'_{n,\delta} B|\xi|^\gamma, \qquad (4.4.6)$$

where we set $\gamma = \min(\frac{1}{2}, \frac{\delta}{2})$. For $\xi = (\xi_1,\ldots,\xi_n) \neq 0$, let j_0 be such that $|\xi_{j_0}| \geq |\xi_k|$ for all $k \in \{1,\ldots,n\}$. We integrate by parts with respect to x_{j_0} in (4.4.5) to obtain

$$\widehat{\Psi}(\xi) = -\int_{\mathbf{R}^n} (-2\pi i \xi_{j_0})^{-1} e^{-2\pi i x \cdot \xi} (\partial_{j_0}\Psi)(x)\, dx, \qquad (4.4.7)$$

where we also used the vanishing of Ψ at infinity, a consequence of (4.4.1). From (4.4.7) we deduce the estimate

$$|\widehat{\Psi}(\xi)| \leq (2\pi|\xi_{j_0}|)^{-1} \int_{\mathbf{R}^n} |\nabla\Psi(x)|\, dx \leq c''_{n,\delta} B|\xi|^{-1}. \qquad (4.4.8)$$

We obtain from (4.4.6) and (4.4.8) that

$$|\widehat{\Psi}(\xi)| \leq \max(c'_{n,\delta}, c''_{n,\delta}) B\big[|\xi|^\gamma \chi_{|\xi|\leq 1} + |\xi|^{-1}\chi_{|\xi|>1}\big]. \qquad (4.4.9)$$

We now break the sum in (4.4.4) into the parts where j satisfies $2^{-j}|\xi| \leq 1$ or $2^{-j}|\xi| > 1$, and use (4.4.9) to deduce (4.4.4). This proves (4.4.2) when $p = 2$.

We now turn our attention to the case $p \neq 2$ in (4.4.2), which we view as a vector-valued inequality. Define an operator $\vec{T}$ acting on functions in $\cup_{1\leq p\leq \infty} L^p(\mathbf{R}^n)$ as follows:

$$\vec{T}(f) = \{\Delta_j^\Psi(f)\}_{j=-N}^N$$

for some fixed $N \in \mathbf{Z}^+$. The inequalities (4.4.2) and (4.4.3) follow from the statements that $\vec{T}$ is a bounded operator from $L^p(\mathbf{R}^n,\mathbf{C})$ to $L^p(\mathbf{R}^n,\ell^2_{2N+1}(\mathbf{C}))$ and from $L^1(\mathbf{R}^n,\mathbf{C})$ to $L^{1,\infty}(\mathbf{R}^n,\ell^2_{2N+1}(\mathbf{C}))$. We just proved that this statement is true when $p = 2$ with constant $B_\star = C_n B$ (which is independent of N), and therefore one hypothesis of Theorem 4.1.1 is satisfied. We observe that $\vec{T}$ can be written in the form

$$\vec{T}(f)(x) = \left\{\int_{\mathbf{R}^n} \Psi_{2^{-j}}(x-y)f(y)\,dy\right\}_{j=-N}^N = \int_{\mathbf{R}^n} \vec{K}(x-y)(f(y))\,dy,$$

where for each $x \in \mathbf{R}^n$, $\vec{K}(x)$ is a bounded linear operator from $\mathbf{C}$ to ℓ^2_{2N+1} given by

$$a \mapsto \vec{K}(x)(a) = \{\Psi_{2^{-j}}(x)a\}_{j=-N}^N. \qquad (4.4.10)$$

We clearly have

$$\|\vec{K}(x)\|_{\mathbf{C}\to\ell^2_{2N+1}} = \Big(\sum_{j=-N}^{N} |\Psi_{2^{-j}}(x)|^2\Big)^{\frac{1}{2}}. \tag{4.4.11}$$

Our goal is to apply Theorem 4.1.1 with parameters N, q, M, s, r (in the notation there) $N = 1$, $q = 2$, $M = 2N+1$, $s = 2$, and $r = 2$. In order to verify hypotheses (4.1.2), (4.1.3), (4.1.5), and (4.1.6), it will suffice to verify that[1]

$$\|\vec{K}(x)\|_{\mathbf{C}\to\ell^2_{2N+1}} \leq c_{n,\delta} B |x|^{-n}, \tag{4.4.12}$$

$$\sup_{y\neq 0} \int_{|x|\geq 2|y|} \|\vec{K}(x-y) - \vec{K}(x)\|_{\mathbf{C}\to\ell^2_{2N+1}} dx \leq c_{n,\delta} B, \tag{4.4.13}$$

$$\sup_{j:|j|\leq N} \sup_{\varepsilon>0} \sup_{R>\varepsilon} \Big|\int_{\varepsilon\leq |y|\leq R} \Psi_{2^{-j}}(y) dy\Big| \leq A \leq c_{n,\delta} B < \infty, \tag{4.4.14}$$

$$\lim_{\delta\downarrow 0} \int_{\delta\leq |y|\leq 1} \vec{K}(y) dy = \Big\{\int_{|y|\leq 1} \Psi_{2^{-j}}(y) dy\Big\}_{j=-N}^{N} \tag{4.4.15}$$

for some constant $c_{n,\delta} > 0$. Of these, (4.4.14) follows immediately with $A = \|\Psi\|_{L^1}$ by passing the absolute value inside the integral and by changing variables, and (4.4.15) is straightforward. So we focus on (4.4.12) and (4.4.13). First we note that

$$\|\vec{K}(x)\|_{\mathbf{C}\to\ell^2_{2N+1}} \leq \sum_{j=-N}^{N} |\Psi_{2^{-j}}(x)| \leq \sum_{j\in\mathbf{Z}} \frac{2^{jn} B}{(1+2^j|x|)^{n+\delta}}.$$

The sum over the indices j with $2^j|x| \leq 1$, produces

$$\sum_{2^j \leq |x|^{-1}} 2^{jn} B \leq c^1_n B |x|^{-n}.$$

The sum over indices j with $2^j|x| > 1$ produces the term

$$\sum_{2^j > |x|^{-1}} 2^{jn} B (2^j|x|)^{-n-\delta} = B|x|^{-n-\delta} \sum_{2^j > |x|^{-1}} 2^{-j\delta} \leq c^2_{n,\delta} B |x|^{-n}.$$

This proves (4.4.12).

We now address (4.4.13). Fix x, y in $\mathbf{R}^n$ such that $|x| \geq 2|y|$. Since Ψ is a $\mathscr{C}^1$ function, by the mean value theorem we may write

$$|\Psi_{2^{-j}}(x-y) - \Psi_{2^{-j}}(x)| \leq 2^{(n+1)j} \int_0^1 |\nabla\Psi(2^j(x-\theta y))| |y| d\theta$$

$$\leq B 2^{(n+1)j} \int_0^1 (1+2^j|x-\theta y|)^{-n-\delta} |y| d\theta$$

$$\leq B 2^{nj} (1+2^{j-1}|x|)^{-n-\delta} 2^j |y|, \tag{4.4.16}$$

since $|x - \theta y| \geq |x| - |\theta y| \geq |x| - |y| \geq \frac{1}{2}|x|$.

---

[1] In view of (4.1.2), the weaker hypothesis $|\Psi_{2^{-j}}(x)| \leq c_{n,\delta} B |x|^{-n}$ would have sufficed for (4.4.12).

4.4 Littlewood–Paley Theory

We also have that

$$|\Psi_{2^{-j}}(x-y) - \Psi_{2^{-j}}(x)| \leq 2^{nj}|\Psi(2^j(x-y))| + 2^{jn}|\Psi(2^j x)|$$
$$\leq B2^{jn}\left(1 + 2^j \tfrac{1}{2}|x|\right)^{-n-\delta} + B2^{nj}\left(1 + 2^j|x|\right)^{-n-\delta}$$
$$\leq 2B2^{nj}\left(1 + 2^{j-1}|x|\right)^{-n-\delta}. \tag{4.4.17}$$

Let $\varepsilon \in [0,1]$. A weighted geometric mean of (4.4.16) and (4.4.17) is

$$|\Psi_{2^{-j}}(x-y) - \Psi_{2^{-j}}(x)| \leq 2^{1-\varepsilon}B2^{nj}(2^j|y|)^\varepsilon \left(1 + 2^{j-1}|x|\right)^{-n-\delta}. \tag{4.4.18}$$

Using this estimate, when $|x| \geq 2|y|$, we write

$$\|\vec{K}(x-y) - \vec{K}(x)\|_{\mathbf{C} \to \ell^2_{2N+1}}$$
$$\leq \left(\sum_{j \in \mathbf{Z}} |\Psi_{2^{-j}}(x-y) - \Psi_{2^{-j}}(x)|^2\right)^{1/2}$$
$$\leq \sum_{j \in \mathbf{Z}} |\Psi_{2^{-j}}(x-y) - \Psi_{2^{-j}}(x)|$$
$$\leq 2B\left(|y| \sum_{2^j < \frac{2}{|x|}} 2^{(n+1)j} + |y|^\gamma \sum_{2^j \geq \frac{2}{|x|}} 2^{nj} 2^{\gamma j}(2^{j-1}|x|)^{-n-\delta}\right) \tag{4.4.19}$$
$$\leq c'_{n,\delta} B\left(|y| |x|^{-n-1} + |y|^\gamma |x|^{\delta-\gamma}|x|^{-n-\delta}\right), \tag{4.4.20}$$

having used (4.4.18) in the first sum in (4.4.19) with $\varepsilon = 1$ and in the second sum with $\varepsilon = \gamma = \min(\frac{1}{2}, \frac{\delta}{2})$. We now deduce (4.4.13) by integrating in polar coordinates (4.4.20) over the region $|x| \geq 2|y|$.

An application of Theorem 4.1.1 concludes the proofs of (4.4.2) and (4.4.3). □

Corollary 4.4.3. *Let Ψ be as in Theorem 4.4.2. Then there is a constant $C_{n,\delta}$ such that for $1 < p < \infty$ and for any functions $f_j \in L^p(\mathbf{R}^n)$ we have*

$$\left\|\left(\sum_{j \in \mathbf{Z}} |\Delta_j^\Psi(f_j)|^2\right)^{\frac{1}{2}}\right\|_{L^p(\mathbf{R}^n)} \leq C_{n,\delta} B \max\left(p, \frac{1}{p-1}\right) \left\|\left(\sum_{j \in \mathbf{Z}} |f_j|^2\right)^{\frac{1}{2}}\right\|_{L^p(\mathbf{R}^n)}. \tag{4.4.21}$$

There also exists a $C'_{n,\delta} < \infty$ such that for all f in $L^1(\mathbf{R}^n)$ we have

$$\left\|\left(\sum_{j \in \mathbf{Z}} |\Delta_j^\Psi(f_j)|^2\right)^{\frac{1}{2}}\right\|_{L^{1,\infty}(\mathbf{R}^n)} \leq C'_{n,\delta} B \left\|\left(\sum_{j \in \mathbf{Z}} |f_j|^2\right)^{\frac{1}{2}}\right\|_{L^1(\mathbf{R}^n)}. \tag{4.4.22}$$

Proof. The proof follows the paradigm of Theorem 4.4.2, except that we define an operator $\vec{S}$ acting on sequences of functions in $\cup_{1 \leq p \leq \infty} L^p(\mathbf{R}^n)$ as follows:

$$\vec{S}(\{f_j\}_{j=-N}^N) = \{\Delta_j^\Psi(f_j)\}_{j=-N}^N.$$

The kernel of $\vec{S}$ is the $(2N+1) \times (2N+1)$ matrix $\vec{L}(x) = \big(L_{ij}(x)\big)_{-N \le i,j \le N}$ with $L_{jj}(x) = \Psi_{2^{-j}}(x)$ and $L_{ij}(x) = 0$ if $i \ne j$. We clearly have

$$\|\vec{L}(x)\|_{\ell^2_{2N+1} \to \ell^2_{2N+1}} \le \sup_{-N \le j \le N} |\Psi_{2^{-j}}(x)| \le \Big(\sum_{j=-N}^{N} |\Psi_{2^{-j}}(x)|^2 \Big)^{\frac{1}{2}}$$

and the last expression coincides with the norm $\|\vec{K}(x)\|_{\mathbf{C} \to \ell^2_{2N+1}}$ of $\vec{K}(x)$ introduced in the proof of Theorem 4.4.2; see (4.4.11). The estimates for $\vec{K}(x)$ are also valid for $\vec{L}(x)$ and another application of Theorem 4.1.1 yields the claimed conclusion. □

In many applications of Theorem 4.4.2, Ψ has the following properties:

$$\begin{aligned} &\widehat{\Psi} \text{ is smooth and nonnegative on } \mathbf{R}^n, \\ &\operatorname{support}(\widehat{\Psi}) \subseteq \Big\{ \xi \in \mathbf{R}^n : 1 - \tfrac{1}{7} \le |\xi| \le 2 \Big\}, \\ &\widehat{\Psi}(\xi) = 1 \quad \text{on} \quad 1 \le |\xi| \le 2 - \tfrac{2}{7}, \\ &\sum_{j \in \mathbf{Z}} \widehat{\Psi}(2^{-j}\xi) = 1 \qquad \text{for all } \xi \ne 0. \end{aligned} \qquad (4.4.23)$$

Such a function can be constructed as follows: Start with a smooth function ϕ with values in $[0,1]$ which is supported in the interval $[\tfrac{6}{7}, 2]$ and is equal to 1 on $[1, \tfrac{12}{7}]$. (Exercise 1.7.3.) Define

$$\widehat{\Psi}(\xi) = \frac{\phi(|\xi|)}{\sum_{k \in \mathbf{Z}} \phi(2^{-k}|\xi|)}, \qquad \xi \ne 0, \qquad (4.4.24)$$

and $\widehat{\Psi}(0) = 0$. Then obviously $\widehat{\Psi}$ is a smooth function with compact support that satisfies (4.4.23). Moreover $\widehat{\Psi}$ is supported in the annulus $1 - \tfrac{1}{7} \le |\xi| \le 2$. Also on the annulus $1 \le |\xi| \le 2 - \tfrac{2}{7}$ equals 1, as all summands in (4.4.24) with $k \ne 0$ vanish. See Figure 4.1.

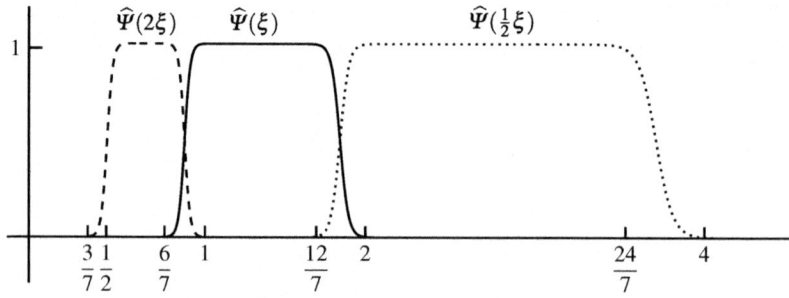

Fig. 4.1 The functions $\widehat{\Psi}(2\xi)$, $\widehat{\Psi}(\xi)$, and $\widehat{\Psi}(\xi/2)$ plotted in one dimension.

4.4 Littlewood–Paley Theory

Corollary 4.4.4. *If Ψ is as in (4.4.23), then estimates (4.4.2) and (4.4.3) are valid.*

Proof. Obviously $\widehat{\Psi}(0) = 0$, which implies $\int_{\mathbf{R}^n} \Psi(y)\,dy = 0$. Also, (4.4.1) holds. □

It is often desirable to group the Δ_j^{Ψ} with $j \leq 0$ together. To achieve this we define a Schwartz function Φ as follows:

$$\widehat{\Phi}(\xi) = \begin{cases} \sum_{j \leq 0} \widehat{\Psi}(2^{-j}\xi) & \text{when } \xi \neq 0, \\ 1 & \text{when } \xi = 0. \end{cases} \qquad (4.4.25)$$

Note that $\widehat{\Phi}(\xi)$ is equal to 1 for $|\xi| \leq 2 - \frac{2}{7}$, vanishes when $|\xi| \geq 2$, and satisfies

$$\widehat{\Phi}(\xi) + \sum_{j=1}^{\infty} \widehat{\Psi}(2^{-j}\xi) = 1$$

for all ξ in $\mathbf{R}^n$. We introduce an operator S_0^{Φ} given by convolution with Φ. There is a version of Theorem 4.4.2 with S_0^{Φ} in place of all Δ_j^{Ψ} for $j \leq 0$; on this see Exercise 4.4.3.

Exercises

4.4.1. Construct a Schwartz function Ψ that satisfies $\sum_{j \in \mathbf{Z}} |\widehat{\Psi}(2^{-j}\xi)|^2 = 1$ for all $\xi \in \mathbf{R}^n \setminus \{0\}$ and whose Fourier transform is supported in the annulus $\frac{6}{7} \leq |\xi| \leq 2$ and is equal to 1 on the annulus $1 \leq |\xi| \leq \frac{12}{7}$.
[*Hint*: Set $\widehat{\Psi}(\xi) = \phi(\xi)\big(\sum_{k \in \mathbf{Z}} |\phi(2^{-k}\xi)|^2\big)^{-1/2}$ for a suitable $\phi \in \mathscr{C}_0^{\infty}(\mathbf{R}^n)$.]

4.4.2. Construct a smooth function Ψ supported in the unit ball of $\mathbf{R}^n$ with integral zero such that

$$\sum_{j \in \mathbf{Z}} \widehat{\Psi}(2^{-j}\xi) = 1, \qquad \xi \neq 0.$$

[*Hint:* Set $\Psi(x) = \Phi(x) - 2^n \Phi(2x)$, where Φ is a smooth function supported in the unit ball of $\mathbf{R}^n$ with integral equal to 1.]

4.4.3. Let $B, \delta > 0$. Let Φ, Ψ be as (4.4.23) and (4.4.25). Prove that there are constants $C_{n,\delta,\Psi}, C'_{n,\delta,\Psi} < \infty$ such that for all $1 < p < \infty$ and all f in $L^p(\mathbf{R}^n)$ we have

$$\left\|\Big(|S_0^{\Phi}(f)|^2 + \sum_{j=1}^{\infty} |\Delta_j^{\Psi}(f)|^2\Big)^{\frac{1}{2}}\right\|_{L^p(\mathbf{R}^n)} \leq C_{n,\delta,\Psi} \max\big(p, (p-1)^{-1}\big) \|f\|_{L^p(\mathbf{R}^n)}$$

and for all f in $L^1(\mathbf{R}^n)$ we have

$$\left\|\Big(|S_0^{\Phi}(f)|^2 + \sum_{j=1}^{\infty} |\Delta_j^{\Psi}(f)|^2\Big)^{\frac{1}{2}}\right\|_{L^{1,\infty}(\mathbf{R}^n)} \leq C'_{n,\delta,\Psi} \|f\|_{L^1(\mathbf{R}^n)}.$$

4.4.4. Let Ψ, Φ be as (4.4.23) and (4.4.25). Prove the operator identities on L^2

$$\sum_{j \in \mathbf{Z}} \Delta_j^\Psi = I \quad \text{and} \quad S_0^\Phi + \sum_{j=1}^\infty \Delta_j^\Psi = I.$$

4.4.5. Let Ψ be a Schwartz function whose Fourier transform vanishes in a neighborhood of the origin and let $\varphi \in \mathscr{S}(\mathbf{R}^n)$. Prove that for any $M > 0$ there is a constant $C_M = C_{M,n,\Psi,\varphi}$ such that

$$\sum_{j \in \mathbf{Z}} |\Delta_j^\Psi(\varphi)(x)| \leq \frac{C_M}{(1+|x|)^M}.$$

Conclude that if Ψ is an (4.4.23), then for all $0 < p \leq \infty$ one has

$$\sum_{|j| \leq N} \Delta_j^\Psi(\varphi) \to \varphi \quad \text{in } L^p \text{ as } N \to \infty.$$

[*Hint:* Use the estimates $|(\Psi_{2^{-j}} * \varphi)(x)| \leq C_{M,n} 2^{\min(0,j)n}(1 + 2^{\min(0,j)n}|x|)^{-M}$ and $|(\Psi_{2^{-j}} * \varphi)(x)| \leq C_{M,L,n} 2^{-Lj}(1+|x|)^{-M}$ for any $L, M \in \mathbf{Z}^+ \cup \{0\}$. These are consequences of Theorems 7.1.1 and 3.3.5 and the second estimate uses that Ψ has vanishing moments of all orders. Last assertion: start with $p = \infty$.]

4.4.6. Let Ψ be a Schwartz function that satisfies (4.4.23). Let $1 < p < \infty$. Prove that for $g \in L^p(\mathbf{R}^n)$ we have

$$\lim_{N \to \infty} \Big\| \sum_{|j| \leq N} \Delta_j^\Psi(g) - g \Big\|_{L^p} = 0.$$

[*Hint:* Use Exercise 4.4.5.]

4.4.7. Let m be a bounded function on $\mathbf{R}^n$ that is supported in the annulus $1 \leq |\xi| \leq 2$ and define $T_j(f) = \big(\widehat{f} m(2^{-j}(\cdot))\big)^\vee$. Suppose that the square function

$$f \mapsto \Big(\sum_{j \in \mathbf{Z}} |T_j(f)|^2 \Big)^{1/2}$$

is bounded on $L^p(\mathbf{R}^n)$ for some $1 < p < \infty$. Show that there is a constant $C_{p,n}$ such that for every finite subset S of the integers and every $f \in L^p(\mathbf{R}^n)$ we have

$$\Big\| \sum_{j \in S} T_j(f) \Big\|_{L^p(\mathbf{R}^n)} \leq C_{p,n} \|f\|_{L^p(\mathbf{R}^n)}.$$

4.4.8. Prove the following generalization of Theorem 4.4.2. Let $A_1, A_2 > 0$. Suppose that K_j, $j \in \mathbf{Z}$ are locally integrable functions on $\mathbf{R}^n \setminus \{0\}$ that satisfy

$$\Big(\sum_{j \in \mathbf{Z}} |K_j(x)|^2 \Big)^{\frac{1}{2}} \leq \frac{A_1}{|x|^n}, \qquad x \neq 0,$$

4.5 Reverse Littlewood–Paley Inequalities

$$\sup_{y \in \mathbf{R}^n \setminus \{0\}} \int_{|x| \geq 2|y|} \Big(\sum_{j \in \mathbf{Z}} |K_j(x-y) - K_j(x)|^2 \Big)^{\frac{1}{2}} dx \leq A < \infty,$$

$$\sup_{j \in \mathbf{Z}} \sup_{\varepsilon > 0} \sup_{R > \varepsilon} \Big| \int_{\varepsilon \leq |y| \leq R} K_j(y) \, dy \Big| \leq A_2 < \infty,$$

and there is a sequence $\varepsilon_k \downarrow 0$ and numbers L_j (for each $j \in \mathbf{Z}$) such that

$$\lim_{\varepsilon_k \downarrow 0} \int_{\varepsilon_k \leq |y| \leq 1} K_j(y) \, dy = L_j.$$

Suppose that the functions K_j coincide with tempered distributions W_j that satisfy

$$\sum_{j \in \mathbf{Z}} |\widehat{W_j}(\xi)|^2 \leq B^2, \qquad \xi \in \mathbf{R}^n.$$

Prove that the operator

$$f \rightarrow \Big(\sum_{j \in \mathbf{Z}} |K_j * f|^2 \Big)^{\frac{1}{2}}$$

maps $L^p(\mathbf{R}^n)$ to itself and is of weak type $(1,1)$ with norms at most $C_{n,p}(A_2 + B)$.
[*Hint:* Notice that (4.4.12), (4.4.13), (4.4.14), and (4.4.15) hold by assumption.]

4.5 Reverse Littlewood–Paley Inequalities

The focus of this section is to study the reverse inequality of that in Theorem 4.4.2. We recall that $\langle f, \varphi \rangle$ denotes the action of a tempered distribution f on a Schwartz function φ, and $\langle f, \varphi \rangle$ coincides with the standard Lebesgue integral $\int_{\mathbf{R}^n} f(x) \varphi(x) \, dx$ if f happens to be an L^p function for some $1 \leq p \leq \infty$.

We can extend the definition of the Littlewood–Paley operator Δ_j^Ω to tempered distributions whenever Ω is a Schwartz function. Precisely, if $\Omega \in \mathscr{S}(\mathbf{R}^n)$ and f in $\mathscr{S}'(\mathbf{R}^n)$, then $\Delta_j^\Omega(f)$ is well defined as the convolution $\Omega_{2^{-j}} * f$. This convolution always produces a smooth function (Theorem 2.7.1).

Recall the reflection $\widetilde{\Omega}$ of a function Ω is given by $\widetilde{\Omega}(x) = \Omega(-x)$ for $x \in \mathbf{R}^n$. We begin by identifying the transpose operator of Δ_j^Ω for a function Ω.

Proposition 4.5.1. (a) *If Ω lies in $L^1(\mathbf{R}^n)$ and f in $L^p(\mathbf{R}^n)$, $1 \leq p \leq \infty$, then we have*

$$\langle f, \Delta_j^\Omega(g) \rangle = \langle \Delta_j^{\widetilde{\Omega}}(f), g \rangle \qquad \text{whenever } g \in L^{p'}(\mathbf{R}^n). \tag{4.5.1}$$

(b) *For any $f \in \mathscr{S}'(\mathbf{R}^n)$ and $\Omega \in \mathscr{S}(\mathbf{R}^n)$ we have*

$$\langle f, \Delta_j^\Omega(\varphi) \rangle = \langle \Delta_j^{\widetilde{\Omega}}(f), \varphi \rangle \qquad \text{whenever } \varphi \in \mathscr{S}(\mathbf{R}^n). \tag{4.5.2}$$

Thus, in these senses, the transpose of the operator Δ_j^Ω is $\Delta_j^{\widetilde{\Omega}}$.

Proof. Fix $f \in L^p(\mathbf{R}^n)$, $1 \leq p \leq \infty$, and $g \in L^{p'}(\mathbf{R}^n)$. By Fubini's theorem we write

$$\int_{\mathbf{R}^n} f(x) \left(\int_{\mathbf{R}^n} g(y) \Omega_{2^{-j}}(x-y) \, dy \right) dx = \int_{\mathbf{R}^n} g(y) \int_{\mathbf{R}^n} f(x) \Omega_{2^{-j}}(x-y) \, dx \, dy$$

$$= \int_{\mathbf{R}^n} g(y) \left(\int_{\mathbf{R}^n} f(x) \widetilde{\Omega}_{2^{-j}}(y-x) \, dx \right) dy,$$

and this proves (4.5.1). The interchange of the integrals is justified from the fact that the double integral converges absolutely, a consequence of Hölder's inequality; note $g, \Delta_j^\Omega(g)$ lie in $L^{p'}(\mathbf{R}^n)$ and $f, \Delta_j^{\widetilde{\Omega}}(f)$ lie in $L^p(\mathbf{R}^n)$.

The identity in (4.5.2) is just a restatement of the definition of the convolution of Schwartz functions and tempered distributions (Definition 2.6.16). □

A reformulation of Theorem 4.4.2 based on duality is as follows:

Proposition 4.5.2. *Suppose that Ψ is an integrable function on $\mathbf{R}^n$ with mean value zero that satisfies (4.4.1). Then there is a constant $C_{n,\delta}$ such that for any $1 < p < \infty$ and any $N \in \mathbf{Z}^+$ we have*

$$\left\| \sum_{|j| \leq N} \Delta_j^\Psi(f_j) \right\|_{L^p(\mathbf{R}^n)} \leq C_{n,\delta} B \max\left(p, (p-1)^{-1}\right) \left\| \left(\sum_{|j| \leq N} |f_j|^2 \right)^{\frac{1}{2}} \right\|_{L^p(\mathbf{R}^n)} \quad (4.5.3)$$

for all L^p functions f_j.

Proof. To verify this assertion by duality, define the operator

$$\vec{T}(f) = \{\Delta_j^\Psi(f)\}_{j=-N}^N, \qquad f \in L^p(\mathbf{R}^n).$$

The transpose operator of $\vec{T}$ is

$$\vec{T}^t(\{g_j\}_{j=-N}^N) = \sum_{j=-N}^N \Delta_j^{\widetilde{\Psi}}(g_j), \qquad \{g_j\}_{j=-N}^N \in L^{p'}(\mathbf{R}^n, \ell_{2N+1}^2),$$

as the following identity based on (4.5.1) indicates:

$$\int_{\mathbf{R}^n} \sum_{j=-N}^N \Delta_j^{\widetilde{\Psi}}(f) g_j \, dx = \sum_{j=-N}^N \int_{\mathbf{R}^n} f \Delta_j^\Psi(g_j) \, dx = \int_{\mathbf{R}^n} f \sum_{j=-N}^N \Delta_j^\Psi(g_j) \, dx.$$

Estimate (4.4.2) (with $\widetilde{\Psi}$ in place of Ψ) says that $\vec{T}$ maps $L^p(\mathbf{R}^n, \mathbf{C})$ to $L^p(\mathbf{R}^n, \ell_{2N+1}^2)$. The dual statement of this is that $\vec{T}^t$ maps $L^{p'}(\mathbf{R}^n, \ell_{2N+1}^2)$ to $L^{p'}(\mathbf{R}^n, \mathbf{C})$. This is exactly the claim in (4.5.3) if p is replaced by p'. Since p is any number in $(1, \infty)$ and $\max(p', (p'-1)^{-1}) \approx \max(p, (p-1)^{-1})$, (4.5.3) is proved. □

We now discuss the converse of Theorem 4.4.2. Before doing so we notice that the converse inequality to (4.4.2) may not hold in general. In fact, if $\widehat{\Psi}$ is supported in the annulus $\frac{9}{7} \leq |\xi| \leq \frac{10}{7}$ and $\widehat{f}$ is supported in $\frac{12}{7} \leq |\xi| \leq \frac{13}{7}$, then $\Delta_j^\Psi(f) = 0$

4.5 Reverse Littlewood–Paley Inequalities

for all $j \in \mathbf{Z}$ but f itself may not be zero. So some condition on Ψ is needed and it turns out that the last condition in (4.4.23) is sufficient.

Theorem 4.5.3. *Let Ψ be a Schwartz function whose Fourier transform is supported in an annulus that does not contain the origin and satisfies*

$$\sum_{j \in \mathbf{Z}} \widehat{\Psi}(2^{-j}\xi) = 1, \qquad \text{for all } \xi \in \mathbf{R}^n \setminus \{0\}. \tag{4.5.4}$$

Let $1 < p < \infty$. Then there is a constant $C_{n,\Psi}$, such that for all $f \in L^p(\mathbf{R}^n)$ we have

$$\|f\|_{L^p(\mathbf{R}^n)} \leq C_{n,\Psi} \max\left(p, (p-1)^{-1}\right) \left\| \left(\sum_{j \in \mathbf{Z}} |\Delta_j^\Psi(f)|^2 \right)^{\frac{1}{2}} \right\|_{L^p(\mathbf{R}^n)}. \tag{4.5.5}$$

Proof. Given Ψ as in the hypothesis of the theorem, pick $0 < c_1 < c_2 < \infty$ such that the support of $\widehat{\Psi}$ is contained in the annulus $c_1 \leq |\xi| \leq c_2$. Pick c_3, c_4 such that $0 < c_3 < c_1 < c_2 < c_4 < \infty$ and fix another Schwartz function Ω whose Fourier transform is supported in the annulus $c_3 \leq |\xi| \leq c_4$ and is equal to 1 on $c_1 \leq |\xi| \leq c_2$. See Figure 4.2.

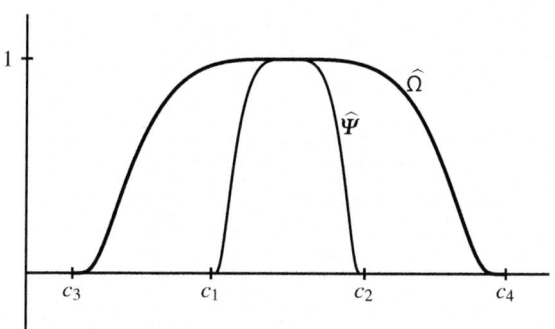

Fig. 4.2 The function $\widehat{\Omega}$ is equal to 1 on the support of $\widehat{\Psi}$.

Note that $\widetilde{\widehat{\Psi}} = \widehat{\widetilde{\Psi}}$, which yields that (4.5.4) is also satisfied for $\widetilde{\Psi}$ in place of Ψ. The key observation is that

$$\Delta_j^{\widetilde{\Psi}} = \Delta_j^{\widetilde{\Psi}} \Delta_j^\Omega,$$

as a quick examination of the Fourier transforms gives

$$\widehat{\widetilde{\Psi}}(2^{-j}\xi) = \widehat{\widetilde{\Psi}}(2^{-j}\xi) \widehat{\Omega}(2^{-j}\xi)$$

for all $\xi \in \mathbf{R}^n$ and all $j \in \mathbf{Z}$.

Let $\widehat{\mathscr{S}_0}(\mathbf{R}^n)$ be the subspace of Schwartz functions whose Fourier transform is compactly supported and does not contain the origin. By Proposition 2.5.4, $\widehat{\mathscr{S}_0}(\mathbf{R}^n)$ is dense in every $L^q(\mathbf{R}^n)$ with $1 < q < \infty$. Notice that for every $\varphi \in \widehat{\mathscr{S}_0}(\mathbf{R}^n)$ we have

$$\varphi = \sum_j \Delta_j^{\widetilde{\Psi}}(\varphi),$$

where the sum contains only finitely many j. This is because any compact set that does not contain the origin is contained in finitely many annuli of the form $c_1 2^j < |\xi| < c_2 2^j$, for different values of j.

Now fix f in $L^p(\mathbf{R}^n)$ for some $1 < p < \infty$. We assume that the expression on the right in (4.5.5) is finite, otherwise there is nothing to prove. Then

$$\begin{aligned}
\langle f, \varphi \rangle &= \left\langle f, \sum_j \Delta_j^{\widetilde{\Psi}} \varphi \right\rangle \\
&= \left\langle f, \sum_j \Delta_j^{\widetilde{\Psi}} \Delta_j^{\Omega}(\varphi) \right\rangle \\
&= \sum_j \langle f, \Delta_j^{\widetilde{\Psi}} \Delta_j^{\Omega}(\varphi) \rangle \\
&= \sum_j \langle \Delta_j^{\Psi}(f), \Delta_j^{\Omega}(\varphi) \rangle \\
&= \sum_j \int_{\mathbf{R}^n} \Delta_j^{\Psi}(f) \Delta_j^{\Omega}(\varphi) \, dx \\
&= \int_{\mathbf{R}^n} \sum_j \Delta_j^{\Psi}(f) \Delta_j^{\Omega}(\varphi) \, dx,
\end{aligned}$$

where the sum consists of only finitely many j, and this justifies the interchanges of different operations that include this sum. Next we apply the Cauchy–Schwarz inequality and then Hölder's inequality to bound the preceding expression by

$$\left\| \left(\sum_j |\Delta_j^{\Psi}(f)|^2 \right)^{\frac{1}{2}} \right\|_{L^p} \left\| \left(\sum_j |\Delta_j^{\Omega}(\varphi)|^2 \right)^{\frac{1}{2}} \right\|_{L^{p'}}.$$

Applying Theorem 4.4.2, we conclude that

$$|\langle f, \varphi \rangle| \leq \left\| \left(\sum_j |\Delta_j^{\Psi}(f)|^2 \right)^{\frac{1}{2}} \right\|_{L^p} C_{n,\Psi} \max\left(p', (p'-1)^{-1}\right) \|\varphi\|_{L^{p'}} < \infty. \quad (4.5.6)$$

Finally, we note that $\max(p', (p'-1)^{-1}) \approx \max(p, (p-1)^{-1})$, so taking the supremum over all $\varphi \in \widehat{\mathscr{S}_0}(\mathbf{R}^n)$ with $\|\varphi\|_{L^{p'}} = 1$ in (4.5.6), we deduce (4.5.5) in view of (1.1.6) in Proposition 1.1.4. At this point we used that f already lies in $L^p(\mathbf{R}^n)$. $\square$

Corollary 4.5.4. *Let Ψ be as in Theorem 4.5.3 and $1 < p < \infty$. Then for any g in $L^p(\mathbf{R}^n)$ we have*

$$\|g\|_{L^p(\mathbf{R}^n)} \approx \left\| \left(\sum_{j \in \mathbf{Z}} |\Delta_j^{\Psi}(g)|^2 \right)^{\frac{1}{2}} \right\|_{L^p(\mathbf{R}^n)}.$$

Proof. This follows from Theorems 4.4.2 and 4.5.3. $\square$

4.5 Reverse Littlewood–Paley Inequalities

A delicate point of Theorem 4.5.3 is that the function f is already assumed to lie in L^p and this assumption is needed in the derivation of (4.5.5); precisely it was needed in the use of (1.1.6) (Proposition 1.1.4). One may wonder if the L^p assumption on f can be relaxed. For instance, could we replace L^p by L^1_{loc}? But a moment's thought gives that the locally integrable function $f = 1$ satisfies $\Delta_j^\Psi(1) = \widehat{\Psi}(0) = 0$ for all j, so (4.5.5) could not possibly hold. The problem with the function 1 is that its Fourier transform is the Dirac mass at zero and identity (4.5.4) fails at zero. So in order to extend the result of Theorem 4.5.3 to locally integrable functions, or even to tempered distributions, we need to take into account that the Fourier transform of this distribution could be supported at the origin.

Our next goal is to investigate which tempered distributions f satisfy

$$C_p(f) = \left\| \left(\sum_{j \in \mathbf{Z}} |\Delta_j^\Psi(f)|^2 \right)^{\frac{1}{2}} \right\|_{L^p} < \infty \qquad (4.5.7)$$

for some $1 < p < \infty$. If for $f \in \mathscr{S}'(\mathbf{R}^n)$ we have $\text{supp}(\widehat{f}) = \{0\}$, then $\Delta_j^\Psi(f) = 0$ for all $j \in \mathbf{Z}$, as the functions $\widehat{\Psi}(2^{-j}(\cdot))$ are supported away from the origin. It turns out that the support of $\widehat{f}$ is $\{0\}$ exactly when f is a polynomial (Lemma 4.5.6). So (4.5.7) holds for distributions that are sums of L^p functions and polynomials. The next theorem says that these are all possible such distributions.

Theorem 4.5.5. *Let $\Psi \in \mathscr{S}(\mathbf{R}^n)$ have Fourier transform supported in an annulus that does not contain the origin and satisfies (4.5.4). Let $1 < p < \infty$. Then there is a constant $C_{n,\Psi}$ such that for any $f \in \mathscr{S}'(\mathbf{R}^n)$ which satisfies (4.5.7) there exists a unique polynomial Q such that the tempered distribution $f - Q$ coincides with an L^p function satisfying*

$$\|f - Q\|_{L^p(\mathbf{R}^n)} \leq C_{n,\Psi} \max\left(p, (p-1)^{-1}\right) \left\| \left(\sum_{j \in \mathbf{Z}} |\Delta_j^\Psi(f)|^2 \right)^{\frac{1}{2}} \right\|_{L^p(\mathbf{R}^n)}. \qquad (4.5.8)$$

Proof. Fix $f \in \mathscr{S}'(\mathbf{R}^n)$ such that $C_p(f) < \infty$ [as defined in (4.5.7)]. Proceeding as in the proof of Theorem 4.5.3, using (4.5.2), we arrive at (4.5.6). We now define a linear functional L on $L^{p'}(\mathbf{R}^n)$ as follows: For a given g in $L^{p'}(\mathbf{R}^n)$ pick a sequence $\varphi_k \in \widehat{\mathscr{S}_0}(\mathbf{R}^n)$ such that $\varphi_k \to g$ in $L^{p'}(\mathbf{R}^n)$ as $k \to \infty$. Applying (4.5.6) to $\varphi_k - \varphi_l$ we obtain that $\{\langle f, \varphi_k \rangle\}_{k=1}^\infty$ is a Cauchy sequence and thus it converges. Set

$$L(g) = \lim_{k \to \infty} \langle f, \varphi_k \rangle$$

and note that $L(g)$ is independent of the choice of φ_k; indeed, if φ_k' is another sequence from $\widehat{\mathscr{S}_0}(\mathbf{R}^n)$ that converges to g in $L^{p'}$, then applying (4.5.6) to $\varphi = \varphi_k - \varphi_k'$ we obtain that $\langle f, \varphi_k - \varphi_k' \rangle \to 0$ as $k \to \infty$. Secondly, using again (4.5.6) with $\varphi = \varphi_k$ and letting $k \to \infty$, as $\|\varphi_k\|_{L^{p'}} \to \|g\|_{L^{p'}}$, we deduce

$$\|L\|_{L^{p'}\to \mathbf{C}} \leq \left\|\left(\sum_j |\Delta_j^\Psi(f)|^2\right)^{\frac{1}{2}}\right\|_{L^p} C_{n,\Psi}\max\left(p',(p'-1)^{-1}\right) < \infty. \qquad (4.5.9)$$

Thus L is a bounded linear functional on $L^{p'}(\mathbf{R}^n)$. By Theorem 1.1.10 (Riesz representation), L can be identified with an $L^p(\mathbf{R}^n)$ function F such that

$$L(g) = \int_{\mathbf{R}^n} F(y)g(y)\,dy \qquad \text{for any } g \in L^{p'}(\mathbf{R}^n)$$

and

$$\|F\|_{L^p} = \|L\|_{L^{p'}\to\mathbf{C}}. \qquad (4.5.10)$$

Consider the tempered distribution $f - F$. For a compactly supported smooth function ψ whose support does not contain the origin we have

$$\langle \widehat{f}-\widehat{F},\psi\rangle = \langle f,\widehat{\psi}\rangle - \langle F,\widehat{\psi}\rangle = L(\widehat{\psi}) - \int_{\mathbf{R}^n} F(y)\widehat{\psi}(y)\,dy = 0,$$

and this yields that the support of $\widehat{f}-\widehat{F}$ is contained in $\{0\}$. Here we made use of the fact that $L(\widehat{\psi}) = \langle f,\widehat{\psi}\rangle$, which is a consequence of the definition of L considering the constant sequence $\widehat{\psi}$ of elements of $\widehat{\mathscr{S}_0}(\mathbf{R}^n)$ which converges to itself in $L^{p'}$. Applying Lemma 4.5.6 (proved below) there is a polynomial Q such that $f - F = Q$. Thus $F = f - Q$. Combining (4.5.9) with (4.5.10), we deduce (4.5.8).

Finally, we obtain the uniqueness of Q. If Q_1 is another polynomial, with $f - Q_1$ in L^p, then $Q - Q_1$ must be an L^p function; but the only polynomial that lies in L^p is the zero polynomial. Thus $Q_1 = Q$. $\square$

Lemma 4.5.6. *If $u \in \mathscr{S}'(\mathbf{R}^n)$ is supported at the origin, then there exist an integer K and complex numbers a_α such that*

$$u = \sum_{|\alpha|\leq K} a_\alpha \partial^\alpha \delta_0,$$

where δ_0 is the Dirac mass at the origin. Consequently, any tempered distribution whose Fourier transform is supported at the origin must be a polynomial.

Proof. As $u \in \mathscr{S}'(\mathbf{R}^n)$, there are $C > 0$, M, and K in $\mathbf{Z}^+ \cup \{0\}$ such that

$$|\langle u,\varphi\rangle| \leq C \sum_{|\alpha|\leq M}\sum_{|\beta|\leq K}\sup_{x\in\mathbf{R}^n}|x^\alpha(\partial^\beta\varphi)(x)| \qquad \text{for all } \varphi \in \mathscr{S}(\mathbf{R}^n).$$

We will first prove that if $\varphi \in \mathscr{S}$ satisfies $(\partial^\gamma\varphi)(0) = 0$ for all $|\gamma|\leq K$ then $\langle u,\varphi\rangle = 0$. To verify this claim, first observe that such a function φ must satisfy

$$|(\partial^\gamma\varphi)(x)| \leq C_\gamma |x|^{K+1-|\gamma|}, \qquad \text{when } |x|\leq 1 \text{ and } |\gamma|\leq K \qquad (4.5.11)$$

for some constant C_γ. This follows by noticing that the Taylor expansion of $\partial^\gamma\varphi$ of order $K - |\gamma|$ at the origin has vanishing coefficients of x^α for all $|\alpha|\leq K - |\gamma|$ and its error is $O(|x|^{K-|\gamma|+1})$ as $|x|\to 0$.

4.5 Reverse Littlewood–Paley Inequalities

Let ζ be a smooth function on $\mathbf{R}^n$ that equals 1 when $|x| \geq 2$ and equals 0 when $|x| \leq 1$. Set $\zeta^\varepsilon(x) = \zeta(x/\varepsilon)$ for $x \in \mathbf{R}^n$ and $\varepsilon > 0$. For $|\alpha| \leq M$ and $|\beta| \leq K$, we write

$$\rho_{\alpha,\beta}(\varphi - \zeta^\varepsilon \varphi) \leq \sum_{0 \leq \gamma \leq \beta} \binom{\beta}{\gamma} \sup_{x \in \mathbf{R}^n} |x^\alpha| \left| \partial^\gamma 1 - \frac{1}{\varepsilon^{|\gamma|}} (\partial^\gamma \zeta)(\tfrac{x}{\varepsilon}) \right| |\partial^{\beta-\gamma} \varphi(x)|.$$

The x in the supremum is restricted in the ball $|x| \leq 2\varepsilon$ (even when $\gamma = 0$) in view of the properties of ζ. So using (4.5.11), we estimate the supremum by a multiple of $(2\varepsilon)^{|\alpha|} \varepsilon^{-|\gamma|} (2\varepsilon)^{K+1-(|\beta|-|\gamma|)} \leq c\varepsilon$, when $\varepsilon > 0$ is sufficiently small. We obtain that $\rho_{\alpha,\beta}(\zeta^\varepsilon \varphi - \varphi) \to 0$ as $\varepsilon \to 0$. Then

$$|\langle u, \varphi \rangle| \leq |\langle u, \zeta^\varepsilon \varphi \rangle| + |\langle u, \varphi - \zeta^\varepsilon \varphi \rangle| \leq 0 + C \sum_{|\alpha| \leq M} \sum_{|\beta| \leq K} \rho_{\alpha,\beta}(\varphi - \zeta^\varepsilon \varphi) \to 0$$

as $\varepsilon \to 0$. Hence $\langle u, \varphi \rangle = 0$, and this proves our claim.

Now for a given $\psi \in \mathscr{S}(\mathbf{R}^n)$ write

$$\psi(x) = (1 - \zeta(x)) \left(\sum_{|\alpha| \leq K} \frac{(\partial^\alpha \psi)(0)}{\alpha!} x^\alpha + h(x) \right) + \zeta(x) \psi(x), \qquad (4.5.12)$$

where $h \in \mathscr{C}^\infty$ and satisfies $h(x) = O(|x|^{K+1})$ as $|x| \to 0$. Then $(1-\zeta)h$ satisfies $\partial^\gamma((1-\zeta)h)(0) = 0$ for all $|\gamma| \leq K$ and thus one has $\langle u, (1-\zeta)h \rangle = 0$ by the previous assertion. Also, $\zeta \psi$ is supported away from the origin, so $\langle u, \zeta \psi \rangle = 0$, by our hypothesis on u. Acting u on both sides of (4.5.12), we deduce

$$\langle u, \psi \rangle = \sum_{|\alpha| \leq K} \frac{(\partial^\alpha \psi)(0)}{\alpha!} \langle u, (\cdot)^\alpha (1-\zeta) \rangle = \sum_{|\alpha| \leq K} a_\alpha \langle \partial^\alpha \delta_0, \psi \rangle,$$

with $a_\alpha = (-1)^{|\alpha|} \langle u, (\cdot)^\alpha (1-\zeta) \rangle / \alpha!$. This concludes the proof of the first assertion.

If the Fourier transform of a tempered distribution v is supported at the origin, using the fact just proved and the result of Example 2.6.9, we conclude that v must be a polynomial. This proves the second assertion of the lemma. $\square$

Exercises

4.5.1. Prove that every tempered distribution u which satisfies *Laplace's equation* $\partial_1^2 u + \cdots + \partial_n^2 u = 0$ must be a polynomial. (Such polynomials are called *harmonic*.) [*Hint:* Use Lemma 4.5.6.]

4.5.2. Let $1 < p < \infty$. Let Φ and Ψ be as in (4.4.23) and (4.4.25). Then for any g in $L^p(\mathbf{R}^n)$ we have

$$\|g\|_{L^p(\mathbf{R}^n)} \approx \left\| \left(|S_0^\Phi(g)|^2 + \sum_{j=1}^\infty |\Delta_j^\Psi(g)|^2 \right)^{\frac{1}{2}} \right\|_{L^p(\mathbf{R}^n)}.$$

[*Hint:* One direction follows from Exercise 4.4.3. For the other direction start with $\langle g, \varphi \rangle = \langle g, S_0^{\Phi}(\varphi) \rangle + \sum_{j=1}^{\infty} \langle g, \Delta_j^{\Psi}(\varphi) \rangle$, where $\varphi \in \mathscr{S}_0(\mathbf{R}^n)$. Pick an even smooth function Θ with compact support that equals 1 on the support of $\widehat{\Phi}$ and a smooth function Ω with compact support in $\mathbf{R}^n \setminus \{0\}$ that equals 1 on the support of $\widehat{\Psi}$.]

4.5.3. Let Ψ, Ω be Schwartz functions whose Fourier transforms are supported in compact annuli that do not contain the origin and satisfy

$$\sum_{j \in \mathbf{Z}} \widehat{\Psi}(2^{-j}\xi) \widehat{\Omega}(2^{-j}\xi) = 1, \qquad \xi \in \mathbf{R}^n \setminus \{0\}.$$

Prove that for any $1 < p < \infty$ there is a constant $C_{\Psi,\Omega,n}$ for any $g \in L^p(\mathbf{R}^n)$ we have

$$\|g\|_{L^p(\mathbf{R}^n)} \le C_{\Psi,\Omega,n} \max\left(p, (p-1)^{-1}\right) \left\| \left(\sum_{j \in \mathbf{Z}} |\Delta_j^{\Psi}(g)|^2 \right)^{\frac{1}{2}} \right\|_{L^p(\mathbf{R}^n)}.$$

In particular, this inequality is valid if $\sum_{j \in \mathbf{Z}} |\widehat{\Psi}(2^{-j}\xi)|^2 = 1$, $\xi \ne 0$ (take $\Omega = \widetilde{\Psi}$).

4.5.4. Let Ψ, Ω be Schwartz functions whose Fourier transforms are supported in compact annuli that do not contain the origin and let Φ, Θ be Schwartz functions whose Fourier transforms are compactly supported and equal to 1 on a neighborhood of the origin. Suppose that

$$\widehat{\Phi}(\xi) \widehat{\Theta}(\xi) + \sum_{j=1}^{\infty} \widehat{\Psi}(2^{-j}\xi) \widehat{\Omega}(2^{-j}\xi) = 1, \qquad \xi \in \mathbf{R}^n.$$

Prove that for any $1 < p < \infty$ there is a constant $C = C_{\Psi,\Omega,\Phi,\Theta,n}$ for any $g \in L^p(\mathbf{R}^n)$ one has

$$\|g\|_{L^p(\mathbf{R}^n)} \le C \max\left(p, (p-1)^{-1}\right) \left\| \left(|S_0^{\Phi}(g)|^2 + \sum_{j=1}^{\infty} |\Delta_j^{\Psi}(g)|^2 \right)^{\frac{1}{2}} \right\|_{L^p(\mathbf{R}^n)}.$$

4.5.5. Fix a nonzero Schwartz function h on the line whose Fourier transform is supported in the interval $\left[-\frac{1}{8}, \frac{1}{8}\right]$. For a finite sequence of numbers $\{a_j\}_{j=1}^{N}$ define

$$f(x) = \sum_{j=1}^{N} a_j e^{2\pi i 2^j x} h(x), \qquad x \in \mathbf{R}.$$

Prove that for all $1 < p < \infty$ there exists a constant C_p independent of N such that

$$\|f\|_{L^p(\mathbf{R})} \le C_p \left(\sum_{j=1}^{N} |a_j|^2 \right)^{\frac{1}{2}} \|h\|_{L^p(\mathbf{R})}.$$

[*Hint:* Fix $\psi \in \mathscr{S}(\mathbf{R})$ with $\widehat{\psi}$ supported in $\left[-\frac{10}{8}, -\frac{6}{8}\right] \cup \left[\frac{6}{8}, \frac{10}{8}\right]$ and equal to 1 on $\left[-\frac{9}{8}, -\frac{7}{8}\right] \cup \left[\frac{7}{8}, \frac{9}{8}\right]$. Then notice $f = \sum_{j=1}^{N} \Delta_j^{\psi}(a_j e^{2\pi i 2^j(\cdot)} h)$ and use (4.5.3).]

4.5.6. Let $1 < p < \infty$ and $0 < c_1 < c_2 < \infty$. Let θ be a Schwartz function whose Fourier transform is supported in $c_1 \leq |\xi| \leq c_2$ and satisfies

$$\sum_{j \in \mathbf{Z}} \widehat{\theta}(2^{-j}\xi) \neq 0, \qquad \text{whenever } \xi \neq 0.$$

Show that there is a constant $C_{p,n,\theta}$ such that for any g in $L^p(\mathbf{R}^n)$ we have

$$\|g\|_{L^p(\mathbf{R}^n)} \leq C_{p,n,\theta} \Big\| \Big(\sum_{j \in \mathbf{Z}} |\Delta_j^\theta(g)|^2 \Big)^{\frac{1}{2}} \Big\|_{L^p(\mathbf{R}^n)}.$$

[*Hint:* Pick $m_1, m_2 \in \mathbf{Z}$ such that $\widehat{\Theta}(\xi) = \sum_{m_1 \leq k \leq m_2} \widehat{\theta}(2^{-k}\xi) \neq 0$ for every ξ in the annulus $\frac{3}{7} \leq |\xi| \leq 4$. Then for Ψ as in (4.4.23) one has $\widehat{\Psi} = \widehat{\Theta}\widehat{\Omega}$, where $\widehat{\Omega}$ is a smooth function with compact support in $\mathbf{R}^n \setminus \{0\}$. Apply Corollary 4.4.3.]

4.5.7. Let $1 < p \leq 2$. Prove that there is a constant $C_{n,p}$ such that for all L^p functions f_j on $\mathbf{R}^n$ whose Fourier transforms are supported in the dyadic annuli $2^j \leq |\xi| \leq 2^{j+1}$, $j \in \mathbf{Z}$, we have

$$\Big\| \sum_j f_j \Big\|_{L^p(\mathbf{R}^n)}^p \leq C_{n,p} \sum_j \|f_j\|_{L^p(\mathbf{R}^n)}^p.$$

[*Hint:* Use Corollary 4.5.2.]

4.6 Littlewood–Paley Theory of Product Type

One may ask whether Theorems 4.4.2 and 4.5.3 still hold if the Littlewood–Paley operators Δ_j^Ψ are replaced by their nonsmooth versions

$$f \mapsto \big(\chi_{2^j \leq |\cdot| < 2^{j+1}} \widehat{f}\big)^\vee. \tag{4.6.1}$$

This question has positive answer in one dimension but negative in higher dimensions. For this reason, a product-type version of a non smooth Littlewood–Paley decomposition provides a substitute in higher dimensions.

We first look at the one-dimensional case. For $j \in \mathbf{Z}$ we consider the interval

$$I_j = [2^j, 2^{j+1}) \cup (-2^{j+1}, -2^j], \tag{4.6.2}$$

and we introduce the one-dimensional *sharp cutoff Littlewood–Paley operator*

$$\Delta_j^\sharp(f) = (\widehat{f}\chi_{I_j})^\vee, \qquad f \in \mathscr{S}(\mathbf{R}). \tag{4.6.3}$$

If Ψ is as in (4.4.23), then $\Delta_j^\sharp$ is a version of Δ_j^Ψ with $\widehat{\Psi}(2^{-j}\xi)$ being replaced by the characteristic function of the set $2^j \leq |\xi| < 2^{j+1}$. We note that although

$\Delta_j^\sharp$ is initially defined on the $\mathscr{S}(\mathbf{R})$, it has an extension on $L^p(\mathbf{R})$ for any $1 < p < \infty$, as the operator given by multiplication by $\chi_{[a,b]}$ on the Fourier transform equals $\frac{i}{2}(M^a HM^{-a} - M^b HM^{-b})$, where H is the Hilbert transform and $M^a(f)(x) = e^{2\pi i x a} f(x)$. A consequence of this identity is the following key inequality, whose proof is based on Corollary 4.2.3 (with $r=2$) and is omitted (see Exercise 4.2.2):

Proposition 4.6.1. *There is a constant C such that for any $1 < p < \infty$ and f_j in $L^p(\mathbf{R})$, $j \in \mathbf{Z}$, we have*

$$\left\|\left(\sum_{j \in \mathbf{Z}} |\Delta_j^\sharp(f_j)|^2\right)^{\frac{1}{2}}\right\|_{L^p(\mathbf{R})} \leq C \max\left(p, \frac{1}{p-1}\right) \left\|\left(\sum_{j \in \mathbf{Z}} |f_j|^2\right)^{\frac{1}{2}}\right\|_{L^p(\mathbf{R})}. \quad (4.6.4)$$

We now extend the operator $\Delta_j^\sharp$ to n dimensions. For $j_1, \ldots, j_n \in \mathbf{Z}$ we set $\boldsymbol{j} = (j_1, \ldots, j_n) \in \mathbf{Z}^n$ and we define a union of 2^n dyadic rectangles $R_{\boldsymbol{j}}$ by setting

$$R_{\boldsymbol{j}} = I_{j_1} \times \cdots \times I_{j_n}, \quad (4.6.5)$$

where I_k is as in (4.6.2). Observe that for different $\boldsymbol{j}, \boldsymbol{j}' \in \mathbf{Z}^n$ the sets $R_{\boldsymbol{j}}$ and $R_{\boldsymbol{j}'}$ have disjoint interiors and that the union of all the $R_{\boldsymbol{j}}$ is equal to $\mathbf{R}^n$ minus the union of all the coordinate planes $x_k = 0$. We call this tiling the *dyadic decomposition* of $\mathbf{R}^n$. We now introduce n-dimensional *sharp cutoff Littlewood–Paley operators*

$$\Delta_{\boldsymbol{j}}^\sharp(f)(x) = (\widehat{f}\chi_{R_{\boldsymbol{j}}})^\vee(x), \qquad f \in \mathscr{S}(\mathbf{R}^n), \quad \boldsymbol{j} \in \mathbf{Z}^n. \quad (4.6.6)$$

Note that if $\boldsymbol{j} = (j_1, \ldots, j_n) \in \mathbf{Z}^n$, then

$$\Delta_{\boldsymbol{j}}^\sharp = \Delta_{j_1}^{\sharp 1} \circ \cdots \circ \Delta_{j_n}^{\sharp n},$$

where $\Delta_{j_r}^{\sharp r}$ is the one-dimensional operator $\Delta_{j_r}^\sharp$ acting on the rth variable, with the remaining variables fixed. As in the one-dimensional case, $\Delta_{\boldsymbol{j}}^\sharp$ admits a well-defined extension on $L^p(\mathbf{R}^n)$ for $1 < p < \infty$ as it is a composition of n L^p-bounded operators. The important property of these operators is the projection identity $\Delta_{\boldsymbol{j}}^\sharp = \Delta_{\boldsymbol{j}}^\sharp \Delta_{\boldsymbol{j}}^\sharp$, which plays a fundamental role in the subsequent main result about them.

Theorem 4.6.2. *Let $\Delta_{\boldsymbol{j}}^\sharp$ be the operators defined in (4.6.6). For each $1 < p < \infty$ there exists a positive constant $C(p,n)$ such that for all $f \in L^p(\mathbf{R}^n)$ we have*

$$\frac{\|f\|_{L^p(\mathbf{R}^n)}}{C(p',n)} \leq \left\|\left(\sum_{\boldsymbol{j} \in \mathbf{Z}^n} |\Delta_{\boldsymbol{j}}^\sharp(f)|^2\right)^{\frac{1}{2}}\right\|_{L^p(\mathbf{R}^n)} \leq C(p,n) \|f\|_{L^p(\mathbf{R}^n)}. \quad (4.6.7)$$

Proof. We start the proof with the one-dimensional case. Pick a Schwartz function ψ on the line whose Fourier transform is supported in the set $2^{-1} \leq |\xi| \leq 2^2$ and is equal to 1 on the set $1 \leq |\xi| \leq 2$. Let Δ_j^ψ be the Littlewood–Paley operators associated with ψ. Observe that

4.6 Littlewood–Paley Theory of Product Type

$$\Delta_j^{\psi}\Delta_j^{\#} = \Delta_j^{\#}\Delta_j^{\psi} = \Delta_j^{\#},$$

since $\widehat{\psi}(2^{-j}\cdot)$ is equal to 1 on the support of $\Delta_j^{\#}(f)\widehat{}$. Proposition 4.6.1 yields the first inequality below:

$$\left\|\Big(\sum_{j\in\mathbf{Z}}|\Delta_j^{\#}(f)|^2\Big)^{\frac{1}{2}}\right\|_{L^p} = \left\|\Big(\sum_{j\in\mathbf{Z}}|\Delta_j^{\#}\Delta_j^{\psi}(f)|^2\Big)^{\frac{1}{2}}\right\|_{L^p}$$

$$\leq C\max(p,(p-1)^{-1})\left\|\Big(\sum_{j\in\mathbf{Z}}|\Delta_j^{\psi}(f)|^2\Big)^{\frac{1}{2}}\right\|_{L^p}$$

$$\leq C'\max(p,(p-1)^{-1})^2\|f\|_{L^p}, \tag{4.6.8}$$

while the second inequality follows from Theorem 4.4.2. We have obtained the upper inequality in (4.6.7) with $C(p,1) \leq C'\max(p,(p-1)^{-1})^2$.

We now turn to the higher-dimensional case of the upper inequality in (4.6.7). A key ingredient for this is the following fundamental property of the Rademacher functions r_j, $j = 0, 1, 2, \ldots$, which are re-indexed by $j \in \mathbf{Z}$ (see [31, Appendix C]):

Lemma 4.6.3. *Let $0 < p < \infty$ and r_j be the sequence of Rademacher functions indexed by the integers. Then there are constants $0 < A_p, B_p < \infty$ such that for any complex-valued sequence $\{c_j\}_{j\in\mathbf{Z}}$ with all but finitely many terms equal to zero, we have*

$$B_p^p\Big(\sum_j |c_j|^2\Big)^{\frac{p}{2}} \leq \int_0^1 \Big|\sum_j c_j r_j(t)\Big|^p dt \leq A_p^p\Big(\sum_j |c_j|^2\Big)^{\frac{p}{2}}. \tag{4.6.9}$$

Moreover, for any complex-valued sequence $\{c_{\mathbf{j}}\}_{\mathbf{j}\in\mathbf{Z}^n}$, $\mathbf{j} = (j_1,\ldots,j_n)$, with all but finitely many terms equal to zero, we have

$$B_p^{np}\Big[\sum_{\mathbf{j}\in\mathbf{Z}^n}|c_{\mathbf{j}}|^2\Big]^{\frac{p}{2}} \leq \int_{[0,1]^n}\Big|\sum_{\mathbf{j}\in\mathbf{Z}^n} c_{\mathbf{j}} r_{j_1}(t_1)\cdots r_{j_n}(t_n)\Big|^p dt_1\cdots dt_n \leq A_p^{np}\Big[\sum_{\mathbf{j}\in\mathbf{Z}^n}|c_{\mathbf{j}}|^2\Big]^{\frac{p}{2}}.$$

With the aid of this lemma we prove the upper inequality in (4.6.7) by induction. Assume this inequality holds when $n = k$ with a constant $C(p,k)$. We look at the dimension $n = k+1$. We write an element $\mathbf{j}$ of $\mathbf{Z}^{k+1}$ as $(j_1, \mathbf{j}')$, where $j_1 \in \mathbf{Z}$ and $\mathbf{j}'$ in $\mathbf{Z}^k$. Analogously we write elements of $\mathbf{R}^{k+1}$ as (x_1, x') and

$$\Delta_{\mathbf{j}}^{\#} = \Delta_{j_1}^{\#1}\Delta_{\mathbf{j}'}^{\#}.$$

We also write

$$\mathbf{r}_{\mathbf{j}'}(t') = r_{j_2}(t_2)\cdots r_{j_{k+1}}(t_{k+1}),$$

where $t' = (t_2,\ldots,t_{k+1})$ and we set

$$S_N^m = [-N,N]^m \cap \mathbf{Z}^m, \quad m = 1,2,\ldots.$$

Using Lemma 4.6.3 (three times), the induction hypothesis, and (4.6.8) we justify the sequence of inequalities:

$$\int_{\mathbf{R}}\int_{\mathbf{R}^k}\Big(\sum_{j\in S_N^{k+1}}|\Delta_j^{\#}(f)(x_1,x')|^2\Big)^{\frac{p}{2}}dx'dx_1$$

$$\leq \frac{1}{B_p^{(k+1)p}}\int_{\mathbf{R}}\int_{\mathbf{R}^k}\int_0^1\int_{[0,1]^k}\Big|\sum_{j_1\in S_N^1}\sum_{j'\in S_N^k}r_{j_1}(t_1)r_{j'}(t')\Delta_{j_1}^{\#1}\Delta_{j'}^{\#}(f)\Big|^p dt'dt_1 dx' dx_1$$

$$= \frac{1}{B_p^{(k+1)p}}\int_{\mathbf{R}}\int_0^1\int_{\mathbf{R}^k}\int_{[0,1]^k}\Big|\sum_{j'\in S_N^k}r_{j'}(t')\Delta_{j'}^{\#}\Big[\sum_{j_1\in S_N^1}r_{j_1}(t_1)\Delta_{j_1}^{\#1}(f)\Big]\Big|^p dt'dx' dt_1 dx_1$$

$$\leq \frac{A_p^{kp}}{B_p^{(k+1)p}}\int_{\mathbf{R}}\int_0^1\int_{\mathbf{R}^k}\Big\{\sum_{j'\in S_N^k}\Big|\Delta_{j'}^{\#}\Big[\sum_{j_1\in S_N^1}r_{j_1}(t_1)\Delta_{j_1}^{\#1}(f)\Big]\Big|^2\Big\}^{\frac{p}{2}}dx'dt_1 dx_1$$

$$\leq \frac{A_p^{kp}C(p,k)^p}{B_p^{(k+1)p}}\int_{\mathbf{R}}\int_0^1\int_{\mathbf{R}^k}\Big|\sum_{j_1\in S_N^1}r_{j_1}(t_1)\Delta_{j_1}^{\#1}(f)\Big|^p dx'dt_1 dx_1$$

$$= \frac{A_p^{kp}C(p,k)^p}{B_p^{(k+1)p}}\int_{\mathbf{R}^k}\int_{\mathbf{R}}\int_0^1\Big|\sum_{j_1\in S_N^1}r_{j_1}(t_1)\Delta_{j_1}^{\#1}(f)\Big|^p dt_1 dx_1 dx'$$

$$\leq \frac{A_p^{(k+1)p}C(p,k)^p}{B_p^{(k+1)p}}\int_{\mathbf{R}^k}\int_{\mathbf{R}}\Big\{\sum_{j_1\in S_N^1}|\Delta_{j_1}^{\#1}(f)|^2\Big\}^{\frac{p}{2}}dx_1 dx'$$

$$\leq \frac{A_p^{(k+1)p}C(p,k)^p C(p,1)^p}{B_p^{(k+1)p}}\int_{\mathbf{R}^k}\int_{\mathbf{R}}|f(x_1,x')|^p dx_1 dx'$$

$$= \frac{A_p^{(k+1)p}C(p,k)^p C(p,1)^p}{B_p^{(k+1)p}}\|f\|_{L^p(\mathbf{R}^{k+1})}^p.$$

Letting S_N increase to $\mathbf{Z}^{k+1}$, the LMCT yields the case $n=k+1$ with

$$C(p,k+1)\leq C(p,k)C(p,1)\Big(\frac{A_p}{B_p}\Big)^{k+1}.$$

This result, combined with (4.6.8), provides the upper inequality in (4.6.7) with constant

$$C(p,n)=C_n\max\Big(p,\frac{1}{p-1}\Big)^{2n}\Big(\frac{A_p}{B_p}\Big)^{2+3+\cdots+n}.$$

We now prove the lower inequality in (4.6.7). Let $f\in L^p(\mathbf{R}^n)$ and φ be a Schwartz function whose Fourier transform support is compact and does not intersect any hyperplane of the form $x_j=0$; such functions are dense in $L^p(\mathbf{R}^n)$, see Exercise 4.6.1. Then we can write

$$\varphi=\sum_{j\in\mathbf{Z}^n}\Delta_j^{\#}(\varphi),$$

where only finitely many $j\in\mathbf{Z}^n$ appear in the sum; this explains the interchange of summation and integration below. Using this information write

4.6 Littlewood–Paley Theory of Product Type

$$\left| \int_{\mathbf{R}^n} f(y)\varphi(y)\,dy \right| = \left| \int_{\mathbf{R}^n} f(y) \sum_{j\in\mathbf{Z}^n} \Delta_j^\sharp \Delta_j^\sharp(\varphi)(y)\,dy \right|$$

$$= \left| \sum_{j\in\mathbf{Z}^n} \int_{\mathbf{R}^n} f(y) \Delta_j^\sharp \Delta_j^\sharp(\varphi)(y)\,dy \right|$$

$$= \left| \sum_{j\in\mathbf{Z}^n} \int_{\mathbf{R}^n} \Delta_j^\sharp(f)(y) \Delta_j^\sharp(\varphi)(y)\,dy \right|$$

$$= \left| \int_{\mathbf{R}^n} \sum_{j\in\mathbf{Z}^n} \Delta_j^\sharp(f)(y) \Delta_j^\sharp(\varphi)(y)\,dy \right|$$

$$\leq \int_{\mathbf{R}^n} \sum_{j\in\mathbf{Z}^n} |\Delta_j^\sharp(f)(y)| |\Delta_j^\sharp(\varphi)(y)|\,dy$$

$$\leq \int_{\mathbf{R}^n} \Big(\sum_{j\in\mathbf{Z}^n} |\Delta_j^\sharp(f)(y)|^2\Big)^{\frac{1}{2}} \Big(\sum_{j\in\mathbf{Z}^n} |\Delta_j^\sharp(\varphi)(y)|^2\Big)^{\frac{1}{2}} dy$$

$$\leq \left\|\Big(\sum_{j\in\mathbf{Z}^n} |\Delta_j^\sharp(f)|^2\Big)^{\frac{1}{2}}\right\|_{L^p} \left\|\Big(\sum_{j\in\mathbf{Z}^n} |\Delta_j^\sharp(\varphi)|^2\Big)^{\frac{1}{2}}\right\|_{L^{p'}}$$

$$\leq C(p',n) \|\varphi\|_{L^{p'}} \left\|\Big(\sum_{j\in\mathbf{Z}^n} |\Delta_j^\sharp(f)|^2\Big)^{\frac{1}{2}}\right\|_{L^p}.$$

We now take the supremum over all Schwartz functions φ whose Fourier transform does not cross any axis and have $L^{p'}$ norm equal to 1. This yields

$$\frac{\|f\|_{L^p(\mathbf{R}^n)}}{C(p',n)} \leq \left\|\Big(\sum_{j\in\mathbf{Z}^n} |\Delta_j^\sharp(f)|^2\Big)^{\frac{1}{2}}\right\|_{L^p(\mathbf{R}^n)},$$

which is the lower inequality in (4.6.7). $\square$

A straightforward modification of Theorem 4.6.2 yields its smooth counterpart.

Theorem 4.6.4. *Fix $\psi \in \mathscr{S}(\mathbf{R})$ whose Fourier transform is supported in a compact subset of $\mathbf{R} \setminus \{0\}$. Define Littlewood–Paley operators $\Delta_j^{\otimes\psi} = \Delta_{j_1}^{\psi,(1)} \circ \cdots \circ \Delta_{j_n}^{\psi,(n)}$, where $j = (j_1, \ldots, j_n)$, and where each $\Delta_{j_k}^{\psi,(k)}$ acts on the kth variable with the remaining variables fixed. Let $1 < p < \infty$. Then there is a positive constant $C_{n,p,\psi}$ such that for any $f \in L^p(\mathbf{R}^n)$ we have*

$$\left\|\Big(\sum_{j\in\mathbf{Z}^n} |\Delta_j^{\otimes\psi}(f)|^2\Big)^{\frac{1}{2}}\right\|_{L^p(\mathbf{R}^n)} \leq C_{n,p,\psi} \|f\|_{L^p(\mathbf{R}^n)}. \tag{4.6.10}$$

Suppose additionally that $\sum_{j\in\mathbf{Z}} \widehat{\psi}(2^{-j}y) = 1$, when $y \in \mathbf{R} \setminus \{0\}$. Then there is another positive constant $c_{n,p,\psi}$ such that for any f in $L^p(\mathbf{R}^n)$ we have

$$\frac{1}{c_{n,p,\psi}} \|f\|_{L^p(\mathbf{R}^n)} \leq \left\|\Big(\sum_{j\in\mathbf{Z}^n} |\Delta_j^{\otimes\psi}(f)|^2\Big)^{\frac{1}{2}}\right\|_{L^p(\mathbf{R}^n)}. \tag{4.6.11}$$

Proof. The proof of (4.6.10) follows the paradigm of the corresponding estimate in Theorem 4.6.2 with only notational changes. The only difference is that in dimension $n = 1$ one directly uses Theorem 4.4.2. We skip the details.

So we focus on inequality (4.6.11). We pick another function ζ in $\mathscr{S}(\mathbf{R})$ whose Fourier transform vanishes in a neighborhood of the origin and which equals 1 on the support of $\widehat{\widetilde{\psi}}$. Define $\Delta_j^{\otimes \zeta}$ analogously, and notice that

$$\Delta_j^{\otimes \widetilde{\psi}} = \Delta_j^{\otimes \zeta} \Delta_j^{\otimes \widetilde{\psi}} \quad \text{for any } \boldsymbol{j} \in \mathbf{Z}^n.$$

Let $f \in L^p(\mathbf{R}^n)$ and let φ be a Schwartz function whose Fourier transform support is compact and does not intersect any plane of the form $x_j = 0$; such functions are dense in $L^p(\mathbf{R}^n)$; see Exercise 4.6.1. Then we write

$$\varphi = \sum_{\boldsymbol{j} \in \mathbf{Z}^n} \Delta_j^{\otimes \widetilde{\psi}}(\varphi),$$

where only finitely many $\boldsymbol{j} \in \mathbf{Z}^n$ appear in the sum. Using this information write

$$\left| \int_{\mathbf{R}^n} f(y)\varphi(y)\,dy \right| = \left| \int_{\mathbf{R}^n} f(y) \sum_{\boldsymbol{j} \in \mathbf{Z}^n} \Delta_j^{\otimes \widetilde{\psi}}(\varphi)(y)\,dy \right|$$

$$= \left| \sum_{\boldsymbol{j} \in \mathbf{Z}^n} \int_{\mathbf{R}^n} f(y) \Delta_j^{\otimes \widetilde{\psi}} \Delta_j^{\otimes \zeta}(\varphi)(y)\,dy \right|$$

$$= \left| \sum_{\boldsymbol{j} \in \mathbf{Z}^n} \int_{\mathbf{R}^n} \Delta_j^{\otimes \psi}(f)(y) \Delta_j^{\otimes \zeta}(\varphi)(y)\,dy \right|$$

$$= \left| \int_{\mathbf{R}^n} \sum_{\boldsymbol{j} \in \mathbf{Z}^n} \Delta_j^{\otimes \psi}(f)(y) \Delta_j^{\otimes \zeta}(\varphi)(y)\,dy \right|$$

$$\leq \int_{\mathbf{R}^n} \sum_{\boldsymbol{j} \in \mathbf{Z}^n} |\Delta_j^{\otimes \psi}(f)| |\Delta_j^{\otimes \zeta}(\varphi)|\,dx$$

$$\leq \int_{\mathbf{R}^n} \Big(\sum_{\boldsymbol{j} \in \mathbf{Z}^n} |\Delta_j^{\otimes \psi}(f)|^2 \Big)^{\frac{1}{2}} \Big(\sum_{\boldsymbol{j} \in \mathbf{Z}^n} |\Delta_j^{\otimes \zeta}(\varphi)|^2 \Big)^{\frac{1}{2}} dx$$

$$\leq \Big\| \Big(\sum_{\boldsymbol{j} \in \mathbf{Z}^n} |\Delta_j^{\otimes \psi}(f)|^2 \Big)^{\frac{1}{2}} \Big\|_{L^p} \Big\| \Big(\sum_{\boldsymbol{j} \in \mathbf{Z}^n} |\Delta_j^{\otimes \zeta}(\varphi)|^2 \Big)^{\frac{1}{2}} \Big\|_{L^{p'}}$$

$$\leq \frac{1}{c_{n,p,\psi}} \|\varphi\|_{L^{p'}} \Big\| \Big(\sum_{\boldsymbol{j} \in \mathbf{Z}^n} |\Delta_j^{\otimes \psi}(f)|^2 \Big)^{\frac{1}{2}} \Big\|_{L^p}.$$

Taking the supremum over all φ as above yields the desired inequality. $\square$

Exercises

4.6.1. Prove that the set
$$\widehat{\mathcal{S}_{0,\ldots,0}} = \{\varphi \in \mathcal{S}(\mathbf{R}^n) : \widehat{\varphi} \in \mathcal{C}_0^\infty \text{ and } \min_{1 \le j \le n} \text{dist}\left[\text{supp}(\widehat{\varphi}), \{x \in \mathbf{R}^n : x_j = 0\}\right] > 0\}$$
is dense in $L^p(\mathbf{R}^n)$ when $1 < p < \infty$. [*Hint:* Mimic the proof of Proposition 2.5.4.]

4.6.2. Let $1 < p < \infty$. Prove that there is a constant $C_{n,p}$ such that for any finite subset S of $\mathbf{Z}^n$ and every f_j in $L^p(\mathbf{R}^n)$, $j \in \mathbf{Z}^n$, we have
$$\left\| \sum_{j \in S} \Delta_j^\sharp(f_j) \right\|_{L^p} \le c_{n,p} \left\| \left(\sum_{j \in S} |f_j|^2 \right)^{\frac{1}{2}} \right\|_{L^p}$$
and
$$\left\| \sum_{j \in S} \Delta_j^\sharp(f_j) \right\|_{L^p} \le c_{n,p} \left\| \left(\sum_{j \in S} |\Delta_j^\sharp(f_j)|^2 \right)^{\frac{1}{2}} \right\|_{L^p}.$$
Conclude that there is a constant $C_{n,p}$ such that for every $f \in L^p(\mathbf{R}^n)$ we have
$$\left\| \sum_{j \in S} \Delta_j^\sharp(f) \right\|_{L^p} \le C_{n,p} \|f\|_{L^p}.$$

4.6.3. Suppose that $\{m_j\}_{j \in \mathbf{Z}^n}$ is a sequence of bounded functions supported in the sets R_j defined in (4.6.5). Let $T_j(f) = (\widehat{f} m_j)^\vee$ be the multiplier operator associated with m_j. Let $1 < p < \infty$. Assume that there is a constant A_p for all sequences of functions $\{f_j\}_{j \in \mathbf{Z}^n}$ with $f_j \in L^p(\mathbf{R}^n)$ the vector-valued inequality
$$\left\| \left(\sum_{j \in \mathbf{Z}^n} |T_j(f_j)|^2 \right)^{\frac{1}{2}} \right\|_{L^p(\mathbf{R}^n)} \le A_p \left\| \left(\sum_{j \in \mathbf{Z}^n} |f_j|^2 \right)^{\frac{1}{2}} \right\|_{L^p(\mathbf{R}^n)}$$
is valid. Prove there is a $C_{p,n} > 0$ such that for all finite subsets S of $\mathbf{Z}$ we have
$$\left\| \sum_{j \in S} m_j \right\|_{\mathcal{M}_p} \le C_{p,n} A_p.$$

4.6.4. Fix $\theta \in \mathbf{S}^{n-1}$. For $j \in \mathbf{Z}$ define sets $S_j^\theta = \{\xi \in \mathbf{R}^n : 2^j \le |\xi \cdot \theta| < 2^{j+1}\}$ and operators $T_{S_j^\theta}(f) = (\widehat{f} \chi_{S_j^\theta})^\vee$ initially on $\mathcal{S}(\mathbf{R}^n)$ and later extended on $L^p(\mathbf{R}^n)$ for $1 < p < \infty$. Prove that for any $g \in L^p(\mathbf{R}^n)$ we have
$$\|g\|_{L^p(\mathbf{R}^n)} \approx \left\| \left(\sum_{j \in \mathbf{Z}} |T_{S_j^\theta}(g)|^2 \right)^{\frac{1}{2}} \right\|_{L^p(\mathbf{R}^n)}.$$
[*Hint:* Consider first the case $\theta = e_1$ and then apply a rotation.]

4.6.5. Show that for $1 < p < \infty$ there is a constant $K(n,p)$ such that for any f_j in $L^p(\mathbf{R}^n)$, $j \in \mathbf{Z}^n$, we have

$$\left\|\left(\sum_{j \in \mathbf{Z}^n} |\Delta_j^\sharp(f_j)|^2\right)^{\frac{1}{2}}\right\|_{L^p(\mathbf{R}^n)} \leq K(n,p) \left\|\left(\sum_{j \in \mathbf{Z}^n} |f_j|^2\right)^{\frac{1}{2}}\right\|_{L^p(\mathbf{R}^n)}.$$

Moreover, $K(n,p) \leq K(1,p)^n (A_p/B_p)^{2n}$, where A_p, B_p are as in Lemma 4.6.3. Finally, construct an example to show that the reverse inequality fails. [*Hint:* For with $n = 1$ appeal to Exercise 4.2.3. Reduce the n-dimensional result to the case $n = 1$ by applying Lemma 4.6.3 multiple times.]

4.6.6. Let ψ and $\Delta_j^{\otimes \psi}$ be as in Theorem 4.6.4. Show that for $1 < p < \infty$ there is a constant $C_{n,p,\psi}$ such that for any $f_j \in L^p(\mathbf{R}^n)$, $j \in \mathbf{Z}^n$, we have

$$\left\|\left(\sum_{j \in \mathbf{Z}^n} |\Delta_j^{\otimes \psi}(f_j)|^2\right)^{\frac{1}{2}}\right\|_{L^p(\mathbf{R}^n)} \leq C_{n,p,\psi} \left\|\left(\sum_{j \in \mathbf{Z}^n} |f_j|^2\right)^{\frac{1}{2}}\right\|_{L^p(\mathbf{R}^n)}.$$

[*Hint:* Apply the method used in Exercise 4.6.5 starting with $n = 1$.]

4.6.7. Use (4.6.9) to prove the following statement: Let T be a linear operator bounded from $L^p(\mathbf{R}^n)$ to itself for some $0 < p < \infty$. Show that for any $f_j \in L^p(\mathbf{R}^n)$ we have

$$\left\|\left(\sum_{j \in \mathbf{Z}} |T(f_j)|^2\right)^{\frac{1}{2}}\right\|_{L^p} \leq \frac{A_p}{B_p} \|T\|_{L^p \to L^p} \left\|\left(\sum_{j \in \mathbf{Z}} |f_j|^2\right)^{\frac{1}{2}}\right\|_{L^p}.$$

As a consequence, derive another proof of Proposition 4.6.1 using (4.6.9).

Chapter 5
Fractional Integrability or Differentiability and Multiplier Theorems

5.1 Powers of the Laplacian and Riesz Potentials

The *Laplacian* is the operator $\Delta = \partial_1^2 + \cdots + \partial_n^2$ initially defined on $\mathscr{C}^2(\mathbf{R}^n)$. The action of the Laplacian is also defined on tempered distributions u on $\mathbf{R}^n$ by duality:

$$\langle \Delta u, \varphi \rangle = \langle u, \Delta \varphi \rangle, \qquad \varphi \in \mathscr{S}(\mathbf{R}^n).$$

Let $\varphi \in \mathscr{S}(\mathbf{R}^n)$. Applying the Fourier transform, we write

$$-\widehat{\Delta \varphi}(\xi) = -\sum_{j=1}^{n} (-2\pi i \xi_j)^2 \widehat{\varphi}(\xi) = (2\pi|\xi|)^2 \widehat{\varphi}(\xi), \qquad \xi \in \mathbf{R}^n$$

The exponent 2 indicates the total number of differentiations on φ. Motivated by this identity, it is tempting to replace the exponent 2 by a complex exponent z and define $(-\Delta)^{z/2}$ as the operator given by the multiplication with the function $(2\pi|\xi|)^z$ on the Fourier transform. Then $(-\Delta)^{z/2}\varphi$ represents in some sense the *total derivative* of φ of *order z*. Precisely, for $z \in \mathbf{C}$ with $\mathrm{Re}\, z > -n$ and Schwartz functions φ we define

$$(-\Delta)^{z/2}\varphi = ((2\pi|\cdot|)^z \widehat{\varphi})^\vee. \tag{5.1.1}$$

If z is an even integer, then clearly $(-\Delta)^{z/2}$ is a derivative of order z. For complex values of z, we call $(-\Delta)^{z/2}$ the *total derivative of order z*. If z is a complex number with real part less than $-n$, then the function $|\xi|^z$ is not locally integrable on $\mathbf{R}^n$ and so (5.1.1) may not be well defined. For this reason, we extend (5.1.1) to $\mathrm{Re}\, z \leq -n$ only to Schwartz functions φ whose Fourier transform vanishes to sufficiently high order at the origin[1] so that the expression $|\xi|^z \widehat{\varphi}(\xi)$ is integrable. Note that the family of operators $(-\Delta)^z$ satisfies the *semigroup property* $(-\Delta)^z (-\Delta)^w = (-\Delta)^{z+w}$ for all $z, w \in \mathbf{C}$ when acting on Schwartz functions whose Fourier transform vanishes to sufficiently high order at the origin.

---
[1] This means that sufficiently many derivatives of $\widehat{\varphi}$ vanish at the origin.

Recall the identity (2.7.5) in Example 2.7.4. For $-n < \operatorname{Re} z < 0$, the locally integrable function
$$u_z(x) = \frac{\pi^{\frac{z+n}{2}}}{\Gamma(\frac{z+n}{2})} |x|^z, \qquad x \in \mathbf{R}^n \setminus \{0\},$$
satisfies $\widehat{u_z} = u_{-n-z}$. This implies that the inverse Fourier transform of $(2\pi|\cdot|)^z$ is
$$\left((2\pi|\cdot|)^z\right)^{\vee}(x) = (2\pi)^z \frac{\pi^{-\frac{z}{2}}}{\pi^{\frac{z+n}{2}}} \frac{\Gamma(\frac{n+z}{2})}{\Gamma(\frac{-z}{2})} |x|^{-z-n}. \tag{5.1.2}$$

When $-n < \operatorname{Re} z < 0$, both $|\xi|^z$ and $|x|^{-z-n}$ are locally integrable functions.

When $s > 0$, then $(-\Delta)^{-s/2} f$ is not really differentiating f, but it is integrating it. For this reason, we introduce a different notation that better reflects the nature of this operator.

Definition 5.1.1. Let z be a complex number with $0 < \operatorname{Re} z < n$. The *Riesz potential operator* of order z is
$$\mathcal{I}_z = (-\Delta)^{-z/2}.$$

In view of identity (5.1.2), we could express $\mathcal{I}_z$ as a convolution operator as follows:
$$\mathcal{I}_z(f)(x) = 2^{-z} \pi^{-\frac{n}{2}} \frac{\Gamma(\frac{n-z}{2})}{\Gamma(\frac{z}{2})} \int_{\mathbf{R}^n} f(x-y)|y|^{-n+z} dy, \qquad f \in \mathscr{S}(\mathbf{R}^n). \tag{5.1.3}$$

Notice that this integral is absolutely convergent for $f \in \mathscr{S}(\mathbf{R}^n)$ for all $\operatorname{Re} z > 0$. Moreover, if f is simply measurable and nonnegative and z is real, then $\mathcal{I}_z(f)(x)$ is well defined, but could be infinite for some (or all) values of $x \in \mathbf{R}^n$.

We begin with a remark concerning the homogeneity of the operator $\mathcal{I}_s$.

Remark 5.1.2. Let $0 < p,q \leq \infty$ and suppose that for some $s \in (0,n)$ we have the estimate
$$\|\mathcal{I}_s\|_{L^p \to L^q} < \infty. \tag{5.1.4}$$
Then the following must be true:
$$\frac{1}{p} - \frac{1}{q} = \frac{s}{n} \quad \text{and} \quad p > 1. \tag{5.1.5}$$

To prove these assertions we consider the function $f = \chi_{B(0,1)}$. Then obviously,
$$0 < \|f\|_{L^p(\mathbf{R}^n)} < \infty. \tag{5.1.6}$$

When $|x| \geq 2$ and $|y| < 1$ we have $|x-y| \approx |x|$, so
$$\mathcal{I}_s(f)(x) = C_{s,n} \int_{|y|<1} |x-y|^{-n+s} dy \geq C'_{s,n} |x|^{-n+s}, \qquad |x| \geq 2, \tag{5.1.7}$$

and assumption (5.1.4) yields that $\|\mathcal{I}_s(f)\|_{L^q} < \infty$. Combining these facts we deduce that $q(-n+s) < -n$ (equivalently $\frac{s}{n} + \frac{1}{q} < 1$) and that

5.1 Powers of the Laplacian and Riesz Potentials

$$\|\mathcal{I}_s(f)\|_{L^q} > 0. \tag{5.1.8}$$

Consider the dilation $f^\lambda(x) = f(\lambda x)$ defined for $\lambda > 0$. Changing variables, we write

$$\begin{aligned}\mathcal{I}_s(f^\lambda)(x) &= C_{s,n} \int_{\mathbf{R}^n} f(\lambda x - \lambda y)|y|^{-n+s}dy \\ &= C_{s,n} \int_{\mathbf{R}^n} f(\lambda x - y')\lambda^{n-s}|y'|^{-n+s}\lambda^{-n}dy' \\ &= \lambda^{-s}\mathcal{I}_s(f)(\lambda x). \end{aligned} \tag{5.1.9}$$

As (5.1.4) is assumed to hold, we obtain

$$\|\mathcal{I}_s(f^\lambda)\|_{L^q(\mathbf{R}^n)} \leq \|\mathcal{I}_s\|_{L^p \to L^q} \|f^\lambda\|_{L^p(\mathbf{R}^n)},$$

which, in view of (5.1.9), can be written as

$$\|\mathcal{I}_s(f)\|_{L^q(\mathbf{R}^n)} \leq \|\mathcal{I}_s\|_{L^p \to L^q} \lambda^{n(\frac{1}{q} - \frac{1}{p} + \frac{s}{n})} \|f\|_{L^p(\mathbf{R}^n)}. \tag{5.1.10}$$

If $\frac{1}{q} - \frac{1}{p} + \frac{s}{n} > 0$, then we let $\lambda \to 0$, whereas if $\frac{1}{q} - \frac{1}{p} + \frac{s}{n} < 0$ we let $\lambda \to \infty$ in (5.1.10). In both cases, recalling (5.1.4), (5.1.6), and (5.1.8), we obtain that a positive quantity is less than or equal to zero. Thus, $\frac{1}{q} - \frac{1}{p} + \frac{s}{n} \neq 0$ is not possible. It follows that $\frac{1}{p} - \frac{1}{q} = \frac{s}{n}$. Combining this identity with the previously obtained relationship $\frac{s}{n} + \frac{1}{q} < 1$, we deduce that $p > 1$. Thus (5.1.5) must necessarily be valid.

It turns out there is a positive estimate under hypothesis (5.1.5).

Theorem 5.1.3. *(Hardy–Littlewood–Sobolev theorem on fractional integration) Let s be a real number, with $0 < s < n$, and let $1 < p < \frac{n}{s}$ and $\frac{n}{n-s} < q < \infty$ be related as in $\frac{1}{p} - \frac{1}{q} = \frac{s}{n}$. Then there exist constants $C(n,s,p)$, $C(s,n) < \infty$ such that for all f in $\mathscr{S}(\mathbf{R}^n)$ we have*

$$\|\mathcal{I}_s(f)\|_{L^q(\mathbf{R}^n)} \leq C(n,s,p) \|f\|_{L^p(\mathbf{R}^n)} \tag{5.1.11}$$

and

$$\|\mathcal{I}_s(f)\|_{L^{\frac{n}{n-s},\infty}(\mathbf{R}^n)} \leq C(n,s) \|f\|_{L^1(\mathbf{R}^n)}. \tag{5.1.12}$$

Thus $\mathcal{I}_s$ has unique bounded extensions from $L^p(\mathbf{R}^n)$ to $L^q(\mathbf{R}^n)$ and from $L^1(\mathbf{R}^n)$ to $L^{\frac{n}{n-s},\infty}(\mathbf{R}^n)$.

Proof. Fix a nonzero function f in the Schwartz class. Write

$$\int_{\mathbf{R}^n} |f(x-y)| |y|^{s-n} dy = I_1(f)(x) + I_2(f)(x),$$

where I_1 and I_2 are defined by

$$I_1(f)(x) = \int_{|y|<\varepsilon} |f(x-y)| \, |y|^{s-n} \, dy,$$

$$I_2(f)(x) = \int_{|y|\geq\varepsilon} |f(x-y)| \, |y|^{s-n} \, dy,$$

for some $\varepsilon = \varepsilon(x) > 0$ to be determined later. Begin by writing

$$I_1(f)(x) = \sum_{j=0}^{\infty} \int_{2^{-j-1}\varepsilon \leq |y| < 2^{-j}\varepsilon} |y|^{-n+s} |f(x-y)| \, dy$$

$$\leq \sum_{j=0}^{\infty} 2^{(j+1)(n-s)} \varepsilon^{-n+s} \int_{|y|<2^{-j}\varepsilon} |f(x-y)| \, dy$$

$$\leq \varepsilon^s M(f)(x) 2^{n-s} v_n \sum_{j=0}^{\infty} 2^{-js}$$

$$\leq \frac{v_n 2^{n-s}}{1-2^{-s}} \varepsilon^s M(f)(x). \tag{5.1.13}$$

Let $1 \leq p < \frac{n}{s}$. Observe that $(n-s)p' = n + \frac{p'n}{q} > n$ and this is valid even when $1' = \infty$. Let $\omega_{n-1} = |\mathbf{S}^{n-1}|$. Hölder's inequality gives that

$$|I_2(f)(x)| \leq \left(\int_{|y|\geq\varepsilon} |y|^{-(n-s)p'} dy \right)^{\frac{1}{p'}} \|f\|_{L^p(\mathbf{R}^n)}$$

$$= \left(\frac{q\omega_{n-1}}{p'n} \right)^{\frac{1}{p'}} \varepsilon^{-\frac{n}{q}} \|f\|_{L^p(\mathbf{R}^n)}, \tag{5.1.14}$$

and note that this estimate is also valid when $p=1$ (in which case $q=\frac{n}{n-s}$), provided the $L^{p'}$ norm is interpreted as the L^∞ norm and the constant $\left(\frac{q\omega_{n-1}}{p'n}\right)^{\frac{1}{p'}}$ is replaced by 1. Combining (5.1.13) and (5.1.14), we obtain that

$$\mathcal{I}_s(f)(x) \leq C'_{n,s,p} \left(\varepsilon^s M(f)(x) + \varepsilon^{-\frac{n}{q}} \|f\|_{L^p} \right). \tag{5.1.15}$$

We choose

$$\varepsilon = \varepsilon(x) = \|f\|_{L^p}^{\frac{p}{n}} \left(M(f)(x) \right)^{-\frac{p}{n}}$$

to minimize the expression on the right-hand side in (5.1.15). We observe that if f is nonzero, then $M(f)(x) > 0$ for all $x \in \mathbf{R}^n$ and thus ε is well defined. This choice of ε yields the estimate

$$\mathcal{I}_s(f)(x) \leq C_{n,s,p} M(f)(x)^{\frac{p}{q}} \|f\|_{L^p}^{1-\frac{p}{q}}. \tag{5.1.16}$$

Now suppose that $p > 1$. We raise (5.1.16) to the power q, we integrate over $\mathbf{R}^n$, and we use the boundedness of the Hardy–Littlewood maximal operator M on $L^p(\mathbf{R}^n)$ (Corollary 1.4.7). This yields (5.1.11).

5.1 Powers of the Laplacian and Riesz Potentials

We are left with the case $p = 1$ when $q = \frac{n}{n-s}$. The result in this case also follows from (5.1.16) by the weak-type $(1,1)$ property of M (Theorem 1.4.6). Indeed for all $\lambda > 0$ we have

$$\left|\left\{C_{n,s,1} M(f)^{\frac{n-s}{n}} \|f\|_{L^1}^{\frac{s}{n}} > \lambda\right\}\right| = \left|\left\{M(f) > \left(\frac{\lambda}{C_{n,s,1} \|f\|_{L^1}^{\frac{s}{n}}}\right)^{\frac{n}{n-s}}\right\}\right|$$

$$\leq 3^n \left(\frac{C_{n,s,1} \|f\|_{L^1}^{\frac{s}{n}}}{\lambda}\right)^{\frac{n}{n-s}} \|f\|_{L^1}$$

$$= \left(C(n,s) \frac{\|f\|_{L^1}}{\lambda}\right)^{\frac{n}{n-s}}.$$

This is the claimed estimate (5.1.12). $\square$

Example 5.1.4. It follows from (5.1.7) that $\mathcal{I}_s$ does not map $L^1(\mathbf{R}^n)$ to $L^{\frac{n}{n-s}}(\mathbf{R}^n)$.

Additionally, $\mathcal{I}_s$ does not map $L^{\frac{n}{s}}(\mathbf{R}^n)$ to $L^\infty(\mathbf{R}^n)$. To see this, let $0 < s < n$ and pick δ such that $0 < \delta < \frac{n-s}{s}$. Consider the function $h(x) = |x|^{-s} (\log \frac{1}{|x|})^{-\frac{s}{n}(1+\delta)}$ for $|x| \leq 1/e$ and zero otherwise. A straightforward substitution based on the convergence of the integral $\int_0^{1/e} r^{-1} [\log(r^{-1})]^{-(1+\delta)} dr$ indicates that h lies in $L^{\frac{n}{s}}(\mathbf{R}^n)$. On the other hand for $|x| \leq \frac{1}{100}$ we have

$$\mathcal{I}_s(h)(x) = c \int_{|y| \leq 1/e} |y|^{-s} \left(\log \frac{1}{|y|}\right)^{-\frac{s}{n}(1+\delta)} |x-y|^{s-n} dy$$

$$\geq c \int_{2|x| \leq |y| \leq 1/e} |y|^{-s} \left(\log \frac{1}{|y|}\right)^{-\frac{s}{n}(1+\delta)} |x-y|^{s-n} dy$$

$$\geq c' \int_{2|x| \leq |y| \leq 1/e} |y|^{-s} \left(\log \frac{1}{2|x|}\right)^{-\frac{s}{n}(1+\delta)} |y|^{s-n} dy$$

$$= c'' \left(\log \frac{1}{2e|x|}\right) \left(\log \frac{1}{2|x|}\right)^{-\frac{s}{n}(1+\delta)}$$

which tends to infinity as $|x| \to 0$, since δ is so small so that $1 - \frac{s}{n}(1+\delta) > 0$. Consequently, $\mathcal{I}_s(h) \notin L^\infty$.

Exercises

5.1.1. Let z_1, z_2 be complex numbers with positive real parts (or z_1, z_2 could be zero). Find a $w \in \mathbf{C}$ such that for all φ in $\mathscr{S}(\mathbf{R}^n)$ we have

$$\int_{\mathbf{R}^n} \mathcal{I}_{z_1}(\varphi)(x) \overline{\mathcal{I}_{z_2}(\varphi)(x)} \, dx = \left\|(-\Delta)^w \varphi\right\|_{L^2(\mathbf{R}^n)}^2.$$

5.1.2. Let $1 \leq q \leq \infty$ and $z \in \mathbf{C}$ have positive real part. Prove that for $\varphi \in \mathscr{S}(\mathbf{R}^n)$ we have

$$\|\varphi\|_{L^2}^2 \leq \|\mathcal{I}_z(\varphi)\|_{L^q} \|(-\Delta)^{\bar{z}/2}\varphi\|_{L^{q'}}.$$

5.1.3. Show that for any $t > 0$ there is a constant $c_{n,t}$ such that

$$\|f\|_{L^1(\mathbf{R}^n)} \leq c_{n,t} \|(-\Delta)^{\frac{n+t}{2}} f\|_{L^2(\mathbf{R}^n)}^{\frac{n}{n+t}} \|f\|_{L^2(\mathbf{R}^n)}^{\frac{t}{n+t}}$$

is valid for all $f \in L^2(\mathbf{R}^n)$. Note that the Fourier transform of $(-\Delta)^{\frac{n+t}{2}} f$ is well defined when $f \in L^2$ and so does its L^2 norm, which could be infinite. [*Hint:* First prove that $\int_{\mathbf{R}^n} |f(x)|\,dx \leq C\big(\int_{\mathbf{R}^n} |f(x)|^2(|x|^{n+t}+1)\,dx\big)^{1/2}$, then apply this inequality to $f(\lambda x)$, and optimize over λ.]

5.1.4. Let s be a real number, with $0 < s < n$, and let $1 < p < \frac{n}{s}$ and $\frac{n}{n-s} < q < \infty$ be related as in (5.1.5). Suppose that $(-\Delta)^{\frac{s}{2}} f \in L^p(\mathbf{R}^n)$ for a given $f \in \mathcal{S}'(\mathbf{R}^n)$. Prove that f coincides with an L^q function whose norm satisfies the estimate

$$\|f\|_{L^q} \leq C(n,p,s) \|(-\Delta)^{\frac{s}{2}} f\|_{L^p},$$

where $C(n,p,s) < \infty$. [*Hint:* Use Theorem 5.1.3 and that $\mathcal{I}_s$ is self-adjoint.]

5.1.5. Let f be a tempered and locally integrable function on $\mathbf{R}^n$. Suppose that either (a) $n \geq 2$ and the distributional derivatives $\partial_j f$ lie in $L^{p_1}(\mathbf{R}^n) \cap L^{p_2}(\mathbf{R}^n)$ for all $j = 1,\ldots,n$, where $1 < p_1 < n < p_2 < \infty$; or (b) $n \geq 3$ and the distributional Laplacian Δf lies in $L^{p_1}(\mathbf{R}^n) \cap L^{p_2}(\mathbf{R}^n)$, where $1 \leq p_1 < \frac{n}{2} < p_2 \leq \infty$. Prove that f lies in $L^\infty(\mathbf{R}^n)$.
[*Hint:* (a) Use the identity $f = \sum_{j=1}^n \mathcal{I}_1(R_j(\partial_j f))$. (b) Write $f = -\mathcal{I}_2(\Delta f)$.]

5.1.6. For $0 < s < n$ define the *fractional maximal function*

$$M^s(f)(x) = \sup_{t>0} \frac{1}{(v_n t^n)^{\frac{n-s}{n}}} \int_{|y| \leq t} |f(x-y)|\,dy, \qquad f \in L^1_{\mathrm{loc}}(\mathbf{R}^n),$$

where $v_n = |B(0,1)|$. Show that for some finite constant $C(n,s)$ we have

$$M^s(f) \leq C(n,s) \mathcal{I}_s(|f|).$$

5.1.7. For continuous functions f on $\mathbf{R}^n$ define the *difference operator* $D_h f(x) = f(x+h) - f(x)$ for $x, h \in \mathbf{R}^n$. Let $0 < s < m$ and $m \in \mathbf{Z}^+$. Prove that for $f \in \mathcal{S}(\mathbf{R}^n)$ we have

$$\int_{\mathbf{R}^n} \int_{\mathbf{R}^n} |\overbrace{D_t \circ \cdots \circ D_t}^{m \text{ times}} f(x)|^2\,dx \frac{dt}{|t|^{n+2s}} = C(m,n,s) \|(|\cdot|^s \widehat{f})^\vee\|_{L^2}^2,$$

where $C(m,n,s) = \int_{\mathbf{R}^n} |e^{2\pi i t_1} - 1|^{2m} |t|^{-n-2s}\,dt < \infty$.
[*Hint:* Use that $(D_t \circ \cdots \circ D_t f)\widehat{}(\xi) = \widehat{f}(\xi)(e^{2\pi i \xi \cdot t} - 1)^m$.]

5.1.8. Prove that $(-\Delta)^{z/2}\varphi$ is a bounded function whenever
(a) $\operatorname{Re} z > -n$ and $\varphi \in \mathscr{S}(\mathbf{R}^n)$; or
(b) $z \in \mathbf{C}$ and the Fourier transform of $\varphi \in \mathscr{S}(\mathbf{R}^n)$ vanishes in a neighborhood of 0.

5.1.9. Fix $w \in \mathbf{C}$ with $\operatorname{Re} w > 0$. Show that for every $\varphi \in \mathscr{C}_0^\infty(\mathbf{R}^n)$ there is a constant $C(n,w,\varphi)$ such that
$$\left|(-\Delta)^{w/2}\varphi(x)\right| \leq C(n,w,\varphi)(1+|x|)^{-n-\operatorname{Re} w}.$$
Moreover, prove that for any $s > 0$ and every nonnegative and nonzero $\varphi \in \mathscr{C}_0^\infty(\mathbf{R}^n)$, there are constants $C, K > 0$ such that for all $|x| > K$ one has
$$\left|(-\Delta)^{s/2}\varphi(x)\right| \geq C|x|^{-n-s}.$$
[*Hint:* Identity (5.1.3) applied to φ can be extended to complex numbers z with $\operatorname{Re} z < 0$ by analytic continuation, for large values of x.]

5.2 Bessel Potentials

In this section we study an *adjustment* of the Riesz potentials in which we replace $-\Delta$ by $I - \Delta$, where I is the identity operator. This simple modification allows one to define the action of $(I - \Delta)^{z/2}$ on $\mathscr{S}(\mathbf{R}^n)$ for all $z \in \mathbf{C}$. Notice that $(1+4\pi^2|\cdot|^2)^{z/2}\widehat{\varphi}$ lies in $\mathscr{S}(\mathbf{R}^n)$, whenever $\varphi \in \mathscr{S}(\mathbf{R}^n)$, thus so does $(I - \Delta)^{z/2}\varphi$, by Proposition 2.6.12. In fact, $(I - \Delta)^{z/2}$ is a one-to-one and onto mapping from $\mathscr{S}(\mathbf{R}^n)$ to $\mathscr{S}(\mathbf{R}^n)$.

Definition 5.2.1. Let z be a complex number satisfying $0 < \operatorname{Re} z < \infty$. The *Bessel potential* of order z is the operator
$$\mathcal{J}_z(f) = (I-\Delta)^{-\frac{z}{2}}(f) = \left((1+4\pi^2|\cdot|^2)^{-\frac{z}{2}}\widehat{f}\,\right)^{\vee},$$
initially acting on Schwartz functions f.

We denote by G_z the kernel of $\mathcal{J}_z$, i.e.,
$$G_z = \left((1+4\pi^2|\cdot|^2)^{-\frac{z}{2}}\right)^{\vee},$$
which a priori is a tempered distribution. The Bessel potential is the operator
$$\mathcal{J}_z(\varphi) = \varphi * G_z, \qquad \varphi \in \mathscr{S}(\mathbf{R}^n).$$

The Bessel potential is obtained by replacing $4\pi^2|\xi|^2$ in the Riesz potential by the smooth term $1 + 4\pi^2|\xi|^2$. This adjustment smooths the function near zero, and this translates into rapid decay for its inverse Fourier transform at infinity.

The next result quantifies the behavior of G_s near zero and near infinity for $s > 0$. To describe this behavior we introduce a function H_s on $\mathbf{R}^n \setminus \{0\}$ by setting

$$H_s(x) = \begin{cases} |x|^{s-n} & \text{for } 0 < s < n, \\ \log \frac{4}{|x|} & \text{for } s = n, \\ 1 & \text{for } s > n. \end{cases}$$

Proposition 5.2.2. *For $s > 0$, G_s is a strictly positive $\mathscr{C}^\infty$ function on $\mathbf{R}^n \setminus \{0\}$ that satisfies $\|G_s\|_{L^1} = 1$. Moreover, there are positive constants $C(s,n)$, $c(s,n)$ such that*

$$G_s(x) \leq C(s,n) e^{-\frac{|x|}{2}} \qquad \text{when } |x| \geq 2 \tag{5.2.1}$$

and

$$\frac{1}{c(s,n)} \leq \frac{G_s(x)}{H_s(x)} \leq c(s,n) \qquad \text{when } 0 < |x| < 2. \tag{5.2.2}$$

Proof. Fix $s > 0$. The definition of the *gamma function*

$$\Gamma(s/2) = \int_0^\infty e^{-u} u^{s/2} \frac{du}{u}$$

yields for $A > 0$ the identity

$$A^{-s/2} = \frac{1}{\Gamma(s/2)} \int_0^\infty e^{-tA} t^{s/2} \frac{dt}{t}$$

via the change of variables $u = tA$. Taking $A = 1 + 4\pi^2 |\xi|^2$ we obtain

$$(1 + 4\pi^2 |\xi|^2)^{-\frac{s}{2}} = \frac{1}{\Gamma(\frac{s}{2})} \int_0^\infty e^{-t} e^{-\pi |2\sqrt{\pi t}\, \xi|^2} t^{\frac{s}{2}} \frac{dt}{t}. \tag{5.2.3}$$

Note that the preceding integral converges at both ends. Let φ be in $\mathscr{S}(\mathbf{R}^n)$. Then

$$\langle G_s, \widehat{\varphi} \rangle = \int_{\mathbf{R}^n} \widehat{G_s}(\xi) \varphi(\xi)\, d\xi = \int_{\mathbf{R}^n} (1 + 4\pi^2 |\xi|^2)^{-\frac{s}{2}} \varphi(\xi)\, d\xi \tag{5.2.4}$$

and using (5.2.3), we rewrite (5.2.4) as

$$\langle G_s, \widehat{\varphi} \rangle = \frac{1}{\Gamma(\frac{s}{2})} \int_0^\infty e^{-t} t^{\frac{s}{2}} \left[\int_{\mathbf{R}^n} e^{-\pi |2\sqrt{\pi t}\,\xi|^2} \varphi(\xi)\, d\xi \right] \frac{dt}{t}, \tag{5.2.5}$$

where the interchange of the integrals on the left in (5.2.5) is justified by the rapid decay of the integrand. Using the fact that $\varepsilon^{-n} e^{-\pi |x/\varepsilon|^2}$ is the inverse Fourier transform of $e^{-\pi |\varepsilon \xi|^2}$, we express (5.2.5) as

$$\langle G_s, \widehat{\varphi} \rangle = \frac{1}{\Gamma(\frac{s}{2})} \int_0^\infty e^{-t} t^{\frac{s}{2}} \left[\int_{\mathbf{R}^n} \frac{1}{(2\sqrt{\pi t})^n} e^{-\pi \left|\frac{x}{2\sqrt{\pi t}}\right|^2} \widehat{\varphi}(x)\, dx \right] \frac{dt}{t}. \tag{5.2.6}$$

As $e^{-\frac{|x|^2}{4t}} \leq C_\beta (4t/|x|^2)^\beta$ for any $\beta > 0$, picking $\beta \in (\frac{n-s}{2}, \frac{n}{2})$, we obtain that the double integral in (5.2.6) converges absolutely, and so interchanging the order of

5.2 Bessel Potentials

integration is allowed. Doing so and using that φ was arbitrary, we deduce that the tempered distribution G_s can be identified with the function

$$G_s(x) = \frac{(2\sqrt{\pi})^{-n}}{\Gamma(\frac{s}{2})} \int_0^\infty e^{-t} e^{-\frac{|x|^2}{4t}} t^{\frac{s-n}{2}} \frac{dt}{t}. \tag{5.2.7}$$

Note that the preceding integral is absolutely convergent for all $x \neq 0$, while for $x = 0$, it converges only when $s > n$. Identity (5.2.7) shows that G_s is smooth on $\mathbf{R}^n \setminus \{0\}$ and that $G_s(x) > 0$ for all $x \in \mathbf{R}^n$. Consequently,

$$\|G_s\|_{L^1} = \int_{\mathbf{R}^n} G_s(x)\,dx = \widehat{G_s}(0) = 1.$$

Now suppose $|x| \geq 2$. Then $t + \frac{|x|^2}{4t} \geq t + \frac{1}{t}$ and $t + \frac{|x|^2}{4t} \geq |x|$. This implies that

$$-t - \frac{|x|^2}{4t} \leq -\frac{t}{2} - \frac{1}{2t} - \frac{|x|}{2} \tag{5.2.8}$$

for all $t > 0$. From this it follows that when $|x| \geq 2$,

$$G_s(x) \leq \frac{(2\sqrt{\pi})^{-n}}{\Gamma(\frac{s}{2})} \left(\int_0^\infty e^{-\frac{t}{2}} e^{-\frac{1}{2t}} t^{\frac{s-n}{2}} \frac{dt}{t} \right) e^{-\frac{|x|}{2}} = \frac{C'(s,n)}{\Gamma(\frac{s}{2})} e^{-\frac{|x|}{2}}, \tag{5.2.9}$$

proving (5.2.1).

Suppose now that $0 < |x| < 2$. Write $G_s(x) = G_s^1(x) + G_s^2(x) + G_s^3(x)$, where

$$G_s^1(x) = \frac{(2\sqrt{\pi})^{-n}}{\Gamma(\frac{s}{2})} \int_0^{|x|^2} e^{-u} e^{-\frac{|x|^2}{4u}} u^{\frac{s-n}{2}} \frac{du}{u}$$

$$= |x|^{s-n} \frac{(2\sqrt{\pi})^{-n}}{\Gamma(\frac{s}{2})} \int_0^1 e^{-t|x|^2} e^{-\frac{1}{4t}} t^{\frac{s-n}{2}} \frac{dt}{t},$$

$$G_s^2(x) = \frac{(2\sqrt{\pi})^{-n}}{\Gamma(\frac{s}{2})} \int_{|x|^2}^4 e^{-t} e^{-\frac{|x|^2}{4t}} t^{\frac{s-n}{2}} \frac{dt}{t},$$

$$G_s^3(x) = \frac{(2\sqrt{\pi})^{-n}}{\Gamma(\frac{s}{2})} \int_4^\infty e^{-t} e^{-\frac{|x|^2}{4t}} t^{\frac{s-n}{2}} \frac{dt}{t}.$$

For $0 \leq t \leq 1$ and $0 < |x| < 2$ we have $e^{-4} \leq e^{-t|x|^2} \leq 1$. Thus, we write[2]

$$G_s^1(x) \approx |x|^{s-n} \frac{(2\sqrt{\pi})^{-n}}{\Gamma(\frac{s}{2})} \int_0^1 e^{-\frac{1}{4t}} t^{\frac{s-n}{2}} \frac{dt}{t} \approx |x|^{s-n}. \tag{5.2.10}$$

For $|x|^2 \leq t \leq 4$ and $|x| < 2$ we have $e^{-4} \leq e^{-t} \leq 1$ and $e^{-\frac{1}{4}} \leq e^{-\frac{|x|^2}{4t}} \leq 1$, so

---

[2] We say $f(x) \approx g(x)$ if there are $0 < c < C < \infty$ such that $c < f(x)/g(x) \leq C$ for all x.

$$G_s^2(x) \approx \int_{|x|^2}^{4} t^{\frac{s-n}{2}} \frac{dt}{t} \approx \begin{cases} |x|^{s-n} - 2^{s-n} & \text{for } s < n, \\ \log \frac{2}{|x|} & \text{for } s = n, \\ 2^{s-n} - |x|^{s-n} & \text{for } s > n. \end{cases} \quad (5.2.11)$$

Finally, when $t \geq 4 > |x|^2$ we have $e^{-\frac{1}{4}} \leq e^{-\frac{|x|^2}{4t}} \leq 1$, which yields

$$G_s^3(x) \approx \frac{(2\sqrt{\pi})^{-n}}{\Gamma(\frac{s}{2})} \int_4^{\infty} e^{-t} t^{\frac{s-n}{2}} \frac{dt}{t} \approx 1. \quad (5.2.12)$$

Combining (5.2.10), (5.2.11), and (5.2.12), for $0 < |x| < 2$ we obtain

$$G_s(x) = G_s^1(x) + G_s^2(x) + G_s^3(x) \approx \begin{cases} |x|^{s-n} + (|x|^{s-n} - 2^{s-n}) + 1 & \text{for } 0 < s < n, \\ |x|^{s-n} + \log \frac{2}{|x|} + 1 & \text{for } s = n, \\ |x|^{s-n} + (2^{s-n} - |x|^{s-n}) + 1 & \text{for } s > n. \end{cases}$$

But this function is comparable to $H_s(x)$ when $0 < |x| < 2$, so we deduce (5.2.2). $\square$

The next proposition is concerned with estimates for the derivatives of G_s.

Proposition 5.2.3. *Let $s > 0$. For each multi-index α there is a positive constant $C_{\alpha,s,n}$ such that for every $x \in \mathbf{R}^n \setminus \{0\}$ one has*

$$|\partial^{\alpha} G_s(x)| \leq C_{\alpha,s,n} \begin{cases} |x|^{s-n-|\alpha|} & \text{when } 0 < |x| < 2 \text{ and } s < n + |\alpha|, \\ \log \frac{4}{|x|} & \text{when } 0 < |x| < 2 \text{ and } s = n + |\alpha|, \\ 1 & \text{when } 0 < |x| < 2 \text{ and } s > n + |\alpha|, \\ e^{-\frac{|x|}{4}} & \text{when } |x| \geq 2. \end{cases} \quad (5.2.13)$$

Proof. We begin by noting that for each $m \in \mathbf{Z}^+ \cup \{0\}$ there is a polynomial p_m of degree m on the real line such that $\frac{d^m}{dt^m} e^{-t^2} = p_m(t) e^{-t^2}$. From this we obtain that for each multi-index $\alpha = (\alpha_1, \ldots, \alpha_n)$ and all $x \in \mathbf{R}^n$ we have

$$\partial^{\alpha} e^{-|x|^2} = p_{\alpha_1}(x_1) \cdots p_{\alpha_n}(x_n) e^{-|x|^2}.$$

Consequently, there is a constant $B_{\alpha,n}$ such that $|\partial^{\alpha} e^{-|x|^2}| \leq B_{\alpha,n}(1+|x|)^{|\alpha|} e^{-|x|^2}$ and thus, by the chain rule, for $t > 0$ and $x \in \mathbf{R}^n$ we have the estimate

$$|\partial_x^{\alpha} e^{-\frac{|x|^2}{4t}}| \leq B_{\alpha,n}\left(1 + \frac{|x|}{2\sqrt{t}}\right)^{|\alpha|} e^{-\frac{|x|^2}{4t}} \frac{1}{(2\sqrt{t})^{|\alpha|}}. \quad (5.2.14)$$

Returning to (5.2.7), for $x \neq 0$, we obtain the identity

$$\partial^{\alpha} G_s(x) = \frac{(2\sqrt{\pi})^{-n}}{\Gamma(\frac{s}{2})} \int_0^{\infty} e^{-t} (\partial^{\alpha} e^{-\frac{|x|^2}{4t}}) t^{\frac{s-n}{2}} \frac{dt}{t}, \quad (5.2.15)$$

5.2 Bessel Potentials

interchanging differentiation and integration, which is justified by the rapid convergence of the integral. For $|x| \geq 2$ we use that $(1+|x|)^{|\alpha|} e^{-|x|^2} \leq B'_\alpha e^{-|x|^2/2}$ and (5.2.14) to obtain

$$e^{-t}\left|\partial_x^\alpha e^{-\frac{|x|^2}{4t}}\right| t^{\frac{s-n}{2}} \leq e^{-\frac{t}{2}} B_{\alpha,n} B'_\alpha e^{-\frac{1}{2}\frac{|x|^2}{4t}} \frac{t^{\frac{s-n}{2}}}{(2\sqrt{t})^{|\alpha|}} \leq B'_\alpha B_{\alpha,n} e^{-\frac{t}{4}-\frac{1}{4t}-\frac{|x|}{4}} \frac{t^{\frac{s-n}{2}}}{(2\sqrt{t})^{|\alpha|}},$$

where the last inequality is a consequence of (5.2.8). Inserting this estimate in (5.2.15) yields (5.2.13) when $|x| \geq 2$, in analogy with (5.2.9).

For $0 < |x| < 2$ we write $\partial^\alpha G_s(x) = \partial^\alpha G_s^1(x) + \partial^\alpha G_s^2(x) + \partial^\alpha G_s^3(x)$, as in the case of no derivatives. The upper estimates for $\partial^\alpha G_s^3(x)$ are similar to those for $G_s^3(x)$ [see (5.2.12)] as the extra term $(1+\frac{|x|}{2\sqrt{t}})^{|\alpha|}(\frac{1}{2\sqrt{t}})^{|\alpha|}$ is bounded for $t \in [4,\infty)$. Now for $t \in [|x|^2, 4)$, the extra term $(1+\frac{|x|}{2\sqrt{t}})^{|\alpha|}(\frac{1}{2\sqrt{t}})^{|\alpha|}$ contributes a factor of $t^{-|\alpha|/2}$ to the integral in (5.2.11). Thus for $0 < |x| < 2$ we obtain

$$|\partial^\alpha G_s^2(x)| \leq c_{\alpha,s,n} \int_{|x|^2}^4 t^{\frac{s-|\alpha|-n}{2}} \frac{dt}{t} \approx \begin{cases} |x|^{s-|\alpha|-n} - 2^{s-|\alpha|-n} & \text{for } s-|\alpha| < n, \\ \log\frac{2}{|x|} & \text{for } s-|\alpha| = n, \\ 2^{s-|\alpha|-n} - |x|^{s-|\alpha|-n} & \text{for } s-|\alpha| > n, \end{cases}$$

for some constant $c_{\alpha,s,n} > 0$. Now we focus on $\partial^\alpha G_s^1(x)$. In view of (5.2.14) we write

$$|\partial^\alpha G_s^1(x)| \leq \frac{(2\sqrt{\pi})^{-n}}{\Gamma(\frac{s}{2})} \int_0^{|x|^2} B_{\alpha,n}\left(1+\frac{|x|}{2\sqrt{u}}\right)^{|\alpha|} \frac{1}{(2\sqrt{u})^{|\alpha|}} e^{-u} e^{-\frac{|x|^2}{4u}} u^{\frac{s-n}{2}} \frac{du}{u}$$

$$\leq \frac{(2\sqrt{\pi})^{-n}}{\Gamma(\frac{s}{2})} \int_0^{|x|^2} B_{\alpha,n}\left(\frac{5|x|}{2\sqrt{u}}\right)^{|\alpha|} \left(\frac{1}{2}\right)^{|\alpha|} e^{-u} e^{-\frac{|x|^2}{4u}} u^{\frac{s-|\alpha|-n}{2}} \frac{du}{u}$$

$$\leq |x|^{s-|\alpha|-n} \frac{(2\sqrt{\pi})^{-n}}{\Gamma(\frac{s}{2})} B_{\alpha,n}\left(\frac{5}{4}\right)^{|\alpha|} \int_0^1 e^{-t|x|^2} e^{-\frac{1}{4t}} t^{\frac{s-2|\alpha|-n}{2}} \frac{dt}{t},$$

and as $e^{-t|x|^2} \leq 1$, we have that $|\partial^\alpha G_s^1(x)| \leq C'_{\alpha,s,n}|x|^{s-|\alpha|-n}$, where $C'_{\alpha,s,n} > 0$. The combined estimate for $|\partial^\alpha G_s^1(x)| + |\partial^\alpha G_s^2(x)| + |\partial^\alpha G_s^3(x)|$ then gives

$$|\partial^\alpha G_s(x)| \leq C''_{\alpha,s,n} \begin{cases} |x|^{s-|\alpha|-n} + (|x|^{s-|\alpha|-n} - 2^{s-|\alpha|-n}) + 1 & \text{for } 0 < s < n+|\alpha|, \\ |x|^{s-|\alpha|-n} + \log\frac{2}{|x|} + 1 & \text{for } s = n+|\alpha|, \\ |x|^{s-|\alpha|-n} + (2^{s-|\alpha|-n} - |x|^{s-|\alpha|-n}) + 1 & \text{for } s > n+|\alpha| \end{cases}$$

when $0 < |x| < 2$. In all cases, this expression is bounded by that in (5.2.13) when $0 < |x| < 2$. This completes the proof. $\square$

Corollary 5.2.4. *Let $0 < s < n$.*
(a) $\mathcal{J}_s$ maps $L^1(\mathbf{R}^n)$ to $L^{\frac{n}{n-s},\infty}(\mathbf{R}^n)$.
(b) $\mathcal{J}_s$ maps $L^1(\mathbf{R}^n)$ to $L^q(\mathbf{R}^n)$ when $1 \leq q < \frac{n}{n-s}$.
(c) $\mathcal{J}_s$ maps $L^p(\mathbf{R}^n)$ to $L^q(\mathbf{R}^n)$ when $1 < p < \frac{n}{s}$ and $q \in [p, \frac{pn}{n-ps}]$.

Proof. When $0 < s < n$ the kernel G_s of $\mathcal{J}_s$ satisfies

$$G_s(x) \leq C_{n,s} \begin{cases} |x|^{-n+s} & \text{when } |x| \leq 2, \\ e^{-\frac{|x|}{2}} & \text{when } |x| \geq 2. \end{cases}$$

It follows that $G_s(x) \leq C'_{n,s}|x|^{-n+s}$ for all $x \in \mathbf{R}^n$. Then Theorem 5.1.3 implies the assertion in (a). Now notice that $G_s \in L^1(\mathbf{R}^n)$ and this gives that $\mathcal{J}_s$ maps $L^1(\mathbf{R}^n)$ to $L^1(\mathbf{R}^n)$. Interpolating between this estimate and the one in (a), via Exercise 1.3.3, we obtain the claim in (b). To obtain the assertion in (c) we note: For $1 < p < \frac{n}{s}$, $\mathcal{J}_s$ maps $L^p(\mathbf{R}^n)$ to $L^p(\mathbf{R}^n)$ (as $G_s \in L^1$) and it also maps $L^p(\mathbf{R}^n)$ to $L^{q_0}(\mathbf{R}^n)$ where $\frac{1}{q_0} = \frac{1}{p} - \frac{s}{n}$ in view of Theorem 5.1.3. By Exercise 1.1.6 we deduce that $\mathcal{J}_s$ maps $L^p(\mathbf{R}^n)$ to $L^q(\mathbf{R}^n)$ when $p \leq q \leq \frac{pn}{n-ps}$. □

Exercises

5.2.1. (Fractional integration by parts) Let $z \in \mathbf{C}$ and $f,g \in \mathscr{S}(\mathbf{R}^n)$. Show that

$$\int_{\mathbf{R}^n} g\,(I-\Delta)^{\frac{z}{2}} f\,dx = \int_{\mathbf{R}^n} f\,(I-\Delta)^{\frac{z}{2}} g\,dx.$$

Moreover, $(I-\Delta)^{\frac{z}{2}}$ could be replaced by $(-\Delta)^{\frac{z}{2}}$ if $\operatorname{Re} z > -n$ or if the Fourier transform of one of $\widehat{f},\widehat{g}$ vanishes in a neighborhood of the origin.

5.2.2. Let $1 \leq p \leq q \leq \infty$. (a) Show that $\mathcal{J}_s$ maps $L^p(\mathbf{R}^n)$ to $L^q(\mathbf{R}^n)$ when $s > n$.
(b) Prove that $\mathcal{J}_n$ maps $L^p(\mathbf{R}^n)$ to $L^q(\mathbf{R}^n)$ when $(p,q) \neq (1,\infty)$.

5.2.3. Let $0 < s < n$. Show that the Bessel potential $\mathcal{J}_s$ maps $L^1(\mathbf{R}^n)$ to $L^{r,\infty}(\mathbf{R}^n)$ when $1 \leq r \leq \frac{n}{n-s}$ and $L^p(\mathbf{R}^n)$ to $L^q(\mathbf{R}^n)$ when $1 < p < \infty$ and $p \leq q \leq \frac{pn}{n-s}$.

5.2.4. Let $1 < s < n$ and $1 < p, r < \infty$ satisfy $\frac{1}{r'} \leq \frac{1}{p} \leq \frac{1}{r'} + \frac{s}{n}$. Prove that there is a constant $C = C(n,s,p,r)$ such that for all $\varphi \in \mathscr{S}(\mathbf{R}^n)$ one has

$$\|\varphi\|_{L^2}^2 \leq C \|\varphi\|_{L^p} \|(I-\Delta)^{\frac{s}{2}} \varphi\|_{L^r}.$$

5.2.5. Prove that for any $s > 0$ there is a constant C_s such that for any f,g be Schwartz functions on $\mathbf{R}^n$ whose Fourier transforms are nonnegative we have

$$\|(I-\Delta)^{\frac{s}{2}}(fg)\|_{L^2(\mathbf{R}^n)} \leq C_s \Big[\|f(I-\Delta)^{\frac{s}{2}} g\|_{L^2(\mathbf{R}^n)} + \|g(I-\Delta)^{\frac{s}{2}} f\|_{L^2(\mathbf{R}^n)} \Big].$$

[*Hint:* Use $(1 + |\xi + \xi'|^2)^{\frac{s}{2}} \leq C_s [(1 + |\xi|^2)^{\frac{s}{2}} + (1 + |\xi'|^2)^{\frac{s}{2}}]$. Note $C_s = 1$ if $s \leq 1$.]

5.2.6. Let $s_1, \ldots, s_n > 0$. Consider the operator $\mathcal{J}_{s_j}^{(j)} = (I - \partial_j^2)^{-\frac{s_j}{2}}$ acting on the jth variable of a function on $\mathbf{R}^n$. (This is basically the one-dimensional $\mathcal{J}_{s_j}$ acting on the jth variable.) Prove that $\mathcal{J}_{s_1}^{(1)} \circ \cdots \circ \mathcal{J}_{s_n}^{(n)}$ maps $L^q(\mathbf{R}^n)$ to itself for any $1 \leq q \leq \infty$.

5.2.7. Fix $0 < s_1 < 1$, $s_1 \leq s_2, \ldots, s_n$, and let G_s denote the kernel of the one-dimensional Bessel potential. (a) Show that $G_{s_1}(x_1) \cdots G_{s_n}(x_n)$ lies in $L^{1/(1-s_1),\infty}(\mathbf{R}^n)$.
(b) Show that $T = \mathcal{J}_{s_1}^{(1)} \circ \cdots \circ \mathcal{J}_{s_n}^{(n)}$ (Exercise 5.2.6) maps $L^1(\mathbf{R}^n)$ to $L^{1/(1-s_1),\infty}(\mathbf{R}^n)$.
(c) Prove that $T: L^p(\mathbf{R}^n) \to L^q(\mathbf{R}^n)$ when $1 < p < \frac{1}{s_1}$, $\frac{1}{1-s_1} < q < \infty$, and $\frac{1}{p} - \frac{1}{q} = s_1$.
[*Hint:* Part (a): Use that $\left|\{x \in \mathbf{R}^n : G_{s_1}(x_1) \cdots G_{s_n}(x_n) > \lambda\}\right|$ is equal to

$$\int_{\mathbf{R}^{n-1}} \left|\left\{x_1 \in \mathbf{R} : G_{s_1}(x_1) > \frac{\lambda}{G_{s_2}(x_2) \cdots G_{s_n}(x_n)}\right\}\right| dx_2 \cdots dx_n.$$

Part (b): Use Exercise 1.6.7. Part (c): Use the version of Young's inequality stated in the footnote of Proposition 2.4.2.]

5.3 The Mikhlin and Hörmander Multiplier Theorems

In this section we obtain a sufficient condition on a function σ in $L^\infty(\mathbf{R}^n)$ to be an L^p *Fourier multiplier*. This means that the operator

$$T_\sigma(f) = (\widehat{f}\sigma)^\vee, \qquad f \in \mathscr{S}(\mathbf{R}^n),$$

admits a bounded extension on $L^p(\mathbf{R}^n)$. Throughout the section we fix a Schwartz function Ψ as in (4.4.23) and we define

$$\widehat{\Theta}(\xi) = \widehat{\Psi}(\xi/2) + \widehat{\Psi}(\xi) + \widehat{\Psi}(2\xi). \tag{5.3.1}$$

Then $\widehat{\Theta}$ is supported in $\{\xi \in \mathbf{R}^n : \frac{3}{7} \leq |\xi| \leq 4\}$ and $\widehat{\Theta} = 1$ on the support of $\widehat{\Psi}$. Recall that for $j \in \mathbf{Z}$, the Littlewood–Paley operator associated with the bump Ψ is

$$\Delta_j^\Psi(f)(x) = \int_{\mathbf{R}^n} f(x-y) 2^{jn} \Psi(2^j y)\, dy.$$

Analogously one defines the Littlewood–Paley operator associated with Θ.

Lemma 5.3.1. *Fix $\Psi \in \mathscr{S}(\mathbf{R}^n)$ as in (4.4.23) and Θ as in (5.3.1). Let $1 \leq \rho < 2$, $s > n/\rho$, and $\sigma \in L^\infty(\mathbf{R}^n)$. Suppose that $(I-\Delta)^{\frac{s}{2}}[\widehat{\Psi}\sigma(2^j \cdot)]$ is an L^ρ function for any $j \in \mathbf{Z}$ and that*

$$K = \sup_{j \in \mathbf{Z}} \left\|(I-\Delta)^{\frac{s}{2}}[\widehat{\Psi}\sigma(2^j \cdot)]\right\|_{L^\rho(\mathbf{R}^n)} < \infty. \tag{5.3.2}$$

Then, for any Schwartz function f on $\mathbf{R}^n$ and any integer j we have

$$|\Delta_j^\Psi T_\sigma(f)| \leq C_{s,n,\rho} K \left[M(|\Delta_j^\Theta(f)|^\rho)\right]^{\frac{1}{\rho}}, \tag{5.3.3}$$

where M denotes the uncentered Hardy–Littlewood maximal operator.

Proof. Fix $f \in \mathscr{S}(\mathbf{R}^n)$. Since $\widehat{\Theta}$ is equal to 1 on the support of $\widehat{\Psi}$, we have that $\widehat{\Psi}(2^{-j}\xi) = \widehat{\Theta}(2^{-j}\xi)\widehat{\Psi}(2^{-j}\xi)$ for all $\xi \in \mathbf{R}^n$. Thus we can write

$$\begin{aligned}
\Delta_j^{\Psi} T_\sigma(f)(x) &= \int_{\mathbf{R}^n} \widehat{f}(\xi)\sigma(\xi)\widehat{\Psi}(2^{-j}\xi)e^{2\pi i x\cdot\xi}\,d\xi \\
&= \int_{\mathbf{R}^n} \widehat{f}(\xi)\widehat{\Theta}(2^{-j}\xi)\widehat{\Psi}(2^{-j}\xi)\sigma(\xi)e^{2\pi i x\cdot\xi}\,d\xi \\
&= \int_{\mathbf{R}^n} \widehat{\Delta_j^{\Theta}(f)}(\xi)\widehat{\Psi}(2^{-j}\xi)\sigma(\xi)e^{2\pi i x\cdot\xi}\,d\xi \\
&= \int_{\mathbf{R}^n} 2^{jn}\widehat{\Delta_j^{\Theta}(f)}(2^j\xi')\widehat{\Psi}(\xi')\sigma(2^j\xi')e^{2\pi i(2^j x\cdot\xi')}\,d\xi' \\
&= \int_{\mathbf{R}^n} [\Delta_j^{\Theta}(f)(2^{-j}\cdot)]^{\widehat{\;}}(\xi')[\widehat{\Psi}\sigma(2^j\cdot)e^{2\pi i(2^j x\cdot(\cdot))}](\xi')\,d\xi' \\
&= \int_{\mathbf{R}^n} \Delta_j^{\Theta}(f)(2^{-j}y')[\widehat{\Psi}\sigma(2^j\cdot)]^{\widehat{\;}}(y' - 2^j x)\,dy' \qquad \text{by (2.2.1)}\\
&= 2^{jn}\int_{\mathbf{R}^n} \Delta_j^{\Theta}(f)(y)[\widehat{\Psi}\sigma(2^j\cdot)]^{\widehat{\;}}(2^j y - 2^j x)\,dy \\
&= \int_{\mathbf{R}^n} \frac{2^{jn}\Delta_j^{\Theta}(f)(y)}{(1+2^j|x-y|)^s}(1+2^j|x-y|)^s[\widehat{\Psi}\sigma(2^j\cdot)]^{\widehat{\;}}(2^j y - 2^j x)\,dy.
\end{aligned}$$

Applying Hölder's inequality, we estimate

$$\begin{aligned}
|\Delta_j^{\Psi} T_\sigma(f)(x)| &\leq \left(\int_{\mathbf{R}^n} 2^{jn}\frac{|\Delta_j^{\Theta}(f)(y)|^\rho}{(1+2^j|x-y|)^{s\rho}}\,dy\right)^{\frac{1}{\rho}} \\
&\quad \cdot \left(\int_{\mathbf{R}^n} 2^{jn}\Big|(1+2^j|x-y|)^s[\widehat{\Psi}\sigma(2^j\cdot)]^{\widehat{\;}}(2^j y - 2^j x)\Big|^{\rho'}\,dy\right)^{\frac{1}{\rho'}},
\end{aligned} \qquad (5.3.4)$$

where the second factor of the product is to be interpreted as an L^∞ norm if $\rho = 1$.

Since $s\rho > n$, Corollary 2.5.2 yields the estimate

$$\left(\int_{\mathbf{R}^n} 2^{jn}\frac{|\Delta_j^{\Theta}(f)(y)|^\rho}{(1+2^j|x-y|)^{s\rho}}\,dy\right)^{\frac{1}{\rho}} \leq C_{s,n,\rho} M\!\left(|\Delta_j^{\Theta}(f)|^\rho\right)^{\frac{1}{\rho}}(x). \qquad (5.3.5)$$

By a change of variables, the second factor in the product in (5.3.4) equals

$$\begin{aligned}
&\left(\int_{\mathbf{R}^n} \Big|(1+|y|)^s\cdot[\widehat{\Psi}\sigma(2^j\cdot)]^{\widehat{\;}}(y)\Big|^{\rho'}\,dy\right)^{\frac{1}{\rho'}} \\
&\leq 2^{\frac{s}{2}}\left(\int_{\mathbf{R}^n} \Big|(1+4\pi^2|y|^2)^{\frac{s}{2}}\cdot[\widehat{\Psi}\sigma(2^j\cdot)]^{\widehat{\;}}(y)\Big|^{\rho'}\,dy\right)^{\frac{1}{\rho'}} \\
&= 2^{\frac{s}{2}}\left(\int_{\mathbf{R}^n} \Big|[(I-\Delta)^{\frac{s}{2}}[\widehat{\Psi}\sigma(2^j\cdot)]]^{\widehat{\;}}(y)\Big|^{\rho'}\,dy\right)^{\frac{1}{\rho'}} \\
&\leq 2^{\frac{s}{2}}\Big\|(I-\Delta)^{\frac{s}{2}}[\widehat{\Psi}\sigma(2^j\cdot)]\Big\|_{L^\rho(\mathbf{R}^n)},
\end{aligned}$$

5.3 The Mikhlin and Hörmander Multiplier Theorems

where we used the Hausdorff–Young inequality (Proposition 2.4.4) in the last step, as $1 \leq \rho < 2$. Combining this estimate with the one in (5.3.5) and inserting them in (5.3.4) yields the claimed conclusion. $\square$

We now prove the main result of this section.

Theorem 5.3.2. *(Mikhlin multiplier theorem)* *If a function σ on $\mathbf{R}^n \setminus \{0\}$ satisfies*

$$|\partial^\beta \sigma(\xi)| \leq C_\beta |\xi|^{-|\beta|} \qquad \text{for all } |\beta| \leq \left[\frac{n}{2}\right] + 2, \tag{5.3.6}$$

for some constants C_β, then for all $1 < p < \infty$, T_σ admits a bounded extension from $L^p(\mathbf{R}^n)$ to itself with norm bounded by $C(n,p) \sup_{|\beta| \leq [\frac{n}{2}]+2} C_\beta$, where $C(n,p)$ depends on p, n.

Proof. Let s be the even integer among the numbers $[\frac{n}{2}]+1$, $[\frac{n}{2}]+2$. Then $s > \frac{n}{2}$ and we have

$$(I-\Delta)^{\frac{s}{2}} = \sum_{|\alpha| \leq s} c_{\alpha,s} \partial^\alpha$$

for some constants $c_{\alpha,s}$ (which vanish when $|\alpha|$ is odd). Let Ψ be as in (4.4.23). Then

$$|\partial^\alpha(\widehat{\Psi}(\xi)\sigma(2^j\xi))| = \left|\sum_{\beta \leq \alpha} \binom{\alpha}{\beta}(\partial^{\alpha-\beta}\widehat{\Psi})(\xi) 2^{j|\beta|}(\partial^\beta\sigma)(2^j\xi)\right|$$

$$\leq \sum_{\beta \leq \alpha} \binom{\alpha}{\beta}|\partial^{\alpha-\beta}\widehat{\Psi}(\xi)| 2^{j|\beta|} C_\beta |2^j\xi|^{-|\beta|}$$

$$\leq K'\left(\sup_{|\beta| \leq s} C_\beta\right) \chi_{\frac{6}{7} \leq |\xi| \leq 2},$$

having used condition (5.3.6) in the first inequality above and the fact that $\frac{6}{7} \leq |\xi| \leq 2$ in the second inequality. It follows that

$$(I-\Delta)^{\frac{s}{2}}[\widehat{\Psi}\sigma(2^j \cdot)]$$

is compactly supported and bounded by a constant. Thus for any $\rho \geq 1$ the constant K in (5.3.2) is finite, i.e.,

$$K = \sup_{j \in \mathbf{Z}} \left\|(I-\Delta)^{\frac{s}{2}}[\widehat{\Psi}\sigma(2^j \cdot)]\right\|_{L^\rho(\mathbf{R}^n)} \leq K'' \sup_{|\beta| \leq s} C_\beta < \infty.$$

As $s > n/2$, we choose $\rho \geq 1$ with $n/s < \rho < 2$, so that the hypotheses on the indices of Lemma 5.3.1 are satisfied.

Suppose first that $p > 2$. In order to be able to apply Theorem 4.5.3 we need to know that $T_\sigma(f)$ lies in $L^p(\mathbf{R}^n)$; this will be the case if $f \in \widehat{\mathscr{S}_0}$, i.e., it is a Schwartz function whose Fourier transform is compactly supported away from the origin. Such functions are dense in $L^p(\mathbf{R}^n)$ for any $1 < p < \infty$; see Proposition 2.5.4. Integrating by parts, we write

$$T_\sigma(f)(x) = \int_{\mathbf{R}^n} \widehat{f}(\xi)\sigma(\xi)e^{2\pi i x\cdot\xi}d\xi$$
$$= \frac{1}{(1+4\pi^2|x|^2)^{\frac{s}{2}}}\int_{\mathbf{R}^n}\widehat{f}(\xi)\sigma(\xi)(I-\Delta)^{\frac{s}{2}}e^{2\pi ix\cdot\xi}d\xi$$
$$= \frac{1}{(1+4\pi^2|x|^2)^{\frac{s}{2}}}\int_{\mathbf{R}^n}\sum_{|\alpha|\le s}c_{\alpha,s}\partial^\alpha\big[\widehat{f}\sigma\big](\xi)e^{2\pi ix\cdot\xi}d\xi,$$

and we notice that hypotheses (5.3.6) combined with the fact that $\widehat{f}$ has compact support that does not contain $\{0\}$, by Leibniz's rule we obtain that

$$|T_\sigma(f)(x)| \le C_f (1+|x|)^{-s}.$$

But this function lies in $L^p(\mathbf{R}^n)$ as $p>2$ and $s>n/2$, which give $ps>n$. Thus $\|T_\sigma(f)\|_{L^p}<\infty$, which allows us to use inequality (4.5.5) in Theorem 4.5.3.

Applying successively inequality (4.5.5) in Theorem 4.5.3, Lemma 5.3.1, and (4.3.17) in Theorem 4.3.3 with $r=2/\rho$ (recall $n/s<\rho<2$), we obtain

$$\|T_\sigma(f)\|_{L^p(\mathbf{R}^n)} \le C_p(n)\Big\|\Big(\sum_{j\in\mathbf{Z}}|\Delta_j^\Psi(T_\sigma(f))|^2\Big)^{\frac{1}{2}}\Big\|_{L^p(\mathbf{R}^n)}$$
$$\le C_p(n)C_{n,p}K''\Big(\sup_{|\beta|\le s}C_\beta\Big)\Big\|\Big(\sum_{j\in\mathbf{Z}}\big[M(|\Delta_j^\Theta(f)|^\rho)\big]^{\frac{2}{\rho}}\Big)^{\frac{1}{2}}\Big\|_{L^p(\mathbf{R}^n)}$$
$$= C_p(n)C_{n,p}K''\Big(\sup_{|\beta|\le s}C_\beta\Big)\Big\|\Big(\sum_{j\in\mathbf{Z}}\big[M(|\Delta_j^\Theta(f)|^\rho)\big]^{\frac{2}{\rho}}\Big)^{\frac{\rho}{2}}\Big\|_{L^{p/\rho}(\mathbf{R}^n)}^{\frac{1}{\rho}}$$
$$\le C'_p(n)C_{n,p}K''\Big(\sup_{|\beta|\le s}C_\beta\Big)\Big\|\Big(\sum_{j\in\mathbf{Z}}|\Delta_j^\Theta(f)|^{\rho\frac{2}{\rho}}\Big)^{\frac{\rho}{2}}\Big\|_{L^{p/\rho}(\mathbf{R}^n)}^{\frac{1}{\rho}}$$
$$= C'_p(n)K''\Big(\sup_{|\beta|\le s}C_\beta\Big)\Big\|\Big(\sum_{j\in\mathbf{Z}}|\Delta_j^\Theta(f)|^2\Big)^{\frac{1}{2}}\Big\|_{L^p(\mathbf{R}^n)}$$
$$\le C''_p(n)K''\Big(\sup_{|\beta|\le s}C_\beta\Big)\|f\|_{L^p(\mathbf{R}^n)},$$

where the last inequality is a consequence of Theorem 4.4.2. Here Θ is as in (5.3.1) and the application of (4.3.17) makes use of the assumptions $1<2/\rho<\infty$ and $1<p/\rho<\infty$ (since $\rho<2<p$). This proves the claimed bound for functions $f\in\mathscr{S}_0$, which is a dense subspace of L^p. By density, there is a bounded extension of T_σ on $L^p(\mathbf{R}^n)$ for $2<p<\infty$ with norm bounded by $C(n,p)\sup_{|\beta|\le s}C_\beta$.

The case $p=2$ is a direct consequence of Plancherel's theorem. Finally, we discuss the case $1<p<2$. Notice that the transpose $(T_\sigma)^t$ of T_σ is equal to $T_{\widetilde{\sigma}}$, where $\widetilde{\sigma}(\xi)=\sigma(-\xi)$. As $\widetilde{\sigma}$ also satisfies (5.3.6), it follows that $T_{\widetilde{\sigma}}=(T_\sigma)^t$ is bounded from $L^p(\mathbf{R}^n)$ to itself for $p>2$, and by duality it follows that T_σ is bounded from $L^p(\mathbf{R}^n)$ to itself for $1<p<2$. $\square$

5.3 The Mikhlin and Hörmander Multiplier Theorems

Corollary 5.3.3. *For any $1 < p < \infty$ there is a constant $C_{n,p}$ such that for any $t \in \mathbf{R}$ we have*

$$\left\|(-\Delta)^{it}\right\|_{L^p(\mathbf{R}^n) \to L^p(\mathbf{R}^n)} + \left\|(I-\Delta)^{it}\right\|_{L^p(\mathbf{R}^n) \to L^p(\mathbf{R}^n)} \leq C_{n,p}(1+|t|)^{[\frac{n}{2}]+2}.$$

Proof. It is tedious but straightforward to verify that for any $k \in \mathbf{Z}^+$ one has

$$\sup_{\xi \in \mathbf{R}^n \setminus \{0\}} \sup_{|\alpha| \leq k} |\xi|^{|\alpha|} \left|\partial^\alpha(|\xi|^{i2t})\right| \leq C(n,k) \prod_{m=0}^{k-1} |it - m| \leq C(n,k)(1+|t|)^k. \quad (5.3.7)$$

A similar bound holds for $(1+4\pi^2|\xi|^2)^{it}$ in place of $|\xi|^{i2t}$. To see this consider the function $(\xi, \xi_{n+1}) \mapsto (|\xi_{n+1}|^2 + |\xi|^2)^{it}$ on $\mathbf{R}^{n+1}$, apply (5.3.7) to a multi-index of the form $(0, \beta)$, where β is an multi-index with n entries, and plug in $\xi_{n+1} = 1/2\pi$. Inserting $k = [\frac{n}{2}] + 2$ in (5.3.7) and in its analog for $(1+4\pi^2|\xi|^2)^{it}$ provides the hypotheses of Theorem 5.3.2, so its conclusion yields our claim.

One may also verify (5.3.7) and the analogous version for $(1+|\xi|^2)^{it}$ by applying the Faà di Bruno formula (Appendix F). $\square$

In fact this corollary is a special case of a more general situation.

Example 5.3.4. Let σ be a smooth function on $\mathbf{R}^n \setminus \{0\}$ that is homogeneous of degree $i\tau$, where τ is real. This means that for all $\lambda > 0$ and all $\xi \neq 0$ we have

$$\sigma(\lambda \xi) = \lambda^{i\tau} \sigma(\xi). \quad (5.3.8)$$

[An explicit example of such a function is $\sigma(\xi) = |\xi|^{i\tau}$.] Then σ is an L^p Fourier multiplier for $1 < p < \infty$. To show this we verify condition (5.3.6). Differentiating both sides of (5.3.8) with respect to ∂_ξ^α, we obtain

$$\lambda^{|\alpha|}(\partial_\xi^\alpha \sigma)(\lambda \xi) = \lambda^{i\tau} \partial_\xi^\alpha \sigma(\xi), \qquad \xi \neq 0.$$

Taking $\lambda = |\xi|^{-1}$, we deduce condition (5.3.6) with $C_\alpha = \sup_{|\theta|=1} |\partial^\alpha \sigma(\theta)|$.

Example 5.3.5. Let z be a complex number. Then for any multi-index β there is a constant C_β such that the function $m(\xi) = (1+|\xi|^2)^{\frac{z}{2}}$ satisfies

$$\left|\partial^\beta m(\xi)\right| \leq C_\beta (1+|\xi|)^{\mathrm{Re}\, z - |\beta|}, \qquad \xi \in \mathbf{R}^n. \quad (5.3.9)$$

This shows that when $\mathrm{Re}\, z \leq 0$, then m is an L^p Fourier multiplier for $1 < p < \infty$. To verify (5.3.9), we introduce the function $M(t, \xi) = (|t|^2 + |\xi|^2)^{\frac{z}{2}}$ on $\mathbf{R}^{n+1} \setminus \{0\}$. Then M is homogeneous of degree z and is smooth on the sphere S^n. Thus,

$$M(\lambda t, \lambda \xi) = \lambda^z M(t, \xi), \qquad \lambda > 0.$$

Differentiating with respect to ∂_ξ^β, we obtain

$$\lambda^{|\beta|}(\partial_\xi^\beta M)(\lambda t, \lambda \xi) = \lambda^z \partial_\xi^\beta M(t, \xi), \qquad \lambda > 0,$$

and choosing $\lambda = |(t,\xi)|^{-1}$, if $(t,\xi) \neq 0$, yields the bound

$$|\partial_\xi^\beta M(t,\xi)| \leq |(t,\xi)|^{-|\beta|+\operatorname{Re} z} \sup_{(t,\xi)' \in S^n} |(\partial_\xi^\beta M)(t,\xi)'|. \tag{5.3.10}$$

From this, plugging in $t = 1$, we obtain (5.3.9) with C_β equal to the supremum on the right in (5.3.10), which is finite as M is smooth on $\mathbf{S}^n$.

The proof of Theorem 5.3.2 provides the following more general result.

Theorem 5.3.6. *(Hörmander multiplier theorem)* *Let Ψ be a Schwartz function as defined in (4.4.23) and let $s > n/2$. Fix $1 \leq \rho < 2$ and $s > n/\rho$. Let K be as in (5.3.2). Then T_σ admits a bounded extension from $L^p(\mathbf{R}^n)$ to itself for all $1 < p < \infty$ with norm bounded by $C(n,p,s,\rho)K$.*

Example 5.3.7. Homogeneity and smoothness yields that for any multi-index α

$$|\partial^\beta \xi^\alpha| \leq c_{\beta,\alpha} |\xi|^{|\alpha|-|\beta|},$$

where $c_{\beta,\alpha}$ vanishes if $\beta_j > \alpha_j$ for some j. This estimate is useful in calculations.

We end this section with a couple more examples of Fourier multipliers.

Example 5.3.8. Let z be a complex number with $\operatorname{Re} z \geq 0$ and let α be a fixed multi-index with $|\alpha| \leq \operatorname{Re} z$. Then the function, defined on $\mathbf{R}^n$,

$$m(\xi) = \frac{\xi^\alpha}{(1+|\xi|^2)^{z/2}}$$

satisfies (5.3.6) (in fact without the restriction on β) and is therefore an L^p Fourier multiplier for $1 < p < \infty$. To prove this assertion we write by Leibniz's rule

$$|\partial^\beta m(\xi)| = \left| \sum_{\gamma \leq \beta} \binom{\beta}{\gamma} (\partial^{\beta-\gamma} \xi^\alpha)(\partial^\gamma (1+|\xi|^2)^{-z/2}) \right|$$

$$\leq C_\beta \sum_{\gamma \leq \beta} \binom{\beta}{\gamma} |\xi|^{|\alpha|-(|\beta|-|\gamma|)} (1+|\xi|)^{-\operatorname{Re} z - |\gamma|}$$

$$\leq C'_\beta \sum_{\gamma \leq \beta} |\xi|^{-|\beta|+|\gamma|} \left[|\xi|^{|\alpha|} (1+|\xi|)^{-\operatorname{Re} z} \right] |\xi|^{-|\gamma|} \leq C''_\beta |\xi|^{-|\beta|},$$

since the expression inside the square brackets is bounded by 1, as $|\alpha| \leq \operatorname{Re} z$.

Example 5.3.9. Let $\theta \in \mathbf{R}$ and $\eta \in \mathbf{R}^n$. On $\mathbf{R}^n$ consider the function

$$\sigma(\xi) = \left(\frac{1+|\xi+\eta|^2}{1+|\xi|^2} \right)^{\frac{\theta}{2}}.$$

Then (5.3.6) holds (without the restriction on β) with a constant that depends on η, precisely,

5.3 The Mikhlin and Hörmander Multiplier Theorems

$$|\partial^\beta \sigma(\xi)| \leq C_{\beta,\theta} \frac{(1+|\eta|)^{\max(|\beta|-\theta,0)}}{(1+|\xi|)^{|\beta|}}. \tag{5.3.11}$$

In view of Leibniz's rule we write

$$|\partial_\xi^\beta \sigma(\xi)| = \left|\sum_{\gamma \leq \beta} \binom{\beta}{\gamma} \partial_\xi^\gamma (1+|\xi+\eta|^2)^{\frac{\theta}{2}} \partial_\xi^{\beta-\gamma}(1+|\xi|^2)^{-\frac{\theta}{2}}\right|$$

$$\leq C_{n,\theta} \sum_{\gamma \leq \beta} \binom{\beta}{\gamma}(1+|\xi+\eta|)^{\theta-|\gamma|}(1+|\xi|)^{-\theta-(|\beta|-|\gamma|)}.$$

Using the estimates

$$(1+|\xi+\eta|)^{\theta-|\gamma|} \leq \begin{cases} (1+|\xi|)^{\theta-|\gamma|}(1+|\eta|)^{\theta-|\gamma|} & \text{if } \theta-|\gamma| \geq 0, \\ (1+|\xi|)^{\theta-|\gamma|}(1+|\eta|)^{|\gamma|-\theta} & \text{if } \theta-|\gamma| < 0, \end{cases}$$

we deduce (5.3.11), which in fact holds for all multi-indices β.

Exercises

5.3.1. Prove that if σ_1 and σ_2 satisfy condition (5.3.6), then so does $\sigma_1 \sigma_2$.

5.3.2. Show that if σ is real-valued and satisfies (5.3.6), then so does $e^{i\sigma}$.

5.3.3. Let m be a function on $\mathbf{R}^2$ which is homogeneous of degree $-\rho \leq 0$ and smooth on the unit circle. Prove that the function $\xi \mapsto m(1,|\xi|)$ lies in $\mathscr{M}_p(\mathbf{R}^n)$ for any $1 < p < \infty$.

5.3.4. Let $0 < c < C < \infty$, $z \in \mathbf{C}$, and $m \in \mathbf{Z}^+$. Let $\sigma: \mathbf{R}^n \to \mathbf{C}$ satisfy (5.3.6).
(a) If $|\sigma(\xi)| \geq c$ for all $\xi \in \mathbf{R}^n$, show that σ^{-m} satisfies condition (5.3.6).
(b) If $|\sigma(\xi)| \leq C$ for all $\xi \in \mathbf{R}^n$ and $m \geq s$, show that σ^m also satisfies (5.3.6).
(c) If $\sigma(\xi) \geq c$ for all $\xi \in \mathbf{R}^n$ and $\operatorname{Re} z \leq 0$, show that σ^z satisfies (5.3.6).
(c) If $0 < \sigma(\xi) \leq C$ for all $\xi \in \mathbf{R}^n$ and $\operatorname{Re} z \geq s$, show that σ^z also satisfies (5.3.6).

5.3.5. Suppose that σ is a complex-valued function on $\mathbf{R}^n$ that satisfies (5.3.6) for all multi-indices α (i.e., without the restriction $|\alpha| \leq s$). Let β be a fixed multi-index. Show that $\nabla \sigma(\xi) \cdot \xi$ and $\xi^\beta \partial^\beta \sigma$ satisfy (5.3.6) for all all multi-indices α.

5.3.6. Prove that the functions $g_t(\xi) = t(t^2 + |\xi|^2)^{-1/2}$, defined for $\xi \in \mathbf{R}^n$ and indexed by $t > 0$, lie in $\mathscr{M}_p(\mathbf{R}^n)$ uniformly in t.

5.3.7. Let $\widehat{\zeta}$ be a smooth function on $\mathbf{R}^n$ that is supported in a compact set that does not contain the origin and let a_j be a bounded sequence of complex numbers. Prove that the function

$$m(\xi) = \sum_{j \in \mathbf{Z}} a_j \widehat{\zeta}(2^{-j}\xi), \quad \xi \in \mathbf{R}^n,$$

lies in $\mathscr{M}_p(\mathbf{R}^n)$ for all $1 < p < \infty$.

5.4 Sobolev Spaces

Just as Lebesgue spaces quantify the integrability of the pth power of functions, Sobolev spaces quantify the L^p integrability of functions and their derivatives.

Definition 5.4.1. Let k be a nonnegative integer, and let $1 < p < \infty$. The *Sobolev space* $L^p_k(\mathbf{R}^n)$ is defined as the space of functions f in $L^p(\mathbf{R}^n)$ such that for all $|\alpha| \leq k$ the distributional derivatives $\partial^\alpha f$ are also $L^p(\mathbf{R}^n)$ functions. This space is normed by the quantity

$$\|f\|_{L^p_k} = \sum_{|\alpha| \leq k} \|\partial^\alpha f\|_{L^p}. \tag{5.4.1}$$

Sobolev space norms quantify smoothness of functions in terms of the integrability of their derivatives. The index k indicates the *degree* of smoothness of a given function in L^p_k. As k increases, the functions become smoother. Equivalently, these spaces form a decreasing sequence $L^p \hookleftarrow L^p_1 \hookleftarrow L^p_2 \hookleftarrow L^p_3 \hookleftarrow \cdots$, meaning that each $L^p_{k+1}(\mathbf{R}^n)$ properly embeds in $L^p_k(\mathbf{R}^n)$. This property, which coincides with our intuition of smoothness, is a consequence of the definition of Sobolev norms.

Next, we extend the definition of Sobolev spaces to the case where the positive integer k is replaced by a real number s. Before we do so, we note that for $s \in \mathbf{R}$ the function $(1 + 4\pi^2 |\cdot|^2)^{\frac{s}{2}}$ lies in $\mathscr{C}^\infty$ and has polynomial growth at infinity, so the product $(1 + 4\pi^2 |\cdot|^2)^{\frac{s}{2}} \widehat{u}$ is a well-defined element of $\mathscr{S}'$. (Definition 2.6.15.) Thus, its inverse Fourier transform $(I - \Delta)^{\frac{s}{2}} u$ is also a well-defined element of $\mathscr{S}'$.

Definition 5.4.2. Let s be a real number and let $1 < p < \infty$. The *Sobolev space* $L^p_s(\mathbf{R}^n)$ is defined as the space of all tempered distributions u in $\mathscr{S}'(\mathbf{R}^n)$ for which $(I - \Delta)^{\frac{s}{2}} u$ is a function in $L^p(\mathbf{R}^n)$. For such distributions u we define

$$\|u\|_{L^p_s(\mathbf{R}^n)} = \|(I - \Delta)^{\frac{s}{2}} u\|_{L^p(\mathbf{R}^n)}.$$

Remark 5.4.3. The function $\left(\frac{1 + 4\pi^2 |\xi|^2}{1 + |\xi|^2}\right)^{s/2}$ and its reciprocal lie in $\mathscr{M}_p(\mathbf{R}^n)$ for any $1 < p < \infty$. This is because of Leibniz's rule and Example 5.3.5. Consequently, a tempered distribution u lies in $L^p_s(\mathbf{R}^n)$ if and only if $\left((1 + |\cdot|^2)^{\frac{s}{2}} \widehat{u}\right)^\vee$ lies in $L^p(\mathbf{R}^n)$; furthermore, in this case we have $\left\|\left((1 + |\cdot|^2)^{\frac{s}{2}} \widehat{u}\right)^\vee\right\|_{L^p} \approx \|(I - \Delta)^{\frac{s}{2}} u\|_{L^p}$.

Remark 5.4.4. (a) $L^p_0 = L^p$. This is straightforward.
(b) For $s > 0$, L^p_s embeds in L^p. Indeed, if $f_s = (I - \Delta)^{\frac{s}{2}} f$, then we have

$$f = (I - \Delta)^{-\frac{s}{2}} f_s = f_s * G_s,$$

where G_s is as in Definition 5.2.1. Theorem 1.6.6 and the fact that $\|G_s\|_{L^1} = 1$ yield

$$\|f\|_{L^p(\mathbf{R}^n)} \leq \|f_s\|_{L^p(\mathbf{R}^n)} = \|f\|_{L^p_s(\mathbf{R}^n)} < \infty.$$

(c) When $s = k \in \mathbf{Z}^+$, the space L^p_s of Definition 5.4.2 coincides with the space L^p_k of Definition 5.4.1 with equivalence of norms. Moreover, as s increases, the functions

5.4 Sobolev Spaces

in L_s^p become smoother. Suppose that $f \in L_k^p$ according to Definition 5.4.2. Then for all $|\alpha| \le k$ we have that the distributional derivatives $\partial^\alpha f$ are equal to

$$\partial^\alpha f = ((2\pi i \cdot)^\alpha \widehat{f})^\vee = \left(\frac{(2\pi i \cdot)^\alpha}{(1+4\pi^2|\cdot|^2)^{\frac{k}{2}}} (1+4\pi^2|\cdot|^2)^{\frac{k}{2}} \widehat{f} \right)^\vee. \quad (5.4.2)$$

The result in Example 5.3.8 gives that when $|\alpha| \le k$, the function

$$\xi \mapsto \frac{(2\pi i \xi)^\alpha}{(1+4\pi^2|\xi|^2)^{k/2}}$$

is an L^p multiplier. By assumption $(I-\Delta)^{\frac{k}{2}} f = \left((1+4\pi^2|\cdot|^2)^{\frac{k}{2}} \widehat{f} \right)^\vee$ lies in $L^p(\mathbf{R}^n)$, and thus it follows from (5.4.2) that the distributional derivatives $\partial^\alpha f$ lie in $L^p(\mathbf{R}^n)$ and that

$$\sum_{|\alpha| \le k} \|\partial^\alpha f\|_{L^p} \le C_{p,n,k} \|(I-\Delta)^{\frac{k}{2}} f\|_{L^p} < \infty.$$

Conversely, suppose that $f \in L_k^p$ according to Definition 5.4.1; then the multinomial identity applied to the expression $(1+4\pi^2|\xi|^2)^k$ yields

$$\left(1+4\pi^2(\xi_1^2 + \cdots + \xi_n^2)\right)^{\frac{k}{2}} = \sum_{|\alpha| \le k} \frac{k!}{(k-|\alpha|)! \alpha_1! \cdots \alpha_n!} \frac{(2\pi \xi)^\alpha}{(1+4\pi^2|\xi|^2)^{\frac{k}{2}}} \frac{(2\pi i \xi)^\alpha}{i^{|\alpha|}}.$$

By Example 5.3.8 the functions $m_\alpha(\xi) = (2\pi \xi)^\alpha (1+4\pi^2|\xi|^2)^{-\frac{k}{2}}$ are L^p Fourier multipliers whenever $|\alpha| \le k$. We have

$$(I-\Delta)^{\frac{k}{2}} f = \sum_{|\alpha| \le k} \frac{k!(-i)^{|\alpha|}}{(k-|\alpha|)! \alpha_1! \cdots \alpha_n!} \left(m_\alpha \widehat{\partial^\alpha f} \right)^\vee,$$

and it follows that

$$\|(I-\Delta)^{\frac{k}{2}} f\|_{L^p} \le C_{p,n,k} \sum_{|\alpha| \le k} \|\partial^\alpha f\|_{L^p} < \infty.$$

This proves the converse direction.

(d) As a consequence of the preceding result for $s > 0$ we deduce that

$$\|f\|_{L_s^p} \approx \sum_{|\alpha| \le [s]} \|\partial^\alpha f\|_{L_{s-[s]}^p}.$$

To see this, simply write $(I-\Delta)^{\frac{s}{2}} f = (I-\Delta)^{\frac{[s]}{2}} (I-\Delta)^{\frac{s-[s]}{2}} f$ and apply the equivalence in (c) with $k = [s]$.

(e) For $-\infty < s < t < \infty$ we have that $L_t^p(\mathbf{R}^n) \hookrightarrow L_s^p(\mathbf{R}^n)$. Indeed, we show that

$$\|(I-\Delta)^{\frac{s}{2}} f\|_{L^p} \le \|(I-\Delta)^{\frac{t}{2}} f\|_{L^p}.$$

Setting $g = (I-\Delta)^{\frac{t}{2}}f \iff f = (I-\Delta)^{-\frac{t}{2}}g$, this is equivalent to showing that

$$\|\mathcal{J}_{t-s}(g)\|_{L^p} = \|(I-\Delta)^{-\frac{t-s}{2}}g\|_{L^p} = \|G_{t-s}*g\|_{L^p} \leq \|g\|_{L^p},$$

which is a consequence of Theorem 1.6.6 and of the fact that $\|G_{t-s}\|_{L^1} = 1$.

(f) The following observation is related to (e): If $f \in L_s^p(\mathbf{R}^n)$, then $\partial^\alpha f \in L_{s-|\alpha|}^p(\mathbf{R}^n)$. The underlying inequality here is that when $1 < p < \infty$ one has

$$\|(I-\Delta)^{\frac{s-|\alpha|}{2}}\partial^\alpha f\|_{L^p} \leq C\|(I-\Delta)^{\frac{s}{2}}f\|_{L^p}.$$

By the "change of variables" $g = (I-\Delta)^{\frac{s}{2}}f$, this is equivalent to

$$\|\partial^\alpha(I-\Delta)^{-\frac{|\alpha|}{2}}g\|_{L^p} = \|(I-\Delta)^{\frac{s-|\alpha|}{2}}\partial^\alpha(I-\Delta)^{-\frac{s}{2}}g\|_{L^p} \leq C\|g\|_{L^p},$$

a valid inequality, as $(2\pi i \xi)^\alpha(1 + 4\pi^2|\xi|^2)^{-|\alpha|/2}$ lies in $\mathcal{M}_p(\mathbf{R}^n)$ (Example 5.3.8).

We now show that, in contrast to the observation (b) in Remark 5.4.4, L_{-s}^p does not embed in any space of functions when $s > 0$.

Example 5.4.5. Consider the Dirac mass at the origin δ_0. Then $\|\delta_0\|_{L_{-s}^p(\mathbf{R}^n)} = \|G_s\|_{L^p}$ when $s > 0$. For $s \geq n$, $\|G_s\|_{L^p} < \infty$, in view of Proposition 5.2.2. For $0 < s < n$ the function $G_s = ((1+4\pi^2|\cdot|^2)^{-\frac{s}{2}})^\vee$ is integrable to the power p as long as $(s-n)p > -n$, that is, exactly when $1 < p < \frac{n}{n-s}$. Thus for $0 < s < n$, δ_0 lies in $L_{-s}^p(\mathbf{R}^n)$ if and only if $1 < p < \frac{n}{n-s}$. For $s \geq n$, δ_0 lies in $L_{-s}^p(\mathbf{R}^n)$ for all $1 < p < \infty$.

Example 5.4.6. Let g be the characteristic function of $[-1,1]$. As $g' = \delta_{-1} - \delta_1$, it follows that g does not lie in $L_1^p(\mathbf{R})$ for any $p > 1$. However, it is conceivable that g lies in $L_s^p(\mathbf{R})$ for some $s < 1$. In fact, we fix $s \in (0,1)$ and we will show that g lies in $L_s^p(\mathbf{R})$ if and only if $1 < p < 1/s$. We pick a smooth function with compact support $\widehat{\varphi}$ equal to 1 on $[-1,1]$ and vanishing on the complement of $[-2,2]$.

Assume first that $1 < p < 1/s$; we will show that $g \in L_s^p(\mathbf{R}^n)$. We write

$$(1+|\xi|^2)^{\frac{s}{2}}\widehat{g}(\xi) = (1+|\xi|^2)^{\frac{s}{2}}\widehat{g}(\xi)\widehat{\varphi}(\xi) + (1+|\xi|^2)^{\frac{s}{2}}\widehat{g}(\xi)(1-\widehat{\varphi}(\xi)). \quad (5.4.3)$$

As $\widehat{g}(\xi) = \sin(2\pi\xi)/\pi\xi$ is smooth, the inverse Fourier transform of the function $(1+|\xi|^2)^{\frac{s}{2}}\widehat{g}(\xi)\widehat{\varphi}(\xi)$ lies in $\mathscr{S}$ and thus in L^p for any p. So we focus on the other term in the sum, which we rewrite as

$$(1+|\xi|^2)^{\frac{s}{2}}\frac{\sin(2\pi\xi)}{\pi\xi}(1-\widehat{\varphi}(\xi)) = \left\{\frac{(1+|\xi|^2)^{\frac{s}{2}}}{2\pi|\xi|^s}(1-\widehat{\varphi}(\xi))\right\}\frac{|\xi|}{i\xi}\frac{e^{2\pi i\xi}-e^{-2\pi i\xi}}{|\xi|^{1-s}}.$$

Notice that the function in the curly brackets is an L^p Fourier multiplier in view of Theorem 5.3.2; see Exercise 5.4.2. Likewise $|\xi|/i\xi$ is an L^p Fourier multiplier as the corresponding operator is the Hilbert transform. It follows that the inverse Fourier transform of the second term to the right in (5.4.3) lies in $L^p(\mathbf{R})$, $1 < p < \infty$, if the inverse Fourier transform of

5.4 Sobolev Spaces

$$\frac{e^{2\pi i \xi} - e^{-2\pi i \xi}}{|\xi|^{1-s}}$$

does so. But this can be calculated using the result of Example 2.7.4 and property (5) in Proposition 2.1.6, and it turns out to be $c_s(|x-1|^{-s} - |x+1|^{-s})$ for some constant c_s. This function decays like $|x|^{-1-s}$ as $|x| \to \infty$ by the mean value theorem and blows up like $|x \pm 1|^{-s}$ as $x \to \pm 1$. Clearly this function lies in $L^p(\mathbf{R})$ if $p < 1/s$. This yields that $g \in L_s^p(\mathbf{R})$ if $1 < p < 1/s$.

Assume now $g \in L_s^p(\mathbf{R})$; we will show that $1 < p < 1/s$. Then the inverse Fourier transform of

$$\left\{ \frac{(1+|\xi|^2)^{\frac{s}{2}}}{2\pi|\xi|^s}(1-\widehat{\varphi}(\xi)) \right\} \frac{|\xi|}{i\xi} \frac{e^{2\pi i\xi} - e^{-2\pi i\xi}}{|\xi|^{1-s}}$$

lies in $L^p(\mathbf{R})$. We now write $1 - \widehat{\varphi}(\xi) = (1-\widehat{\varphi}(\xi))(1-\widehat{\varphi}(2\xi))$ and we use that both $\xi/i|\xi|$ and $(1-\widehat{\varphi}(2\xi))|\xi|^s/(1+|\xi|^2)^{\frac{s}{2}}$ lie in $\mathscr{M}_p(\mathbf{R})$, to obtain that the inverse Fourier transform of

$$(1-\widehat{\varphi}(\xi))\frac{e^{2\pi i\xi} - e^{-2\pi i\xi}}{|\xi|^{1-s}}$$

must lie in $L^p(\mathbf{R})$. But this equals

$$c_s(|x-1|^{-s} - |x+1|^{-s}) - c_s\big((|\cdot -1|^{-s} - |\cdot +1|^{-s}) * \varphi\big)(x),$$

and as the second term lies in $L^p(\mathbf{R})$ when $p > 1$ and $s < 1$, then the first term must lie in $L^p(\mathbf{R})$; consequently, $\int_{|x\pm 1|<1} |x\pm 1|^{-sp} dx < \infty$, which yields $s < 1/p$.

Next we show that Sobolev spaces are Banach spaces. Before we do so, we notice that if $\varphi \in \mathscr{S}(\mathbf{R}^n)$, then $(I-\Delta)^{s/2}\varphi$ also lies in $\mathscr{S}(\mathbf{R}^n)$, a fact that is easily seen by examining the Fourier transforms.

Theorem 5.4.7. *For any $s \in \mathbf{R}$ and $1 < p < \infty$, $L_s^p(\mathbf{R}^n)$ is a complete normed vector space, i.e., a Banach space.*

Proof. Suppose that $\{f_k\}_{k=1}^\infty$ is a Cauchy sequence in $L_s^p(\mathbf{R}^n)$. This means that $\{g_k\}_{k=1}^\infty = \{(I-\Delta)^{s/2}f_k\}_{k=1}^\infty$ is a Cauchy sequence in L^p. As L^p is complete g_k converges to an element g in L^p. Now $f = (I-\Delta)^{-s/2}g$ is well defined as a tempered distribution[3]. We claim that $f \in L_s^p(\mathbf{R}^n)$ and that $f_k \to f$ in L_s^p. Obviously, $(I-\Delta)^{s/2}f = g \in L^p$, thus f is an element of L_s^p. Moreover,

$$\|f_k - f\|_{L_s^p} = \|g_k - g\|_{L^p} \to 0$$

as $k \to \infty$. This shows that $L_s^p(\mathbf{R}^n)$ is a complete normed vector space. □

Theorem 5.4.8. *For any $s \in \mathbf{R}$ and $1 < p < \infty$, $\mathscr{S}(\mathbf{R}^n)$ is dense in $L_s^p(\mathbf{R}^n)$.*

Proof. Let $f \in L_s^p(\mathbf{R}^n)$. As $(I-\Delta)^{s/2}f$ lies in L^p, we find a sequence of Schwartz functions ψ_j converging to $(I-\Delta)^{s/2}f$ in L^p as $j \to \infty$. Define $\varphi_j = (I-\Delta)^{-s/2}\psi_j$.

---

[3] f is in fact an L^p function if $s > 0$.

Clearly, $\widehat{\varphi_j} = (1+4\pi^2|\cdot|^2)^{-s/2}\widehat{\psi_j}$ is a Schwartz function, so by Proposition 2.6.12 we have $\varphi_j \in \mathscr{S}(\mathbf{R}^n)$. We will show that $\varphi_j \to f$ in L_s^p. This is equivalent to proving that $\left\|(I-\Delta)^{\frac{s}{2}}(f-\varphi_j)\right\|_{L^p} \to 0$ as $j \to \infty$. But this is obvious by the choice of ψ_j, as $\left\|(I-\Delta)^{\frac{s}{2}}(f-\varphi_j)\right\|_{L^p} = \left\|(I-\Delta)^{\frac{s}{2}}f - \psi_j\right\|_{L^p}$ which tends to 0 as $j \to \infty$. □

We discuss embedding of Sobolev spaces in other function spaces. If $(X, \|\cdot\|_X)$ and $(Y, \|\cdot\|_Y)$ are normed vector spaces, we write $X \hookrightarrow Y$ if X can be identified with a subspace of Y and there is a constant C such that $\|f\|_Y \leq C\|f\|_X$ for all $f \in X$.

Theorem 5.4.9. *(Sobolev embedding theorem)* Let $0 < s < \infty$ and $1 < p < \infty$.
(a) If $0 < s < \frac{n}{p}$, then for any $q \in \left(\frac{n}{n-s}, \infty\right)$ satisfying $\frac{1}{p} - \frac{1}{q} = \frac{s}{n}$ we have

$$L_s^p(\mathbf{R}^n) \hookrightarrow L^q(\mathbf{R}^n).$$

(b) If $s = \frac{n}{p}$, then for any q satisfying $p = \frac{n}{s} \leq q < \infty$ we have

$$L_s^p(\mathbf{R}^n) \hookrightarrow L^q(\mathbf{R}^n).$$

(c) If $s > \frac{n}{p}$, let M denote the largest integer strictly less than $s - \frac{n}{p}$. Then $L_s^p(\mathbf{R}^n)$ embeds in the space of functions whose partial derivatives up to and including order M exist, are continuous uniformly on $\mathbf{R}^n$, and are bounded on $\mathbf{R}^n$; moreover there is a constant $C_{n,p,s}$ such that for all $f \in L_s^p(\mathbf{R}^n)$ it holds that

$$\|f\|_{L^\infty} \leq \sum_{|\alpha|\leq M} \|\partial^\alpha f\|_{L^\infty} \leq C_{n,p,s}\|f\|_{L_s^p}.$$

Now, if $N = s - \frac{n}{p}$ happens to be an integer, then every function in $L_s^p(\mathbf{R}^n)$ has partial derivatives of order N in $L^q(\mathbf{R}^n)$ for any q satisfying $p \leq q < \infty$; precisely,

$$L_s^p(\mathbf{R}^n) \hookrightarrow L_N^q(\mathbf{R}^n).$$

Remark 5.4.10. Part (c) essentially says: If $s - \frac{n}{p} \in \mathbf{R}^+ \setminus \mathbf{Z}^+$, then elements of $L_s^p(\mathbf{R}^n)$ have derivatives up to and including order $[s - \frac{n}{p}]$ in L^q for $p \leq q \leq \infty$; if $s - \frac{n}{p} \in \mathbf{Z}^+$, then they have derivatives up to and including order $[s - \frac{n}{p}] - 1$ in L^q for $q \leq p \leq \infty$, and have derivatives of order $[s - \frac{n}{p}]$ in L^q for $p \leq q < \infty$.

Proof. (a) If $f \in L_s^p$, then we write

$$f = (I-\Delta)^{-\frac{s}{2}}(I-\Delta)^{\frac{s}{2}}f = G_s * (I-\Delta)^{\frac{s}{2}}f,$$

where G_s is the Bessel potential. Since $s < \frac{n}{p} < n$, Proposition 5.2.2 gives that

$$|G_s(x)| \leq C_{s,n}|x|^{s-n}$$

for all $x \in \mathbf{R}^n \setminus \{0\}$. This implies that $|f| \leq C_{s,n}\mathcal{I}_s\left(|(I-\Delta)^{\frac{s}{2}}f|\right)$. Theorem 5.1.3 now yields the required conclusion:

5.4 Sobolev Spaces

$$\|f\|_{L^q} \leq C'_{s,n} \| \mathcal{I}_s(|(I-\Delta)^{\frac{s}{2}}f|)\|_{L^q} \leq C''_{s,n}\|f\|_{L^p_s}.$$

(b) Given q satisfying $\frac{n}{s} = p \leq q < \infty$, there is an r in $[1,\infty)$ such that

$$1 + \frac{1}{q} = \frac{1}{p} + \frac{1}{r}.$$

Then $1 < \frac{s}{n} + \frac{1}{r}$, which implies that $(-n+s)r > -n$. Thus, the function $|x|^{-n+s}\chi_{|x|\leq 2}$ is integrable to the rth power, which implies that G_s is in $L^r(\mathbf{R}^n)$. As f in L^p_s can be written $G_s * (I-\Delta)^{\frac{s}{2}}f$, Young's inequality (Proposition 2.4.2) gives that

$$\|f\|_{L^q(\mathbf{R}^n)} \leq \|G_s\|_{L^r(\mathbf{R}^n)}\|(I-\Delta)^{\frac{s}{2}}f\|_{L^p(\mathbf{R}^n)} = C_{n,s}\|f\|_{L^p_s}.$$

(c) Let $s > n/p$ and M be the largest integer strictly less than $s - \frac{n}{p}$. Then for all $|\alpha| \leq M$, by Proposition 5.2.3, $\partial^\alpha G_s(x)$ decays exponentially when $|x| \geq 2$ and it is bounded by a constant multiple of $\max(|x|^{-n+s-|\alpha|}, \log\frac{2}{|x|}, 1)$ when $0 < |x| < 2$. This function lies in $L^{p'}(B(0,2))$ when $|\alpha| < s - n/p$ and thus $\partial^\alpha G_s$ lies in $L^{p'}(\mathbf{R}^n)$ for all $|\alpha| \leq M$. Let $f \in L^p_s(\mathbf{R}^n)$. Applying Theorem 1.7.1 (with $g = (I-\Delta)^{s/2}f$ and $\varphi = G_s$) we obtain that $f = G_s * (I-\Delta)^{s/2}f$ lies in $\mathscr{C}^M(\mathbf{R}^n)$ and satisfies

$$\partial^\alpha f = \partial^\alpha(G_s * (I-\Delta)^{s/2}f) = (\partial^\alpha G_s) * (I-\Delta)^{s/2}f$$

for all $|\alpha| \leq M$. The boundedness and uniform continuity of this function are consequences of Theorem 1.6.7. Hölder's inequality now yields

$$\sum_{|\alpha|\leq M}\|\partial^\alpha f\|_{L^\infty} \leq \sum_{|\alpha|\leq M}\|\partial^\alpha G_s\|_{L^{p'}}\|(I-\Delta)^{s/2}f\|_{L^p} = \left(\sum_{|\alpha|\leq M}\|\partial^\alpha G_s\|_{L^{p'}}\right)\|f\|_{L^p_s}$$

and thus the claimed embedding is valid.

In the event that $N = s - \frac{n}{p}$ is an integer, for $p \leq q < \infty$ we notice that

$$\|f\|_{L^q_N} = \|(I-\Delta)^{\frac{N}{2}}f\|_{L^q} \leq C\|(I-\Delta)^{\frac{N}{2}}f\|_{L^p_{s-N}} = C\|f\|_{L^p_s},$$

where the preceding inequality is a consequence of the assertion in part (b). $\square$

Exercises

5.4.1. Let $s \in \mathbf{Z}^+$ and $f \in L^p_s(\mathbf{R}^n)$ for $1 < p < \infty$. Suppose that g is a $\mathscr{C}^s$ function on $\mathbf{R}^n$ with the property $\partial^\alpha g \in L^\infty(\mathbf{R}^n)$ for all $|\alpha| \leq s$. Prove the $fg \in L^p_s(\mathbf{R}^n)$.

5.4.2. Let $\widehat{\Phi}$ be a $\mathscr{C}^\infty_0$ function on $\mathbf{R}^n$ that equals 1 in a neighborhood of the origin. Prove that the function $\xi \mapsto (1+|\xi|^2)^{s/2}|\xi|^{-s}(1-\widehat{\Phi}(\xi))$ lies in $\mathscr{M}_p(\mathbf{R}^n)$ for all $1 < p < \infty$.

5.4.3. Let $s > 0$ and $1 < p < \infty$. Let $\phi \in \mathscr{C}_0^\infty(\mathbf{R}^n)$ be equal to 1 on the unit ball and vanishing outside the double of the unit ball. Let $\phi^\varepsilon(x) = \phi(\varepsilon x)$. Prove that for any $f \in \mathscr{S}(\mathbf{R}^n)$, the sequence $f\phi^\varepsilon$ converges to f in $L_s^p(\mathbf{R}^n)$. Then use Theorem 5.4.8 to conclude that smooth functions with compact support are dense in $L_s^p(\mathbf{R}^n)$. [*Hint:* Show that $\{f\phi^\varepsilon\}_{\varepsilon>0}$ is Cauchy in $L_s^p(\mathbf{R}^n)$ and converges to f in $L^p(\mathbf{R}^n)$ as $\varepsilon \to 0$.]

5.4.4. Let $1 < p < n$, $\frac{n}{n-1} < q < \infty$ and $\frac{1}{p} - \frac{1}{q} = \frac{1}{n}$. Suppose that $f, \Delta f$ lie in $L^p(\mathbf{R}^n)$. Prove that f lies in $L_1^q(\mathbf{R}^n)$.

5.4.5. Let $1 < p < \infty$ and $s \in \mathbf{R}$.
(a) Let $f \in L_s^p(\mathbf{R}^n)$ and $g \in L^1(\mathbf{R}^n)$. Prove that $f * g$ lies in $L_s^p(\mathbf{R}^n)$.
(b) Let $f \in L_s^p(\mathbf{R}^n)$ and let $g \in L^r(\mathbf{R}^n)$ for some r in $(1, \infty)$. Prove that $f * g$ lies in $L_s^q(\mathbf{R}^n)$ when $1 < q < \infty$ and $1 + 1/q = 1/p + 1/r$.

5.4.6. (Fractional integration by parts for Sobolev spaces) Let $s > 0$, $f \in L_s^p(\mathbf{R}^n)$, and $g \in L_s^{p'}(\mathbf{R}^n)$, $1 < p < \infty$. Show that for any $t \in \mathbf{R}$ we have
$$\int_{\mathbf{R}^n} g\,(I-\Delta)^{\frac{s}{2}+it} f\, dx = \int_{\mathbf{R}^n} f\,(I-\Delta)^{\frac{s}{2}+it} g\, dx.$$
[*Hint:* Use Exercise 5.2.1, Corollary 5.3.3, and density.]

5.4.7. Show that translations, dilations, and modulations $M^a f(x) = f(x) e^{2\pi i x \cdot a}$ preserve $L_s^p(\mathbf{R}^n)$ for any $s \in \mathbf{R}$. [*Hint:* Use the result of Example 5.3.9.]

5.4.8. Let $\varphi \in \mathscr{S}(\mathbf{R}^n)$ and $f \in L_s^p(\mathbf{R}^n)$ for $1 < p < \infty$ and $s \in \mathbf{R}$. Prove that $\varphi f \in L_s^p$. [*Hint:* Write $(I-\Delta)^{\frac{s}{2}}(\varphi f)(x)$ as
$$\int_{\mathbf{R}^n} \widehat{\varphi}(\eta) e^{2\pi i x \cdot \eta} \left[\int_{\mathbf{R}^n} \left(\frac{1+4\pi^2|\xi+\eta|^2}{1+4\pi^2|\xi|^2} \right)^{\frac{s}{2}} [(I-\Delta)^{\frac{s}{2}} f]^{\wedge}(\xi) e^{2\pi i x \cdot \xi} d\xi \right] d\eta$$
and then use Example 5.3.9.]

5.4.9. Let u be the inverse Fourier transform of $t^{-1}(\log t)^{-1} \chi_{t \geq 2}$ on the real line. Show that u lies in $L_s^2(\mathbf{R})$ for all $s \leq 1/2$ but $u \notin L^\infty(\mathbf{R})$. How is this example related to Theorem 5.4.9?

5.4.10. Let $s, t \in \mathbf{R}$. Suppose that $|\partial^\alpha \sigma(\xi)| \leq C_\alpha (1+|\xi|)^{s-t} |\xi|^{-|\alpha|}$ for all multi-indices α and all $\xi \in \mathbf{R}^n \setminus \{0\}$. Prove that the operator $T_\sigma(\varphi) = (\widehat{\varphi}\sigma)^\vee$, initially defined for $\varphi \in \mathscr{S}(\mathbf{R}^n)$, admits a bounded extension from $L_s^p(\mathbf{R}^n)$ to $L_t^p(\mathbf{R}^n)$.

5.5 Interpolation of Analytic Families of Operators

In this section we prove an interpolation result for families of operators indexed by a complex parameter in which they depend analytically. We begin with a lemma that allows us to approximate a general $\mathscr{C}_0^\infty$ function by a family of $\mathscr{C}_0^\infty$ functions which are analytic in an auxiliary variable.

5.5 Interpolation of Analytic Families of Operators

Lemma 5.5.1. *Let $0 < p_0 \le p_1 \le \infty$ satisfy $p_0 < \infty$. Define p in terms of the identity $1/p = (1-\theta)/p_0 + \theta/p_1$, where $0 < \theta < 1$. Given $f \in \mathscr{C}_0^\infty(\mathbf{R}^n)$ and $\varepsilon > 0$, there exist $N_\varepsilon \in \mathbf{Z}^+$, smooth functions h_j^ε, $j = 1, \ldots, N_\varepsilon$, supported in cubes with (pairwise) disjoint interiors, and nonzero complex constants c_j^ε, such that the functions*

$$f_z^\varepsilon = \sum_{j=1}^{N_\varepsilon} |c_j^\varepsilon|^{\frac{p}{p_0}(1-z) + \frac{p}{p_1}z} h_j^\varepsilon \tag{5.5.1}$$

satisfy

$$\|f_\theta^\varepsilon - f\|_{L^{p_0}} < \varepsilon, \quad \begin{cases} \|f_\theta^\varepsilon - f\|_{L^{p_1}} < \varepsilon & \text{if } p_1 < \infty, \\ \|f_\theta^\varepsilon\|_{L^\infty} \le \|f\|_{L^\infty} + \varepsilon & \text{if } p_1 = \infty, \end{cases} \tag{5.5.2}$$

and for any real number t they also satisfy

$$\|f_{it}^\varepsilon\|_{L^{p_0}}^{p_0} \le \|f\|_{L^p}^p + \varepsilon', \quad \|f_{1+it}^\varepsilon\|_{L^{p_1}} \le (\|f\|_{L^p}^p + \varepsilon')^{\frac{1}{p_1}}, \tag{5.5.3}$$

where ε' depends on $\varepsilon, p, \|f\|_{L^p}$ and tends to zero as $\varepsilon \to 0$.

Proof. Given $f \in \mathscr{C}_0^\infty(\mathbf{R}^n)$ and $\varepsilon > 0$, by uniform continuity, there is a mesh of cubes of diameters at most 1, such that if x, y belong to the same cube in the mesh, then $|f(x) - f(y)| \le \varepsilon/C$, for some $C > 1$ to be chosen later. Let Q_j^ε, $j = 1, \ldots, N_\varepsilon$, be those cubes in the mesh whose interior intersects the support of f. There are $x_j \in Q_j^\varepsilon$ with $f(x_j) \ne 0$ and we define $c_j^\varepsilon = f(x_j)$. By construction we have

$$\Big\| f - \sum_{j=1}^{N_\varepsilon} c_j^\varepsilon \chi_{Q_j^\varepsilon} \Big\|_{L^\infty} \le \frac{\varepsilon}{C} \tag{5.5.4}$$

and by choosing C large enough (depending on f) we ensure

$$\Big\| f - \sum_{j=1}^{N_\varepsilon} c_j^\varepsilon \chi_{Q_j^\varepsilon} \Big\|_{L^{p_\kappa}}^{\min(1,p_\kappa)} < \frac{\varepsilon^{\min(1,p_\kappa)}}{2}, \quad \kappa \in \{0,1\}, \tag{5.5.5}$$

and

$$\Big\| f - \sum_{j=1}^{N_\varepsilon} c_j^\varepsilon \chi_{Q_j^\varepsilon} \Big\|_{L^p} < \varepsilon. \tag{5.5.6}$$

Now pick $g_j^\varepsilon \in \mathscr{C}_0^\infty$ satisfying $0 \le g_j^\varepsilon \le \chi_{Q_j^\varepsilon}$ such that

$$\Big\| \sum_{j=1}^{N_\varepsilon} c_j^\varepsilon (\chi_{Q_j^\varepsilon} - g_j^\varepsilon) \Big\|_{L^{p_\kappa}}^{\min(1,p_\kappa)} < \frac{\varepsilon^{\min(1,p_\kappa)}}{2}, \quad \kappa \in \{0,1\}, \text{ when } p_1 < \infty. \tag{5.5.7}$$

We set $h_j^\varepsilon = e^{i\phi_j^\varepsilon} g_j^\varepsilon$, where ϕ_j^ε is the argument of the complex number c_j^ε. Then h_j^ε is that function claimed in (5.5.1). Combining (5.5.5) and (5.5.7) and using the subadditivity of the expression $\|\cdot\|_{L^p}^{\min(1,p)}$ we obtain that the function

$$f_\theta^\varepsilon = \sum_{j=1}^{N_\varepsilon} |c_j^\varepsilon| h_j^\varepsilon = \sum_{j=1}^{N_\varepsilon} c_j^\varepsilon g_j^\varepsilon$$

satisfies (5.5.2) when $p_1 < \infty$. Additionally, we have

$$|f_\theta^\varepsilon| \leq \sum_{j=1}^{N_\varepsilon} |c_j^\varepsilon| \chi_{Q_j^\varepsilon} = \left| \sum_{j=1}^{N_\varepsilon} c_j^\varepsilon \chi_{Q_j^\varepsilon} \right| \leq \left| \sum_{j=1}^{N_\varepsilon} c_j^\varepsilon \chi_{Q_j^\varepsilon} - f \right| + |f| \leq \frac{\varepsilon}{C} + |f| \leq \varepsilon + \|f\|_{L^\infty},$$

so (5.5.2) also holds when $p_1 = \infty$. We now notice that

$$\|f_{it}^\varepsilon\|_{L^{p_0}}^{p_0} \leq \sum_{j=1}^{N_\varepsilon} |c_j^\varepsilon|^p |Q_j^\varepsilon| = \left\| \sum_{j=1}^{N_\varepsilon} c_j^\varepsilon \chi_{Q_j^\varepsilon} \right\|_{L^p}^p \leq \left(\varepsilon^{\min(1,p)} + \|f\|_{L^p}^{\min(1,p)} \right)^{\frac{p}{\min(1,p)}},$$

where we made use of (5.5.6) and of the subadditivity of $\|\cdot\|_{L^p}^{\min(1,p)}$.

We set $\varepsilon' = \varepsilon^p$ if $p \leq 1$ and $\varepsilon' = (\varepsilon + \|f\|_{L^p})^p - \|f\|_{L^p}^p$ when $1 < p < \infty$. Then $\varepsilon' \to 0$ as $\varepsilon \to 0$ and this proves (5.5.3) for p_0 and analogously for p_1 when $p_1 < \infty$; now if $p_1 = \infty$, then $\|f_{1+it}^\varepsilon\|_{L^\infty} \leq 1$ and the right-hand side of the second inequality in (5.5.3) is equal to 1, so the inequality is still valid. □

We discuss an extension of Theorem 2.4.1 in which the operators are allowed to vary analytically in a complex variable in the unit strip $\mathbf{S} = \{z \in \mathbf{C} : 0 < \mathrm{Re}(z) < 1\}$.

Definition 5.5.2. Suppose that for every $z \in \overline{\mathbf{S}} = \{z \in \mathbf{C} : 0 \leq \mathrm{Re}\, z \leq 1\}$ there is an associated linear operator T_z defined on $\mathscr{C}_0^\infty(\mathbf{R}^n)$ and taking values in $L^1_{\mathrm{loc}}(\mathbf{R}^n)$. We call $\{T_z\}_z$ an *analytic family* if for all φ, ψ in $\mathscr{C}_0^\infty(\mathbf{R}^n)$ the function

$$z \mapsto \int_{\mathbf{R}^n} T_z(\varphi)\, \psi\, dx \tag{5.5.8}$$

is analytic in the open strip $\mathbf{S} = \{z \in \mathbf{C} : 0 < \mathrm{Re}\, z < 1\}$ and continuous on its closure. The analytic family $\{T_z\}_z$ is called of *admissible growth* if there is a constant γ with $0 \leq \gamma < \pi$ and an s satisfying $1 < s \leq \infty$, such that for any φ in $\mathscr{C}_0^\infty(\mathbf{R}^n)$ and every compact subset K of $\mathbf{R}^n$ there is constant $C(\varphi, K)$ such that

$$\log \|T_z(\varphi)\|_{L^s(K)} \leq C(\varphi, K) e^{\gamma |\mathrm{Im}\, z|}, \qquad \text{for all } z \in \overline{\mathbf{S}}. \tag{5.5.9}$$

Examples of such families are given at the end of this section.

Theorem 5.5.3. (Stein's interpolation theorem for analytic families) *For $z \in \overline{\mathbf{S}}$, let T_z be linear operators on $\mathscr{C}_0^\infty(\mathbf{R}^n)$ with values in $L^1_{\mathrm{loc}}(\mathbf{R}^n)$ that form an analytic family of admissible growth. Let $0 < p_0, p_1 \leq \infty$, $0 < q_0, q_1 \leq \infty$, fix $0 < \theta < 1$, and define p, q by the equations*

5.5 Interpolation of Analytic Families of Operators

$$\frac{1}{p} = \frac{1-\theta}{p_0} + \frac{\theta}{p_1} \quad \text{and} \quad \frac{1}{q} = \frac{1-\theta}{q_0} + \frac{\theta}{q_1}. \tag{5.5.10}$$

Let $B_0, B_1 > 0$ and M_0 and M_1 be nonnegative continuous functions on the real line that satisfy

$$M_0(y) + M_1(y) \leq e^{c e^{\tau|y|}} \tag{5.5.11}$$

for some $c, \tau \geq 0$ with $\tau < \pi$ and all $y \in \mathbf{R}$. Suppose that for all $f \in \mathscr{C}_0^\infty(\mathbf{R}^n)$ we have

$$\|T_{iy}(f)\|_{L^{q_0}} \leq B_0 M_0(y) \|f\|_{L^{p_0}}, \tag{5.5.12}$$

$$\|T_{1+iy}(f)\|_{L^{q_1}} \leq B_1 M_1(y) \|f\|_{L^{p_1}}, \tag{5.5.13}$$

for all $y \in \mathbf{R}$. Then for all f in $\mathscr{C}_0^\infty(\mathbf{R}^n)$ we have

$$\|T_\theta(f)\|_{L^q} \leq M(\theta) B_0^{1-\theta} B_1^\theta \|f\|_{L^p}, \tag{5.5.14}$$

where

$$M(\theta) = \exp\left\{ \frac{\sin(\pi\theta)}{2} \int_{-\infty}^{\infty} \left[\frac{\log M_0(y)}{\cosh(\pi y) - \cos(\pi\theta)} + \frac{\log M_1(y)}{\cosh(\pi y) + \cos(\pi\theta)} \right] dy \right\}.$$

Thus, by density, T_θ has a unique bounded extension from L^p to L^q when $p < \infty$.

We observe that assumption (5.5.11) guarantees the absolute convergence of the integral defining $M(\theta)$.

Proof. Case I: $\min(q_0, q_1) > 1$. This forces $q_0', q_1' < \infty$ and so $q' < \infty$ as well. Given T_z as in the statement of the theorem, for $f, g \in \mathscr{C}_0^\infty$ one may be tempted to consider the family of operators $H(z) = \int_{\mathbf{R}^n} T_z(f) g \, dx$ which is analytic in $\mathbf{S}$, continuous and bounded in $\overline{\mathbf{S}}$ and satisfies the hypotheses of Proposition C.0.2 with bounds $|H(iy)| \leq B_0 M_0(y) \|f\|_{L^{p_0}} \|g\|_{L^{q_0'}}$ and $|H(1+iy)| \leq B_1 M_1(y) \|f\|_{L^{p_1}} \|g\|_{L^{q_1'}}$ for all real y. Applying the result of Proposition C.0.2 and identity (C.0.2) (with $x = 1 - \theta$ and $x = \theta$) yields for all $f, g \in \mathscr{C}_0^\infty(\mathbf{R}^n)$

$$\left| \int_{\mathbf{R}^n} T_\theta(f) g \, dx \right| \leq M(\theta) \left(B_0 \|f\|_{L^{p_0}} \|g\|_{L^{q_0'}} \right)^{1-\theta} \left(B_1 \|f\|_{L^{p_1}} \|g\|_{L^{q_1'}} \right)^\theta. \tag{5.5.15}$$

Unfortunately this estimate does not provide the claimed assertion; it supplies, however, a useful continuity estimate for the operator T_θ.

To improve (5.5.15), let us first consider the situation where $\min(p_0, p_1) < \infty$; this forces $p < \infty$. Without loss of generality assume that $p_0 \leq p_1$. Fix $f, g \in \mathscr{C}_0^\infty$ and $\varepsilon > 0$. By Lemma 5.5.1 we can find f_z^ε and g_z^ε such that

$$f_z^\varepsilon = \sum_{j=1}^{N_\varepsilon} |c_j^\varepsilon|^{\frac{p}{p_0}(1-z) + \frac{p}{p_1} z} u_j^\varepsilon, \quad g_z^\varepsilon = \sum_{k=1}^{M_\varepsilon} |d_k^\varepsilon|^{\frac{q'}{q_0'}(1-z) + \frac{q'}{q_1'} z} v_k^\varepsilon,$$

where $u_j^\varepsilon, v_k^\varepsilon$ are in $\mathscr{C}_0^\infty(\mathbf{R}^n)$ and (for t real)

$$\|f_\theta^\varepsilon - f\|_{L^{p_0}} < \varepsilon, \quad \begin{cases} \|f_\theta^\varepsilon - f\|_{L^{p_1}} < \varepsilon & \text{if } p_1 < \infty \\ \|f_\theta^\varepsilon\|_{L^\infty} \le \|f\|_{L^\infty} + \varepsilon & \text{if } p_1 = \infty \end{cases}, \quad \begin{cases} \|g_\theta^\varepsilon - g\|_{L^{q_1'}} < \varepsilon, \\ \|g_\theta^\varepsilon - g\|_{L^{q_0'}} < \varepsilon, \end{cases} \tag{5.5.16}$$

$$\|f_{it}^\varepsilon\|_{L^{p_0}} \le (\|f\|_{L^p} + \varepsilon')^{\frac{p}{p_0}}, \quad \|g_{it}^\varepsilon\|_{L^{q_0'}} \le (\|g\|_{L^{q'}} + \varepsilon')^{\frac{q'}{q_0'}}, \tag{5.5.17}$$

$$\|f_{1+it}^\varepsilon\|_{L^{p_1}} \le (\|f\|_{L^p} + \varepsilon')^{\frac{p}{p_1}}, \quad \|g_{1+it}^\varepsilon\|_{L^{q_1'}} \le (\|g\|_{L^{q'}} + \varepsilon')^{\frac{q'}{q_1'}}. \tag{5.5.18}$$

Now consider the function defined on the closure of the unit strip

$$F(z) = \int_{\mathbf{R}^n} T_z(f_z^\varepsilon) g_z^\varepsilon \, dx = \sum_{j=1}^{N_\varepsilon} \sum_{k=1}^{M_\varepsilon} |c_j^\varepsilon|^{\frac{p}{p_0}(1-z) + \frac{p}{p_1}z} |d_k^\varepsilon|^{\frac{q'}{q_0'}(1-z) + \frac{q'}{q_1'}z} \int_{\mathbf{R}^n} T_z(u_j^\varepsilon) v_k^\varepsilon \, dx.$$

Applying Hölder's inequality with exponents s and s' to $\int_{\mathbf{R}^n} T_z(u_j^\varepsilon) v_k^\varepsilon \, dx$ and using condition (5.5.9) we obtain for any z in $\overline{\mathbf{S}}$

$$|F(z)| \le \left[\sum_{j=1}^{N_\varepsilon} \sum_{k=1}^{M_\varepsilon} (1 + |c_j^\varepsilon|)^{\frac{p}{p_0} + \frac{p}{p_1}} (1 + |d_k^\varepsilon|)^{\frac{q'}{q_0'} + \frac{q'}{q_1'}} \|v_k^\varepsilon\|_{L^{s'}} \right] e^{[\max_{j,k} C(u_j^\varepsilon, \text{supp } v_k^\varepsilon)] e^{\gamma |\text{Im} z|}}$$
$$\le e^{C' e^{\gamma |\text{Im} z|}},$$

where C' equals $\max_{j,k} C(u_j^\varepsilon, \text{supp } v_k^\varepsilon)$ plus the logarithm of the double sum in the square brackets. Thus F satisfies the hypothesis of Proposition C.0.2, as $\gamma < \pi$.

Hölder's inequality, hypothesis (5.5.12) and (5.5.17) give for y real

$$|F(iy)| \le B_0 M_0(y) \|f_{iy}^\varepsilon\|_{L^{p_0}} \|g_{iy}^\varepsilon\|_{L^{q_0'}} \le B_0 M_0(y) (\|f\|_{L^p} + \varepsilon')^{\frac{p}{p_0}} (\|g\|_{L^{q'}} + \varepsilon')^{\frac{q'}{q_0'}}.$$

Likewise, Hölder's inequality, the hypothesis (5.5.13) and (5.5.18) imply for y real

$$|F(1+iy)| \le B_1 M_1(y) \|f_{1+iy}^\varepsilon\|_{L^{p_1}} \|g_{1+iy}^\varepsilon\|_{L^{q_1'}} \le B_1 M_1(y) (\|f\|_{L^p} + \varepsilon')^{\frac{p}{p_1}} (\|g\|_{L^{q'}} + \varepsilon')^{\frac{q'}{q_1'}}.$$

As $\log |F|$ is subharmonic in $\mathbf{S}$, applying Proposition C.0.2 in Appendix C we obtain

$$\log |F(\theta)| \le \int_{-\infty}^{+\infty} \Omega(1-\theta, t) \log[M_0(t) Q_0] \, dt + \int_{-\infty}^{+\infty} \Omega(\theta, t) \log[M_1(t) Q_1] \, dt,$$

where Ω is the Poisson kernel on the strip [defined in (C.0.1)] and

$$Q_0 = B_0 (\|f\|_{L^p} + \varepsilon')^{\frac{p}{p_0}} (\|g\|_{L^{q'}} + \varepsilon')^{\frac{q'}{q_0'}}, \quad Q_1 = B_1 (\|f\|_{L^p} + \varepsilon')^{\frac{p}{p_1}} (\|g\|_{L^{q'}} + \varepsilon')^{\frac{q'}{q_1'}}.$$

Using identity (C.0.2) (with $x = 1 - \theta$ and $x = \theta$) and the fact that

$$Q_0^{1-\theta} Q_1^\theta = B_0^{1-\theta} B_1^\theta (\|f\|_{L^p} + \varepsilon') (\|g\|_{L^{q'}} + \varepsilon')$$

we obtain [with $M(\theta)$ as in the statement of the theorem] that

5.5 Interpolation of Analytic Families of Operators

$$\left| \int_{\mathbf{R}^n} T_\theta(f_\theta^\varepsilon) g_\theta^\varepsilon \, dx \right| = |F(\theta)| \leq M(\theta) B_0^{1-\theta} B_1^\theta \left(\|f\|_{L^p} + \varepsilon' \right) \left(\|g\|_{L^{q'}} + \varepsilon' \right). \quad (5.5.19)$$

An application of the triangle inequality gives

$$\left| \int_{\mathbf{R}^n} T_\theta(f) g \, dx - \int_{\mathbf{R}^n} T_\theta(f_\theta^\varepsilon) g_\theta^\varepsilon \, dx \right| \quad (5.5.20)$$

$$\leq \left| \int_{\mathbf{R}^n} T_\theta(f - f_\theta^\varepsilon) g \, dx \right| + \left| \int_{\mathbf{R}^n} T_\theta(f_\theta^\varepsilon)(g - g_\theta^\varepsilon) \, dx \right|.$$

We now apply (5.5.15) in each of the terms on the right side of this inequality and we use (5.5.16) to deduce that (5.5.20) tends to zero as $\varepsilon \to 0$. We now return to (5.5.19) and let $\varepsilon \to 0$. Using that $\varepsilon' \to 0$, we conclude

$$\left| \int_{\mathbf{R}^n} T_\theta(f) g \, dx \right| \leq M(\theta) B_0^{1-\theta} B_1^\theta \|f\|_{L^p} \|g\|_{L^{q'}}. \quad (5.5.21)$$

Finally, we obtain (5.5.14) by taking the supremum in (5.5.21) over all g in $\mathscr{C}_0^\infty(\mathbf{R}^n)$ with $L^{q'}$ norm equal to 1.

Suppose now that $p_0 = p_1 = \infty$, which forces $p = \infty$. In this case we work directly with the analytic function

$$F(z) = \int_{\mathbf{R}^n} T_z(f) g_z^\varepsilon \, dx$$

on $\mathbf{S}$, which is continuous on $\overline{\mathbf{S}}$, it satisfies

$$|F(\kappa + iy)| \leq B_\kappa M_\kappa(y) \|f\|_{L^\infty} \left(\|g\|_{L^{q'}} + \varepsilon' \right)^{\frac{q'}{q_\kappa'}}, \quad \kappa \in \{0, 1\}, \quad y \in \mathbf{R},$$

and is bounded by $e^{C'e^{\gamma|\operatorname{Im} z|}}$ for all $z \in \overline{\mathbf{S}}$. Proposition C.0.2 yields the bound

$$|F(\theta)| = \left| \int_{\mathbf{R}^n} T_\theta(f) g_\theta^\varepsilon \, dx \right| \leq M(\theta) B_0^{1-\theta} B_1^\theta \|f\|_{L^\infty} \left(\|g\|_{L^{q'}} + \varepsilon' \right). \quad (5.5.22)$$

At this point we make use of the inequality

$$\left| \int_{\mathbf{R}^n} T_\theta(f) g \, dx \right| \leq \left| \int_{\mathbf{R}^n} T_\theta(f) g_\theta^\varepsilon \, dx \right| + \left| \int_{\mathbf{R}^n} T_\theta(f)(g_\theta^\varepsilon - g) \, dx \right| \quad (5.5.23)$$

and the auxiliary estimate (5.5.15) with $p_0 = p_1 = \infty$; this implies that the second term on the right in (5.5.23) is bounded by a constant times ε. Inserting a limsup as $\varepsilon \to 0$ in both (5.5.22) and (5.5.23) allows one to replace $F(\theta)$ by $\int_{\mathbf{R}^n} T_\theta(f) g \, dx$ in (5.5.22). After doing this, we take the supremum over all g in $\mathscr{C}_0^\infty(\mathbf{R}^n)$ with $L^{q'}$ norm equal to 1 to deduce (5.5.14).

Case II: $\min(q_0, q_1) \leq 1$. Assume first that $\min(p_0, p_1) < \infty$ and as before suppose $p_0 \leq p_1$, so that $p_0, p < \infty$. Choose $r > 1$ such that $r \min(q_0, q_1) > q$. Let us fix a nonnegative step function g with $\|g\|_{L^{r'}} = 1$. Assume that $g = \sum_{k=1}^{K} a_k \chi_{E_k}$, where $a_k > 0$ and E_k are cubes of finite measure and with disjoint interiors. For $z \in \mathbf{C}$ define

$$g^z = \sum_{k=1}^{K} a_k^{R(z)} \chi_{E_k},$$

where we set

$$R(z) = r'\left[1 - \frac{q}{rq_0}(1-z) - \frac{q}{rq_1}z\right].$$

Notice that $R(\theta) = 1$, $R(it) = r'(1 - \frac{q}{rq_0})$, $R(1+it) = r'(1 - \frac{q}{rq_1})$ for $t \in \mathbf{R}$ and that

$$\left\|g^{it}\right\|_{L^{(rq_0/q)'}}^{(rq_0/q)'} = \left\|g^{1+it}\right\|_{L^{(rq_1/q)'}}^{(rq_1/q)'} = \|g\|_{L^{r'}}^{r'} = 1. \tag{5.5.24}$$

We fix $f \in \mathscr{C}_0^\infty$ and $\varepsilon > 0$. Let f_z^ε be as in Case I obtained by Lemma 5.5.1. Define the function

$$G(z) = \int_{\mathbf{R}^n} |T_z(f_z^\varepsilon)(x)|^{\frac{q}{r}} |g^z(x)| \, dx = \sum_{k=1}^{K} \int_{E_k} |F_k(x,z)|^{\frac{q}{r}} dx, \tag{5.5.25}$$

where

$$F_k(x,z) = a_k^{\frac{r}{q}R(z)} \sum_{j=1}^{N_\varepsilon} |c_j^\varepsilon|^{\frac{p}{p_0}(1-z) + \frac{p}{p_1}z} T_z(u_j^\varepsilon)(x).$$

If we knew that each term of the sum on the right in (5.5.25) is log-subharmonic, it would follow from Lemma B.0.2 that so is G. To achieve this we use Lemma B.0.5, which requires knowing that for each k, the mapping $z \mapsto F_k(\cdot, z)$ is analytic from $\mathbf{S}$ to the Banach space $L^1(E_k)$. To prove this, in view of Theorem B.0.3, it suffices to show that for $w \in L^\infty(E_k)$ the function

$$z \mapsto \int_{E_k} F_k(x,z) w(x) \, dx$$

is analytic in $\mathbf{S}$ and continuous on its closure; on this see[4] Exercise 5.5.1.

To apply Proposition C.0.2 to G we verify its hypotheses. Using Hölder's inequality with indices $\frac{rq_0}{q}$ and $\left(\frac{rq_0}{q}\right)'$, (5.5.12), (5.5.3), and (5.5.24) we obtain

$$G(it) \leq \left\{\int_{\mathbf{R}^n} |T_{it}(f_{it}^\varepsilon)(x)|^{q_0} dx\right\}^{\frac{q}{rq_0}} \|g^{it}\|_{L^{(\frac{rq_0}{q})'}} \leq \left(B_0 M_0(t)(\|f\|_{L^p}^p + \varepsilon')^{\frac{1}{p_0}}\right)^{\frac{q}{r}}$$

when $t \in \mathbf{R}$. Similarly, we obtain the estimate

$$G(1+it) \leq \left\{\int_{\mathbf{R}^n} |T_{1+it}(f_{1+it}^\varepsilon)(x)|^{q_1} dx\right\}^{\frac{q}{rq_1}} \|g^{1+it}\|_{L^{(\frac{rq_1}{q})'}} \leq \left[B_1 M_1(t)(\|f\|_{L^p}^p + \varepsilon')^{\frac{1}{p_1}}\right]^{\frac{q}{r}}.$$

---
[4] The condition $s > 1$ in Definition 5.5.2 is used here. Case I only requires $s \geq 1$.

5.5 Interpolation of Analytic Families of Operators

Finally, we verify condition (C.0.5) for G. Let E be a compact set that contains all E_k. We apply Hölder's inequality with indices $\frac{rs}{q}$ and $\left(\frac{rs}{q}\right)'$ to obtain for $z \in \overline{\mathbf{S}}$

$$G(z) \leq \|T_z(f_z^\varepsilon)\|_{L^s(E)}^{\frac{q}{r}} \|g^z\|_{L^{(\frac{rs}{q})'}}$$

$$\leq \left[\sum_{j=1}^{N_\varepsilon} (1+|c_j^\varepsilon|)^{\frac{p}{p_0}+\frac{p}{p_1}} \|T_z(u_j^\varepsilon)\|_{L^s(E)}\right]^{\frac{q}{r}} \left[\sum_{k=1}^{K} (1+|a_k|)^{r'[1+\frac{q}{r}(\frac{1}{q_0}+\frac{1}{q_1})]} \|\chi_{E_k}\|_{L^{(\frac{rs}{q})'}}\right]$$

$$\leq e^{\frac{q}{r}\sup_j C(u_j^\varepsilon,E)} e^{\gamma|\mathrm{Im}\,z|} \left[\sum_{j=1}^{N_\varepsilon}(1+|c_j^\varepsilon|)^{\frac{p}{p_0}+\frac{p}{p_1}}\right]^{\frac{q}{r}}$$

$$\cdot \left[\sum_{k=1}^{K}(1+|a_k|)^{r'[1+\frac{q}{r}(\frac{1}{q_0}+\frac{1}{q_1})]} \|\chi_{E_k}\|_{L^{(\frac{rs}{q})'}}\right]$$

having used (5.5.9). Taking the logarithm, we deduce condition (C.0.5) for G.

As $g^\theta = g$, by Proposition C.0.2 we conclude

$$\int_{\mathbf{R}^n} |T_\theta(f_\theta^\varepsilon)(x)|^{\frac{q}{r}} g(x)\,dx = G(\theta) \leq \left(B_0^{1-\theta} B_1^\theta M(\theta)(\|f\|_{L^p}^p + \varepsilon')^{\frac{1}{p}}\right)^{\frac{q}{r}}. \quad (5.5.26)$$

Inequality (5.5.26) implies that

$$\|T_\theta(f_\theta^\varepsilon)\|_{L^q} = \left\||T_\theta(f_\theta^\varepsilon)|^{\frac{q}{r}}\right\|_{L^r}^{\frac{r}{q}}$$

$$= \sup\left\{\int |T_\theta(f_\theta^\varepsilon)(x)|^{\frac{q}{r}} g(x)\,dx :\ g \geq 0,\ g \text{ step function},\ \|g\|_{L^{r'}} = 1\right\}^{\frac{r}{q}}$$

$$\leq B_0^{1-\theta} B_1^\theta M(\theta)(\|f\|_{L^p}^p + \varepsilon')^{\frac{1}{p}}. \quad (5.5.27)$$

We also note that a similar (but simpler) argument, applying Proposition C.0.2 to the log-subharmonic function $H(z) = \int_{\mathbf{R}^n} |T_z(f)(x)|^{\frac{q}{r}} |g^z(x)|\,dx$, yields the estimate

$$|H(\theta)| = \left|\int_{\mathbf{R}^n} |T_\theta(f)(x)|^{\frac{q}{r}} g(x)\,dx\right| \leq \left(B_0^{1-\theta} B_1^\theta M(\theta) \|f\|_{L^{p_0}}^{1-\theta} \|f\|_{L^{p_1}}^{\theta}\right)^{\frac{q}{r}}.$$

It follows from this that

$$\|T_\theta(f)\|_{L^q} \leq B_0^{1-\theta} B_1^\theta M(\theta) \|f\|_{L^{p_0}}^{1-\theta} \|f\|_{L^{p_1}}^{\theta} \quad (5.5.28)$$

via a duality argument similar to that leading to (5.5.27).

We now make use of the triangle inequality

$$\|T_\theta(f)\|_{L^q}^{\min(1,q)} \leq \|T_\theta(f - f_\theta^\varepsilon)\|_{L^q}^{\min(1,q)} + \|T_\theta(f_\theta^\varepsilon)\|_{L^q}^{\min(1,q)}.$$

For the second term on the right above we use (5.5.27), while the first term is bounded by a constant multiple of $(\varepsilon^{1-\theta})^{\min(1,q)}$ in view of (5.5.28), and hence it tends to zero as $\varepsilon \to 0$. We deduce (5.5.14) by letting $\varepsilon \to 0$.

Finally, if $p_0 = p_1 = \infty$, then we must have $p = \infty$, and the claimed assertion is contained in (5.5.28), which is valid even when $p_0 = p_1 = \infty$. □

Example 5.5.4. We examine some families for which Theorem 5.5.3 applies.

1. The family $T_z(\varphi)(x) = \int_{|y|\leq 1} \varphi(x-y)|y|^{-nz} dy$, defined for $\varphi \in \mathscr{C}_0^\infty(\mathbf{R}^n)$, satisfies the analyticity condition of Definition 5.5.2 in $\mathbf{S}$ but is not continuous on $\overline{\mathbf{S}}$. In fact, the analyticity assertion can be reduced to Lemma 2.7.6 by Fubini's theorem. The continuity on the boundary fails at $z = 1$, as this is easily seen by an example.

2. The operators $\{(1-z)T_z\}_z$, where T_z is as in the previous example, form an analytic family of admissible growth. To verify the continuity on $\overline{\mathbf{S}}$, we write

$$(1-z)T_z(\varphi)(x) = \varphi(x)(1-z)\int_{|y|\leq 1} \frac{dy}{|y|^{nz}} + (1-z)\int_{|y|\leq 1} (\varphi(x-y) - \varphi(x)) \frac{dy}{|y|^{nz}}$$

and we notice that the first integral on the right equals $\omega_{n-1} n^{-1}(1-z)^{-1}$. So the factor $1-z$ cancels the singularity caused by this integral. Also, using that $|\varphi(x-y) - \varphi(x)| \leq \|\nabla\varphi\|_{L^\infty}|y|$ we see that the second integral converges absolutely and produces a continuous function of z on $\overline{\mathbf{S}}$. These calculations also show that $(1-z)T_z(\varphi)(x)$ is bounded on compact sets independently of z, so it satisfies the admissibility condition (5.5.9).

3. The operators $V_z = (I - \Delta)^{z^2+(a+ib)z}$, $z \in \overline{\mathbf{S}}$, $a, b \in \mathbf{R}$, form an analytic family of admissible growth. The analyticity can be derived from Lemma 2.7.5 and is omitted. To verify that V_z is of admissible growth, we notice that the real part of $z^2 + (a+ib)z$ equals $x^2 + ax - (y + \frac{b}{2})^2 + \frac{1}{4}b^2$ for $z = x + iy \in \overline{\mathbf{S}}$, so for $\varphi \in \mathscr{C}_0^\infty$ we have

$$\|V_z(\varphi)\|_{L^\infty} \leq \int_{\mathbf{R}^n} (1 + 4\pi^2|\xi|^2)^{1+|a|+b^2/4} |\widehat{\varphi}(\xi)| d\xi < \infty,$$

and this constant is independent of $\operatorname{Im} z$.

4. Let G be a nonnegative function in $L^1(\mathbf{R}^n)$. Then the family $W_z(\varphi)(x) = \int_{\mathbf{R}^n} \varphi(x-y) G(y)^z dy$, $z \in \overline{\mathbf{S}}$, $\varphi \in \mathscr{C}_0^\infty$, is analytic of admissible growth. We only verify the assertion of analyticity. To see this, we fix a point z_0 in the unit strip $\mathbf{S}$ and we pick $\delta > 0$ such that $2\delta < \min(\operatorname{Re} z_0, 1 - \operatorname{Re} z_0)$. Then for $|z| < \delta$, by Lemma 2.7.5, the integrand of

$$\frac{W_{z+z_0}(\varphi)(x) - W_{z_0}(\varphi)(x)}{z} = \int_{\mathbf{R}^n} \varphi(x-y) G(y)^{z_0} \frac{G(y)^z - 1}{z} dy$$

is bounded by $\frac{2}{\delta}|\varphi(x-y)|G(y)^{\operatorname{Re} z_0} \max(G(y)^{2\delta}, G(y)^{-2\delta})$, which lies in $L^1(dy)$ by the choice of δ. So the LDCT allows the passing of the limit inside the integral and the existence of a complex derivative of $W_z(\varphi)(x)$ follows.

Exercises

5.5.1. Suppose that $\{T_z\}_z$ is an analytic family of admissible growth according to Definition 5.5.2. Prove that for any bounded function g with compact support and any $\varphi \in \mathscr{C}_0^\infty(\mathbf{R}^n)$, the mapping
$$z \mapsto \int_{\mathbf{R}^n} T_z(\varphi) g\, dx$$
is analytic in the unit strip and continuous on its closure. [*Hint:* Approximate g in $L^{s'}$ by $\mathscr{C}_0^\infty$ functions. The condition $s > 1$ in Definition 5.5.2 is needed here.]

5.5.2. (Kato–Ponce inequality) Let $0 < s < 2N$, $N \in \mathbf{Z}^+$, and $\psi \in \mathscr{S}(\mathbf{R}^n)$. Show that for any $1 < p < \infty$ there is a constant $C = C_{p,n,s,N}$ such that for all f in $\mathscr{S}(\mathbf{R}^n)$ we have
$$\|\psi f\|_{L^p_s(\mathbf{R}^n)} \leq C \|f\|_{L^p_s(\mathbf{R}^n)} \sum_{|\alpha| \leq 2N} \|\partial^\alpha \psi\|_{L^\infty}.$$
Then extend this inequality by density to all $f \in L^p_s(\mathbf{R}^n)$ (Theorem 5.4.8).
[*Hint:* Note that $f \in \mathscr{S}$ if and only if $(I - \Delta)^{\frac{s}{2}} f \in \mathscr{S}$. Apply Theorem 5.5.3 to the family of operators $T_z(f) = (I - \Delta)^{Nz}[\psi (I - \Delta)^{-Nz} f]$, $z \in \overline{\mathbf{S}}$.]

5.5.3. (Kato–Ponce-type inequality) Let $1 < p, q, r < \infty$ satisfy $1/p + 1/q = 1/r$ and let $s > 0$. Prove that when f lies in $L^p_s(\mathbf{R}^n)$ and g lies in $L^q_s(\mathbf{R}^n)$, then fg is an element of $L^r_s(\mathbf{R}^n)$ and there is a constant $C = C_{p,q,s,n}$ such that
$$\|fg\|_{L^r_s} \leq C \|f\|_{L^p_s} \|g\|_{L^q_s}.$$
[*Hint:* Prove the inequality for $f, g \in \mathscr{C}_0^\infty$ and use density. Apply Theorem 5.5.3 to the family $T_z(f) = (I - \Delta)^{zN}\big[\big((I-\Delta)^{-zN} f\big)\big((I-\Delta)^{-zN} g\big)\big]$, $N = [\frac{s}{2}] + 1$, for $g \in \mathscr{C}_0^\infty$ fixed. For the density argument use Exercise 5.4.3.]

5.5.4. (Interpolation between Sobolev spaces) Let $1 < p_0, q_0, p_1, q_1 < \infty$ and s_0, s_1, t_0, t_1 be real numbers. Suppose that σ is a tempered distribution whose Fourier transform is a locally integrable function and tempered at infinity; cf. Example 2.6.2. Define $T(\varphi) = \varphi * \sigma$, for $\varphi \in \mathscr{S}(\mathbf{R}^n)$. Assume that for some constants $M_0, M_1 > 0$ we have $\|T(\varphi)\|_{L^{q_i}_{t_i}(\mathbf{R}^n)} \leq M_i \|\varphi\|_{L^{p_i}_{s_i}(\mathbf{R}^n)}$ for all $\varphi \in \mathscr{S}(\mathbf{R}^n)$ and $i = 0, 1$. Show that there is a constant C depending on all the preceding parameters, such that $\|T(\varphi)\|_{L^q_t(\mathbf{R}^n)} \leq C \|\varphi\|_{L^p_s(\mathbf{R}^n)}$ for all $\varphi \in \mathscr{S}(\mathbf{R}^n)$. Here $1/p = (1-\theta)/p_0 + \theta/p_1$, $1/q = (1-\theta)/q_0 + \theta/q_1$, $s = (1-\theta)s_0 + \theta s_1$, $t = (1-\theta)t_0 + \theta t_1$, and $0 < \theta < 1$.
[*Hint:* Apply Theorem 5.5.3 to the family $T_z = (I - \Delta)^{\frac{1}{2}(1-z)(t_0-s_0) + \frac{1}{2}z(t_1-s_1)} T$.]

5.5.5. Let $N \in \mathbf{Z}^+$. Suppose $\sigma, g \in L^2(\mathbf{R}^n)$ satisfy
$$\int_{\mathbf{R}^n} |\widehat{\sigma}(\xi)| |\widehat{g}(\xi)| |\xi|^{2N+\frac{1}{2}} d\xi < \infty.$$

(This condition implies that $\partial^\alpha(\sigma * (-\Delta)^{it} g)$ exists for all $|\alpha| \leq 2N$ and all t real.) Suppose that there is a constant B such that for any multi-index α with $|\alpha| = 2N$ we have

$$\left|\partial^\alpha(\sigma * (-\Delta)^{it} g)(x)\right| \leq \frac{B\|g\|_{L^2}}{(1+|x|)^{2N}}, \qquad t \in \mathbf{R}, \, x \in \mathbf{R}^n.$$

Prove that for any $0 \leq \theta \leq 2N$, $(-\Delta)^{\frac{\theta}{2}}(\sigma * g)$ is a well-defined function and there is a constant $C_{N,n}$ (depending only on N, n) such that

$$\left|(-\Delta)^{\frac{\theta}{2}}(\sigma * g)(x)\right| \leq C_{N,n} \frac{B^{\frac{\theta}{2N}} \|\sigma\|_{L^2}^{1-\frac{\theta}{2N}} \|g\|_{L^2}}{(1+|x|)^{\theta}}, \qquad x \in \mathbf{R}^n.$$

[*Hint:* Show that the mapping $z \mapsto (1+|x|)^{2Nz}(-\Delta)^{Nz}(\sigma * g)(x)$ is analytic on the unit strip using (2.7.8). Then apply Corollary C.0.3.]

5.6 The Calderón–Torchinsky Multiplier Theorem

Theorem 5.3.6 improves Theorem 5.3.2 in allowing the multiplier to have fractional derivatives. In this section we adjust the number of derivatives to depend on p. For $p = 2$, naturally, no derivatives are needed of the multiplier, but this number gradually grows as p moves away from 2.

Let σ be a complex-valued bounded function on $\mathbf{R}^n$. Associated with σ we define a linear operator

$$T_\sigma(\varphi) = (\widehat{\varphi}\sigma)^\vee$$

initially defined for $\varphi \in \mathscr{S}(\mathbf{R}^n)$. The next result provides a weaker but more useful formulation of Theorem 5.3.6, especially interesting when $r = 2$, although, for our purposes it will be useful for r near infinity.

Theorem 5.6.1. *Let $\Psi \in \mathscr{S}(\mathbf{R}^n)$ be as in (4.4.23). Let $1 < r < \infty$ and s be a real number such that $s > \max(\frac{n}{2}, \frac{n}{r})$. Suppose that $\sigma \in L^\infty(\mathbf{R}^n)$ satisfies*

$$K_0 = \sup_{j \in \mathbf{Z}} \left\|(I-\Delta)^{\frac{s}{2}}\left[\sigma(2^j \cdot)\widehat{\Psi}\right]\right\|_{L^r(\mathbf{R}^n)} < \infty. \tag{5.6.1}$$

Then T_σ admits a bounded extension from $L^p(\mathbf{R}^n)$ to itself for all $1 < p < \infty$.

Proof. As 1 and n/s are smaller than both 2 and r, we pick ρ such that

$$\max\left(1, \tfrac{n}{s}\right) < \rho < \min(2, r).$$

The statement of Theorem 5.3.6 yields

$$\|T_\sigma(\varphi)\|_{L^p(\mathbf{R}^n)} \leq C_{p,n} K \|\varphi\|_{L^p(\mathbf{R}^n)}, \qquad \varphi \in \mathscr{S}(\mathbf{R}^n),$$

where K, defined in (5.3.2) in Lemma 5.3.1, is

5.6 The Calderón–Torchinsky Multiplier Theorem

$$K = \sup_{j \in \mathbf{Z}} \left\| (I-\Delta)^{\frac{s}{2}} [\sigma(2^j \cdot)\widehat{\Psi}] \right\|_{L^p(\mathbf{R}^n)},$$

where $1 \leq \rho < 2$ and $s > n/\rho$. So to prove our result, it will suffice to show that

$$\left\| (I-\Delta)^{\frac{s}{2}} [\sigma(2^j \cdot)\widehat{\Psi}] \right\|_{L^p(\mathbf{R}^n)} \leq C \left\| (I-\Delta)^{\frac{s}{2}} [\sigma(2^j \cdot)\widehat{\Psi}] \right\|_{L^r(\mathbf{R}^n)} \tag{5.6.2}$$

for every $j \in \mathbf{Z}$. Define Θ as in (5.3.1), so that $\widehat{\Theta}$ equals 1 on the support of $\widehat{\Psi}$. Replacing $\widehat{\Psi}$ by $\widehat{\Psi}\widehat{\Theta}$ on the left, (5.6.2) is derived by the following lemma. □

Lemma 5.6.2. *Let $\widehat{\Theta} \in \mathscr{C}_0^\infty(\mathbf{R}^n)$ and $s > 0$. Suppose $1 < \rho < r < \infty$. Then there is a constant $C = C(n,\rho,r,s,\Theta)$ such that for any $\omega \in L_s^r$ we have*

$$\left\| (I-\Delta)^{\frac{s}{2}} [\omega \widehat{\Theta}] \right\|_{L^p(\mathbf{R}^n)} \leq C \left\| (I-\Delta)^{\frac{s}{2}} \omega \right\|_{L^r(\mathbf{R}^n)}, \tag{5.6.3}$$

[Taking $\omega = \sigma(2^j \cdot)\widehat{\Psi}$, we obtain (5.6.2) from (5.6.3).]

Proof. It will suffice to prove (5.6.3) for $\omega \in \mathscr{C}_0^\infty(\mathbf{R}^n)$. Indeed, if this is known, given $\omega \in L_s^r$ pick a sequence $\varphi_j \in \mathscr{C}_0^\infty$ converging to ω in L_s^r (Exercise 5.4.3). Then $\{\varphi_j\}_j$ is a Cauchy sequence in L_s^r and (5.6.3) yields that $\{\varphi_j \widehat{\Theta}\}_j$ is a Cauchy sequence in L_s^p. But this sequence converges in L_s^p (Theorem 5.4.8) and the limit coincides with the L^p limit of the sequence $\{\varphi_j \widehat{\Theta}\}_j$, which is $\omega \widehat{\Theta}$. This implies (5.6.3) for $\omega \in L_s^r$.

So, in proving (5.6.3), let us work with functions $\omega \in \mathscr{C}_0^\infty$. We pick a positive integer m such that $s/2 < m$. Consider the family of operators

$$T_z(\omega) = (I-\Delta)^{mz} \left[\widehat{\Theta}(I-\Delta)^{\frac{s}{2}-mz} \omega \right], \quad \omega \in \mathscr{C}_0^\infty, \tag{5.6.4}$$

defined on the strip $\mathbf{S} = \{z \in \mathbf{C} : 0 < \operatorname{Re} z < 1\}$. Notice that for $\omega, g \in \mathscr{C}_0^\infty$,

$$\int_{\mathbf{R}^n} g T_z(\omega) dx = \int_{\mathbf{R}^n} \int_{\mathbf{R}^n} \widehat{g}(\xi)(1+4\pi^2|\xi|^2)^{mz} \Theta(\xi-\eta)(1+4\pi^2|\eta|^2)^{\frac{s}{2}-mz} \omega^\vee(\eta) d\eta d\xi,$$

which converges absolutely. This function is continuous and bounded on $\overline{\mathbf{S}}$; it is also analytic in $\mathbf{S}$, which can be obtained by the LDCT using (2.7.8) and the rapid decay of the integrand. Additionally, one can write

$$T_z(\omega)(x) = \int_{\mathbf{R}^n} \int_{\mathbf{R}^n} (1+4\pi^2|\xi|^2)^{mz} \Theta(\xi-\eta)(1+4\pi^2|\eta|^2)^{\frac{s}{2}-mz} \omega^\vee(\eta) d\eta \, e^{-2\pi i x \cdot \xi} d\xi,$$

and from this one obtains that $T_z(\omega)$ is bounded on compact sets by a constant independent of $|\operatorname{Im} z|$. This verifies the admissibility condition (5.5.9).

Let $C(n,t) = (1+|t|)^{[\frac{n}{2}]+2}$. Then

$$\left\| T_{it}(\omega) \right\|_{L^p} \leq C' C(n,mt) \left\| \widehat{\Theta}(I-\Delta)^{\frac{s}{2}-mit} \omega \right\|_{L^p}$$

$$\leq C' C(n,mt) \left| \operatorname{supp}(\widehat{\Theta}) \right|^{\frac{1}{p}-\frac{1}{r}} \left\| (I-\Delta)^{-mit+\frac{s}{2}} \omega \right\|_{L^r}$$

$$\leq C' C(n,mt)^2 \left| \operatorname{supp}(\widehat{\Theta}) \right|^{\frac{1}{p}-\frac{1}{r}} \left\| (I-\Delta)^{\frac{s}{2}} \omega \right\|_{L^r},$$

having used Corollary 5.3.3 twice and Hölder's inequality. Likewise, we obtain

$$\begin{aligned}\left\|T_{1+it}(\omega)\right\|_{L^p} &\leq C'C(n,mt)\left\|(I-\Delta)^m\left[\widehat{\Theta}(I-\Delta)^{-m-mit+\frac{s}{2}}\omega\right]\right\|_{L^p}\\&\leq C'C(n,mt)\left\|\sum_{|\alpha|\leq 2m}\sum_{\beta\leq\alpha}C_{\alpha,\beta}(\partial^{\alpha-\beta}\widehat{\Theta})\partial^\beta\left[(I-\Delta)^{-m-mit+\frac{s}{2}}\omega\right]\right\|_{L^p}\\&\leq C''C(n,mt)|\mathrm{supp}(\widehat{\Theta})|^{\frac{1}{p}-\frac{1}{r}}\sum_{|\beta|\leq 2m}\left\|\partial^\beta(I-\Delta)^{-m-mit+\frac{s}{2}}\omega\right\|_{L^r}.\end{aligned}$$

Now notice that $\partial^\beta(I-\Delta)^{-m}$ is an operator with symbol $(2\pi i\xi)^\beta(1+4\pi^2|\xi|^2)^{-m}$, which satisfies the conditions of Theorem 5.3.2 when $|\beta|\leq 2m$ (Example 5.3.8). Finally, applying Corollary 5.3.3 to $(I-\Delta)^{-mit}$ we obtain

$$\left\|T_{1+it}(\omega)\right\|_{L^p}\leq C'''C(n,mt)^2\left\|(I-\Delta)^{\frac{s}{2}}\omega\right\|_{L^r}.$$

As the constants $C(n,t)$ grow at most polynomially in $|t|$, Theorem 5.5.3 (with $q_0 = q_1 = p$ and $p_0 = p_1 = r$) yields

$$\left\|T_{s/2m}(\omega)\right\|_{L^p}\leq C(n,p,r,s,\Theta)\left\|(I-\Delta)^{\frac{s}{2}}\omega\right\|_{L^r},$$

which is exactly (5.6.3). □

Example 5.6.3. Let $\sigma(\xi) = (1-|\xi|)\chi_{[-1,1]}(\xi)$. We show that σ lies in $\mathscr{M}_p(\mathbf{R})$ for any $1 < p < \infty$. Let $0 < t < 1$. We have that $\sigma' = \chi_{[-1,0]} - \chi_{[0,1]}$ and the functions[5] $(I-\partial^2)^{\frac{t}{2}}\chi_{[-1,0]}$ and $(I-\partial^2)^{\frac{t}{2}}\chi_{[0,1]}$ lie in $L^r(\mathbf{R})$ when $1 < r < 1/t$ by the work contained in Example 5.4.6. Thus, so does $(I-\partial^2)^{\frac{t}{2}}\sigma'$.

Using the Fourier transform one verifies the identity,

$$(I-\partial^2)^{\frac{1+t}{2}}\sigma = (I-\partial^2)^{\frac{t-1}{2}}\sigma + T_m\!\left((I-\partial^2)^{\frac{t}{2}}\sigma'\right),$$

where T_m is the Fourier multiplier associated with $m(\xi) = -2\pi i\xi(1+4\pi^2|\xi|^2)^{-1/2}$. As T_m and $(I-\partial^2)^{\frac{t-1}{2}}$ preserve L^r, it follows that $(I-\partial^2)^{\frac{1+t}{2}}\sigma$ lies in $L^r(\mathbf{R})$ when $1 < r < \frac{1}{t}$. So we fix r such that $\max(1,\frac{1}{t+1}) = 1 < r < \frac{1}{t} < \infty$, set $s = 1+t$, and consider the dilated and translated version

$$\tau(\xi) = \left(1 - 8\left|\xi - \frac{3}{2}\right|\right)\chi_{[-\frac{1}{8},\frac{1}{8}]}\!\left(\xi - \frac{3}{2}\right)$$

of $\sigma(\xi)$. Notice that τ is supported in $[\frac{11}{8},\frac{13}{8}]$ and if Ψ is as in (4.4.23), then $\widehat{\Psi}\tau = \tau$. Moreover, $\widehat{\Psi}\tau(2^j\cdot) = 0$ if $j \neq 0$. This shows that the function τ satisfies condition (5.6.1) for our choices of s and r, which are related as follows: $\frac{1}{s} < 1 < r < \frac{1}{s-1}$. Applying Theorem 5.6.1 we obtain that $\tau \in \mathscr{M}_p(\mathbf{R})$ and so is σ by (2.8.2) and (2.8.3).

---

[5] The symbol ∂^2 denotes the Laplacian in dimension 1, i.e., the second derivative.

5.6 The Calderón–Torchinsky Multiplier Theorem

To be able to reduce the number of derivatives required of a multiplier for p near 2, we will need the following result.

Theorem 5.6.4. *Fix $1 < r_0, r_1 < \infty$, $0 < p_0, p_1, q_0, q_1, s_0, s_1 < \infty$, $0 < B_0, B_1 < \infty$. Suppose that $r_0 s_0 > n$ and $r_1 s_1 > n$. Let $\widehat{\Psi} \in \mathscr{C}_0^\infty(\mathbf{R}^n)$ be supported in $1/2 \leq |\xi| \leq 2$ and satisfy $\sum_{j \in \mathbf{Z}} \widehat{\Psi}(2^{-j}\xi) = 1$ when $\xi \neq 0$. Assume that for $\kappa \in \{0,1\}$ we have*

$$\|T_\sigma(f)\|_{L^{q_\kappa}(\mathbf{R}^n)} \leq B_\kappa \sup_{j \in \mathbf{Z}} \|\sigma(2^j \cdot)\widehat{\Psi}\|_{L^{r_\kappa}_{s_\kappa}(\mathbf{R}^n)} \|f\|_{L^{p_\kappa}(\mathbf{R}^n)} \qquad (5.6.5)$$

for all $f \in \mathscr{C}_0^\infty(\mathbf{R}^n)$ and for all $\sigma \in L^\infty$ that satisfy

$$\sup_{j \in \mathbf{Z}} \|\sigma(2^j \cdot)\widehat{\Psi}\|_{L^{r_\kappa}_{s_\kappa}(\mathbf{R}^n)} < \infty. \qquad (5.6.6)$$

For $0 < \theta < 1$ let

$$\frac{1}{p} = \frac{1-\theta}{p_0} + \frac{\theta}{p_1}, \quad \frac{1}{q} = \frac{1-\theta}{q_0} + \frac{\theta}{q_1}, \quad \frac{1}{r} = \frac{1-\theta}{r_0} + \frac{\theta}{r_1}, \quad s = (1-\theta)s_0 + \theta s_1.$$

Then there is a constant $C_* = C_*(r_0, r_1, s_0, s_1, n, \theta)$ such that for all $f \in \mathscr{C}_0^\infty(\mathbf{R}^n)$ and all $\sigma \in L^\infty(\mathbf{R}^n)$ that satisfy

$$\sup_{j \in \mathbf{Z}} \|\sigma(2^j \cdot)\widehat{\Psi}\|_{L^r_s(\mathbf{R}^n)} < \infty \qquad (5.6.7)$$

we have

$$\|T_\sigma(f)\|_{L^q(\mathbf{R}^n)} \leq C_* B_0^{1-\theta} B_1^\theta \sup_{j \in \mathbf{Z}} \|\sigma(2^j \cdot)\widehat{\Psi}\|_{L^r_s} \|f\|_{L^p(\mathbf{R}^n)}. \qquad (5.6.8)$$

Proof. Fix $\widehat{\Phi} \geq 0$ in $\mathscr{C}_0^\infty$ supported in $\frac{1}{4} \leq |\xi| \leq 4$ and $\widehat{\Phi} \equiv 1$ on the support of $\widehat{\Psi}$. Also fix $\sigma \in L^\infty$ satisfying (5.6.7). For $j \in \mathbf{Z}$ introduce the function

$$\varphi_j = (I - \Delta)^{\frac{s}{2}}[\sigma(2^j \cdot)\widehat{\Psi}]$$

which lies in $L^r(\mathbf{R}^n)$; note that the hypotheses imply that $1 < r < \infty$ and $s > 0$. For $0 \leq \operatorname{Re} z \leq 1$, we define a function σ_z on $\mathbf{R}^n$ by setting

$$\sigma_z(\xi) = \sum_{j \in \mathbf{Z}} (I - \Delta)^{-\frac{s_0}{2}(1-z) - \frac{s_1}{2}z}\left[|\varphi_j|^{\frac{r}{r_0}(1-z) + \frac{r}{r_1}z} e^{i\operatorname{Arg}\varphi_j}\right](2^{-j}\xi)\widehat{\Phi}(2^{-j}\xi). \qquad (5.6.9)$$

For any $\xi \in \mathbf{R}^n$, this sum has at most four terms, in view of the support properties of $\widehat{\Phi}$. Moreover, notice that $\sigma_\theta = \sigma$ since $\widehat{\Phi}$ is equal to 1 on the support of $\widehat{\Psi}$. Let us momentarily assume that σ_z is a bounded function; this will be shown shortly.

We examine a few properties of σ_z. First we claim that for $\kappa \in \{0,1\}$ and $t \in \mathbf{R}$,

$$\|(I-\Delta)^{\frac{s_\kappa}{2}}[\sigma_{\kappa+it}(2^k \cdot)\widehat{\Psi}]\|_{L^{r_\kappa}}$$
$$\leq C(1+|t|)^{[\frac{n}{2}]+2} \sum_{j:|j-k|\leq 3} \|(I-\Delta)^{\frac{s}{2}}[\sigma(2^j \cdot)\widehat{\Psi}]\|_{L^r}^{\frac{r}{r_\kappa}}, \qquad (5.6.10)$$

where $C = C(n, r_\kappa, |s_1 - s_0|, \Psi, \Phi)$. Indeed, notice first that in the sum defining $\sigma_z(2^k \cdot)\widehat{\Psi}$ contains only the terms with $j \in \{k-3, k-2, k-1, k, k+1, k+2, k+3\}$. Then we obtain (5.6.10) using Exercise 5.5.2 and then Corollary 5.3.3 and the observation that when $z = \kappa + it$, $t \in \mathbf{R}$, the L^{r_κ} norm of the expression inside the square brackets in (5.6.9) equals $\|\varphi_j\|_{L^r}^{r/r_\kappa}$. Thus (5.6.6) holds with $\sigma_{\kappa+it}$ in place of σ when $\kappa = 0, 1$. This implies the validity of (5.6.5) with $\sigma_{\kappa+it}$ in place of σ when $\kappa = 0, 1$. Combining (5.6.5) with (5.6.10) for $\kappa = 0, 1$ yields for all $t \in \mathbf{R}$

$$\left\|T_{\sigma_{\kappa+it}}(f)\right\|_{L^{q_\kappa}} \leq B_\kappa C (1+|t|)^{[\frac{n}{2}]+2} \sup_{k \in \mathbf{Z}} \left\|(I-\Delta)^{\frac{s}{2}}\left[\sigma(2^k \cdot)\widehat{\Psi}\right]\right\|_{L^r}^{\frac{r}{r_\kappa}} \|f\|_{L^{p_\kappa}} \quad (5.6.11)$$

for all $f \in \mathscr{C}_0^\infty(\mathbf{R}^n)$, where T_{σ_z} is the multiplier operator associated with σ_z.

We now estimate the L^∞ norm of σ_z. Fix $\xi \in \mathbf{R}^n \setminus \{0\}$. Then there is a j_0 such that $|\xi| \approx 2^{j_0}$ and there are at most four terms in the sum in (5.6.9). For these terms we estimate the L^∞ norm of

$$(I-\Delta)^{-\frac{s_0}{2}(1-z) - \frac{s_1}{2}z}\left[|\varphi_j|^{\frac{r}{r_0}(1-z) + \frac{r}{r_1}z} e^{i\operatorname{Arg}\varphi_j}\right].$$

For $z = \tau + it$ with $0 \leq \tau \leq 1$, let $s_\tau = (1-\tau)s_0 + \tau s_1$ and

$$\frac{1}{r_\tau} = \frac{1-\tau}{r_0} + \frac{\tau}{r_1}.$$

By Theorem 5.4.9 (c) ($s_0 > n/r_0$, $s_1 > n/r_1 \implies s_\tau > n/r_\tau$) we have

$$\left\|(I-\Delta)^{-\frac{s_0}{2}(1-z)-\frac{s_1}{2}z}\left[|\varphi_j|^{\frac{r}{r_0}(1-z)+\frac{r}{r_1}z}e^{i\operatorname{Arg}\varphi_j}\right]\right\|_{L^\infty}$$

$$\leq C(r_\tau, s_\tau, n)\left\|(I-\Delta)^{-\frac{s_0}{2}(1-z)-\frac{s_1}{2}z}\left[|\varphi_j|^{\frac{r}{r_0}(1-z)+\frac{r}{r_1}z}e^{i\operatorname{Arg}\varphi_j}\right]\right\|_{L^{r_\tau}_{s_\tau}}$$

$$\leq C(r_\tau, s_\tau, n)\left\|(I-\Delta)^{it\frac{s_0-s_1}{2}}\left[|\varphi_j|^{\frac{r}{r_0}(1-z)+\frac{r}{r_1}z}e^{i\operatorname{Arg}\varphi_j}\right]\right\|_{L^{r_\tau}}$$

$$\leq C'(r_\tau, s_\tau, n)(1+|s_0-s_1||t|)^{[\frac{n}{2}]+2}\left\||\varphi_j|^{\frac{r}{r_0}(1-z)+\frac{r}{r_1}z}e^{i\operatorname{Arg}\varphi_j}\right\|_{L^{r_\tau}}$$

$$\leq C''(r_0, r_1, s_0, s_1, \tau, n)(1+|t|)^{[\frac{n}{2}]+2}\left\||\varphi_j|^{\frac{r}{r_0}(1-z)+\frac{r}{r_1}z}\right\|_{L^{r_\tau}}$$

$$= C''(r_0, r_1, s_0, s_1, \tau, n)(1+|t|)^{[\frac{n}{2}]+2}\|\varphi_j\|_{L^r}^{\frac{r}{r_\tau}},$$

having used Corollary 5.3.3. It follows from this that $\|\sigma_{\tau+it}\|_{L^\infty} < \infty$, precisely, that

$$\|\sigma_{\tau+it}\|_{L^\infty} \leq C''(r_0, r_1, s_0, s_1, \tau, n)(1+|t|)^{[\frac{n}{2}]+2}\left(\sup_{j \in \mathbf{Z}}\|\sigma(2^j \cdot)\widehat{\Psi}\|_{L^r_s}\right)^{\frac{r}{r_\tau}}. \quad (5.6.12)$$

Finally, we show that $\{T_{\sigma_z}\}_z$ is an analytic family of operators of admissible growth. Obviously, for $\varphi \in \mathscr{C}_0^\infty(\mathbf{R}^n)$ one has

$$\|T_{\sigma_z}(\varphi)\|_{L^2} \leq \|\sigma_z\|_{L^\infty}\|\varphi\|_{L^2},$$

5.6 The Calderón–Torchinsky Multiplier Theorem

and using (5.6.12) we obtain the admissibility condition (5.5.9). To verify the analyticity of the family, for any $f, g \in \mathscr{C}_0^\infty$, we must show that the function

$$F(z) = \int_{\mathbf{R}^n} T_{\sigma_z}(f)(x)g(x)\,dx = \int_{\mathbf{R}^n} \sigma_z(\xi)\widehat{f}(\xi)g^\vee(\xi)\,d\xi$$

is analytic on the unit strip $\mathbf{S} = \{z \in \mathbf{C}: 0 < \operatorname{Re} z < 1\}$ and continuous on its closure. Introduce a $\mathscr{C}_0^\infty$ function Ω equal to 1 on the unit ball and vanishing outside the double of the unit ball. Then if we can show that the functions

$$F_m(z) = \int_{\mathbf{R}^n} \sigma_z(\xi)\widehat{f}(\xi)g^\vee(\xi)\bigl(1 - \Omega(2^m\xi)\bigr)\Omega(2^{-m}\xi)\,d\xi, \qquad m \in \mathbf{Z}^+, \quad (5.6.13)$$

are analytic on $\mathbf{S}$ and continuous on $\overline{\mathbf{S}}$, the same conclusion will follow for F, as $F_m \to F$ uniformly on compact subsets of $\overline{\mathbf{S}}$; this last assertion follows from the LDCT and (5.6.12). The advantage of working with F_m is that only finitely many j (depending on m) appear in the definition of σ_z in the integral in (5.6.13) in view of the support of Ω. So we fix such a j, we set

$$G_j = |\varphi_j|^r \in L^1,$$
$$H_j = \widehat{\Phi}\widehat{f}(2^j \cdot)g^\vee g(2^j \cdot)\bigl(1 - \Omega(2^{j+m} \cdot)\bigr)\Omega(2^{j-m} \cdot),$$
$$s(z) = -\frac{s_0}{2}(1 - z) - \frac{s_1}{2}z,$$

and matters reduce to showing the analyticity of the function

$$\int_{\mathbf{R}^n} (I - \Delta)^{s(z)}\bigl[G_j^{\frac{1-z}{r_0}+\frac{z}{r_1}}e^{i\operatorname{Arg}\varphi_j}\bigr](2^{-j}\xi)\widehat{\Phi}(2^{-j}\xi)\widehat{f}(\xi)g^\vee(\xi)\bigl(1 - \Omega(2^m\xi)\bigr)\Omega(\tfrac{\xi}{2^m})\,d\xi$$

$$= \int_{\mathbf{R}^n} (I - \Delta)^{s(z)}\bigl[G_j^{\frac{1-z}{r_0}+\frac{z}{r_1}}e^{i\operatorname{Arg}\varphi_j}\bigr](\xi)H_j(\xi)2^{jn}\,d\xi$$

$$= \int_{\mathbf{R}^n} G_j(\xi)^{\frac{1-z}{r_0}+\frac{z}{r_1}}e^{i\operatorname{Arg}\varphi_j(\xi)}\bigl[(I-\Delta)^{s(z)}H_j\bigr](\xi)2^{jn}\,d\xi \qquad \text{(Exercise 5.4.6)}$$

$$= \int_{\mathbf{R}^n} \frac{G_j(\xi)^{\frac{1-z}{r_0}+\frac{z}{r_1}}e^{i\operatorname{Arg}\varphi_j(\xi)}}{(1+4\pi^2|\xi|^2)^n}\Biggl(\sum_{|\alpha|\le 2n} c_\alpha(-2\pi i\xi)^\alpha\Biggr)\bigl[(I-\Delta)^{s(z)}H_j\bigr](\xi)2^{jn}\,d\xi$$

$$= 2^{jn}\sum_{\substack{|\alpha|\le 2n \\ \beta \le \alpha}} c_\alpha \binom{\alpha}{\beta}\int_{\mathbf{R}^n}\int_{\mathbf{R}^n} \frac{G_j(\xi)^{\frac{1-z}{r_0}+\frac{z}{r_1}}e^{i\operatorname{Arg}\varphi_j(\xi)}}{(1+4\pi^2|\xi|^2)^n}\times$$

$$\times \partial^\beta(1+4\pi^2|\eta|^2)^{s(z)}\bigl(\partial^{\alpha-\beta}\widehat{H_j}\bigr)(\eta)e^{2\pi i\xi\cdot\eta}\,d\eta\,d\xi, \qquad (5.6.14)$$

where, after writing

$$(1+4\pi^2|\xi|^2)^n = \sum_{|\alpha|\le 2n} c_\alpha(-2\pi i\xi)^\alpha$$

via the identity in Remark 5.4.4 (c), we used Fourier inversion and we expanded the αth derivative via Leibniz's rule. The double integral converges absolutely, and the continuity of this function on $\overline{\mathbf{S}}$ is straightforward. Suppose now that we wanted to show that the function of z in (5.6.14) is analytic at a point $w_0 \in \mathbf{S}$. Pick $\delta > 0$ such that

$$2\delta < \min\left(\operatorname{Re}\left(\frac{1-w_0}{r_0} + \frac{w_0}{r_1}\right), 1 - \operatorname{Re}\left(\frac{1-w_0}{r_0} + \frac{w_0}{r_1}\right)\right). \tag{5.6.15}$$

Note that this is possible, since by assumption we have $1 < r_0, r_1 < \infty$. By Lemma 2.7.5, when $G_j(\xi) \neq 0$, for

$$0 < |w| < \left|\frac{1}{r_1} - \frac{1}{r_0}\right|^{-1} \delta$$

we obtain

$$\left|\frac{G_j(\xi)^{(\frac{1}{r_1} - \frac{1}{r_0})w} - 1}{w}\right| \leq \frac{2\left|\frac{1}{r_1} - \frac{1}{r_0}\right|}{\delta} \max\left(|G_j(\xi)|^{2\delta}, |G_j(\xi)|^{-2\delta}\right). \tag{5.6.16}$$

Then for

$$0 < |w| < \left|\frac{1}{r_1} - \frac{1}{r_0}\right|^{-1} \delta,$$

combining (5.6.16) and (5.6.15) gives

$$\left|\frac{G_j(\xi)^{\frac{1-w_0-w}{r_0} + \frac{w_0+w}{r_1}} - G_j(\xi)^{\frac{1-w_0}{r_0} + \frac{w_0}{r_1}}}{w}\right| \leq \frac{2\left|\frac{1}{r_1} - \frac{1}{r_0}\right|}{\delta} \max(G_j(\xi), 1).$$

As $G_j = |\varphi_j|^r \in L^1$, it follows that $\max(G_j, 1)(1 + 4\pi^2 |\cdot|^2)^{-n}$ also lies in $L^1(\mathbf{R}^n)$. So the LDCT can now be justified when the z derivative hits the term involving $G_j(\xi)$. When the z derivative hits $\partial^\beta (1 + 4\pi^2 |\eta|^2)^{s(z)}$, a similar (but easier) argument applies, and the rapid convergence of the η integral in (5.6.14) allows the use of the LDCT. This argument yields the analyticity of the family $\{T_{\sigma_z}\}_z$.

Now that we know that T_{σ_z} is an analytic family of operators on the strip, estimates (5.6.11) and Theorem 5.5.3 allow us to deduce (5.6.8). $\square$

We now prove a result that extends Theorem 5.3.6 to the range $s \leq \frac{n}{2}$.

Theorem 5.6.5. *(Calderón–Torchinsky multiplier theorem) Fix Ψ as in (4.4.23). Let $1 < r, p < \infty$ and $0 < s \leq \frac{n}{2}$ satisfy $rs > n$ and $|\frac{1}{p} - \frac{1}{2}| < \frac{s}{n}$. Then there is a constant $C(n,p,s,r)$ such that for any $\sigma \in L^\infty(\mathbf{R}^n)$ that satisfies*

$$\sup_{j \in \mathbf{Z}} \|\sigma(2^j \cdot)\widehat{\Psi}\|_{L^r_s(\mathbf{R}^n)} < \infty \tag{5.6.17}$$

and for every $f \in \mathscr{C}_0^\infty(\mathbf{R}^n)$ we have

$$\|T_\sigma(f)\|_{L^p(\mathbf{R}^n)} \leq C(p,n,s,r) \sup_{j \in \mathbf{Z}} \|\sigma(2^j \cdot)\widehat{\Psi}\|_{L^r_s} \|f\|_{L^p(\mathbf{R}^n)}. \tag{5.6.18}$$

5.6 The Calderón–Torchinsky Multiplier Theorem

Proof. Let us fix $1 < p < 2$, $1 < r < \infty$, $0 < s \leq \frac{n}{2}$, $r > \frac{n}{s}$ such that $\frac{1}{p} - \frac{1}{2} < \frac{s}{n}$. Select parameters as follows:

Let $s_1 = s - n(\frac{1}{p} - \frac{1}{2}) > 0$. Consider the dotted line passing through $(\frac{1}{2}, s_1)$ and $(\frac{1}{p}, s)$ (see Figure 5.1) and select a point $(\frac{1}{p_0}, s_0)$ on this line satisfying $s_0 > \frac{n}{2}$ and $p_0 > 1$, i.e.,

$$s_0 = s_1 + \left(\frac{1}{p_0} - \frac{1}{2}\right)\frac{s - s_1}{\frac{1}{p} - \frac{1}{2}} > \frac{n}{2}$$

and

$$\frac{1}{2} + \frac{\frac{n}{2} - s_1}{s - s_1}\left(\frac{1}{p} - \frac{1}{2}\right) < \frac{1}{p_0} < 1.$$

Define $\theta \in (0, 1)$ as follows:

$$\theta = \frac{\frac{1}{p_0} - \frac{1}{p}}{\frac{1}{p_0} - \frac{1}{2}} = \frac{s_0 - s}{s_0 - s_1}.$$

As $n/r < s$, notice that

$$\frac{\frac{1}{r} - \frac{s_1}{n}}{1 - \theta} < \min\left(\frac{\frac{s_0}{n} - \frac{1}{r}}{\theta}, \frac{\frac{1}{r}}{1 - \theta}\right).$$

Pick $1/q$ strictly between the above numbers and define:

$$\frac{1}{r_0} = \frac{1}{r} + \frac{\theta}{q}, \quad \frac{1}{r_1} = \frac{1}{r} - \frac{1-\theta}{q}.$$

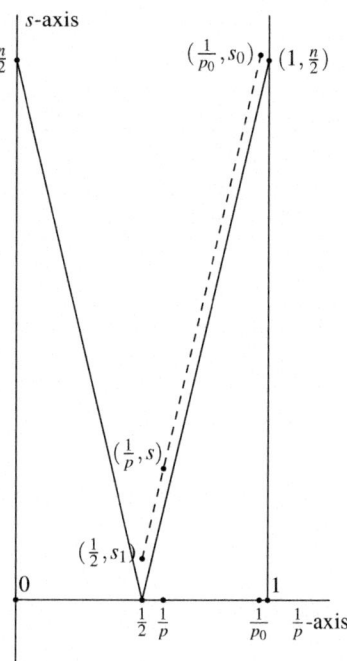

Fig. 5.1 By construction, the point $(\frac{1}{p}, s)$ lies on the dotted line segment joining $(\frac{1}{p_0}, s_0)$ to $(\frac{1}{2}, s_1)$.

After selecting these parameters, we make a few observations. First of all, we have $0 < \frac{1}{r_0} < \frac{s_0}{n} < 1$ and $0 < \frac{1}{r_1} < \frac{s_1}{n} < \frac{1}{2}$. Thus the conditions $1 < r_0, r_1 < \infty$, $r_0 s_0 > n$, and $r_1 s_1 > n$ are satisfied. Also, one has

$$\frac{1}{p} = \frac{1-\theta}{p_0} + \frac{\theta}{p_1}, \quad s = (1-\theta)s_0 + \theta s_1, \quad \frac{1}{r} = \frac{1-\theta}{r_0} + \frac{\theta}{r_1}.$$

As $s_0 > \max(\frac{n}{2}, \frac{n}{r_0})$, for any function that satisfies (5.6.6) with $\kappa = 0$ one obtains now from Theorem 5.6.1 that (5.6.5) holds when $\kappa = 0$ with $q_0 = p_0$.

Now fix an L^∞ function σ that satisfies (5.6.6) with $\kappa = 1$. As $r_1 s_1 > n$, Theorem 5.4.9 (c) (Sobolev embedding) gives that $\sigma(2^k \cdot)\widehat{\Psi}$ is a bounded continuous function which satisfies for any $\xi \in \mathbf{R}^n$ and any $k \in \mathbf{Z}$

$$|\sigma(2^k \xi)\widehat{\Psi}(\xi)| \leq C \|\sigma(2^k \cdot)\widehat{\Psi}\|_{L^{r_1}_{s_1}(\mathbf{R}^n)} \leq C \sup_{j \in \mathbf{Z}} \|\sigma(2^j \cdot)\widehat{\Psi}\|_{L^{r_1}_{s_1}(\mathbf{R}^n)} < \infty. \quad (5.6.19)$$

We now replace ξ by $2^{-k}\xi$ in (5.6.19) and we write $\sigma(\xi) = \sigma(\xi)\sum_k \widehat{\Psi}(2^{-k}\xi)$, $\xi \neq 0$. But for a given $\xi \neq 0$, at most three terms in the sum are nonzero, so we obtain from (5.6.19) that

$$\|\sigma\|_{L^\infty} \leq 3C \sup_{j \in \mathbf{Z}} \|\sigma(2^j \cdot)\widehat{\Psi}\|_{L^{r_1}_{s_1}(\mathbf{R}^n)} < \infty.$$

As $\|T_\sigma\|_{L^2 \to L^2} \leq \|\sigma\|_{L^\infty}$, it follows that any σ that satisfies (5.6.6) with $\kappa = 1$ also satisfies (5.6.5) with $\kappa = 1$ and $p_1 = q_1 = 2$. We have now verified all hypotheses of Theorem 5.6.4. An application of Theorem 5.6.4 yields (5.6.18) when $1 < p < 2$. Figure 5.1 displays pictorially the interpolation: It is between the endpoints of the dotted line which contains $(\frac{1}{p}, s)$. In that picture the solid slanted lines represent the equation $|\frac{1}{p} - \frac{1}{2}| = \frac{s}{n}$ while the region above that is the set of points $(\frac{1}{p}, s)$ that satisfy $|\frac{1}{p} - \frac{1}{2}| < \frac{s}{n}$.

The case $2 < p < \infty$ follows by duality via Proposition 2.8.6, while the case $p = 2$ is contained in the following argument

$$\|T_\sigma\|_{L^p(\mathbf{R}^n) \to L^p(\mathbf{R}^n)} = \|\sigma\|_{L^\infty} \leq 2 \sup_{j \in \mathbf{Z}} \|\sigma(2^j \cdot)\widehat{\Psi}\|_{L^\infty} \leq C \sup_{j \in \mathbf{Z}} \|\sigma(2^j \cdot)\widehat{\Psi}\|_{L^r_s},$$

where the last inequality is due to Theorem 5.4.9 (c) as $rs > n$. □

Exercises

5.6.1. Let $s > \frac{n}{r}$ and $1 < r \leq 2$. Prove that there is a constant $C = C(n, r, s)$ such that for any p satisfying $1 \leq p \leq \infty$ we have

$$\|T_\sigma\|_{L^p \to L^p} \leq C \|\sigma\|_{L^r_s}.$$

5.6.2. Let $0 < \delta < \frac{n}{2}$, $\varphi \in \mathscr{C}^\infty_0$ be supported in $\frac{11}{8} < |\xi| < \frac{13}{8}$ and let a_k lie in a bounded subset of $\mathbf{R}^n$ for $k \in \mathbf{Z}$. Prove that the function

$$\sigma(\xi) = \sum_{k \in \mathbf{Z}} \varphi(2^{-k}\xi)|2^{-k}\xi - a_k|^\delta$$

lies in $\mathscr{M}_p(\mathbf{R}^n)$ for any $1 < p < \infty$. [Hint: Let $N = [\frac{n}{2}] + 1$ and pick s such that $\frac{n}{2} < s < N$ and r satisfying $\max(\frac{n}{s}, 1) < r < \frac{n}{s-\delta}$. Let $\psi \in \mathscr{C}^\infty_0$ have $L^{r'}$ norm equal to 1. Apply Corollary C.0.4 to the function $z \mapsto \int_{\mathbf{R}^n}(I - \Delta)^{\frac{Nz}{2}}[\varphi|\cdot - a_j|^{\delta + Nz - s}]\psi \, dx$ for $\theta = \frac{s}{N}$. Note that $\widehat{\Psi}\varphi = \varphi$ if Ψ is as in (4.4.23), and apply Theorem 5.6.1.]

5.6.3. Let $a, b > 0$ satisfy $\frac{a}{b} < \frac{1}{2}$. Let φ be a smooth function supported in the interval $[-\frac{1}{8}, \frac{1}{8}]$. Prove that $\sigma(x) = |x|^a e^{i|x|^{-b}} \varphi(x)$ lies in $\mathscr{M}_p(\mathbf{R})$ when $|\frac{1}{p} - \frac{1}{2}| < \frac{a}{b}$. [Hint: Given p satisfying $|\frac{1}{p} - \frac{1}{2}| < \frac{a}{b}$, choose s such that $|\frac{1}{p} - \frac{1}{2}| < s < \frac{a}{b}$ and then pick r with $\frac{1}{s} < r < \frac{1}{s(1+b)-a}$. Write the L^r norm of $(I - \partial^2)^{z/2}[\sigma(x)|x|^{(z-s)(1+b)}]$ as a supremum of integrals against $\mathscr{C}_0^\infty$ functions of $L^{r'}$ norm 1 and use Corollary C.0.4 to prove that this norm is finite when $\theta = s$. Then verify condition (5.6.17).]

5.6.4. Let $a, b > 0$ satisfy $\frac{a}{b} < \frac{n}{2}$. Let φ be a smooth function supported in the ball $|x| \leq \frac{1}{8}$ in $\mathbf{R}^n$. Prove that $\sigma(x) = |x|^a e^{i|x|^{-b}} \varphi(x)$ lies in $\mathscr{M}_p(\mathbf{R}^n)$ when $|\frac{1}{p} - \frac{1}{2}| < \frac{a}{nb}$. [Hint: Construct an analytic function as in the preceding exercise.]

5.7 The Marcinkiewicz Multiplier Theorem

To motivate our discussion, let us consider the following function in $\mathbf{R}^3 \setminus \{0\}$:

$$(\xi_1, \xi_2, \xi_3) \mapsto \frac{\xi_2 \xi_3^3}{i\xi_1 + \xi_2^2 + \xi_3^6}.$$

This function is certainly not homogeneous of degree zero, but it is smooth away from the origin and is invariant under the set of dilations:

$$(\xi_1, \xi_2, \xi_3) \mapsto (\lambda^6 \xi_1, \lambda^3 \xi_2, \lambda \xi_3), \qquad \lambda > 0.$$

Let us examine this situation a bit more generally. Suppose that there exist $k_1, \ldots, k_n$ in $\mathbf{R}^+$ and $\tau \in \mathbf{R}$ such that the smooth function σ on $\mathbf{R}^n \setminus \{0\}$ satisfies

$$\sigma(\lambda^{k_1} \xi_1, \ldots, \lambda^{k_n} \xi_n) = \lambda^{i\tau} \sigma(\xi_1, \ldots, \xi_n)$$

for all $\xi_1, \ldots, \xi_n \in \mathbf{R} \setminus \{0\}$ and $\lambda > 0$. Then differentiation gives

$$\lambda^{\alpha_1 k_1 + \cdots + \alpha_n k_n} \partial^\alpha \sigma(\lambda^{k_1} \xi_1, \ldots, \lambda^{k_n} \xi_n) = \lambda^{i\tau} \partial^\alpha \sigma(\xi_1, \ldots, \xi_n)$$

for every multi-index $\alpha = (\alpha_1, \ldots, \alpha_n)$. Now for a fixed $\xi \in \mathbf{R}^n \setminus \{0\}$ pick the unique $\lambda(\xi) > 0$ such that $(\lambda(\xi)^{k_1} \xi_1, \ldots, \lambda(\xi)^{k_n} \xi_n) \in \mathbf{S}^{n-1}$. Then $\lambda(\xi)^{k_j \alpha_j} |\xi_j|^{\alpha_j} \leq 1$, and it follows that

$$|\partial^\alpha \sigma(\xi_1, \ldots, \xi_n)| \leq \left[\sup_{\mathbf{S}^{n-1}} |\partial^\alpha \sigma|\right] \lambda(\xi)^{\alpha_1 k_1 + \cdots + \alpha_n k_n} \leq C_\alpha |\xi_1|^{-\alpha_1} \cdots |\xi_n|^{-\alpha_n},$$

where C_α is the maximum of $\partial^\alpha \sigma$ on $\mathbf{S}^{n-1}$.

So functions σ that are homogeneous of a purely imaginary degree with respect to general groups of dilations and are smooth on $\mathbf{S}^{n-1}$ satisfy

$$|\partial^\alpha \sigma(\xi_1, \ldots, \xi_n)| \leq C_\alpha |\xi_1|^{-\alpha_1} \cdots |\xi_n|^{-\alpha_n} \tag{5.7.1}$$

for all multi-indices α and all $(\xi_1, \ldots, \xi_n) \in \mathbf{R}^n$ with $\xi_j \neq 0$ for all j. In this section we show that such functions are L^p Fourier multipliers.

Next we define product-type Sobolev spaces.

Definition 5.7.1. For $s_j \geq 0$ define the operator $(I - \partial_j^2)^{\frac{s_j}{2}}$ acting on an element $f \in \mathscr{S}'(\mathbf{R}^n)$ as follows:

$$(I - \partial_j^2)^{\frac{s_j}{2}}(f) = \left((1 + 4\pi^2 |(\cdot)_j|^2)^{\frac{s_j}{2}} \widehat{f} \right)^{\vee},$$

where $(\cdot)_j$ denotes the jth variable ξ_j of $\xi = (\xi_1, \ldots, \xi_n)$. For $s_1, \ldots, s_n > 0$ define the *product-type Sobolev space*

$$L^p_{s_1,\ldots,s_n}(\mathbf{R}^n) = \{ f \in L^p(\mathbf{R}^n) : \|f\|_{L^p_{s_1,\ldots,s_n}} = \left\| (I - \partial_1^2)^{\frac{s_1}{2}} \ldots (I - \partial_n^2)^{\frac{s_n}{2}} f \right\|_{L^p} < \infty \}.$$

Proposition 5.7.2. For $s_1, \ldots, s_n$ in $\mathbf{Z}^+ \cup \{0\}$ and $1 < p < \infty$ and $f \in L^p_{s_1,\ldots,s_n}(\mathbf{R}^n)$ we have

$$\|f\|_{L^p_{s_1,\ldots,s_n}(\mathbf{R}^n)} \approx \sum_{l_1=0}^{s_1} \cdots \sum_{l_n=0}^{s_n} \left\| \partial_1^{l_1} \cdots \partial_n^{l_n} f \right\|_{L^p(\mathbf{R}^n)}.$$

Proof. In view of the calculation in Remark 5.4.4, we know this result when $n = 1$. We now prove it for $n = 2$. We have

$$\|f\|_{L^p_{s_1,s_2}(\mathbf{R}^2)}^p = \int_{\mathbf{R}} \left[\int_{\mathbf{R}} \left| (I - \partial_1^2)^{\frac{s_1}{2}} (I - \partial_2^2)^{\frac{s_2}{2}} f(x_1,x_2) \right|^p dx_1 \right] dx_2$$

$$\approx \int_{\mathbf{R}} \left[\sum_{l_1=0}^{s_1} \int_{\mathbf{R}} \left| \partial_1^{l_1} (I - \partial_2^2)^{\frac{s_2}{2}} f(x_1,x_2) \right|^p dx_1 \right] dx_2$$

$$\approx \sum_{l_1=0}^{s_1} \int_{\mathbf{R}} \left[\int_{\mathbf{R}} \left| (I - \partial_2^2)^{\frac{s_2}{2}} \partial_1^{l_1} f(x_1,x_2) \right|^p dx_2 \right] dx_1$$

$$\approx \sum_{l_1=0}^{s_1} \int_{\mathbf{R}} \left[\sum_{l_2=0}^{s_2} \int_{\mathbf{R}} \left| \partial_2^{l_2} \partial_1^{l_1} f(x_1,x_2) \right|^p dx_2 \right] dx_1$$

$$= \sum_{l_1=0}^{s_1} \sum_{l_2=0}^{s_2} \left\| \partial_1^{l_1} \partial_2^{l_2} f \right\|_{L^p(\mathbf{R}^2)}^p.$$

This proves the assertion when $n = 2$ and a straightforward adaptation works in higher dimensions. $\square$

Our goal in this section is to prove the following theorem.

Theorem 5.7.3. (Marcinkiewicz multiplier theorem) Let σ be a $\mathscr{C}^n$ function defined on $(\mathbf{R} \setminus \{0\})^n$ that satisfies

$$\left| \partial_1^{m_1} \cdots \partial_n^{m_n} \sigma(\xi_1, \ldots, \xi_n) \right| \leq A_{m_1,\ldots,m_n} |\xi_1|^{-m_1} \cdots |\xi_n|^{-m_n} \quad (5.7.2)$$

for all $m_j \in \{0,1\}$ and all $\xi_j \neq 0$, $j = 1, \ldots, n$. Let $1 < p < \infty$. Then σ lies in $\mathscr{M}_p(\mathbf{R}^n)$ and there is a constant $C_{n,p}$ such that

$$\|\sigma\|_{\mathscr{M}_p} \leq C_{n,p} \sup_{m_1,\ldots,m_n \in \{0,1\}} A_{m_1,\ldots,m_n}.$$

5.7 The Marcinkiewicz Multiplier Theorem

Before discussing the proof we set up some notation. We use ψ to denote a Schwartz function on the real line whose Fourier transform is supported in $[-2, -\frac{6}{7}] \cup [\frac{6}{7}, 2]$, equals 1 on $[-\frac{12}{7}, -1] \cup [1, \frac{12}{7}]$ and satisfies $\sum_{j \in \mathbf{Z}} \widehat{\psi}(2^{-j}\eta) = 1$ when $\eta \neq 0$. We define θ as follows:

$$\widehat{\theta}(\eta) = \widehat{\psi}(\eta/2) + \widehat{\psi}(\eta) + \widehat{\psi}(2\eta), \qquad \eta \in \mathbf{R}.$$

Then $\widehat{\theta}$ is supported in $\{\frac{3}{7} \leq |\xi| \leq 4\}$ and $\widehat{\theta} = 1$ on the support of $\widehat{\psi}$. To simplify the notation, if $\xi = (\xi_1, \ldots, \xi_n) \in \mathbf{R}^n$ and $\boldsymbol{J} = (j_1, \ldots, j_n) \in \mathbf{Z}^n$, we write

$$2^{\boldsymbol{J}}\xi = (2^{j_1}\xi_1, \ldots, 2^{j_n}\xi_n)$$

and we define functions $\otimes\psi$ and $\otimes\theta$ on $\mathbf{R}^n$ by setting

$$\otimes\psi(x) = \prod_{\ell=1}^{n} \psi(x_\ell), \qquad \otimes\theta(x) = \prod_{\ell=1}^{n} \theta(x_\ell). \tag{5.7.3}$$

Then we have

$$\widehat{\otimes\psi}(\xi) = \otimes\widehat{\psi}(\xi) = \prod_{\ell=1}^{n} \widehat{\psi}(\xi_\ell), \qquad \widehat{\otimes\theta}(\xi) = \otimes\widehat{\theta}(\xi) = \prod_{\ell=1}^{n} \widehat{\theta}(\xi_\ell).$$

Let $k \in \{1, \ldots, n\}$. For $j \in \mathbf{Z}$ we define the Littlewood–Paley operators associated to the bumps ψ and θ by

$$\Delta_j^{\psi,k}(f)(x) = \int_{\mathbf{R}} f(x_1, \ldots, x_{k-1}, x_k - y, x_{k+1}, \ldots, x_n) 2^j \psi(2^j y) dy$$

and

$$\Delta_j^{\theta,k}(f)(x) = \int_{\mathbf{R}} f(x_1, \ldots, x_{k-1}, x_k - y, x_{k+1}, \ldots, x_n) 2^j \theta(2^j y) dy.$$

Finally, when $\boldsymbol{J} = (j_1, \ldots, j_n)$, we write

$$\Delta_{\boldsymbol{J}}^{\otimes\psi} = \Delta_{j_1}^{\psi,1} \circ \cdots \circ \Delta_{j_n}^{\psi,n}, \qquad \Delta_{\boldsymbol{J}}^{\otimes\theta} = \Delta_{j_1}^{\theta,1} \circ \cdots \circ \Delta_{j_n}^{\theta,n}.$$

Then

$$\Delta_{\boldsymbol{J}}^{\otimes\psi}(f) = f * 2^{|\boldsymbol{J}|} \prod_{\ell=1}^{n} \psi(2^{j_\ell}(\cdot)_\ell).$$

Next we have the following lemma.

Lemma 5.7.4. *Suppose that σ is defined on $(\mathbf{R} \setminus \{0\})^n$ and satisfies (5.7.2) for all $m_j \in \{0,1\}$. Let $\otimes\psi$ be as in (5.7.3). Then for p satisfying $1 < p < \infty$ we have*

$$\sup_{\boldsymbol{J} \in \mathbf{Z}^n} \left\| \sigma(2^{\boldsymbol{J}} \cdot) \otimes \widehat{\psi} \right\|_{L^p_{1,\ldots,1}} \leq C_{n,\psi,p} \sup_{m_1,\ldots,m_n \in \{0,1\}} A_{m_1,\ldots,m_n}. \tag{5.7.4}$$

Proof. By Proposition 5.7.2 the expression on the left in (5.7.4) is comparable to

$$\sum_{m_1=0}^{1}\cdots\sum_{m_n=0}^{1}\left\|\partial_1^{m_1}\cdots\partial_n^{m_n}[\sigma(2^J\,\cdot\,)\otimes\widehat{\psi}]\right\|_{L^\rho}$$

and by Leibniz's rule this can be estimated by

$$\sum_{m_1=0}^{1}\cdots\sum_{m_n=0}^{1}\sum_{l_1=0}^{m_1}\cdots\sum_{l_n=0}^{m_n}\left\|\partial_1^{l_1}\cdots\partial_n^{l_n}[\sigma(2^J\,\cdot\,)]\partial_1^{m_1-l_1}\cdots\partial_n^{m_n-l_n}\otimes\widehat{\psi}\right\|_{L^\rho}$$

$$=\sum_{m_1=0}^{1}\cdots\sum_{m_n=0}^{1}\sum_{l_1=0}^{m_1}\cdots\sum_{l_n=0}^{m_n}2^{j_1 l_1}\cdots 2^{j_n l_n}\left\|(\partial_1^{l_1}\cdots\partial_n^{l_n}\sigma)(2^J\,\cdot\,)(\partial_1^{m_1-l_1}\widehat{\psi})\cdots(\partial_n^{m_n-l_n}\widehat{\psi})\right\|_{L^\rho}.$$

Inserting the estimate

$$|\partial_1^{l_1}\cdots\partial_n^{l_n}\sigma(2^J\xi)|\le A_{l_1,\ldots,l_n}|2^{j_1}\xi_1|^{-l_1}\cdots|2^{j_n}\xi_n|^{-l_n},$$

coming from (5.7.2), we obtain that the expression on the left in (5.7.4) is at most

$$\sum_{m_1=0}^{1}\cdots\sum_{m_n=0}^{1}\sum_{l_1=0}^{m_1}\cdots\sum_{l_n=0}^{m_n}A_{l_1,\ldots,l_n}\left\||(\cdot)_1|^{-l_1}\cdots|(\cdot)_n|^{-l_n}(\partial_1^{m_1-l_1}\widehat{\psi})\cdots(\partial_n^{m_n-l_n}\widehat{\psi})\right\|_{L^\rho},$$

and this is bounded by $C_{n,\psi,\rho}\sup_{m_i\in\{0,1\}}A_{m_1,\ldots,m_n}$, as the function $\otimes\widehat{\psi}$ is supported in $([-2,-\frac{6}{7}]\cup[\frac{6}{7},2])^n$. $\square$

In the following lemma we denote by $M^{(\ell)}$ the one-dimensional Hardy–Littlewood maximal operator acting only in the ℓth coordinate.

Lemma 5.7.5. Let $1<\rho<2$. Let ψ,θ be as above. Then, for any $f\in\mathscr{S}(\mathbf{R}^n)$ and for all $J=(j_1,\ldots,j_n)\in\mathbf{Z}^n$ we have

$$|\Delta_J^{\otimes\psi}T_\sigma(f)|\le CK\left[M^{(1)}\cdots M^{(n)}(|\Delta_J^{\otimes\theta}(f)|^\rho)\right]^{\frac{1}{\rho}},\tag{5.7.5}$$

where

$$K=\sup_{J\in\mathbf{Z}^n}\left\|\sigma(2^J\,\cdot\,)\otimes\widehat{\psi}\right\|_{L^\rho_{1,\ldots,1}}.$$

Proof. Since $\widehat{\theta}$ is equal to 1 on the support of $\widehat{\psi}$, $\otimes\widehat{\theta}(2^{-J}\xi)$ is equal to 1 on the support of $\otimes\widehat{\psi}(2^{-J}\xi)$ for any $J=(j_1,\ldots,j_n)\in\mathbf{Z}^n$. Using this for $x\in\mathbf{R}^n$ we write

$$\Delta_J^{\otimes\psi}T_\sigma(f)(x_1,\ldots,x_n)$$

$$=\int_{\mathbf{R}^n}\widehat{f}(\xi)\otimes\widehat{\psi}(2^{-J}\xi)\sigma(\xi)e^{2\pi ix\cdot\xi}d\xi$$

$$=\int_{\mathbf{R}^n}\widehat{f}(\xi)\otimes\widehat{\theta}(2^{-J}\xi)\sigma(\xi)\otimes\widehat{\psi}(2^{-J}\xi)e^{2\pi ix\cdot\xi}d\xi$$

$$=\int_{\mathbf{R}^n}(\Delta_J^{\otimes\theta}(f))^{\widehat{\,}}(\xi)\sigma(\xi)\otimes\widehat{\psi}(2^{-J}\xi)e^{2\pi ix\cdot\xi}d\xi$$

$$=\int_{\mathbf{R}^n}2^{j_1+\cdots+j_n}(\Delta_J^{\otimes\theta}(f))^{\widehat{\,}}(2^J\xi')\sigma(2^J\xi')\otimes\widehat{\psi}(\xi')e^{2\pi i(2^J x)\cdot\xi'}d\xi'$$

5.7 The Marcinkiewicz Multiplier Theorem

$$= \int_{\mathbf{R}^n} (\Delta_J^{\otimes \theta}(f)(2^{-J} \cdot))\widehat{}(\xi') \left[\sigma(2^J \cdot) \otimes \widehat{\psi} \, e^{2\pi i (2^J x) \cdot (\cdot)}\right](\xi') d\xi'$$

$$= \int_{\mathbf{R}^n} \Delta_J^{\otimes \theta}(f)(2^{-J} y') [\sigma(2^J \cdot) \otimes \widehat{\psi}]\widehat{}(y' - 2^J x) \, dy'$$

$$= 2^{j_1 + \cdots + j_n} \int_{\mathbf{R}^n} \Delta_J^{\otimes \theta}(f)(y) [\sigma(2^J \cdot) \otimes \widehat{\psi}]\widehat{}(2^J y - 2^J x) \, dy$$

$$= \int_{\mathbf{R}^n} \frac{2^{j_1 + \cdots + j_n} \Delta_J^{\otimes \theta}(f)(y)}{\prod_{\ell=1}^n (1 + 2^{j_\ell} |x_\ell - y_\ell|)} \cdot \prod_{\ell=1}^n (1 + 2^{j_\ell} |x_\ell - y_\ell|) [\sigma(2^J \cdot) \otimes \widehat{\psi}]\widehat{}(2^J y - 2^J x) \, dy.$$

Hölder's inequality now gives that $|\Delta_J^{\otimes \psi} T_\sigma(f)(x)|$ is bounded by

$$\left(\int_{\mathbf{R}^n} 2^{j_1 + \cdots + j_n} \frac{|\Delta_J^{\otimes \theta}(f)(y)|^\rho}{\prod_{\ell=1}^n (1 + 2^{j_\ell}|x_\ell - y_\ell|)^\rho} \, dy\right)^{\frac{1}{\rho}}$$

$$\times \left(\int_{\mathbf{R}^n} 2^{j_1 + \cdots + j_n} \left|\prod_{\ell=1}^n (1 + 2^{j_\ell}|x_\ell - y_\ell|) \cdot [\sigma(2^J \cdot) \otimes \widehat{\psi}]\widehat{}(2^J y - 2^J x)\right|^{\rho'} dy\right)^{\frac{1}{\rho'}}.$$

As $\rho > 1$, n consecutive applications of the one-dimensional version of Corollary 2.5.2 yield the estimate

$$\left(\int_{\mathbf{R}^n} 2^{j_1 + \cdots + j_n} \frac{|\Delta_J^{\otimes \theta}(f)(y)|^\rho}{\prod_{\ell=1}^n (1 + 2^{j_\ell}|x_\ell - y_\ell|)^\rho} \, dy\right)^{\frac{1}{\rho}} \leq C \left[M^{(1)} \cdots M^{(n)} (|\Delta_J^{\otimes \theta}(f)|^\rho)(x)\right]^{\frac{1}{\rho}}.$$

We now write

$$\left(\int_{\mathbf{R}^n} 2^{j_1 + \cdots + j_n} \left|\prod_{\ell=1}^n (1 + 2^{j_\ell}|x_\ell - y_\ell|) [\sigma(2^J \cdot) \otimes \widehat{\psi}]\widehat{}(2^J y - 2^J x)\right|^{\rho'} dy\right)^{\frac{1}{\rho'}}$$

$$= \left(\int_{\mathbf{R}^n} \left|\prod_{\ell=1}^n (1 + |y_\ell|) [\sigma(2^J \cdot) \otimes \widehat{\psi}]\widehat{}(y)\right|^{\rho'} dy\right)^{\frac{1}{\rho'}}$$

$$\leq 2^{\frac{n}{2}} \left(\int_{\mathbf{R}^n} \left|\prod_{\ell=1}^n (1 + 4\pi^2 |y_\ell|^2)^{\frac{1}{2}} [\sigma(2^J \cdot) \otimes \widehat{\psi}]\widehat{}(y)\right|^{\rho'} dy\right)^{\frac{1}{\rho'}}$$

$$\leq 2^{\frac{n}{2}} \left\|(I - \partial_1^2)^{\frac{1}{2}} \cdots (I - \partial_n^2)^{\frac{1}{2}} [\sigma(2^J \cdot) \otimes \widehat{\psi}]\right\|_{L^\rho} \tag{5.7.6}$$

$$\leq CK.$$

This yields (5.7.5). $\square$

We now have the ingredients needed to prove Theorem 5.7.3.

Proof. Suppose first that $p > 2$. For technical reasons we will need to know that $T_\sigma(f)$ lies a priori in L^p. To achieve this, we assume that f lies in $\widehat{\mathscr{S}_{0,\ldots,0}}$, the space of Schwartz functions whose Fourier transform is compact and does not intersect any plane of the form $\xi_k = 0$. Such functions are dense in $L^p(\mathbf{R}^n)$ for any $1 < p < \infty$ (Exercise 4.6.1). Fix such a function f. For fixed $x = (x_1, \ldots, x_n) \in \mathbf{R}^n$ we write

$$T_\sigma(f)(x) = \int_{\mathbf{R}^n} \widehat{f}(\xi)\sigma(\xi)e^{2\pi i x\cdot\xi}d\xi$$
$$= \frac{1}{\prod_{j=1}^n(1+2\pi i x_j)} \int_{\mathbf{R}^n} \widehat{f}(\xi)\sigma(\xi)(I+\partial_1)\cdots(I+\partial_n)e^{2\pi i x\cdot\xi}d\xi$$
$$= \frac{1}{\prod_{j=1}^n(1+2\pi i x_j)} \int_{\mathbf{R}^n} (I-\partial_1)\cdots(I-\partial_n)[\widehat{f}\sigma](\xi)e^{2\pi i x\cdot\xi}d\xi.$$

In view of Leibniz's rule, hypotheses (5.7.2), and the fact that $\widehat{f}$ has compact support that does not intersect the planes $\xi_k = 0$, we obtain

$$|T_\sigma(f)(x_1,\ldots,x_n)| \leq C_f (1+|x_1|)^{-1}\cdots(1+|x_n|)^{-1}.$$

But this function lies in $L^p(\mathbf{R}^n)$ for all $p > 1$. Thus $\|T_\sigma(f)\|_{L^p} < \infty$, which allows the use of inequality (4.6.11) in Theorem 4.6.4.

Applying first the inequality (4.6.11) of Theorem 4.6.4, then Lemma 5.7.5, then Exercise 4.3.7 (with $r = 2/\rho$), and finally inequality (4.6.10) in Theorem 4.6.4, we obtain

$$\|T_\sigma(f)\|_{L^p(\mathbf{R}^n)} \leq C_p(n) \bigg\|\Big(\sum_{J\in\mathbf{Z}^n} |\Delta_J^{\otimes\psi}(T_\sigma(f))|^2\Big)^{\frac{1}{2}}\bigg\|_{L^p}$$
$$\leq C'_p(n)K \bigg\|\Big(\sum_{J\in\mathbf{Z}^n} \big[M^{(1)}\cdots M^{(n)}(|\Delta_J^{\otimes\theta}(f)|^\rho)\big]^{\frac{2}{\rho}}\Big)^{\frac{1}{2}}\bigg\|_{L^p(\mathbf{R}^n)}$$
$$= C'_p(n)K \bigg\|\Big(\sum_{J\in\mathbf{Z}^n} \big[M^{(1)}\cdots M^{(n)}(|\Delta_J^{\otimes\theta}(f)|^\rho)\big]^{\frac{2}{\rho}}\Big)^{\frac{\rho}{2}}\bigg\|_{L^{p/\rho}}^{\frac{1}{\rho}}$$
$$\leq C''_p(n)K \bigg\|\Big(\sum_{J\in\mathbf{Z}^n} |\Delta_J^{\otimes\theta}(f)|^2\Big)^{\frac{\rho}{2}}\bigg\|_{L^{p/\rho}}^{\frac{1}{\rho}}$$
$$= C''_p(n)K \bigg\|\Big(\sum_{J\in\mathbf{Z}^n} |\Delta_J^{\otimes\theta}(f)|^2\Big)^{\frac{1}{2}}\bigg\|_{L^p(\mathbf{R}^n)}$$
$$\leq C'''_p(n)K \|f\|_{L^p(\mathbf{R}^n)}.$$

Exercise 4.3.7 makes use of the assumptions $1 < 2/\rho < \infty$ and $1 < p/\rho < \infty$. This proves the claimed bound for functions $f \in \mathscr{S}_{0,\ldots,0}$, which is a dense subspace of L^p. By density, there is a bounded extension of T_σ on $L^p(\mathbf{R}^n)$ for $2 < p < \infty$ with norm bounded by $C'''_p(n)K$. We recall that by Lemma 5.7.4, the constant K is bounded by a multiple of

$$\sup_{m_1,\ldots,m_n\in\{0,1\}} A_{m_1,\ldots,m_n},$$

as claimed.

Finally, the case $1 < p < 2$ follows by a duality argument, while $p = 2$ is straightforward. $\square$

Exercises

5.7.1. Suppose that σ_1, σ_2 are complex-valued functions that satisfy (5.7.1).
(a) Prove that $\sigma_1 \sigma_2$ satisfies (5.7.1).
(b) Let β be a fixed multi-index and suppose that σ_3 is a complex-valued function that satisfies $|\partial^\gamma \sigma_3(\xi)| \leq c_{\beta,\gamma} |\xi_1|^{\beta_1 - \gamma_1} \cdots |\xi_n|^{\beta_n - \gamma_n}$ for all multi-indices γ. Show that $\partial^\beta \sigma_3$ also satisfies (5.7.1).
(c) Verify that $\nabla \sigma_1(\xi) \cdot \xi$ and $\xi^\beta \partial^\beta \sigma_2(\xi)$ satisfy (5.7.1).

5.7.2. Let σ be a real-valued function that satisfies (5.7.1). Show that $e^{i\sigma}$ also satisfies (5.7.1). [*Hint:* Use the Faà di Bruno formula (Appendix F).]

5.7.3. Let $s_1, \ldots, s_n > 0$ and $1 < p < \infty$. Show that the function
$$\frac{(1+4\pi^2|\xi_1|^2)^{\frac{s_1}{2}} \cdots (1+4\pi^2|\xi_n|^2)^{\frac{s_n}{2}}}{(1+4\pi^2|\xi|^2)^{\frac{s_1 + \cdots + s_n}{2}}}$$
is an L^p Fourier multiplier. Conclude that $L^p_{s_1 + \cdots + s_n}(\mathbf{R}^n)$ continuously embeds in the product-type Sobolev space $L^p_{s_1, \ldots, s_n}(\mathbf{R}^n)$.

5.7.4. Let $s_1, \ldots, s_n > 0$ and $1 < p < \infty$. Prove that $L^p_{s_1, \ldots, s_n}(\mathbf{R}^n)$ is a complete normed vector space and $\mathscr{S}(\mathbf{R}^n)$ is a dense subspace of it. [*Hint:* Mimic the proofs of Theorems 5.4.7 and 5.4.8.]

5.7.5. Suppose that $k_1, \ldots, k_n \in \mathbf{Z}^+$ are such that $k = (\frac{1}{k_1} + \cdots + \frac{1}{k_n})^{-1} \in \mathbf{Z}^+$. Show that the function
$$M(\xi_1, \ldots, \xi_n) = \frac{|\xi_1 \cdots \xi_n|^{2k}}{|\xi_1|^{2k_1} + \cdots + |\xi_n|^{2k_n}},$$
defined on $\mathbf{R}^n \setminus \{0\}$, is an L^p Fourier multiplier for $1 < p < \infty$.

5.7.6. Let τ be a real number and let $\rho_1, \ldots, \rho_n$ be positive integers. Prove that the following functions are L^p multipliers on $\mathbf{R}^n$ for $1 < p < \infty$:
$$\sigma_m(\xi_1, \ldots, \xi_n) = \left(\sum_{\substack{S \subseteq \{1,\ldots,n\} \\ |S|=m}} \prod_{j \in S} |\xi_j|^{2\rho_j} \right)^{i\tau}, \quad 1 \leq m \leq n.$$

5.7.7. Let τ be a real number and let $\rho_1, \ldots, \rho_n$ be positive integers. Prove that the following function defined on $(\mathbf{R} \setminus \{0\})^n$ lies in $\mathscr{M}_p(\mathbf{R}^n)$ for $1 < p < \infty$:
$$(|\xi_1|^{-2\rho_1} + \cdots + |\xi_n|^{-2\rho_n})^{i\tau}.$$
[*Hint:* Use Exercise 5.7.6 with $m = n-1$ and Exercise 5.7.1.]

5.7.8. Let $b > 0$. Prove that the following function lies in $\mathscr{M}_p(\mathbf{R}^n)$ for $1 < p < \infty$:
$$\frac{\xi_1}{i\xi_1 + (|\xi|^2 - |\xi_1|^2)^b}, \quad \xi = (\xi_1, \ldots, \xi_n) \in \mathbf{R}^n \setminus \{0\}.$$

5.7.9. Consider the differential operators

$$L_1 = \partial_1 - (\partial_2^2 + \cdots + \partial_n^2),$$
$$L_2 = \partial_1 + \partial_2^2 + \cdots + \partial_n^2.$$

Prove that for every $1 < p < \infty$ there exists a constant $C_p < \infty$ such that for all $f \in \mathscr{S}(\mathbf{R}^n)$ we have

$$\|\partial_1 f\|_{L^p} \leq C_p \min\left(\|L_1(f)\|_{L^p}, \|L_2(f)\|_{L^p}\right).$$

[*Hint:* Use the previous exercise.]

5.7.10. Suppose that $m_1, \ldots, m_n$ are positive integers and $c_j > 0$ if m_j is even while $c_j < 0$ if m_j is odd. Consider the differential operator $L = c_1 \partial_1^{2m_1} + \cdots + c_n \partial_n^{2m_n}$.
(a) Show that for any $1 < p < \infty$ there is a constant C (that depends only on p, n, m_j, c_j) such that for all Schwartz functions f on $\mathbf{R}^n$ and $1 \leq j, k \leq n$ we have

$$\|\partial_k^{m_k} \partial_j^{m_j} f\|_{L^p(\mathbf{R}^n)} \leq C \|L(f)\|_{L^p(\mathbf{R}^n)}.$$

(b) Let m be an odd positive integer. Prove that for any $\varphi \in \mathscr{S}(\mathbf{R}^{n+1})$ we have

$$\|\partial_k^{m_k} \partial_j^{m_j} \partial_{n+1}^{m-1} \varphi\|_{L^p(\mathbf{R}^{n+1})} \leq C \|L(\varphi) + \partial_{n+1}^m \varphi\|_{L^p(\mathbf{R}^{n+1})},$$

where $1 \leq j, k \leq n$ and C depends only on p, n, m_j, c_j and m.

Chapter 6
Bounded Mean Oscillation

6.1 Basic Properties of Functions of Bounded Mean Oscillation

The *mean* (or average) of an L^1_{loc} function f over a measurable subset K of $\mathbf{R}^n$ (with positive measure) is

$$f_K = \frac{1}{|K|} \int_K f(y)\,dy.$$

Let us call $|f - f_K|$ the *oscillation* of f over K. Then the *mean oscillation* of f over K is the quantity

$$\frac{1}{|K|} \int_K |f(y) - f_K|\,dy.$$

In this chapter we study functions whose mean oscillation over all cubes is bounded.

Definition 6.1.1. For f a complex-valued locally integrable function on $\mathbf{R}^n$, define

$$\|f\|_{BMO} = \sup_Q \frac{1}{|Q|} \int_Q |f(x) - f_Q|\,dx,$$

where the supremum is taken over all cubes Q in $\mathbf{R}^n$ with sides parallel to the axes.[1] The function f is of bounded mean oscillation if $\|f\|_{BMO} < \infty$ and $BMO(\mathbf{R}^n)$ is the set of all locally integrable functions f on $\mathbf{R}^n$ with $\|f\|_{BMO} < \infty$.

If the pair $\big(BMO, \|\cdot\|_{BMO}\big)$ were a normed linear space, then we would have

$$\|f+g\|_{BMO} \leq \|f\|_{BMO} + \|g\|_{BMO}, \tag{6.1.1}$$
$$\|\lambda f\|_{BMO} = |\lambda|\,\|f\|_{BMO}, \tag{6.1.2}$$
$$\|f\|_{BMO} = 0 \implies f = 0 \quad \text{a.e.} \tag{6.1.3}$$

Although properties (6.1.1) and (6.1.2) can be easily verified, one notes that (6.1.3) fails. If $\|f\|_{BMO} = 0$, then f would have to be a constant C_Q over every cube

---

[1] All cubes in this text have sides parallel to the axes, unless stated otherwise.

Q. Covering $\mathbf{R}^n$ by a union of overlapping cubes, we conclude that $C_Q = C_{Q'}$ for all cubes Q and Q'. So if $\|f\|_{BMO} = 0$, then f is almost everywhere equal to a constant (possibly nonzero). Thus $\|\cdot\|_{BMO}$ is only a seminorm on BMO, even though we often refer to it as a norm. To rectify (6.1.3), instead of considering classes of functions that are equal a.e., we consider equivalence classes of functions formed by the binary relation $f \equiv g \iff f - g$ is a constant a.e. Under this adjustment, the pair $(BMO, \|\cdot\|_{BMO})$ becomes a normed linear space.

Next we observe that BMO is invariant under translations just like every L^p space. But BMO is closer to L^∞ than all other L^p spaces. The reason is that it remains invariant under dilations, just like L^∞ does; to verify this, let $f^\lambda(x) = f(\lambda x)$, $\lambda > 0$, then $(f^\lambda)_Q = f_{\lambda Q}$, $\lambda Q = \{\lambda x : x \in Q\}$, and so we obtain

$$\frac{1}{|Q|} \int_Q |f^\lambda(x) - (f^\lambda)_Q| \, dx = \frac{1}{|\lambda Q|} \int_{\lambda Q} |f(x) - f_{\lambda Q}| \, dx,$$

so taking the supremum of both sides we deduce $\|f^\lambda\|_{BMO} = \|f\|_{BMO}$.

Our last observation is that BMO in fact contains L^∞. Indeed,

$$\|f\|_{BMO} = \sup_Q |f - f_Q|_Q \leq \sup_Q \left[|f|_Q + |f_Q|_Q\right] \leq 2\|f\|_{L^\infty}.$$

Although it is more natural to define BMO in terms of cubes, one can define another BMO space, replacing cubes by balls in the definition.

Definition 6.1.2. For f a complex-valued locally integrable function on $\mathbf{R}^n$, define

$$\|f\|_{BMO_{\text{balls}}} = \sup_B \frac{1}{|B|} \int_B |f(x) - f_B| \, dx,$$

where the supremum is taken over all balls B in $\mathbf{R}^n$.

The following proposition provides one of the most useful criteria to verify that a function lies in BMO.

Proposition 6.1.3. *Let $f \in L^1_{\text{loc}}(\mathbf{R}^n)$. Suppose that there exists an $A > 0$ such that for all cubes K (respectively, balls K) in $\mathbf{R}^n$ there exists a constant c_K such that*

$$\frac{1}{|K|} \int_K |f(x) - c_K| \, dx \leq A. \tag{6.1.4}$$

Then $f \in BMO$ (resp., $f \in BMO_{\text{balls}}$) and $\|f\|_{BMO} \leq 2A$ (resp., $\|f\|_{BMO_{\text{balls}}} \leq 2A$).

Proof. We note that

$$|f - f_K| \leq |f - c_K| + |f_K - c_K| \leq |f - c_K| + \frac{1}{|K|} \int_K |f(x) - c_K| \, dx.$$

Averaging over cubes K (resp., balls K) and using (6.1.4), we obtain that $\|f\|_{BMO} \leq 2A$, and analogously $\|f\|_{BMO_{\text{balls}}} \leq 2A$, if the sets K are balls. □

6.1 Basic Properties of Functions of Bounded Mean Oscillation

We now show that the seminorms $\|f\|_{BMO_{\text{balls}}}$ and $\|f\|_{BMO}$ are in fact comparable; thus the spaces $BMO(\mathbf{R}^n)$ and $\|f\|_{BMO_{\text{balls}}}(\mathbf{R}^n)$ contain the same functions.

Given any cube Q in $\mathbf{R}^n$, we let B be the smallest ball that contains it. Let v_n be the volume of the unit ball. Then

$$\frac{1}{|Q|}\int_Q |f(x)-f_B|\,dx \leq \frac{|B|}{|Q|}\frac{1}{|B|}\int_B |f(x)-f_B|\,dx \leq \frac{v_n\sqrt{n}^n}{2^n}\|f\|_{BMO_{\text{balls}}}.$$

It follows from Proposition 6.1.3 that $\|f\|_{BMO} \leq 2^{1-n}v_n\sqrt{n}^n\|f\|_{BMO_{\text{balls}}}$. Now given a ball B find the smallest cube Q that contains it. Then write

$$\frac{1}{|B|}\int_B |f(x)-f_Q|\,dx \leq \frac{|Q|}{|B|}\frac{1}{|Q|}\int_Q |f(x)-f_Q|\,dx \leq \frac{2^n}{v_n}\|f\|_{BMO},$$

and this implies $\|f\|_{BMO_{\text{balls}}} \leq 2^{n+1}v_n^{-1}\|f\|_{BMO}$. We conclude that the spaces BMO and BMO_{balls} have comparable seminorms, hence they are isomorphic.

Proposition 6.1.4. *If $f \in BMO$, then $|f| \in BMO$. f,g are real-valued BMO functions, then so are $\max(f,g)$ and $\min(f,g)$. Moreover,*

$$\||f|\|_{BMO} \leq 2\|f\|_{BMO}, \tag{6.1.5}$$

$$\|\max(f,g)\|_{BMO} \leq \frac{3}{2}\|f\|_{BMO} + \frac{3}{2}\|g\|_{BMO}, \tag{6.1.6}$$

$$\|\min(f,g)\|_{BMO} \leq \frac{3}{2}\|f\|_{BMO} + \frac{3}{2}\|g\|_{BMO}. \tag{6.1.7}$$

Proof. To prove (6.1.5), note that for each cube Q we have $\big||f|-|f_Q|\big| \leq |f-f_Q|$, which implies

$$\big||f|-|f_Q|\big|_Q \leq |f-f_Q|_Q \leq \|f\|_{BMO}. \tag{6.1.8}$$

Thus, for each cube Q there is a constant $C_Q = |f_Q|$ such that (6.1.8) holds. Appealing to Proposition 6.1.3 we deduce (6.1.5). Next, note that

$$\max(f,g) = \frac{f+g+|f-g|}{2} \quad\text{and}\quad \min(f,g) = \frac{f+g-|f-g|}{2}.$$

Then we obtain the estimate

$$\|\max(f,g)\|_{BMO} \leq \frac{\|f\|_{BMO}+\|g\|_{BMO}+\||f-g|\|_{BMO}}{2}$$
$$\leq \frac{\|f\|_{BMO}+\|g\|_{BMO}+2\|f-g\|_{BMO}}{2}$$
$$\leq \frac{\|f\|_{BMO}+\|g\|_{BMO}+2\|f\|_{BMO}+2\|g\|_{BMO}}{2},$$

from which we obtain (6.1.6). Likewise we obtain (6.1.7). □

Example 6.1.5. We show that the unbounded function $\log|x|$ lies in $BMO(\mathbf{R}^n)$. Hence $L^\infty(\mathbf{R}^n)$ is a proper subspace of $BMO(\mathbf{R}^n)$.

Indeed, we will show that $\log|x|$ lies in $BMO_{\text{balls}}(\mathbf{R}^n)$. Let $B(x_0,R)$ be a ball. If $|x_0| > 2R$, then for $|x - x_0| \leq R$ we have $\frac{1}{2}|x_0| \leq |x| \leq \frac{3}{2}|x_0|$, hence

$$\frac{1}{v_n R^n} \int_{|x-x_0| \leq R} \big|\log|x| - \log|x_0|\big| dx = \frac{1}{v_n R^n} \int_{|x-x_0| \leq R} \left|\log\frac{|x|}{|x_0|}\right| dx$$

$$\leq \max\left(\log\frac{3}{2}, \left|\log\frac{1}{2}\right|\right) = \log 2.$$

Also, if $|x_0| \leq 2R$, then

$$\frac{1}{v_n R^n} \int_{|x-x_0| \leq R} \big|\log|x| - \log R\big| dx = \frac{1}{v_n R^n} \int_{|x-x_0| \leq R} \left|\log\frac{|x|}{R}\right| dx$$

$$\leq \frac{1}{v_n R^n} \int_{|x| \leq 3R} \left|\log\frac{|x|}{R}\right| dx$$

$$= \frac{1}{v_n} \int_{|x| \leq 3} \big|\log|x|\big| dx = \frac{3^n(n\log 3 - 1) + 2}{n}.$$

We apply Proposition 6.1.3 with $C_{B(x_0,R)}$ being $\log R$ or $\log|x_0|$ to deduce that $\log|x|$ lies in $BMO_{\text{balls}}(\mathbf{R}^n)$ and hence in $BMO(\mathbf{R}^n)$.

The function $\log|x|$ turns out to be a typical element of BMO, but we will make this statement a bit more precise in the next section. It is interesting, however, to notice that BMO does not remain invariant under abrupt cutoffs.

Example 6.1.6. The function $h(x) = \chi_{x>0} \log\frac{1}{x}$ is not in $BMO(\mathbf{R})$. Indeed, the problem is at the origin. Consider the intervals $(-\varepsilon, \varepsilon)$, where $0 < \varepsilon < \frac{1}{2}$. We have that

$$h_{(-\varepsilon,\varepsilon)} = \frac{1}{2\varepsilon} \int_{-\varepsilon}^{+\varepsilon} h(x)\, dx = \frac{1}{2\varepsilon} \int_0^\varepsilon \log\frac{1}{x}\, dx = \frac{1 + \log\frac{1}{\varepsilon}}{2}.$$

But then

$$\frac{1}{2\varepsilon} \int_{-\varepsilon}^{+\varepsilon} \big|h(x) - h_{(-\varepsilon,\varepsilon)}\big| dx \geq \frac{1}{2\varepsilon} \int_{-\varepsilon}^0 \big|h_{(-\varepsilon,\varepsilon)}\big| dx = \frac{1 + \log\frac{1}{\varepsilon}}{4},$$

and the latter is clearly unbounded as $\varepsilon \to 0$. This discussion also reveals examples of two BMO functions whose product is not in BMO ($\chi_{(0,\infty)}$ and $\log\frac{1}{|x|}$).

Proposition 6.1.7. *Under the identification of functions whose difference is a constant a.e., BMO is a complete normed linear space, i.e., a Banach space.*

Proof. Let $\{f_k\}_{k=1}^\infty$ be a Cauchy sequence in BMO. Let $Q_N = [-N,N]^n$, $N = 1, 2, \ldots$. Then

6.1 Basic Properties of Functions of Bounded Mean Oscillation

$$\frac{1}{|Q_N|} \int_{Q_N} |f_k - f_m|\, dx \leq \|f_k - f_m\|_{BMO},$$

and this gives that $\{f_k\}_{k=1}^\infty$ is a Cauchy sequence in $L^1(Q_N)$ for any N. By completeness, there is a function $F^N \in L^1(Q_N)$ such that $f_k \to F^N$ in $L^1(Q_N)$ as $k \to \infty$.

As $f_k \to F^{N+1}$ in $L^1(Q_{N+1})$, it follows that $F^N = F^{N+1}$ a.e. on Q_N by the uniqueness of the limit. We define a function F by setting $F = F^N$ on Q_N, $N \in \mathbf{Z}^+$. Clearly F is well defined and lies in $L^1_{\text{loc}}(\mathbf{R}^n)$. Moreover, $f_k \to F$ in $L^1(K)$ for any compact set K. Next we show that F lies in BMO and that $f_k \to F$ in BMO.

As $\{f_k\}_{k=1}^\infty$ is Cauchy in BMO, for any $\varepsilon > 0$ there is a $k_0 \in \mathbf{Z}^+$ such that for $k, m \geq k_0$ we have

$$\sup_Q \frac{1}{|Q|} \int_Q |f_k - f_m|\, dx < \varepsilon.$$

Letting $m \to \infty$ yields

$$\frac{1}{|Q|} \int_Q |f_k - F|\, dx \leq \varepsilon \qquad \text{for any cube } Q. \tag{6.1.9}$$

It follows from this that for any cube Q we have

$$\frac{1}{|Q|} \int_Q |F - F_Q|\, dx \leq \frac{1}{|Q|} \int_Q |F - f_{k_0}|\, dx + \frac{1}{|Q|} \int_Q |f_{k_0} - (f_{k_0})_Q|\, dx + |(f_{k_0})_Q - F_Q|$$

$$\leq \varepsilon + \|f_{k_0}\|_{BMO} + \varepsilon.$$

This shows that F lies in BMO. Now taking the supremum in (6.1.9) over all cubes $Q \subset \mathbf{R}^n$ and using Proposition 6.1.3 we obtain that $\|f_k - F\|_{BMO} \leq 2\varepsilon$, for all $k \geq k_0$; i.e., $f_k \to F$ in BMO. □

We now examine some basic properties of BMO functions. For a ball B and $a > 0$, we denote by aB the ball that is concentric with B and whose radius is a times the radius of B.

Proposition 6.1.8. *Let f be in $BMO_{\text{balls}}(\mathbf{R}^n)$ and let B and B' be balls in $\mathbf{R}^n$.*
(i) *If $B \subset B'$, then*

$$|f_B - f_{B'}| \leq \frac{|B'|}{|B|} \|f\|_{BMO_{\text{balls}}}. \tag{6.1.10}$$

(ii) *Let $B = B_0 \subset B_1 \subset \cdots \subset B_m = B'$, where B_i is a ball of radius at most twice that of the ball B_{i-1} for each $i = 1, \ldots, m$. Then we have*

$$|f_B - f_{B'}| \leq 2^n m \|f\|_{BMO_{\text{balls}}}. \tag{6.1.11}$$

(iii) *For any $\delta > 0$ there is a constant $C_{n,\delta}$ such that if B is centered at $x_0 \in \mathbf{R}^n$ and has radius R, then we have*

$$R^\delta \int_{\mathbf{R}^n} \frac{|f(x) - f_B|}{(R + |x - x_0|)^{n+\delta}}\, dx \leq C_{n,\delta} \|f\|_{BMO_{\text{balls}}}. \tag{6.1.12}$$

An analogous estimate holds for cubes with center x_0 and side length R.

Proof. (i) We write

$$\left|f_B - f_{B'}\right| \leq \frac{1}{|B|}\int_B |f - f_{B'}|\,dx \leq \frac{1}{|B|}\int_{B'} |f - f_{B'}|\,dx \leq \frac{|B'|}{|B|}\|f\|_{BMO_{\text{balls}}}.$$

(ii) By (6.1.10), for each $i = 1,\ldots,m$ we write

$$\left|f_{B_i} - f_{B_{i-1}}\right| \leq \frac{|B_i|}{|B_{i-1}|}\|f\|_{BMO_{\text{balls}}} \leq 2^n\|f\|_{BMO_{\text{balls}}}$$

and we use this inequality to derive (6.1.11) by introducing intermediate terms:

$$\left|f_B - f_{B'}\right| \leq \left|f_{B_0} - f_{B_1}\right| + \left|f_{B_1} - f_{B_2}\right| + \cdots + \left|f_{B_{m-1}} - f_{B_m}\right| \leq 2^n m\|f\|_{BMO_{\text{balls}}}.$$

(iii) In the proof below we assume $x_0 = 0$ and $R = 1$. Once this case is known, given a ball $B(x_0,R)$, we replace $f(x)$ by the function $x \mapsto f(Rx + x_0)$ to obtain (6.1.12) in general. Then setting $B = B(0,1)$, we write

$$\int_{\mathbf{R}^n} \frac{|f(x) - f_B|}{(1+|x|)^{n+\delta}}\,dx$$

$$= \int_B \frac{|f(x) - f_B|}{(1+|x|)^{n+\delta}}\,dx + \sum_{k=0}^\infty \int_{2^{k+1}B \setminus 2^k B} \frac{|f(x) - f_{2^{k+1}B} + f_{2^{k+1}B} - f_B|}{(1+|x|)^{n+\delta}}\,dx$$

$$\leq \int_B |f(x) - f_B|\,dx + \sum_{k=0}^\infty 2^{-k(n+\delta)} \int_{2^{k+1}B} \left(|f(x) - f_{2^{k+1}B}| + |f_{2^{k+1}B} - f_B|\right)dx$$

$$\leq v_n\|f\|_{BMO_{\text{balls}}} + \sum_{k=0}^\infty 2^{-k(n+\delta)}(1+2^n(k+1))(2^{k+1})^n v_n\|f\|_{BMO_{\text{balls}}}$$

$$= C'_{n,\delta}\|f\|_{BMO_{\text{balls}}},$$

where we used (6.1.11) in the last inequality. This completes the proof. □

Finally, we note there is a completely analogous version of Proposition 6.1.8 with cubes in place of balls.

Exercises

6.1.1. Show that for all $f \in L^1_{\text{loc}}(\mathbf{R}^n)$ we have

$$\frac{1}{2}\|f\|_{BMO} \leq \sup_Q \frac{1}{|Q|} \inf_{c_Q} \int_Q |f(x) - c_Q|\,dx \leq \|f\|_{BMO}.$$

6.1.2. Let f be a real-valued *BMO* function on $\mathbf{R}^n$. Prove that the sequence

6.1 Basic Properties of Functions of Bounded Mean Oscillation

$$f_N(x) = \begin{cases} N & \text{if } f(x) > N, \\ f(x) & \text{if } |f(x)| \le N, \\ -N & \text{if } f(x) < -N \end{cases}$$

satisfies $\|f_N\|_{BMO} \le \frac{9}{4}\|f\|_{BMO}$. [*Hint:* Write $f_N = \max(-N, \min(f, N))$.]

6.1.3. Let $1 \le p \le \infty$. Find functions F in $L^p(\mathbf{R}^n)$ and $G \in BMO(\mathbf{R}^n)$ such that FG does not lie in $L^p(\mathbf{R}^n)$.

6.1.4. Show that for all f in $BMO_{\text{balls}}(\mathbf{R}^n)$ and all $r > 0$ we have

$$|f_{rB} - f_B| \le 2^n \left(1 + \log_2 \max(r, \tfrac{1}{r})\right) \|f\|_{BMO_{\text{balls}}}.$$

6.1.5. Let $a > 0$ and let $f \in BMO(\mathbf{R}^n)$. Let B and B' be balls in $\mathbf{R}^n$ both of radius r whose centers have distance ar (these balls could be overlapping). Prove that

$$|f_B - f_{B'}| \le 2^{n+1} \log_2(a+2) \|f\|_{BMO_{\text{balls}}}.$$

Also show that $\sup_{B, B' \text{ balls with } |B|=|B'|} |f_B - f_{B'}|/\|f\|_{BMO}$ may be unbounded.
[*Hint:* Pick $m \in \mathbf{Z}$ such that $2^m \le a+2 < 2^{m+1}$ and let x_0 be the midpoint of the line segment joining the centers of B and B'. Consider the ball $B'' = B(x_0, 2^m r)$ and estimate $|f_B - f_{B''}|$ and $|f_{B'} - f_{B''}|$ via telescoping sums, using (6.1.10).]

6.1.6. Let $f \in BMO(\mathbf{R}^n)$ and $N \in \mathbf{Z}^+$. Verify the following assertions:
(a) For any two cubes Q and Q' of side length 1 contained in $[0, 2^N]^n$ we have

$$|f_Q - f_{Q'}| \le N 2^{n+1} \|f\|_{BMO}.$$

(b) Let $-\infty < l < L < \infty$. Conclude that for any two cubes Q, Q' of side length 2^l both contained in a cube of side length 2^L the following estimate is valid:

$$|f_Q - f_{Q'}| \le (L - l + 1) 2^{n+1} \|f\|_{BMO}.$$

[*Hint:* Part (a). For any interval I of length 1 contained in $[0, 2^N]$ there is a sequence of intervals $I = I_0 \subset I_1 \subset \cdots \subset I_{N-1} \subset [0, 2^N]$ with $|I_j| = 2^j$. Then use (6.1.10).]

6.1.7. Let Φ be a concave strictly increasing function from $[0, \infty)$ to $[0, \infty)$ that satisfies $\Phi(0) = 0$, $\lim_{t \to \infty} \Phi(t) = \infty$, and $\Phi(t+s) \le \Phi(t) + \Phi(s)$ for all $t, s \ge 0$. Prove that if $f \in BMO$, then $\Phi(|f|)$ lies also in BMO and

$$\|\Phi(|f|)\|_{BMO} \le 2\Phi(\|f\|_{BMO}).$$

Let $0 < p < 1$. Two important examples of such functions Φ are

$$\Phi(t) = t^p, \quad \Phi(t) = \log(t+1), \quad \Phi(t) = [\log(t+1)]^p.$$

Conclude that $|\log|x||^p, \log(|\log|x||+1), \log(|\log(|\log|x||+1)|+1)$ lie in $BMO(\mathbf{R}^n)$.
[*Hint:* Apply Proposition 6.1.3 with $c_Q = \Phi(|f_Q|)$ and Jensen's inequality (Exercise 1.1.5) to the convex function Φ^{-1}. Note that $\Phi \circ \Phi$ has similar properties.]

6.1.8. Let $0 < p < 1$ and $A > 0$. Assume that Φ is an increasing function from $[0, \infty)$ to $[0, \infty)$ that satisfies $\Phi(0) = 0$, $\lim_{t \to \infty} \Phi(t) = \infty$, and

$$|\Phi(t) - \Phi(s)| \leq A\left(|t-s| + |t-s|^p\right)$$

for all $t, s \geq 0$. Prove that if $f \in BMO$, then $\Phi(|f|)$ lies also in BMO and

$$\|\Phi(|f|)\|_{BMO} \leq 2A\left(\|f\|_{BMO} + \|f\|_{BMO}^p\right).$$

An example of such a function is $\Phi(t) = t^p \log(t+1)$.

6.1.9. (a) Let $z \in \mathbf{C}$. Prove that the function $x \mapsto \log|x-z|$ lies in $BMO(\mathbf{R})$.
(b) Let $P(x)$ be a polynomial with complex coefficients of degree d. Show that the function $x \mapsto \log|P(x)|$ has $BMO(\mathbf{R})$ norm bounded by $7d$. [*Hint:* Part (a). Use the idea of Example 6.1.5. Part (b). Express P as a product of linear factors.]

6.2 The John–Nirenberg Theorem

A measurable function g is called *exponentially integrable* over any compact subset K of $\mathbf{R}^n$ if there is a positive constant c such that

$$\int_K e^{c|g(x)|} dx < \infty. \tag{6.2.1}$$

In Example 6.1.5 we verified that the function $g(x) = \log|x|$ lies in $BMO(\mathbf{R}^n)$. This function is exponentially integrable, i.e., it satisfies (6.2.1) with any constant $c < n$. It turns out that exponential integrability is a general property of BMO functions, as a consequence of the next theorem.

Theorem 6.2.1. (*John–Nirenberg theorem*) *For all* $f \in BMO(\mathbf{R}^n)$, *all cubes* Q, *and all* $\alpha > 0$ *we have*

$$\left|\{x \in Q : |f(x) - f_Q| > \alpha \|f\|_{BMO}\}\right| \leq e|Q|e^{-\frac{\alpha}{2^n e}}. \tag{6.2.2}$$

Proof. If $\|f\|_{BMO} = 0$, then (6.2.2) is valid as the set on the left in (6.2.2) has measure zero. So we may assume that $\|f\|_{BMO} \neq 0$. As (6.2.2) remains unchanged if we replace f by $f/\|f\|_{BMO}$, it suffices to assume that $\|f\|_{BMO} = 1$. We fix a closed cube Q and we introduce the following selection criterion for a subcube R of Q:

$$\frac{1}{|R|} \int_R |f(x) - f_Q| dx > e. \tag{6.2.3}$$

Since

6.2 The John–Nirenberg Theorem

$$\frac{1}{|Q|}\int_Q |f(x) - f_Q|\,dx \le \|f\|_{BMO} = 1 < e,$$

the cube Q itself does not satisfy (6.2.3). Set $Q^0 = Q$ and subdivide Q^0 into 2^n equal closed subcubes of side length equal to half of the side length of Q. Select such a subcube R if it satisfies criterion (6.2.3). Now subdivide all unselected cubes into 2^n equal subcubes of half their side length by bisecting the sides, and select among these subcubes those that satisfy (6.2.3). Continuing this process indefinitely we obtain a countable collection of cubes $\{Q_j^1\}_j$. We call the cubes Q_j^1 of *first generation*. (We use the superscript k to denote the generation of cubes.)

We now fix a selected first-generation cube Q_j^1 and we introduce the following selection criterion for subcubes R of Q_j^1:

$$\frac{1}{|R|}\int_R |f(x) - f_{Q_j^1}|\,dx > e. \tag{6.2.4}$$

Observe that Q_j^1 does not satisfy the selection criterion (6.2.4). We apply a similar stopping time selection argument to the function $f - f_{Q_j^1}$ inside the cube Q_j^1. Subdivide Q_j^1 into 2^n equal closed subcubes of side length equal to half of the side length of Q_j^1 by bisecting the sides, and select such a subcube R if it satisfies the selection criterion (6.2.4). Continue this process indefinitely. Also repeat this process for any other cube Q_j^1 of the first generation. The collection of all selected subcubes of all cubes of the first generation that satisfy (6.2.4) is denoted by $\{Q_l^2\}_l$; these are called cubes *of second generation*. Every cube of second generation is contained in a unique cube of first generation.

For a fixed selected cube Q_l^2 of second generation, introduce the selection criterion for subcubes R of Q_l^2

$$\frac{1}{|R|}\int_R |f(x) - f_{Q_l^2}|\,dx > e.$$

We repeat the previously outlined process to obtain a collection of cubes of third generation inside Q_l^2. Repeat this procedure for any other cube Q_j^2 of the second generation. Denote by $\{Q_s^3\}_s$ the thus obtained collection of all cubes of the *third generation*.

We iterate this procedure indefinitely to obtain a countable family of cubes $\{Q_j^k\}_j$ for each generation k. We claim that these cubes satisfy the following properties:

(A-k) The interior of every Q_j^k is contained in a unique $Q_{j'}^{k-1}$.

(B-k) $e < |Q_j^k|^{-1}\int_{Q_j^k}|f(x) - f_{Q_{j'}^{k-1}}|\,dx \le 2^n e.$

(C-k) $|f_{Q_j^k} - f_{Q_{j'}^{k-1}}| \le 2^n e.$

(D-k) $\sum_j |Q_j^k| \le \dfrac{1}{e}\sum_{j'}|Q_{j'}^{k-1}|.$

(E-k) $|f - f_{Q_{j'}^{k-1}}| \le e$ a.e. on the set $Q_{j'}^{k-1}\setminus \bigcup_j Q_j^k.$

We prove properties (A-k)–(E-k). Note that (A-k) and the lower inequality in (B-k) are satisfied by construction. The upper inequality in (B-k) is a consequence of the fact that the unique cube R_{j_0} with double the side length of Q_j^k that contains Q_j^k and is contained in $Q_{j'}^{k-1}$ was not selected in the process. Indeed, we have

$$e \geq \frac{1}{|R_{j_0}|} \int_{R_{j_0}} |f(x) - f_{Q_{j'}^{k-1}}|\, dx \geq \frac{1}{2^n |Q_j^k|} \int_{Q_j^k} |f(x) - f_{Q_{j'}^{k-1}}|\, dx.$$

Now (C-k) follows from the upper inequality in (B-k). To prove (E-k) we note that for every point in $Q_{j'}^{k-1} \setminus \bigcup_j Q_j^k$ there is a sequence of cubes shrinking to it and the averages of $|f - f_{Q_{j'}^{k-1}}|$ over all these cubes is at most e. Then $|f - f_{Q_{j'}^{k-1}}| \leq e$ a.e. on $Q_{j'}^{k-1} \setminus \bigcup_j Q_j^k$ by the Lebesgue differentiation theorem (Corollary 1.5.6). It remains to prove (D-k). By (A-k), given a cube Q_j^k of generation k there is unique cube $Q_{j'}^{k-1}$ of generation $k-1$ that contains it. Let us denote by $I_{j'}$ all indices i of cubes of generation k such that $i' = j'$. Then all cubes Q_i^k with $i \in I_{j'}$ have disjoint interiors and are contained in $Q_{j'}^{k-1}$. Using this we write

$$\sum_j |Q_j^k| < \frac{1}{e} \sum_j \int_{Q_j^k} |f(x) - f_{Q_{j'}^{k-1}}|\, dx$$

$$= \frac{1}{e} \sum_{j'} \sum_{i \in I_{j'}} \int_{Q_i^k} |f(x) - f_{Q_{j'}^{k-1}}|\, dx$$

$$\leq \frac{1}{e} \sum_{j'} \int_{Q_{j'}^{k-1}} |f(x) - f_{Q_{j'}^{k-1}}|\, dx$$

$$\leq \frac{1}{e} \sum_{j'} |Q_{j'}^{k-1}|,$$

as the BMO norm of f equals 1. We have now established (A-k)–(E-k) and we turn our attention to some consequences. Applying (D-k) successively $k-1$ times, we obtain

$$\sum_j |Q_j^k| \leq e^{-k} |Q^0|. \tag{6.2.5}$$

For a cube Q_j^1 of generation 1 we have $|f_{Q_j^1} - f_{Q^0}| \leq 2^n e$ by (C-1) and $|f - f_{Q_j^1}| \leq e$ a.e. on $Q_j^1 \setminus \bigcup_l Q_l^2$ by (E-2). These two facts give

$$|f - f_{Q^0}| \leq 2^n e + e \qquad \text{a.e. on} \quad Q_j^1 \setminus \bigcup_l Q_l^2,$$

which, combined with (E-1), yields

$$|f - f_{Q^0}| \leq 2^n 2 e \qquad \text{a.e. on} \quad Q^0 \setminus \bigcup_l Q_l^2. \tag{6.2.6}$$

6.2 The John–Nirenberg Theorem

For a cube Q_l^2 of second generation we have $|f - f_{Q_l^2}| \leq e$ a.e. on $Q_l^2 \setminus \bigcup_s Q_s^3$ by (E-3). Combining this with $|f_{Q_l^2} - f_{Q_{l'}^1}| \leq 2^n e$ and $|f_{Q_{l'}^1} - f_{Q^0}| \leq 2^n e$, which follow from (C-2) and (C-1) respectively, yields

$$|f - f_{Q^0}| \leq 2^n 3e \qquad \text{a.e.} \quad \text{on} \quad Q_l^2 \setminus \bigcup_s Q_s^3.$$

In view of (6.2.6), the same estimate is valid on $Q^0 \setminus \bigcup_s Q_s^3$. Continuing this reasoning, we obtain by induction that for all $k \geq 1$ we have

$$|f - f_{Q^0}| \leq 2^n k e \qquad \text{a.e.} \quad \text{on} \quad Q^0 \setminus \bigcup_s Q_s^k. \qquad (6.2.7)$$

This proves the almost everywhere inclusion

$$\{x \in Q : |f(x) - f_Q| > 2^n k e\} \subseteq \bigcup_j Q_j^k$$

for all $k = 1, 2, 3, \ldots$. (This also holds when $k = 0$ with the understanding that there is only one cube in the family on the right, the cube $Q^0 = Q$.) We now use (6.2.5) and (6.2.7) to prove (6.2.2). We fix an $\alpha > 0$. If $2^n k e < \alpha \leq 2^n (k+1) e$ for some $k \geq 0$, then

$$\begin{aligned}
|\{x \in Q : |f(x) - f_Q| > \alpha\}| &\leq |\{x \in Q : |f(x) - f_Q| > 2^n k e\}| \\
&\leq \sum_j |Q_j^k| \\
&\leq \frac{1}{e^k} |Q^0| \\
&\leq |Q| e e^{-\alpha/(2^n e)},
\end{aligned}$$

since $-k \leq 1 - \frac{\alpha}{2^n e}$. This yields (6.2.2). $\qquad \square$

Having proven the important distribution inequality (6.2.2), we are now in a position to deduce from it a few corollaries.

Corollary 6.2.2. *Every BMO function is exponentially integrable over any cube. Precisely, for any $0 < \gamma < (2^n e)^{-1}$, for all $f \in BMO(\mathbf{R}^n)$, and any cube Q we have*

$$\frac{1}{|Q|} \int_Q e^{\gamma \frac{|f(x) - f_Q|}{\|f\|_{BMO}}} dx \leq 1 + \frac{2^n e^2 \gamma}{1 - 2^n e \gamma}.$$

Proof. We use (1.2.2) in Proposition 1.2.3 with $\varphi(t) = e^t - 1$. Certainly (Q, dx) is a σ-finite measure space and φ is an increasing continuously differentiable function on $[0, \infty)$ with $\varphi(0) = 0$. Then we write

$$\frac{1}{|Q|} \int_Q e^{|h|} dx = 1 + \frac{1}{|Q|} \int_Q (e^{|h|} - 1) dx = 1 + \frac{1}{|Q|} \int_0^\infty e^\lambda |\{x \in Q : |h(x)| > \lambda\}| d\lambda$$

for a measurable function h on $\mathbf{R}^n$. We fix $\gamma < (2^n e)^{-1}$ and set $h = \gamma \|f\|_{BMO}^{-1} |f - f_Q|$. Then

$$\left|\{x \in Q : |h(x)| > \lambda\}\right| = \left|\{x \in Q : |f(x) - f_Q| > \tfrac{\lambda}{\gamma}\|f\|_{BMO}\}\right|.$$

So we apply (6.2.2) with $\alpha = \lambda/\gamma$. Combining these inequalities gives

$$\frac{1}{|Q|} \int_Q e^{\gamma \frac{|f(x)-f_Q|}{\|f\|_{BMO}}} dx \le 1 + \frac{1}{|Q|} \int_0^\infty e^\lambda\, e|Q| e^{-\frac{1}{2^n e}\frac{\lambda}{\gamma}} d\lambda = 1 + \frac{2^n e^2 \gamma}{1 - 2^n e \gamma},$$

noting that the integral converges since $1 - (2^n e \gamma)^{-1} < 0$. $\square$

As a consequence of Corollary 6.2.2 we deduce the exponential integrability of BMO functions.

Corollary 6.2.3. *Let $f \in BMO(\mathbf{R}^n)$. Then for any compact subset K of $\mathbf{R}^n$ we have*

$$\int_K e^{c|f(x)|} dx < \infty, \tag{6.2.8}$$

whenever $c < (2^n e \|f\|_{BMO})^{-1}$.

Proof. Given a compact set K pick a cube Q that contains it. For $c < (2^n e \|f\|_{BMO})^{-1}$ set $\gamma = c\|f\|_{BMO} < (2^n e)^{-1}$ and use that

$$c|f(x)| \le c|f_Q| + c|f(x) - f_Q| \le c|f_Q| + \gamma \frac{|f(x) - f_Q|}{\|f\|_{BMO}}$$

and Corollary 6.2.2 to obtain that $e^{c|f|}$ is integrable over Q. $\square$

Another important corollary of Theorem 6.2.1 is the following.

Corollary 6.2.4. *For all $0 < p < \infty$ and for all $f \in L^1_{\text{loc}}(\mathbf{R}^n)$ we have*

$$\sup_Q \left(\frac{1}{|Q|} \int_Q |f(x) - f_Q|^p dx\right)^{\frac{1}{p}} \le e\, 2^n (ep\Gamma(p))^{\frac{1}{p}} \|f\|_{BMO(\mathbf{R}^n)}. \tag{6.2.9}$$

Consequently, BMO is contained in $L^p_{\text{loc}}(\mathbf{R}^n) = \{f : |f|^p \in L^1_{\text{loc}}(\mathbf{R}^n)\}$ for all $p < \infty$.

Proof. If $f \in L^1_{\text{loc}} \setminus BMO$, then (6.2.9) holds. So we assume that $f \in BMO$. Write

$$\begin{aligned}
\frac{1}{|Q|} \int_Q |f(x) - f_Q|^p dx &= \frac{1}{|Q|} \int_0^\infty p\lambda^{p-1} |\{x \in Q : |f(x) - f_Q| > \lambda\}| d\lambda \\
&\le \frac{p}{|Q|} e|Q| \int_0^\infty \lambda^p e^{-\frac{\lambda}{2^n e\|f\|_{BMO}}} \frac{d\lambda}{\lambda} \\
&= p\Gamma(p)\, e\big(2^n e \|f\|_{BMO}\big)^p,
\end{aligned}$$

having used (6.2.2) in the inequality. This proves (6.2.9). It follows that $|f - f_Q|^p$ is integrable over any cube Q, thus so is $|f|^p$. Then $|f|^p \in L^1_{\text{loc}}(\mathbf{R}^n)$ for any $p < \infty$. $\square$

6.2 The John–Nirenberg Theorem

Since the inequality in Corollary 6.2.4 can be reversed when $p > 1$ via Hölder's inequality, we obtain the following important L^p characterization of *BMO* norms.

Corollary 6.2.5. *For all $1 \leq p < \infty$ and f in $L^1_{\text{loc}}(\mathbf{R}^n)$ we have*

$$\|f\|_{BMO} \leq \sup_Q \left(\frac{1}{|Q|} \int_Q |f(x) - f_Q|^p \, dx\right)^{\frac{1}{p}} \leq e^2 \cdot 2^n \, p \, \|f\|_{BMO}. \quad (6.2.10)$$

Proof. The left inequality in (6.2.10) is obtained by Hölder's inequality and the definition of the *BMO* norm of f. The other direction follows from Corollary 6.2.4, which provides the constant $(p\,\Gamma(p))^{\frac{1}{p}} e^{\frac{1}{p}+1} 2^n$. Note that

$$\frac{\Gamma(p)}{p^p} = \int_0^\infty e^{-t}\left(\frac{t}{p}\right)^p \frac{dt}{t} = \int_0^\infty e^{-pt} t^p \frac{dt}{t} = \int_0^\infty e^{-p(t-\log t)}\frac{dt}{t} \leq \int_0^\infty e^{-(t-\log t)}\frac{dt}{t} = 1$$

if $p \geq 1$. This yields that $\Gamma(p)^{\frac{1}{p}}(pe)^{\frac{1}{p}}e\,2^n \leq p \cdot e \cdot e\,2^n$ for $p \geq 1$; thus the upper inequality in (6.2.10) holds. $\square$

Exercises

6.2.1. Let $A, B > 0$ and let $f \in L^1_{\text{loc}}(\mathbf{R}^n)$. Suppose that for each cube Q there is a constant c_Q such that for every $\lambda > 0$ it holds that

$$|\{x \in Q : |f(x) - c_Q| > \lambda\}| \leq B|Q|e^{-A\lambda}.$$

Prove that f lies in *BMO* with norm at most $2B/A$.

6.2.2. Given $1 \leq p < \infty$ and f locally integrable on $\mathbf{R}^n$ prove that

$$\frac{1}{2}\|f\|_{BMO} \leq \sup_Q \left(\inf_{c_Q} \frac{1}{|Q|} \int_Q |f(x) - c_Q|^p \, dx\right)^{\frac{1}{p}} \leq e^2 p 2^n \|f\|_{BMO}.$$

6.2.3. Let $1 < p < \infty$. Let Q be a cube in $\mathbf{R}^n$ and let Q' be another cube that contains Q and has side length 2^m times that of Q. Prove that for any $f \in BMO(\mathbf{R}^n)$ we have

$$\left(\frac{1}{|Q|} \int_Q |f(y) - f_{Q'}|^p \, dx\right)^{\frac{1}{p}} \leq 2^n \left(e^2 p + m\right) \|f\|_{BMO}.$$

6.2.4. Let g_K be locally integrable on $\mathbf{R}^n$ associated with a measurable subset K of $\mathbf{R}^n$ with positive measure. Prove that $(a) \implies (b) \implies (c) \implies (d) \implies (a)$.
(a) There is a constant $B > 0$ such that for all $\lambda > 0$ and all $p \geq 1$ we have

$$\sup_K \lambda \left(\frac{1}{|K|}|\{x \in K : |g_K(x)| > \lambda\}|\right)^{\frac{1}{p}} \leq Bp.$$

(b) There is a constant $c > 0$ such that for all $\lambda > 0$ we have

$$\sup_K \frac{1}{|K|}\left|\{x \in K : |g_K(x)| > \lambda\}\right| \leq 2e^{-c\lambda}.$$

[*Hint:* Try $p = \lambda/2B$ if $\lambda \geq 2B$. The value $c = (2B)^{-1}\log 2$ works.]

(c) For any c' satisfying $0 < c' < c$ one has

$$\sup_K \frac{1}{|K|} \int_K e^{c'|g_K(x)|} dx \leq \frac{c+c'}{c-c'} < \infty.$$

(d) There is a constant $A > 0$ such that for all $p \geq 1$ we have

$$\sup_K \left(\frac{1}{|K|} \int_K |g_K(x)|^p dx\right)^{\frac{1}{p}} < Ap.$$

What does this set of equivalences say about *BMO* functions?

6.2.5. Let $p > 1$. Prove that $\left|\log|x|\right|^p$ and $\left|\log|\log|x|\right|\,\log|x|$ are not in $BMO(\mathbf{R}^n)$. [*Hint:* Otherwise estimate (6.2.8) would be violated.]

6.3 Dyadic Maximal Functions and Dyadic BMO

In this section we discuss dyadic analogs of certain ideas we have explored so far. We begin by recalling notions related to dyadic cubes.

Definition 6.3.1. A *dyadic cube* is a set of the form $\prod_{j=1}^n [m_j 2^{-k}, (m_j + 1)2^{-k})$, where $m_1, \ldots, m_n, k \in \mathbf{Z}$. A *dyadic child* of a dyadic cube Q is any of the 2^n dyadic cubes obtained by bisecting each of its sides by hyperplanes parallel to the faces of the cube. We denote by $\mathscr{D}$ the set of all dyadic cubes in $\mathbf{R}^n$. An *ancestor* of a dyadic cube Q is any dyadic cube that contains it. A *descendant* of a dyadic cube Q is any dyadic cube contained in it.

Naturally each dyadic cube has 2^n dyadic children, 2^{2n} dyadic grandchildren, and in general 2^{nk} dyadic subcubes of length $2^{-k}L$, where L is the side length of the original cube. In fact, all dyadic cubes contained in a fixed dyadic cube Q_0 are dyadic descendants of it.

By construction, two dyadic cubes of the same length are disjoint. Given two dyadic cubes of different size, there is a unique ancestor of the smaller one of the same size as the bigger one. This ancestor is either the bigger cube or is disjoint from it. This implies that two dyadic cubes are either disjoint or one contains the other.

It is useful to split dyadic cubes in generations. Let us call dyadic cubes of generation zero to be all dyadic cubes of side length 1 and denote by $\mathscr{D}_0$ the set of all such cubes. We denote by $\mathscr{D}_1$ the dyadic cubes of generation 1, that is, all the dyadic

6.3 Dyadic Maximal Functions and Dyadic BMO

children of cubes in $\mathscr{D}_0$. Continuing in this way, we denote by $\mathscr{D}_k$ all dyadic cubes of side length 2^{-k} and we call these dyadic cubes of generation k. As the cubes are shrinking, k increases toward infinity, but we can also consider $k < 0$ in this terminology. In specific problems, however, it is useful to set generation zero to be a fixed dyadic cube and define the future generations in relation to this. A paradigm of this situation appears in the dyadic Calderón–Zygmund decomposition.

Definition 6.3.2. Given a measurable function f on $\mathbf{R}^n$, we define the *dyadic maximal function* $M_d(f)$ of f by

$$M_d(f)(x) = \sup_{\substack{Q \ni x \\ Q \text{ dyadic cube}}} \frac{1}{|Q|} \int_Q |f(t)|\, dt.$$

The supremum is taken over all dyadic cubes Q in $\mathbf{R}^n$ that contain a given point x.

The dyadic maximal function shares many properties with the classical Hardy–Littlewood maximal function. But it is different in a significant way. It can vanish on a portion of the space. For instance, if h is supported in $[0, \infty)$, then $M_d(h)$ vanishes on $(-\infty, 0)$ as no dyadic cubes that contain negative numbers reach the support of h.

Theorem 6.3.3. (a) *For all $\lambda > 0$ and all measurable functions f on $\mathbf{R}^n$ we have*

$$|\{x \in \mathbf{R}^n : M_d(f)(x) > \lambda\}| \leq \frac{1}{\lambda} \int_{\{M_d(f) > \lambda\}} |f(t)|\, dt. \tag{6.3.1}$$

(b) *The operator M_d maps $L^1(\mathbf{R}^n)$ to $L^{1,\infty}(\mathbf{R}^n)$ with constant at most 1.*
(c) *For $1 < p < \infty$, M_d maps $L^p(\mathbf{R}^n)$ to itself with constant at most $p/(p-1)$.*
(d) *For $1 < p < \infty$, M_d maps $L^{p,\infty}(\mathbf{R}^n)$ to itself with constant at most $p/(p-1)$.*

Proof. (a) Fix $\lambda > 0$. For a measurable function f on $\mathbf{R}^n$ consider the set

$$E_f = \{x \in \mathbf{R}^n : M_d(f)(x) > \lambda\}.$$

We first prove (6.3.1) under the additional assumption that $f \in L^1(\mathbf{R}^n)$.

For each $x \in E_f$ there is a dyadic cube Q_x that contains x such that the average of $|f|$ over Q_x is strictly bigger than λ. Then each Q_x is contained in E_f as for every $y \in Q_x$, the average of $|f|$ over Q_x gives that $M_d(f)(y) > \lambda$. Consequently we have $E_f = \bigcup_{x \in E_f} Q_x$. Now all cubes Q_x have measure bounded by $\|f\|_{L^1}/\lambda$. Thus for each $x \in E_f$ there is a unique maximal dyadic cube $Q_x^{\max}$ of the form Q_y (for some $y \in E_f$) that contains Q_x. Say that all distinct cubes $Q_x^{\max}$ are indexed by a subset E_f' of E_f; then $E_f = \bigcup_{x \in E_f} Q_x = \bigcup_{x \in E_f'} Q_x^{\max}$. Moreover, all distinct cubes of the form $Q_x^{\max}$ are disjoint as they are maximal with respect to inclusion. Let us denote by $\{Q_j, j \in \mathbf{Z}\}$ the collection $\{Q_x^{\max} : x \in E_f'\}$. Then we have

$$\sum_j |Q_j| = |E_f|$$

and also

$$\int_{Q_j} |f(y)|\,dy > \lambda |Q_j|.$$

It follows that

$$|E_f| = \sum_j |Q_j| \le \frac{1}{\lambda} \sum_j \int_{Q_j} |f(y)|\,dy = \frac{1}{\lambda} \int_{E_f} |f(y)|\,dy. \qquad (6.3.2)$$

To remove the assumption that $f \in L^1(\mathbf{R}^n)$, for each $N \in \mathbf{Z}^+$ consider the truncations

$$f_N = f \chi_{|f| \le N} \chi_{[-2^N, 2^N]^n}. \qquad (6.3.3)$$

We use that $M_d(f_N)$ increases monotonically to $M_d(f)$ as $N \to \infty$; this fact follows from Exercise 1.4.2 for the classical Hardy–Littlewood maximal operator, but the same proof also holds for M_d. Then $|E_{f_N}|$ increases monotonically to $|E_f|$ and $\chi_{E_{f_N}} |f_N|$ also increases monotonically to $\chi_{E_f} |f|$. Letting $N \to \infty$ in (6.3.2) and using the LMCT yields (6.3.1).

(b) This is a direct consequence of part (a).

(c) This assertion follows from interpolation. One may use the result of Exercise 1.3.4, but here we provide a direct proof. For $f \in L^p(\mathbf{R}^n)$, $1 < p < \infty$, we write

$$\begin{aligned}
\|M_d(f)\|_{L^p}^p &= p \int_0^\infty \lambda^{p-1} |E_\lambda|\,d\lambda \\
&\le p \int_0^\infty \lambda^{p-2} \int_{E_\lambda} |f(x)|\,dx\,d\lambda \\
&= p \int_{\mathbf{R}^n} |f(x)| \int_0^{M_d(f)(x)} \lambda^{p-2}\,d\lambda\,dx \\
&= \frac{p}{p-1} \int_{\mathbf{R}^n} |f(x)| M_d(f)(x)^{p-1}\,dx \\
&\le \frac{p}{p-1} \left(\int_{\mathbf{R}^n} M_d(f)(x)^p\,dx \right)^{\frac{p-1}{p}} \left(\int_{\mathbf{R}^n} |f(x)|^p\,dx \right)^{\frac{1}{p}},
\end{aligned}$$

where we used (6.3.2), Tonelli's theorem, and Hölder's inequality. We would like to divide both sides by $\|M_d(f)\|_{L^p}^{p-1}$ but we don't know that $0 < \|M_d(f)\|_{L^p} < \infty$. In order to ensure this, we assume that f is nonzero and bounded and supported in a cube of the form $[-2^N, 2^N]^n$ for some $N \in \mathbf{Z}^+$. Then $M_d(f)$ is also bounded by $\|f\|_{L^\infty}$ and decays like $C_f |x|^{-n}$ when $x \notin [-2^N, 2^N]^n$ for some constant $C_f > 0$. These facts imply that $\|M_d(f)\|_{L^p} < \infty$, so we can divide by $\|M_d(f)\|_{L^p}^{p-1}$ and obtain

$$\|M_d(f)\|_{L^p} \le \frac{p}{p-1} \|f\|_{L^p}. \qquad (6.3.4)$$

For a general function f in $L^p(\mathbf{R}^n)$, we apply (6.3.4) to the truncations f_N defined in (6.3.3). Then we make use of the LMCT in view of the fact that $M_d(f_N)$ increases monotonically to $M_d(f)$ as $N \to \infty$ (Exercise 1.4.2). This proves the assertion in part (c); note that the bound obtained is better than that provided by Theorem 1.3.3.

6.3 Dyadic Maximal Functions and Dyadic BMO

(d) Let $1 < p < \infty$. Suppose that $f \in L^p(\mathbf{R}^n)$. We write (6.3.2) as

$$\lambda \big|\{M_d(f) > \lambda\}\big|^{\frac{1}{p}} \leq \big|\{M_d(f) > \lambda\}\big|^{\frac{1}{p}-1} \int_{\{M_d(f) > \lambda\}} |f(y)|\, dy.$$

As $f \in L^p(\mathbf{R}^n)$ we have that $\big|\{M_d(f) > \lambda\}\big| < \infty$; thus the right-hand side is bounded by $\frac{p}{p-1}\|f\|_{L^{p,\infty}}$ by Theorem 1.2.10. Thus we have

$$\|M_d(f)\|_{L^{p,\infty}} \leq \frac{p}{p-1}\|f\|_{L^{p,\infty}} \qquad \text{for } f \in L^p(\mathbf{R}^n). \tag{6.3.5}$$

Now given $f \in L^{p,\infty}$, consider the increasing sequence of nonnegative functions $|f_N|$, where f_N are defined in (6.3.3); these functions satisfy $|f_N| \uparrow |f|$ and $|f_N| \in L^p$.

Clearly we have $\|f_N\|_{L^{p,\infty}} \uparrow \|f\|_{L^{p,\infty}}$. Exercise 1.4.2 gives that $M(f_N) \uparrow M(f)$ and $\|M(f_N)\|_{L^{p,\infty}} \uparrow \|M(f)\|_{L^{p,\infty}}$. The same proofs yield

$$\|M_d(f_N)\|_{L^{p,\infty}} \uparrow \|M_d(f)\|_{L^{p,\infty}} \qquad \text{as } N \to \infty.$$

Using estimate (6.3.5) for f_N and taking the limit as $N \to \infty$ we obtain the claimed conclusion. $\square$

We now define dyadic *BMO*. We do so by taking the cubes Q in Definition 6.1.1 to be dyadic.

Definition 6.3.4. For f a complex-valued locally integrable function on $\mathbf{R}^n$, define

$$\|f\|_{BMO_d} = \sup_{Q \text{ dyadic cube}} \frac{1}{|Q|} \int_Q |f(x) - f_Q|\, dx,$$

where the supremum is taken over all dyadic cubes Q in $\mathbf{R}^n$. The function f is of *dyadic bounded mean oscillation* if $\|f\|_{BMO_d} < \infty$. The space $BMO_d(\mathbf{R}^n)$ is the set of all locally integrable functions f on $\mathbf{R}^n$ with $\|f\|_{BMO_d} < \infty$.

Functions whose difference is a constant are identified in BMO_d. Under this identification, BMO_d becomes a normed vector space, which is also complete; on this see Exercise 6.3.5. Almost all properties of *BMO* are also shared by its dyadic analog BMO_d. For instance we have the following:

Proposition 6.3.5. Let $f \in L^1_{\text{loc}}(\mathbf{R}^n)$. Suppose that there exists an $A > 0$ such that for all dyadic cubes Q there exists a constant c_Q such that

$$\sup_Q \frac{1}{|Q|} \int_Q |f(x) - c_Q|\, dx \leq A. \tag{6.3.6}$$

Then $f \in BMO_d$ and $\|f\|_{BMO_d} \leq 2A$.

Proof. We note that

$$|f - f_Q| \leq |f - c_Q| + |f_Q - c_Q| \leq |f - c_Q| + \frac{1}{|Q|} \int_Q |f(x) - c_Q|\, dx.$$

Averaging over dyadic cubes Q and using (6.3.6), we obtain that $\|f\|_{BMO_d} \leq 2A$. $\square$

Example 6.3.6. The function $h(t) = \log t \, \chi_{t>0}$ lies in $BMO_d(\mathbf{R})$ but not in $BMO(\mathbf{R})$. The fact that this function does not lie in $BMO(\mathbf{R})$ was shown in Example 6.1.6. So we only show that $h \in BMO_d(\mathbf{R})$. Let $[a,b)$ be a dyadic interval with $b > a > 0$. We need to choose a constant $C_{[a,b)} = \log c$ such that for all a,b

$$\frac{1}{b-a} \int_a^b |\log t - \log c|\, dt = \frac{1}{\frac{b}{c} - \frac{a}{c}} \int_{\frac{a}{c}}^{\frac{b}{c}} |\log t|\, dt$$

remains bounded. Now if $b - a \geq \frac{b}{2}$ we choose $c = b$. In this case $1 - \frac{a}{b} \geq \frac{1}{2}$ so

$$\frac{1}{1 - \frac{a}{b}} \int_{\frac{a}{b}}^{1} |\log t|\, dt \leq 2 \int_0^1 |\log t|\, dt = 2.$$

If $b - a < \frac{b}{2}$ we choose $c = a$. Then $\frac{b}{a} < 2$, and hence

$$\frac{1}{\frac{b}{a} - 1} \int_1^{\frac{b}{a}} |\log t|\, dt \leq \frac{1}{\frac{b}{a} - 1} \int_1^{\frac{b}{a}} |\log 2|\, dt = |\log 2| \leq 2.$$

Thus $\|h\|_{BMO_d} \leq 4$ by Proposition 6.3.5.

Proposition 6.3.7. *Let f be in BMO_d. Then for any dyadic cubes Q, Q' with $Q \subset Q'$ we have*

$$|f_Q - f_{Q'}| \leq \frac{|Q'|}{|Q|} \|f\|_{BMO_d}.$$

Let $N \in \mathbf{Z}^+$. If the side length of Q' is 2^N times larger than that of Q, then

$$|f_Q - f_{Q'}| \leq N 2^n \|f\|_{BMO_d}.$$

Proof.

$$|f_Q - f_{Q'}| \leq \frac{1}{|Q|} \int_Q |f - f_{Q'}|\, dx \leq \frac{1}{|Q|} \int_{Q'} |f - f_{Q'}|\, dx \leq \frac{|Q'|}{|Q|} \|f\|_{BMO_d}.$$

For the second assertion we fix a sequence of dyadic cubes $Q = Q_0 \subset Q_1 \subset \cdots \subset Q_N = Q'$ such that the side length of Q_j is 2^j times that of Q. Then we have $|Q_{j+1}|/|Q_j| = 2^{(j+1)n}|Q|/2^{jn}|Q| = 2^n$ and we write

$$|f_{Q_0} - f_{Q_N}| \leq |f_{Q_0} - f_{Q_1}| + |f_{Q_1} - f_{Q_2}| + \cdots + |f_{Q_{N-1}} - f_{Q_N}| \leq N 2^n \|f\|_{BMO_d}.$$

This concludes the proof. $\square$

Deeper properties of *BMO* area also shared by BMO_d.

6.3 Dyadic Maximal Functions and Dyadic BMO

Theorem 6.3.8. *For all $f \in BMO_d(\mathbf{R}^n)$, for all $Q \in \mathscr{D}$, and all $\alpha > 0$ we have*

$$\left|\{x \in Q: |f(x) - f_Q| > \alpha \|f\|_{BMO_d}\}\right| \leq e|Q|e^{-\frac{\alpha}{2^n e}}. \tag{6.3.7}$$

Proof. The proof of Theorem 6.2.1 in the previous section starts with a cube Q^0 (of generation 0) and constructs several sequences of cubes Q_j^k of generation k by subdividing the unselected cubes of the previous generation into 2^n pieces. If the original cube Q^0 was dyadic, then so would be all of its descendants; hence in the proof of Theorem 6.2.1 we can replace all cubes Q_j^k that appear with dyadic cubes. Then we obtain (6.3.7) just as we did in the non-dyadic case. □

Similar straightforward adaptations provide the following corollaries:

Corollary 6.3.9. *Every BMO_d function is exponentially integrable in the following sense. For any $0 < \gamma < (2^n e)^{-1}$, for all $f \in BMO_d(\mathbf{R}^n)$, and any $Q \in \mathscr{D}$ we have*

$$\frac{1}{|Q|} \int_Q e^{\gamma \frac{|f(x) - f_Q|}{\|f\|_{BMO_d}}} dx \leq 1 + \frac{2^n e^2 \gamma}{1 - 2^n e \gamma}. \tag{6.3.8}$$

Corollary 6.3.10. *Let $f \in BMO_d(\mathbf{R}^n)$. The for any compact subset K of $\mathbf{R}^n$ we have*

$$\int_K e^{c|f(x)|} dx < \infty, \tag{6.3.9}$$

provided $c < (2^n e \|f\|_{BMO_d})^{-1}$.

Proof. To verify (6.3.9) we simply cover a compact set by a finite sum of dyadic cubes and we apply (6.3.8) on each such cube. □

Corollary 6.3.11. *For all $0 < p < \infty$ and for all $f \in BMO_d$ we have*

$$\sup_{Q \in \mathscr{D}} \left(\frac{1}{|Q|} \int_Q |f(x) - f_Q|^p dx\right)^{\frac{1}{p}} \leq e \, 2^n (ep\Gamma(p))^{\frac{1}{p}} \|f\|_{BMO_d(\mathbf{R}^n)}. \tag{6.3.10}$$

Corollary 6.3.12. *For all $1 \leq p < \infty$ and f in $L^1_{\text{loc}}(\mathbf{R}^n)$ we have*

$$\|f\|_{BMO_d} \leq \sup_{Q \in \mathscr{D}} \left(\frac{1}{|Q|} \int_Q |f(x) - f_Q|^p dx\right)^{\frac{1}{p}} \leq e^2 \cdot 2^n p \|f\|_{BMO_d}. \tag{6.3.11}$$

Exercises

6.3.1. Write $\mathbf{R}^n$ as a union of orthants[2] $H_\omega = \{(x_1, \ldots, x_n) : \xi_j \geq 0 \iff j \in \omega\}$, indexed by subsets ω of $\{1, 2 \ldots, n\}$. Show that two dyadic cubes have a common ancestor if only if they lie in the same orthant H_ω.

---

[2] Higher-dimensional octants.

6.3.2. Prove that every cube of length L in $\mathbf{R}^n$ is contained in the union of 3^n dyadic cubes, each having length less than L.

6.3.3. Let $k \in \mathbf{Z}$. Given a cube Q in $\mathbf{R}^n$ of side length L satisfying $2^{k-1} \leq L < 2^k$ (if Q is open, we could assume that $2^{k-1} \leq L \leq 2^k$), prove that there is a dyadic cube D_Q of side length 2^k and a $\sigma = (\sigma_1, \ldots, \sigma_n)$, where $\sigma_j \in \{0, 1/2, -1/2\}$ such that such that $Q \subseteq 2^k \sigma + D_Q$.

6.3.4. Prove that $\|M_d\|_{L^1(\mathbf{R}^n) \to L^{1,\infty}(\mathbf{R}^n)} = 1$.

6.3.5. Show that $BMO_d(\mathbf{R}^n)$ is a complete.

6.3.6. Prove that the function $\log|x| \chi_{x_1>0} \cdots \chi_{x_n>0}$ lies in $BMO_d(\mathbf{R}^n)$ but not in $BMO(\mathbf{R}^n)$.

6.3.7. Prove that the function $\big|\log|x|\big|^p \log\big(|\log|x|| \chi_{x>0} + 1\big) \chi_{x>0}$ lies in $BMO_d(\mathbf{R})$ when $0 < p < 1$. [*Hint:* Use Exercise 6.1.8.]

6.3.8. Given a locally integrable function f on $\mathbf{R}^n$, consider the *dyadic average operator* $E_k(f)(x) = \sum_{Q \in \mathscr{D}_k} f_Q \chi_Q(x)$ and the *martingale maximal function* $G(f) = \sup_{k \in \mathbf{Z}} |E_k(f)|$. Prove the following assertions:

(a) G is of weak type $(1,1)$ with constant at most 1.
(b) $E_N(f) \to f$ a.e. as $N \to \infty$ for all $f \in L^1_{\text{loc}}(\mathbf{R}^n)$. (Relate this to Exercise 1.5.5.)
(c) $E_N(f) \to 0$ as $N \to -\infty$ for all $f \in L^p(\mathbf{R}^n)$ with $1 \leq p < \infty$.

6.4 The Sharp Maximal Function

Recall the Hardy–Littlewood maximal function with respect to cubes:

$$M_c(f)(x) = \sup_{Q \ni x} \frac{1}{|Q|} \int_Q |f(y)|\, dy$$

defined for $f \in L^1_{\text{loc}}(\mathbf{R}^n)$. In this section we introduce a related maximal operator that controls the mean oscillation of a function near any point.

Definition 6.4.1. Given a locally integrable function f on $\mathbf{R}^n$, we define its *sharp maximal function* $M_c^\#(f)$ as

$$M_c^\#(f)(x) = \sup_{Q \ni x} \frac{1}{|Q|} \int_Q |f(t) - f_Q|\, dt, \tag{6.4.1}$$

where the supremum is taken over all cubes Q in $\mathbf{R}^n$ that contain the given point x.

The sharp maximal function is also related to the space BMO. In fact,

$$BMO(\mathbf{R}^n) = \{f \in L^1_{\text{loc}}(\mathbf{R}^n) : M_c^\#(f) \in L^\infty(\mathbf{R}^n)\}$$

and $\|f\|_{BMO} = \|M_c^\#(f)\|_{L^\infty}$.

6.4 The Sharp Maximal Function

There is also a dyadic version of the sharp maximal function.

Definition 6.4.2. The *dyadic sharp maximal function* is defined as

$$M_d^\#(f)(x) = \sup_{Q \in \mathscr{D}, Q \ni x} \frac{1}{|Q|} \int_Q |f(t) - f_Q|\, dt, \qquad (6.4.2)$$

where $\mathscr{D}$ is the set of all dyadic cubes in $\mathbf{R}^n$.

Obviously, the $M_d^\#$ is pointwise smaller than $M_c^\#$. Also, in analogy with the non-dyadic case we have $BMO_d(\mathbf{R}^n) = \{f \in L^1_{\text{loc}}(\mathbf{R}^n) : M_d^\#(f) \in L^\infty(\mathbf{R}^n)\}$ and $\|f\|_{BMO_d} = \|M_d^\#(f)\|_{L^\infty}$.

Proposition 6.4.3. *Let $f \in L^1_{\text{loc}}(\mathbf{R}^n)$. Suppose that there exists an $A > 0$ such that for all cubes Q (resp., dyadic cubes) that contain a fixed x in $\mathbf{R}^n$ there exists a constant c_Q such that*

$$\frac{1}{|Q|}\int_Q |f(y) - c_Q|\, dy \le A.$$

Then $M_c^\#(f)(x) \le 2A$ (resp., $M_d^\#(f)(x) \le 2A$).

Proof. Fix $x \in \mathbf{R}^n$. For a cube Q (resp., dyadic cube) that contains x we have

$$|f - f_Q|_Q \le |f - c_Q|_Q + |f_Q - c_Q|_Q \le |f - c_Q|_Q + \frac{1}{|Q|}\int_Q |f(x) - c_Q|\, dx \le 2A.$$

Taking the supremum over all cubes Q (resp., dyadic cubes) that contain x, we obtain $M_c^\#(f)(x) \le 2A$ [resp., $M_d^\#(f)(x) \le 2A$]. $\square$

We discuss some properties of $M_c^\#$, $M_d^\#$ and their connections with M_c, M_d.

Proposition 6.4.4. *Let f, g be locally integrable functions on $\mathbf{R}^n$. Then*

(1) We have $M_c^\#(f+g) \le M_c^\#(f) + M_c^\#(g)$ and $M_d^\#(f+g) \le M_d^\#(f) + M_d^\#(g)$.
(2) $M_c^\#(f) \le 2M_c(f)$ and $M_d^\#(f) \le 2M_d(f)$.
(3) $M_c^\#(|f|) \le 2M_c^\#(f)$ and $M_d^\#(|f|) \le 2M_d^\#(f)$.

Proof. We skip the proof of (1) as it is straightforward. The proof of (2) follows directly from (6.4.1) and (6.4.2) using the inequality $|f - f_Q| \le |f| + |f|_Q$. To prove (3) for each cube Q (resp., dyadic cube) that contains a given x in $\mathbf{R}^n$ we let $c_Q = |f_Q|$. Then the inequality $||f| - |f_Q|| \le |f - f_Q|$ implies that

$$\frac{1}{|Q|}\int_Q ||f(y)| - |f_Q||\, dy \le M_c^\#(f)(x)$$

[or analogously with $M_d^\#(f)(x)$ on the right if the cube Q is dyadic]. It follows from Proposition 6.4.3 that $M_c^\#(|f|) \le 2M_c^\#(f)$ [resp., $M_d^\#(|f|) \le 2M_d^\#(f)$]. $\square$

The next result lays the foundation of norm comparability between M_d and $M_d^\#$.

Theorem 6.4.5. *(A good lambda distributional inequality for M_d)* For all $\gamma > 0$, all $\lambda > 0$, and all locally integrable functions f on $\mathbf{R}^n$, we have the estimate

$$\left|\{x \in \mathbf{R}^n : M_d(f)(x) > 2\lambda, M_d^\#(f)(x) \leq \gamma\lambda\}\right| \leq 2^n \gamma \left|\{x \in \mathbf{R}^n : M_d(f)(x) > \lambda\}\right|.$$

Proof. We may suppose that the set $\Omega_\lambda = \{x \in \mathbf{R}^n : M_d(f)(x) > \lambda\}$ has finite measure; otherwise, there is nothing to prove. Then for each $x \in \Omega_\lambda$ there is a maximal (with respect to inclusion) dyadic cube Q^x containing x such that

$$\frac{1}{|Q^x|} \int_{Q^x} |f(y)|\,dy > \lambda; \qquad (6.4.3)$$

otherwise, Ω_λ would have infinite measure. Notice that the entire Q^x is contained in Ω_λ, since given $z \in Q_x$ the average (6.4.3) shows that $z \in \Omega_\lambda$. Let Q_j be the collection of all such maximal dyadic cubes containing all x in Ω_λ, i.e., $\{Q_j\}_j = \{Q^x : x \in \Omega_\lambda\}$. Maximal dyadic cubes are disjoint; hence any two different Q_j are disjoint. Moreover, we note that if $x, y \in Q_j$, then $Q_j = Q^x = Q^y$. It follows that $\Omega_\lambda = \bigcup_j Q_j$ and also this set contains

$$\{x \in \mathbf{R}^n : M_d(f)(x) > 2\lambda, M_d^\#(f)(x) \leq \gamma\lambda\}.$$

Thus, in order to prove the required estimate, it will suffice to show that for all Q_j we have

$$\left|\{x \in Q_j : M_d(f)(x) > 2\lambda, M_d^\#(f)(x) \leq \gamma\lambda\}\right| \leq 2^n \gamma |Q_j|, \qquad (6.4.4)$$

for once (6.4.4) is established, the conclusion follows by summing on j.

For each cube Q_j, we let Q_j' be the unique dyadic parent of Q_j, i.e., the unique ancestor with double side length. We will show that for each j we have

$$\{x \in Q_j : M_d(f)(x) > 2\lambda\} \subseteq \{x \in Q_j : M_d((f - f_{Q_j'})\chi_{Q_j})(x) > \lambda\}. \qquad (6.4.5)$$

We fix $x \in Q_j$ such that $M_d(f)(x) > 2\lambda$. Then the supremum

$$\sup_{R \ni x} \frac{1}{|R|} \int_R |f(y)|\,dy = M_d(f)(x) \qquad (6.4.6)$$

is taken over all dyadic cubes R that either contain Q_j or are contained in Q_j (since $Q_j \cap R \neq \emptyset$). If $R \supsetneq Q_j$, the maximality of Q_j implies that (6.4.3) does not hold for R; thus the average of $|f|$ over R is at most λ. Thus, if $M_d(f)(x) > 2\lambda$, then the average in (6.4.6) is bigger than 2λ for some dyadic cube R contained (not properly) in Q_j. Therefore, if $x \in Q_j$ and $M_d(f)(x) > 2\lambda$, then we can replace f by $f\chi_{Q_j}$ in (6.4.6) and we must have $M_d(f\chi_{Q_j})(x) > 2\lambda$. Then for $x \in Q_j$ we have

$$M_d\big((f - f_{Q_j'})\chi_{Q_j}\big)(x) \geq M_d(f\chi_{Q_j})(x) - |f_{Q_j'}| > 2\lambda - \lambda = \lambda,$$

6.4 The Sharp Maximal Function

since $|f_{Q'_j}| \leq |f|_{Q'_j} \leq \lambda$ because of the maximality of Q_j. This proves (6.4.5), hence

$$\left|\{x \in Q_j : M_d(f)(x) > 2\lambda\}\right| \leq \left|\{x \in Q_j : M_d((f - f_{Q'_j})\chi_{Q_j})(x) > \lambda\}\right|, \quad (6.4.7)$$

and using the fact that M_d is of weak type $(1,1)$ with constant 1, by Theorem 6.3.3 (b), we control the last expression in (6.4.7) by

$$\begin{aligned}
\frac{1}{\lambda}\int_{Q_j}|f(y) - f_{Q'_j}|\,dy &= \frac{1}{\lambda}\int_{\mathbf{R}^n}|f(y) - f_{Q'_j}|\chi_{Q_j}\,dy \\
&\leq \frac{2^n|Q_j|}{\lambda}\frac{1}{|Q'_j|}\int_{Q'_j}|f(y) - f_{Q'_j}|\,dy \quad (6.4.8) \\
&\leq \frac{2^n|Q_j|}{\lambda}M_d^\#(f)(\xi_j)
\end{aligned}$$

for all $\xi_j \in Q_j$. In proving (6.4.4) we may assume that for some $\xi_j \in Q_j$ we have $M_d^\#(f)(\xi_j) \leq \gamma\lambda$; otherwise, the set on the left in (6.4.4) is empty and has zero measure. For this ξ_j, using (6.4.7) and (6.4.8) we obtain (6.4.4). $\square$

We now use the distributional inequality of Theorem 6.4.5 to deduce that several norms of $M_d(f)$ are controlled by the corresponding norms of $M_d^\#(f)$.

Theorem 6.4.6. *Fix $0 < p_0 < \infty$ and let p satisfy $p_0 < p < \infty$. Then for all functions f in $L^1_{\text{loc}}(\mathbf{R}^n)$ with the property*

$$\text{for every } B > 0 \implies C_B(f) = \sup_{0 < \lambda \leq B} \lambda^{p_0}\left|\{M_d(f) > \lambda\}\right| < \infty, \quad (6.4.9)$$

we have

$$\left\|M_d(f)\right\|_{L^p(\mathbf{R}^n)} \leq 2^{n+p+2+\frac{1}{p}}\left\|M_d^\#(f)\right\|_{L^p(\mathbf{R}^n)} \quad (6.4.10)$$

and also

$$\left\|M_d(f)\right\|_{L^{p,\infty}(\mathbf{R}^n)} \leq 2^{n+p+2+\frac{1}{p}}\left\|M_d^\#(f)\right\|_{L^{p,\infty}(\mathbf{R}^n)}. \quad (6.4.11)$$

Proof. Fix $p_0 < p < \infty$. For a positive real number N we set

$$I_N = \int_0^N p\lambda^{p-1}\left|\{x \in \mathbf{R}^n : M_d(f)(x) > \lambda\}\right|d\lambda.$$

We note that I_N is finite, as it is bounded by

$$\int_0^N p\lambda^{p-p_0-1}C_N(f)\,d\lambda = \frac{pN^{p-p_0}}{p-p_0}C_N(f) < \infty,$$

where $C_N(f)$ is defined in (6.4.9). We now write

$$I_N = 2^p\int_0^{\frac{N}{2}} p\lambda^{p-1}\left|\{x \in \mathbf{R}^n : M_d(f)(x) > 2\lambda\}\right|d\lambda$$

and we use Theorem 6.4.5 to obtain the following sequence of inequalities:

$$
\begin{aligned}
I_N &\leq 2^p \int_0^{\frac{N}{2}} p\lambda^{p-1} \big|\{x \in \mathbf{R}^n : M_d(f)(x) > 2\lambda,\, M_d^\#(f)(x) \leq \gamma\lambda\}\big| \, d\lambda \\
&\quad + 2^p \int_0^{\frac{N}{2}} p\lambda^{p-1} \big|\{x \in \mathbf{R}^n : M_d^\#(f)(x) > \gamma\lambda\}\big| \, d\lambda \\
&\leq 2^p 2^n \gamma \int_0^{\frac{N}{2}} p\lambda^{p-1} \big|\{x \in \mathbf{R}^n : M_d(f)(x) > \lambda\}\big| \, d\lambda \\
&\quad + 2^p \int_0^{\frac{N}{2}} p\lambda^{p-1} \big|\{x \in \mathbf{R}^n : M_d^\#(f)(x) > \gamma\lambda\}\big| \, d\lambda \\
&\leq 2^p 2^n \gamma I_N + \frac{2^p}{\gamma^p} \int_0^{\frac{N\gamma}{2}} p\lambda^{p-1} \big|\{x \in \mathbf{R}^n : M_d^\#(f)(x) > \lambda\}\big| \, d\lambda.
\end{aligned}
$$

At this point we pick a γ such that $2^p 2^n \gamma = 1/2$. Since I_N is finite, we can subtract from both sides of the inequality the quantity $\frac{1}{2} I_N$ to obtain

$$
I_N \leq 2^{p+1} 2^{p(n+p+1)} \int_0^{\frac{N\gamma}{2}} p\lambda^{p-1} \big|\{x \in \mathbf{R}^n : M_d^\#(f)(x) > \lambda\}\big| \, d\lambda,
$$

from which we obtain (6.4.10) by letting $N \to \infty$.

Using Theorem 6.4.5 again we write

$$
\begin{aligned}
&\big|\{x \in \mathbf{R}^n : M_d(f)(x) > 2\lambda\}\big| \\
&\leq \big|\{x \in \mathbf{R}^n : M_d(f)(x) > 2\lambda,\, M_d^\#(f)(x) \leq \gamma\lambda\}\big| + \big|\{x \in \mathbf{R}^n : M_d^\#(f)(x) > \gamma\lambda\}\big| \\
&\leq 2^n \gamma \big|\{x \in \mathbf{R}^n : M_d(f)(x) > \lambda\}\big| + \big|\{x \in \mathbf{R}^n : M_d^\#(f)(x) > \gamma\lambda\}\big|.
\end{aligned}
$$

Multiplying by λ^p and taking the supremum over $\lambda \leq N$, we obtain

$$
\sup_{0 < \lambda \leq N} \lambda^p \big|\{x \in \mathbf{R}^n : M_d(f)(x) > 2\lambda\}\big|
$$
$$
\leq 2^n \gamma \sup_{0 < \lambda \leq N} \lambda^p \big|\{x \in \mathbf{R}^n : M_d(f)(x) > \lambda\}\big| + \frac{1}{\gamma^p} \|M_d^\#(f)\|_{L^{p,\infty}}^p,
$$

or equivalently,

$$
2^{-p} \sup_{0 < t \leq 2N} t^p \big|\{x \in \mathbf{R}^n : M_d(f)(x) > t\}\big|
$$
$$
\leq 2^n \gamma \sup_{0 < \lambda \leq N} \lambda^p \big|\{x \in \mathbf{R}^n : M_d(f)(x) > \lambda\}\big| + \frac{1}{\gamma^p} \|M_d^\#(f)\|_{L^{p,\infty}}^p
$$
$$
\leq 2^n \gamma \sup_{0 < \lambda \leq 2N} \lambda^p \big|\{x \in \mathbf{R}^n : M_d(f)(x) > \lambda\}\big| + \frac{1}{\gamma^p} \|M_d^\#(f)\|_{L^{p,\infty}}^p.
$$

Notice that (6.4.9) implies that

6.4 The Sharp Maximal Function

$$\sup_{0<\lambda\leq 2N} \lambda^p \big|\{x\in \mathbf{R}^n : M_d(f)(x) > \lambda\}\big| \leq (2N)^{p-p_0} C_{2N}(f) < \infty,$$

where $C_N(f)$ was defined in the proof of (6.4.10). Choosing γ such that $2^n\gamma = \tfrac{1}{2}2^{-p}$, we obtain

$$\frac{1}{2}2^{-p} \sup_{0<t\leq 2N} t^p \big|\{x\in \mathbf{R}^n : M_d(f)(x) > t\}\big| \leq 2^{p(n+p+1)} \|M_d^\#(f)\|_{L^{p,\infty}}^p.$$

Letting $N \to \infty$, we deduce (6.4.11). $\square$

Remark 6.4.7. Let $1 < p_0 < \infty$. Functions in $L^{p_0,\infty}$ satisfy condition (6.4.9), so Theorem 6.4.6 applies to functions f in $L^{p_0,\infty}$.

Exercises

6.4.1. Construct examples to show that M_d and M_c are not pointwise comparable.

6.4.2. Prove that for all cubes Q in $\mathbf{R}^n$ and all points x we have

$$\frac{1}{2} M_c^\#(f)(x) \leq \sup_{Q\ni x} \inf_{a\in \mathbf{C}} \frac{1}{|Q|} \int_Q |f(y) - a|\, dy \leq M_c^\#(f)(x).$$

6.4.3. For $f \in L^1_{\mathrm{loc}}(\mathbf{R}^n)$ define *the sharp maximal function with respect to balls*

$$M^\#(f)(x) = \sup_B \frac{1}{|B|} \int_B |f(y) - f_B|\, dy,$$

where the supremum is taken over all balls that contain a given point x. Prove a version of Proposition 6.4.3 for $M^\#$. Deduce that $M^\#$ is pointwise equivalent to $M_c^\#$.

6.4.4. Let $f \in L^1_{\mathrm{loc}}(\mathbf{R}^n)$ and $x \in \mathbf{R}^n$. Prove that

$$M_c^\#(f)(x) \leq \sup_{Q\ni x} \frac{1}{|Q|^2} \int_Q \int_Q |f(y) - f(z)|\, dy\, dz \leq 2 M_c^\#(f)(x),$$

where the supremum is taken over all cubes Q that contain x. Obtain that

$$\|f\|_{BMO} \approx \sup_Q \frac{1}{|Q|^2} \int_Q \int_Q |f(y) - f(z)|\, dy\, dz.$$

6.4.5. Let $0 < s < n$. Show that the Riesz potential $\mathcal{I}_s$ maps $L^{n/s}(\mathbf{R}^n)$ to BMO. [*Hint:* Show that for any $R > 0$ and $x \in \mathbf{R}^n$ we have

$$\frac{1}{|B(0,R)|^2} \int_{B(0,R)} \int_{B(0,R)} \left| \frac{1}{|x-y|^{n-s}} - \frac{1}{|x-z|^{n-s}} \right| dy\, dz \leq \frac{CR}{(|x|+R)^{n-s+1}}$$

by considering the cases $|x| \leq 2R$ and $|x| > 2R$. Use this estimate to prove

$$\frac{1}{|B(0,R)|^2} \int_{B(0,R)} \int_{B(0,R)} |\mathcal{I}_s(f)(y) - \mathcal{I}_s(f)(z)| \, dy \, dz \leq C' \|f\|_{L^{n/s}(\mathbf{R}^n)}$$

by Hölder's inequality, where C' is independent of f and R. By translation, this estimate is valid over any ball and thus over any cube. Then use Exercise 6.4.4.]

6.4.6. Suppose $f_k, f \in L^1_{\text{loc}}(\mathbf{R}^n)$, and $|f_k| \leq C|f|$ a.e. for $k = 1, 2, \ldots$, where C is a positive constant. Suppose that $f_k \to f$ a.e. as $k \to \infty$. Show that

$$M_c^\#(f) \leq \liminf_{k \to \infty} M_c^\#(f_k).$$

6.4.7. (a) Let $f, f_k, k = 1, 2, \ldots$ be functions in $L^\infty(\mathbf{R}^n)$ such that $\|f_k - f\|_{L^\infty} \to 0$ as $k \to \infty$. Show that $M_c^\#(f_k) \to M_c^\#(f)$ pointwise everywhere.
(b) Let $f, f_k, k = 1, 2, \ldots$, be functions in $L^1(\mathbf{R}^n)$. Suppose that $f_k \to f$ in $L^1(\mathbf{R}^n)$ and the sequence $\{f_k\}_{k=1}^\infty$ is *equicontinuous* in L^1 in the following sense: For every $\varepsilon > 0$ there is a $\delta > 0$ such that

$$|A| < \delta \implies \int_A |f_k| \, dy < \varepsilon \qquad \text{for all } k.$$

Prove that $M_c^\#(f_k) \to M_c^\#(f)$ pointwise everywhere.

6.5 Interpolation Using BMO

In this section we discuss an interpolation theorem in which the space L^∞ is replaced by *BMO*. The sharp function plays a key role in this result. Before doing so, we state a corollary of Theorem 6.4.6 that is quite useful in this type of interpolation.

Proposition 6.5.1. *Let $0 < p_0 < \infty$. Then for any p with $p_0 < p < \infty$ and for all f in $L^1_{\text{loc}}(\mathbf{R}^n)$ satisfying (6.4.9) [in particular if $M_d(f) \in L^{p_0,\infty}$] we have*

$$\begin{aligned}
\|f\|_{L^p(\mathbf{R}^n)} &\leq \|M_d(f)\|_{L^p(\mathbf{R}^n)} \\
&\leq 2^{n+p+2+\frac{1}{p}} \|M_d^\#(f)\|_{L^p(\mathbf{R}^n)} \\
&\leq 2^{n+p+2+\frac{1}{p}} \|M_c^\#(f)\|_{L^p(\mathbf{R}^n)} \\
&\leq 2^{n+p+3+\frac{1}{p}} \|M_c(f)\|_{L^p(\mathbf{R}^n)}.
\end{aligned} \qquad (6.5.1)$$

Analogously, we have

6.5 Interpolation Using BMO

$$\begin{aligned}\|f\|_{L^{p,\infty}(\mathbf{R}^n)} &\leq \|M_d(f)\|_{L^{p,\infty}(\mathbf{R}^n)} \\ &\leq 2^{n+p+2+\frac{1}{p}}\|M_d^{\#}(f)\|_{L^{p,\infty}(\mathbf{R}^n)} \\ &\leq 2^{n+p+2+\frac{1}{p}}\|M_c^{\#}(f)\|_{L^{p,\infty}(\mathbf{R}^n)} \\ &\leq 2^{n+p+3+\frac{1}{p}}\|M_c(f)\|_{L^{p,\infty}(\mathbf{R}^n)}.\end{aligned} \qquad (6.5.2)$$

If in addition $p > 1$, then all of the above inequalities become equivalences.

Proof. Since for every point in $\mathbf{R}^n$ there is a sequence of dyadic cubes shrinking to it, the Lebesgue differentiation theorem yields that for almost every point x in $\mathbf{R}^n$ the averages of the locally integrable function f over the dyadic cubes containing x converge to $f(x)$. Consequently, $|f| \leq M_d(f)$ a.e., hence the first inequalities in (6.5.1) and (6.5.2) hold. Theorem 6.4.6 provides the second inequalities in both (6.5.1) and (6.5.2). The third inequalities are trivial, while the last inequalities are consequences of Proposition 6.4.4 (2).

Finally, if $p > 1$, then (1.4.7) and Exercise 1.4.8 provide the missing estimates that reverse all inequalities in (6.5.1) and (6.5.2), respectively. □

For the purposes of the next result we set L^{∞}_{fin} to be the space of all L^{∞} functions that are supported in a set of finite measure.

Theorem 6.5.2. *Let $1 \leq p_0 < \infty$. Let T be a linear operator that maps $L^{p_0}(\mathbf{R}^n)$ to $L^{p_0,\infty}(\mathbf{R}^n)$ with bound A_0 and $L^{\infty}_{\text{fin}}(\mathbf{R}^n)$ to $BMO(\mathbf{R}^n)$ with bound A_1. Then for all p with $p_0 < p < \infty$ there is an extension of T on $L^p(\mathbf{R}^n)$ and a constant C_{n,p,p_0} such that for all $f \in L^p(\mathbf{R}^n)$ we have*

$$\|T(f)\|_{L^p(\mathbf{R}^n)} \leq C_{n,p,p_0} A_0^{\frac{p_0}{p}} A_1^{1-\frac{p_0}{p}} \|f\|_{L^p(\mathbf{R}^n)}. \qquad (6.5.3)$$

Proof. We consider two cases according to the value of p_0.

Case I: $p_0 > 1$. Define the operator $S(f) = M_c^{\#}(T(f))$ for $f \in L^{p_0} + L^{\infty}_{\text{fin}}$. It is easy to see that S is a subadditive operator. We first prove that S maps L^{∞}_{fin} to L^{∞}. Indeed, for $f \in L^{\infty}_{\text{fin}}$ we have

$$\|S(f)\|_{L^{\infty}} = \|M_c^{\#}(T(f))\|_{L^{\infty}} = \|T(f)\|_{BMO} \leq A_1 \|f\|_{L^{\infty}}.$$

Next we show that S maps L^{p_0} to $L^{p_0,\infty}$. For $f \in L^{p_0}$ we have

$$\begin{aligned}\|S(f)\|_{L^{p_0,\infty}} &= \|M_c^{\#}(T(f))\|_{L^{p_0,\infty}} \\ &\leq 2\|M_c(T(f))\|_{L^{p_0,\infty}} \\ &\leq 2\frac{3^n p_0}{p_0-1}\|T(f)\|_{L^{p_0,\infty}} \qquad \text{(by Exercise 1.4.8)} \\ &\leq 2\frac{3^n p_0}{p_0-1} A_0 \|f\|_{L^{p_0}}.\end{aligned}$$

Interpolating between these estimates using Theorem 1.3.4, with $\mathscr{F}$ being the space of finitely simple functions, we deduce

$$\left\|M_c^\#(T(f))\right\|_{L^p} = \left\|S(f)\right\|_{L^p} \leq 2\left(\frac{p}{p-p_0}\right)^{\frac{1}{p}}\left(2\frac{3^n p_0}{p_0-1}A_0\right)^{\frac{p_0}{p}} A_1^{1-\frac{p_0}{p}} \left\|f\right\|_{L^p}$$

for all $f \in \mathscr{F}$, where $p_0 < p < \infty$. Consider now a function h in $\mathscr{F}$. Then $h \in L^{p_0}(\mathbf{R}^n)$ and by assumption $T(h) \in L^{p_0,\infty}(\mathbf{R}^n)$. This gives that $M_d(T(h)) \in L^{p_0,\infty}(\mathbf{R}^n)$ by Theorem 6.3.3 (d); then Proposition 6.5.1 [in particular (6.5.1)] is applicable and gives

$$\left\|T(h)\right\|_{L^p} \leq 2^{n+p+2+\frac{1}{p}} \left\|M_c^\#(T(h))\right\|_{L^p} \leq C_{n,p,p_0} A_0^{\frac{p_0}{p}} A_1^{1-\frac{p_0}{p}} \left\|h\right\|_{L^p}.$$

From this and the density of $\mathscr{F}$ in L^p we derive (6.5.3) for all $f \in L^p(\mathbf{R}^n)$.

Case II: $p_0 = 1$. We fix a $\delta \in (0,1)$ and we define the following version of S

$$S_\delta(f) = M_c^\#(|T(f)|^\delta)^{\frac{1}{\delta}}$$

on $L^1 + L^\infty$. We prove that S_δ maps L_{fin}^∞ to L^∞. Indeed, for $f \in L_{\text{fin}}^\infty$ one has

$$\begin{aligned}
\left\|S_\delta(f)\right\|_{L^\infty} &= \left\|M_c^\#(|T(f)|^\delta)^{\frac{1}{\delta}}\right\|_{L^\infty} \\
&= \left\|M_c^\#(|T(f)|^\delta)\right\|_{L^\infty}^{\frac{1}{\delta}} \\
&= \left\||T(f)|^\delta\right\|_{BMO}^{\frac{1}{\delta}} \\
&\leq 2^{\frac{1}{\delta}} \left\|T(f)\right\|_{BMO} \qquad \text{(by Exercise 6.1.7)} \\
&\leq 2^{\frac{1}{\delta}} A_1 \left\|f\right\|_{L^\infty}.
\end{aligned}$$

At the other endpoint for $f \in L^1$ we write

$$\begin{aligned}
\left\|S_\delta(f)\right\|_{L^{1,\infty}} &= \left\|M_c^\#(|T(f)|^\delta)^{\frac{1}{\delta}}\right\|_{L^{1,\infty}} \\
&= \left\|M_c^\#(|T(f)|^\delta)\right\|_{L^{\frac{1}{\delta},\infty}}^{\frac{1}{\delta}} \\
&\leq \left(\frac{3^{n\delta}\frac{2}{\delta}}{\frac{1}{\delta}-1}\right)^{\frac{1}{\delta}} \left\||T(f)|^\delta\right\|_{L^{\frac{1}{\delta},\infty}}^{\frac{1}{\delta}} \qquad \text{[by (1.4.7)]} \\
&= \left(\frac{3^{n\delta}2}{1-\delta}\right)^{\frac{1}{\delta}} \left\|T(f)\right\|_{L^{1,\infty}} \\
&\leq \left(\frac{3^{n\delta}2}{1-\delta}\right)^{\frac{1}{\delta}} A_0 \left\|f\right\|_{L^1}.
\end{aligned}$$

We interpolate between the estimates $S_\delta : L_{\text{fin}}^\infty \to L^\infty$ and $S_\delta : L^1 \to L^{1,\infty}$, applying Theorem 1.3.4 with $\mathscr{F}$, the space of all finitely simple functions on $\mathbf{R}^n$. We deduce that for all $1 < p < \infty$ and all $f \in \mathscr{F}$ one has

6.5 Interpolation Using BMO

$$\left\|M_c^\#(|T(f)|^\delta)\right\|_{L^{p/\delta}}^{\frac{1}{\delta}} = \left\|S_\delta(f)\right\|_{L^p}$$

$$\leq 2\left(\frac{p}{p-1}\right)^{\frac{1}{p}} \left[\left(\frac{3^n\delta 2}{1-\delta}\right)^{\frac{1}{\delta}} A_0\right]^{\frac{1}{p}} \left(2^{\frac{1}{\delta}}A_1\right)^{1-\frac{1}{p}} \|f\|_{L^p}. \quad (6.5.4)$$

In order to prove (6.5.3) we fix p with $1 < p < \infty$ and consider a function h in $\mathscr{F}$. Then $h \in L^1(\mathbf{R}^n)$ and thus by assumption $T(h)$ lies in $L^{1,\infty}(\mathbf{R}^n)$. It follows that $|T(h)|^\delta$ lies in $L^{1/\delta,\infty}(\mathbf{R}^n)$ hence $M_d(|T(h)|^\delta) \in L^{1/\delta,\infty}(\mathbf{R}^n)$ by Theorem 6.3.3 (d). Finally Proposition 6.5.1 applies (with $p_0 = 1/\delta$) and gives

$$\|T(h)\|_{L^p} = \left(\||T(h)|^\delta\|_{L^{p/\delta}}\right)^{\frac{1}{\delta}} \leq \left(2^{n+\frac{p}{\delta}+2+\frac{\delta}{p}} \|M_c^\#(|T(h)|^\delta)\|_{L^{p/\delta}}\right)^{\frac{1}{\delta}}. \quad (6.5.5)$$

Inserting the estimate in (6.5.4) to (6.5.5) we obtain (6.5.3) with h in place of f, where $C_{n,p,1}$ also depends on δ (but we could take $\delta = 1/2$). Finally, by the density of $\mathscr{F}$ in $L^p(\mathbf{R}^n)$ we deduce (6.5.3) for all $f \in L^p(\mathbf{R}^n)$. $\square$

Remark 6.5.3. Theorem 6.5.2 is also valid if *BMO* is replaced by the bigger space BMO_d. In fact one replaces $M_c^\#$ by $M_d^\#$ and uses a version of Exercise 6.1.7 with BMO_d in place of *BMO*.

Next, we turn to an application of Theorem 6.5.2.

Theorem 6.5.4. *Let K be a locally integrable function on $\mathbf{R}^n \setminus \{0\}$ which satisfies (3.3.3) and (3.3.4) (with constant A_2), and associated with K there is a $W \in \mathscr{S}'(\mathbf{R}^n)$ and a sequence $\delta_k \downarrow 0$ as in (3.3.7). Assume that the operator T given by convolution with W admits a bounded extension from L^2 to L^2. Then for all $f \in L^\infty_{\text{fin}}$ we have*

$$\|T(f)\|_{BMO} \leq 2\big(A_2 + (2\sqrt{n})^{\frac{n}{2}} \|T\|_{L^2 \to L^2}\big) \|f\|_{L^\infty}. \quad (6.5.6)$$

Proof. Let us fix a cube Q centered at a point c_Q and let Q^* be the cube with the same center and with sides parallel to those of Q and side length $\ell(Q^*) = 2\sqrt{n}\ell(Q)$, where $\ell(Q)$ is the side length of Q. Given a function f bounded whose support has finite measure we split $f = f_0 + f_\infty$, where $f_0 = f\chi_{Q^*}$ and $f_\infty = f\chi_{(Q^*)^c}$. For fixed $x \in Q$ we claim that $T(f_\infty)(x)$ is well defined and is equal to

$$T(f_\infty)(x) = \lim_{\delta_k \downarrow 0} \int_{|x-y| \geq \delta_k} K(x-y) f(y) \chi_{(Q^*)^c}(y) dy = \int_{(Q^*)^c} K(x-y) f(y) dy,$$

and the reason is that the last integral is absolutely convergent; indeed, for $y \in (Q^*)^c$ and $x \in Q$ we have $|x-y| \geq \sqrt{n}\ell(Q)$, thus

$$|K(x-y)| \leq \frac{A_1}{|x-y|^n} \leq \frac{A_1}{(\sqrt{n}\ell(Q))^n}$$

and the function f is bounded and supported in a set of finite measure.

We set

$$C_Q = T(f_\infty)(c_Q).$$

Then we write

$$\frac{1}{|Q|}\int_Q |T(f)(x) - C_Q|\,dx$$
$$\leq \frac{1}{|Q|}\int_Q |T(f_0)(x)|\,dx + \frac{1}{|Q|}\int_Q |T(f_\infty)(x) - T(f_\infty)(c_Q)|\,dx$$
$$= \frac{1}{|Q|}\int_Q |T(f_0)(x)|\,dx + \frac{1}{|Q|}\int_Q \left|\int_{(Q^*)^c} [K(x-y) - K(c_Q - y)]f(y)\,dy\right|\,dx$$
$$\leq \left(\frac{1}{|Q|}\int_Q |T(f_0)(x)|^2\,dx\right)^{\frac{1}{2}} + \frac{\|f\|_{L^\infty}}{|Q|}\int_Q \int_{(Q^*)^c} |K(x-y) - K(c_Q - y)|\,dy\,dx$$
$$\leq \|T\|_{L^2 \to L^2}\left(\frac{1}{|Q|}\int_{Q^*} |f_0(x)|^2\,dx\right)^{\frac{1}{2}} + \frac{\|f\|_{L^\infty}}{|Q|}\int_Q A_2\,dx$$
$$\leq ((2\sqrt{n})^{\frac{n}{2}}\|T\|_{L^2\to L^2} + A_2)\|f\|_{L^\infty},$$

where we just need to explain why

$$\int_{(Q^*)^c} |K(x - c_Q + (c_Q - y)) - K(c_Q - y)|\,dy \leq A_2.$$

This is because

$$|c_Q - y| \geq \sqrt{n}\,\ell(Q)$$

as $y \notin Q^*$ but

$$|x - c_Q| \leq \frac{\sqrt{n}}{2}\ell(Q)$$

since $x \in Q$. Thus

$$|c_Q - y| \geq 2|x - c_Q|$$

and condition (3.3.4) applies. See Figure 6.1.

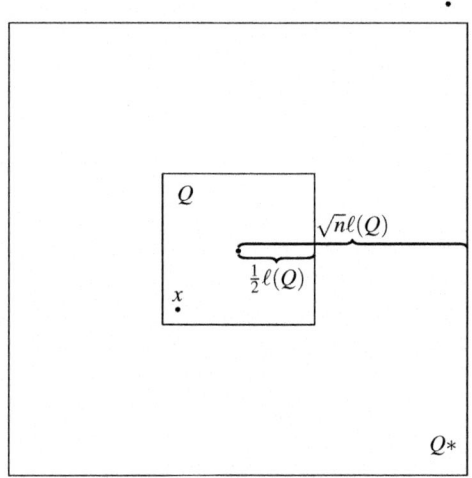

Fig. 6.1 A cube Q and the augmented cube Q^* with $\ell(Q^*) = 2\sqrt{n}\,\ell(Q)$. Here $x \in Q$ and $y \notin Q^*$.

Finally, we obtain (6.5.6) using Proposition 6.1.3. □

Theorem 6.5.4 can be used to obtain L^p bounds for the singular integrals discussed in Chapter 3 via an approach that is not based on Calderón–Zygmund theory. In fact, if T is a singular integral operator as in the statement of Theorem 6.5.4, then T maps L^2 to L^2 and L^∞_{fin} to BMO. Theorem 6.5.2 implies that T admits a

6.5 Interpolation Using BMO

bounded extension on $L^p(\mathbf{R}^n)$ for $2 < p < \infty$. By duality one obtains boundedness for $1 < p < 2$.

Exercises

6.5.1. Let $1 < q < p < \infty$ and define θ by $\frac{1}{p} = \frac{1-\theta}{q}$. Prove that there is a constant $C_{n,p,q}$ such that for all locally integrable functions f on $\mathbf{R}^n$ we have

$$\|f\|_{L^p} \leq C_{n,p,q} \|f\|_{L^{q,\infty}}^{1-\theta} \|f\|_{BMO}^{\theta}.$$

[*Hint:* Assuming $f \in L^{q,\infty}(\mathbf{R}^n)$ implies $M_d(f) \in L^{q,\infty}(\mathbf{R}^n)$ and allows the use of Proposition 6.5.1. You may also need Proposition 1.3.5.]

6.5.2. Let $\delta < p < \infty$ and $C = 2^{\frac{1}{\delta}} 3^{\frac{n}{\delta}} \left(\frac{p}{p-\delta}\right)^{\frac{1}{\delta}}$. Prove that for all functions g such that $|g|^\delta \in L^1_{\text{loc}}(\mathbf{R}^n)$ we have

$$\|M_c^\#(|g|^\delta)^{1/\delta}\|_{L^p} \leq C \|g\|_{L^p}$$

and

$$\|M_c^\#(|g|^\delta)^{1/\delta}\|_{L^{p,\infty}} \leq C \|g\|_{L^{p,\infty}}.$$

[*Hint:* Use the results of Exercises 1.3.4 and 1.4.8.]

6.5.3. Let $0 < \delta < p_0 < p < \infty$. Prove that there is a constant $C = C_{p,\delta,n,p_0}$ such that for all functions $g \in L^{p_0,\infty}(\mathbf{R}^n)$ we have

$$\|g\|_{L^p} \leq C \|M_c^\#(|g|^\delta)^{1/\delta}\|_{L^p}$$

and

$$\|g\|_{L^{p,\infty}} \leq C \|M_c^\#(|g|^\delta)^{1/\delta}\|_{L^{p,\infty}}.$$

Combine this result with the preceding exercise to deduce that for all functions $g \in L^{p_0,\infty}(\mathbf{R}^n)$ we have

$$\|g\|_{L^p} \approx \|M_c^\#(|g|^\delta)^{1/\delta}\|_{L^p} \quad \text{and} \quad \|g\|_{L^{p,\infty}} \approx \|M_c^\#(|g|^\delta)^{1/\delta}\|_{L^{p,\infty}}.$$

6.5.4. (Extension of Exercise 6.5.1) Let $0 < q < p < \infty$ and define θ by $\frac{1}{p} = \frac{1-\theta}{q}$. Prove that there is a constant $C_{n,p,q}$ such that for all $f \in L^1_{\text{loc}}(\mathbf{R}^n)$ we have

$$\|f\|_{L^p} \leq C_{n,p,q} \|f\|_{L^{q,\infty}}^{1-\theta} \|f\|_{BMO}^{\theta}.$$

[*Hint:* Use Exercises 6.5.1 and 6.5.3.]

Chapter 7
Hardy Spaces

7.1 Smoothness and Cancellation

Hardy spaces are spaces of functions or distributions that have vanishing integral or moments, i.e., objects that contain cancellation. To properly exploit cancellation we need to understand how it pairs with smoothness. We achieve this by studying the action of functions with smoothness on other ones with cancellation and assume that these functions are scaled differently.

We denote by v_n the volume of the unit ball in $\mathbf{R}^n$. An inequality that will appear in the sequel is the following:

$$\int_{\mathbf{R}^n} \frac{dx}{(1+|x|)^M} \leq \frac{v_n M}{M-n}, \qquad \text{when } M > n. \tag{7.1.1}$$

To verify (7.1.1), using polar coordinates we write

$$\int_{\mathbf{R}^n} \frac{dx}{(1+|x|)^M} = nv_n \int_0^\infty \frac{r^{n-1}}{(1+r)^M} dr \leq nv_n \int_0^1 r^{n-1} dr + nv_n \int_1^\infty r^{n-1-M} dr,$$

and the conclusion follows by calculating these integrals and adding the outputs.

Theorem 7.1.1. *Suppose that Φ and Ψ are functions on $\mathbf{R}^n$ that satisfy*

$$|\Phi(x)| \leq \frac{A}{(1+|x|)^M}, \qquad |\Psi(x)| \leq \frac{B}{(1+|x|)^K}$$

where $M, K > n$. Then for $t, s > 0$ we have

$$\left| \int_{\mathbf{R}^n} \Phi_t(x-a) \Psi_s(x-b) \, dx \right| \leq \frac{C_0 A B \max(t,s)^{-n}}{(1+\max(t,s)^{-1}|a-b|)^{\min(M,K)}}, \tag{7.1.2}$$

for some $M, K > n$ and all $x \in \mathbf{R}^n$. Then for $t, s > 0$ we have

$$C_0 = v_n \left(\frac{M 4^K}{M-n} + \frac{K 4^M}{K-n} \right).$$

Consequently, for each $y \in \mathbf{R}^n$ we have

$$|(\Phi_t * \Psi_s)(y)| \leq \frac{C_0 A B \max(t,s)^{-n}}{(1+\max(t,s)^{-1}|y|)^{\min(M,K)}}. \tag{7.1.3}$$

Proof. By symmetry we may assume that $t \leq s$. In the case $s^{-1}|a-b| \leq 1$ we use that

$$\frac{s^{-n}}{(1+s^{-1}|x-b|)^K} \leq s^{-n} \leq \frac{s^{-n} 2^{\min(M,K)}}{(1+s^{-1}|a-b|)^{\min(M,K)}}.$$

Then (7.1.2) is a consequence of the estimate

$$\int_{\mathbf{R}^n} \frac{A t^{-n}}{(1+t^{-1}|x-a|)^M} \frac{B s^{-n}}{(1+s^{-1}|x-b|)^K} dx$$

$$\leq \frac{A B s^{-n} 2^{\min(M,K)}}{(1+s^{-1}|a-b|)^{\min(M,K)}} \int_{\mathbf{R}^n} \frac{t^{-n}}{(1+t^{-1}|x-a|)^M} dx$$

$$\leq \frac{A B s^{-n}}{(1+s^{-1}|a-b|)^{\min(M,K)}} \frac{v_n M 2^K}{M-n},$$

where the last inequality follows from (7.1.1) and the fact that $\min(M,K) \leq K$.

We now consider the case $s^{-1}|a-b| \geq 1$. Let H_a and H_b be the two half-spaces, containing the points a and b, respectively, formed by the hyperplane perpendicular to the line segment $[a,b]$ at its midpoint. Write

$$\int_{\mathbf{R}^n} \frac{A t^{-n}}{(1+t^{-1}|x-a|)^M} \frac{B s^{-n}}{(1+s^{-1}|x-b|)^K} dx = \int_{H_a} \cdots dx + \int_{H_b} \cdots dx.$$

For $x \in H_a$ use that $|x-b| \geq \frac{1}{2}|a-b|$ to obtain

$$\int_{H_a} \frac{A t^{-n}}{(1+t^{-1}|x-a|)^M} \frac{B s^{-n}}{(1+s^{-1}|x-b|)^K} dx$$

$$\leq \frac{B s^{-n}}{(1+s^{-1}\frac{1}{2}|a-b|)^K} \int_{\mathbf{R}^n} \frac{A t^{-n}}{(1+t^{-1}|x-a|)^M} dx$$

$$\leq \frac{A B s^{-n}}{(1+s^{-1}|a-b|)^K} \frac{v_n M 2^K}{M-n}.$$

For $x \in H_b$ we use that $|x-a| \geq \frac{1}{2}|a-b|$ to write

$$\int_{H_b} \frac{A t^{-n}}{(1+t^{-1}|x-a|)^M} \frac{B s^{-n}}{(1+s^{-1}|x-b|)^K} dx$$

$$\leq \frac{A t^{-n}}{(1+t^{-1}\frac{1}{2}|a-b|)^M} \int_{\mathbf{R}^n} \frac{B s^{-n}}{(1+s^{-1}|x-b|)^K} dx$$

7.1 Smoothness and Cancellation

$$\begin{aligned}
&\leq \frac{A t^{-n} 2^M}{(t^{-1}|a-b|)^M} \frac{B v_n K}{K-n} \\
&= \frac{A t^{-n} 2^M (t/s)^M}{(s^{-1}|a-b|)^M} \frac{B v_n K}{K-n} \\
&\leq \frac{A B (t/s)^{(M-n)} s^{-n} 4^M}{(1+s^{-1}|a-b|)^M} \frac{v_n K}{K-n} \qquad (\text{as } s^{-1}|a-b| \geq 1) \\
&\leq \frac{A B s^{-n}}{(1+s^{-1}|a-b|)^M} \frac{v_n K 4^M}{K-n}.
\end{aligned}$$

Combining the estimates over H_a and H_b we deduce (7.1.2). Taking $a=0$, $y=b$, and considering the reflection of Ψ about 0, we derive (7.1.3) from (7.1.2). □

Theorem 7.1.2. *Fix $a, b \in \mathbf{R}^n$ and $N \in \mathbf{Z}^+ \cup \{0\}$. Let $\Phi \in \mathscr{C}^{N+1}(\mathbf{R}^n)$ be such that*

$$\sum_{|\beta|=N+1} |\partial^\beta \Phi(x)| \leq \frac{A}{(1+|x|)^M},$$

where $M \geq 0$. Let Ψ be another function on $\mathbf{R}^n$ that satisfies

$$|\Psi(x)| \leq \frac{B}{(1+|x|)^K},$$

where $K > N + M + n + 1$ and

$$\int_{\mathbf{R}^n} \Psi(x) x^\beta \, dx = 0 \qquad \text{for all } |\beta| \leq N. \tag{7.1.4}$$

Then when $0 < s \leq t < \infty$ we have

$$\left| \int_{\mathbf{R}^n} \Phi_t(x-a) \Psi_s(x-b) \, dx \right| \leq \frac{C_N C(K-N-M,n) A B t^{-n} \left(\frac{s}{t}\right)^{N+1}}{(1+t^{-1}|a-b|)^M}, \tag{7.1.5}$$

where $C(K-N-M,n) = \frac{v_n(K-N-M-1)}{K-N-M-n-1}$ and $C_N = \sum_{|\beta|=N+1} \frac{1}{\beta!}$. Consequently, for all $y \in \mathbf{R}^n$, it holds that

$$|(\Phi_t * \Psi_s)(y)| \leq \frac{C_N C(K-N-M,n) A B t^{-n} \left(\frac{s}{t}\right)^{N+1}}{(1+t^{-1}|y|)^M}. \tag{7.1.6}$$

Proof. To prove this assertion, we make use of the *Taylor expansion formula*

$$F(x_0 + h) = \sum_{|\alpha| \leq N} \frac{\partial^\alpha F(x_0)}{\alpha!} h^\alpha + R(h, x_0, N), \tag{7.1.7}$$

where $x_0, h \in \mathbf{R}^n$, $N \in \mathbf{Z}^+ \cup \{0\}$, and

$$R(h, x_0, N) = (N+1) \sum_{|\beta|=N+1} \frac{h^\beta}{\beta!} \int_0^1 (1-\theta)^N \partial^\beta F(x_0 + \theta h) \, d\theta. \tag{7.1.8}$$

is the remainder in integral form. Here F is a $\mathscr{C}^{N+1}$ function on $\mathbf{R}^n$.

We set $x_0 = b - a$ and $h = x - b$ so that $x_0 + h = x - a$. Using (7.1.7), (7.1.8), and the cancellation property (7.1.4) (which is also valid for Ψ_s), we write

$$\left| \int_{\mathbf{R}^n} \Phi_t(x-a) \Psi_s(x-b) \, dx \right|$$

$$= \left| \int_{\mathbf{R}^n} \left[\Phi_t(x-a) - \sum_{|\gamma| \leq N} \frac{\partial^\gamma \Phi_t(b-a)}{\gamma!} (x-b)^\gamma \right] \Psi_s(x-b) \, dx \right|$$

$$= \left| \int_{\mathbf{R}^n} \left[(N+1) \sum_{|\beta|=N+1} \frac{(x-b)^\beta}{\beta!} \int_0^1 (1-\theta)^N \frac{\partial^\beta \Phi(\frac{b-a+\theta(x-b)}{t})}{t^{n+N+1}} \, d\theta \right] \Psi_s(x-b) \, dx \right|$$

$$\leq \sum_{|\beta|=N+1} \frac{N+1}{\beta!} \int_{\mathbf{R}^n} \int_0^1 \frac{t^{-n} t^{-N-1}(1-\theta)^N}{(1+t^{-1}|(1-\theta)b+\theta x - a|)^M} \frac{ABs^{-n}|x-b|^{N+1}}{(1+s^{-1}|x-b|)^K} \, d\theta \, dx$$

$$\leq \sum_{|\beta|=N+1} \frac{N+1}{\beta!} \int_{\mathbf{R}^n} \int_0^1 \frac{t^{-n}(s/t)^{N+1}(1-\theta)^N}{(1+t^{-1}|\xi^\theta_{b,x} - a|)^M} \frac{ABs^{-n}}{(1+s^{-1}|x-b|)^{K-N-1}} \, d\theta \, dx, \quad (7.1.9)$$

where we set $\xi^\theta_{b,x} = (1-\theta)b + \theta x$. Using $s \leq t$ and $|\xi^\theta_{b,x} - b| \leq |x-b|$ we write

$$1 + t^{-1}|a-b| \leq 1 + t^{-1}|a - \xi^\theta_{b,x}| + t^{-1}|\xi^\theta_{b,x} - b|$$
$$\leq 1 + t^{-1}|a - \xi^\theta_{b,x}| + s^{-1}|x-b|$$
$$\leq (1 + t^{-1}|\xi^\theta_{b,x} - a|)(1 + s^{-1}|x-b|);$$

consequently, as $M \geq 0$,

$$\frac{1}{(1+t^{-1}|\xi^\theta_{b,x} - a|)^M} \leq \left(\frac{1+s^{-1}|x-b|}{1+t^{-1}|a-b|} \right)^M.$$

Inserting this estimate in (7.1.9) we obtain

$$\left| \int_{\mathbf{R}^n} \Phi_t(x-a) \Psi_s(x-b) \, dx \right| \leq C_N \int_{\mathbf{R}^n} \frac{(1+s^{-1}|x-b|)^M}{(1+t^{-1}|a-b|)^M} \frac{ABt^{-n}(s/t)^{N+1}s^{-n}}{(1+s^{-1}|x-b|)^{K-N-1}} \, dx$$

$$\leq C_N \frac{ABt^{-n}\left(\frac{s}{t}\right)^{N+1}}{(1+t^{-1}|a-b|)^M} \frac{v_n(K-N-M-1)}{K-N-M-n-1},$$

in view of the assumption $K > N + M + n + 1$, (7.1.1), and the fact that the integral in θ produces a factor of $(N+1)^{-1}$. (Recall $C_N = \sum_{|\beta|=N+1} \frac{1}{\beta!}$.) This yields (7.1.5). Finally, assertion (7.1.6) is obtained by taking $a = 0$, $y = b$, and by considering the reflection of Ψ about the origin (which satisfies the same assumptions as Ψ). □

Example 7.1.3. Let $\Phi, \Psi \in \mathscr{S}(\mathbf{R}^n)$ and assume that $\int_{\mathbf{R}^n} \Psi(x) x^\alpha \, dx = 0$ for all multi-indices α with $|\alpha| \leq N$ (for some fixed integer N). Then for any $M > n$ there is a constant C_M such that for any $y \in \mathbf{R}^n$ we have

7.2 Definition of Hardy Spaces and Preliminary Estimates

$$|(\Phi_t * \Psi)(y)| \leq \begin{cases} \dfrac{C_M}{(2+|y|)^M} & \text{when } t < 1, \\ \dfrac{C_M t^{-n-N-1}}{(2+t^{-1}|y|)^M} & \text{when } t \geq 1. \end{cases} \qquad (7.1.10)$$

The first of these estimates is a consequence of Theorem 7.1.1 while the second one is due to Theorem 7.1.2, in both cases with $s=1$.

Exercises

7.1.1. Fix $N \in \mathbf{Z}^+ \cup \{0\}$ and $a, b \in \mathbf{R}^n$. Let Φ, Ψ be Schwartz functions that satisfy

$$\int_{\mathbf{R}^n} \Psi(x) x^\beta \, dx = \int_{\mathbf{R}^n} \Phi(x) x^\beta \, dx = 0$$

for all $|\beta| \leq N$. Prove that for any $t, s > 0$ and any $M > 0$ there is a constant C_M such that

$$\left| \int_{\mathbf{R}^n} \Phi_t(x-a) \Psi_s(x-b) \, dx \right| \leq \frac{C_M \max(t,s)^{-n} \min\left(\frac{s}{t}, \frac{t}{s}\right)^{N+1}}{(1+\max(t,s)^{-1}|a-b|)^M}.$$

7.1.2. Suppose that g is a bounded measurable function supported in a cube of side length ℓ centered at $c \in \mathbf{R}^n$ and let $\Phi \in \mathscr{S}(\mathbf{R}^n)$. Show that for every $M > 0$ there is a constant C_M such that

$$|(\Phi_t * g)(x)| \leq \frac{C_M \|g\|_{L^\infty}}{(1+\ell^{-1}|x-c|)^M} \qquad \text{when } t < \ell,$$

and if g has vanishing integral then

$$|(\Phi_t * g)(x)| \leq \frac{C_M \|g\|_{L^\infty} (\ell/t)^{n+1}}{(1+t^{-1}|x-c|)^M} \qquad \text{when } t \geq \ell.$$

7.2 Definition of Hardy Spaces and Preliminary Estimates

Several boundedness results in analysis hold on L^p for $p > 1$ but break down on L^1. The Hardy space H^1 provides a good substitute for L^1 in several ways. The main focus of this chapter is the study of H^1 and, to a certain extent, of other Hardy spaces H^p for $p < 1$.

Definition 7.2.1. Let $\Phi \in \mathscr{S}(\mathbf{R}^n)$ and let f be a tempered distribution on $\mathbf{R}^n$. We define the *nontangential maximal function* of f with respect to Φ as

$$M^*(f;\Phi)(x) = \sup_{t>0} \sup_{\substack{y\in \mathbf{R}^n \\ |y-x|<t}} |(\Phi_t * f)(y)|. \qquad (7.2.1)$$

The term *nontangential* stems from the fact that the two suprema in (7.2.1) are taken over points (y,t) in the cone $\Gamma_x = \{(y,t): |x-y|<t\}$ in $\mathbf{R}^{n+1}_+$, which touches $\mathbf{R}^n$ only at the point x, i.e., it is nontangential to $\mathbf{R}^n$.

Remark 7.2.2. Let M be the Hardy–Littlewood maximal function. We verify that for any Schwartz function Φ there is a constant C_Φ such that

$$M^*(f;\Phi) \le C_\Phi M(f), \qquad (7.2.2)$$

when f is locally integrable on $\mathbf{R}^n$ and tempered at infinity; the latter means that $|f(x)| \le C(1+|x|)^K$ when $|x| \ge R$ for some $K,R,C>0$ (Example 2.6.2). To see this we pick $N > K+n$ and $C'_\Phi > 0$ such that $|\Phi(x)| \le C'_\Phi (2+|x|)^{-N}$ for all $x \in \mathbf{R}^n$. The integral defining the convolution $|f| * |\Phi_t|$ converges absolutely due to the choice of N, and thus $M^*(f;\Phi)(x)$ is well defined for any $x \in \mathbf{R}^n$. Then for all $x \in \mathbf{R}^n$ we have

$$|\Phi_t(x)| = \frac{1}{t^n}\left|\Phi\left(\frac{x}{t}\right)\right| \le \frac{1}{t^n}\frac{C'_\Phi}{(2+|\frac{x}{t}|)^N}.$$

For any $y \in \mathbf{R}^n$ satisfying $|y-x|<t$ we obtain

$$2 + \frac{|y-z|}{t} \ge 2 + \frac{|z-x|}{t} - \frac{|y-x|}{t} \ge 2 + \frac{|z-x|}{t} - 1 = 1 + \frac{|z-x|}{t}.$$

This gives that

$$M^*(f;\Phi)(x) \le \sup_{t>0} \sup_{\substack{y\in\mathbf{R}^n \\ |y-x|<t}} \int_{\mathbf{R}^n} \frac{C'_\Phi |f(z)|}{(2+\frac{|y-z|}{t})^N} \frac{dz}{t^n} \le \sup_{t>0} \int_{\mathbf{R}^n} \frac{C'_\Phi |f(z)|}{(1+\frac{|x-z|}{t})^N} \frac{dz}{t^n},$$

and from this we obtain (7.2.2) using Corollary 2.5.2.

We define the Hardy space H^p in terms of the specific nontangential maximal function associated with the Gaussian $\Phi(x) = e^{-\pi|x|^2}$ which gives rise to the approximate identity $\Phi_t(x) = t^{-n}\Phi(t^{-1}x)$, $t>0$.

Definition 7.2.3. Let $0 < p < \infty$ and $\Phi(x) = e^{-\pi|x|^2}$ for $x \in \mathbf{R}^n$. The *Hardy space* $H^p(\mathbf{R}^n)$ is the set of all tempered distributions f such that $M^*(f;\Phi)$ lies in $L^p(\mathbf{R}^n)$, and in this case we set

$$\|f\|_{H^p} = \|M^*(f;\Phi)\|_{L^p}.$$

This expression is a norm when $p \ge 1$ and a quasi-norm when $p < 1$ (Exercise 7.2.2). It is not clear from Definition 7.2.3 whether the H^p spaces coincide with any other known spaces for some values of p. In the next theorem we show that this is the case when $1 < p < \infty$.

7.2 Definition of Hardy Spaces and Preliminary Estimates

Theorem 7.2.4. (a) *Let $1 < p < \infty$. Then there is a constant $C_{n,p}$ such that for all $f \in L^p(\mathbf{R}^n)$ we have*

$$\|f\|_{H^p} \leq C_{n,p} \|f\|_{L^p}.$$

Moreover, for all $f \in H^p(\mathbf{R}^n)$ we have

$$\|f\|_{L^p} \leq \|f\|_{H^p}.$$

In other words, $H^p(\mathbf{R}^n)$ coincides with (is isomorphic to) $L^p(\mathbf{R}^n)$.

(b) *When $p = 1$, every element of H^1 is an integrable function. In other words, $H^1(\mathbf{R}^n) \subseteq L^1(\mathbf{R}^n)$ and for all $f \in H^1$ we have*

$$\|f\|_{L^1} \leq \|f\|_{H^1}. \tag{7.2.3}$$

Proof. (a) Let $1 < p < \infty$ and $f \in H^p(\mathbf{R}^n)$. Fix $\Phi(x) = e^{-\pi|x|^2}$ for $x \in \mathbf{R}^n$. The set $\{\Phi_t * f : t > 0\}$ lies in a multiple of the unit ball of $L^p(\mathbf{R}^n)$, which is the dual space of the separable Banach space $L^{p'}(\mathbf{R}^n)$. By the Banach–Alaoglu theorem this set is weakly* sequentially compact. Therefore, there exists a sequence $t_j \to 0$ such that $\Phi_{t_j} * f$ converges to some $f_0 \in L^p$ in the weak* topology of L^p. This means

$$\int_{\mathbf{R}^n} (\Phi_{t_j} * f) h\, dx \to \int_{\mathbf{R}^n} f_0 h\, dx, \qquad h \in L^{p'}(\mathbf{R}^n). \tag{7.2.4}$$

On the other hand, $\Phi_{t_j} * f \to f$ in $\mathscr{S}'(\mathbf{R}^n)$ as $t_j \to 0$ (Exercise 2.7.1), and thus the tempered distribution f coincides with the L^p function f_0. Since the family $\{\Phi_t\}_{t>0}$ is an approximate identity, Theorem 1.9.4 (a) gives that

$$\|\Phi_t * f - f\|_{L^p} \to 0 \qquad \text{as } t \to 0. \tag{7.2.5}$$

It follows from this that for any $\varepsilon > 0$ there is a $t_\varepsilon > 0$ such that for $0 < t < t_\varepsilon$ one has $\|f\|_{L^p} \leq \|\Phi_t * f\|_{L^p} + \varepsilon$. As $\varepsilon > 0$ was arbitrary, it follows that

$$\|f\|_{L^p} \leq \big\| \sup_{t>0} |\Phi_t * f| \big\|_{L^p} \leq \|M^*(f;\Phi)\|_{L^p} = \|f\|_{H^p}. \tag{7.2.6}$$

The converse inequality is a consequence of (7.2.2) and the boundedness of the Hardy–Littlewood maximal operator on $L^p(\mathbf{R}^n)$ for $p > 1$.

(b) Let us denote by $\mathscr{C}_{00}(U)$ the space of continuous functions $g(x)$ that are supported in an open set U and tend to zero as $|x| \to \infty$. We embed L^1 in the space of complex Borel measures $\mathscr{M}$ whose total variation is finite; this space is the dual of the separable space $\mathscr{C}_{00}(\mathbf{R}^n)$. By the Banach–Alaoglu theorem, the unit ball of $\mathscr{M}$ is weakly* sequentially compact, and we can extract a sequence $t_j \to 0$ such that $\Phi_{t_j} * f$ converges to a complex Borel measure μ in $\mathscr{M}$ in the sense

$$\int_{\mathbf{R}^n} (\Phi_{t_j} * f) h\, dx \to \int_{\mathbf{R}^n} h\, d\mu, \qquad h \in \mathscr{C}_{00}(\mathbf{R}^n). \tag{7.2.7}$$

As $\Phi_{t_j} * f \to f$ in $\mathscr{S}'(\mathbf{R}^n)$ (Exercise 2.7.1), the distribution f can be identified with the measure μ. If we can show that $d\mu = f_0\, dx$ for some $f_0 \in L^1(\mathbf{R}^n)$, it will follow that $f = f_0$ a.e.; thus the given $f \in \mathscr{S}'$ can be identified with an L^1 function.

Next we prove that for all subsets E of $\mathbf{R}^n$ we have $|E| = 0 \implies \mu(E) = 0$. Since $\sup_{t>0} |\Phi_t * f|$ lies in $L^1(\mathbf{R}^n)$, given $\varepsilon > 0$, there exists a $\delta > 0$ such that for any measurable subset F of $\mathbf{R}^n$ we have

$$|F| < \delta \implies \int_F \sup_{t>0} |\Phi_t * f|\, dx < \varepsilon.$$

Given E with $|E| = 0$, we can find an open set U such that $E \subseteq U$ and $|U| < \delta$.
Then for any g in $\mathscr{C}_{00}(U)$ we have

$$\begin{aligned}
\left| \int_{\mathbf{R}^n} g\, d\mu \right| &= \lim_{j \to \infty} \left| \int_{\mathbf{R}^n} g(x)(\Phi_{t_j} * f)(x)\, dx \right| \\
&\leq \|g\|_{L^\infty} \int_U \sup_{t>0} |(\Phi_t * f)(x)|\, dx \\
&< \varepsilon \|g\|_{L^\infty}.
\end{aligned}$$

Let $|\mu|$ be the variation measure of measure μ. Then we have {see [37] (20.49)}

$$|\mu|(U) = \int_U 1\, d|\mu| = \sup\left\{ \left| \int_{\mathbf{R}^n} g\, d\mu \right| : g \in \mathscr{C}_{00}(U),\ \|g\|_{L^\infty} \leq 1 \right\},$$

which implies $|\mu|(U) < \varepsilon$. Thus $|\mu|(E) < \varepsilon$ and as ε was arbitrary, it follows that $|\mu|(E) = 0$ and thus $\mu(E) = 0$. This argument shows that μ is absolutely continuous with respect to Lebesgue measure. By the Radon–Nikodym theorem we obtain the existence of a function f_0 in $L^1(\mathbf{R}^n)$ such that $d\mu = f_0\, dx$. Inserting this in (7.2.7) we obtain the analog of (7.2.4) in the case $p = 1$. As Theorem 1.9.4 (a) also applies when $p = 1$, it follows that (7.2.5) also holds when $p = 1$. Finally, (7.2.3) is a consequence of (7.2.6), which is also valid for $p = 1$. $\square$

Remark 7.2.5. One may wonder whether $H^1(\mathbf{R}^n)$ coincides with $L^1(\mathbf{R}^n)$. We provide an example showing that $L^p(\mathbf{R}^n)$ is not contained in $H^p(\mathbf{R}^n)$ for any $p \leq 1$. Set $\Phi(x) = e^{-\pi|x|^2}$ on $\mathbf{R}^n$. One sees that for $|x| \geq 1$

$$M^*(\chi_{B(0,1)}; \Phi)(x) \geq (\chi_{B(0,1)} * \Phi_{|x|})(x) = \frac{1}{|x|^n} \int_{|y| \leq 1} e^{-\pi\left(\frac{|x-y|}{|x|}\right)^2} dy \geq v_n \frac{e^{-4\pi}}{|x|^n},$$

since

$$\frac{|x-y|}{|x|} \leq \frac{|x|+|y|}{|x|} \leq \frac{|x|+1}{|x|} \leq 2.$$

As the function $|x|^{-n}$ is not integrable to any power $p \leq 1$ over $B(0,1)^c$, we have $\chi_{B(0,1)} \in L^p \setminus H^p$. Thus H^1 is a proper subspace of L^1 and $L^p(\mathbf{R}^n)$ is not contained in $H^p(\mathbf{R}^n)$ for any $p < 1$. Also, $H^p(\mathbf{R}^n)$ is not contained in $L^p(\mathbf{R}^n)$ when $p < 1$ as

7.2 Definition of Hardy Spaces and Preliminary Estimates

certain distributions that are not functions are members of $H^p(\mathbf{R}^n)$; see for instance Exercise 7.3.4.

Next we show that H^1 functions must have integral zero.

Theorem 7.2.6. *Suppose that $g \in H^1(\mathbf{R}^n)$. Then $\int_{\mathbf{R}^n} g(x)\,dx = 0$.*

Proof. By Theorem 7.2.4 (b), g lies in L^1 and so its integral is well defined. Set $c = \int_{\mathbf{R}^n} g(x)\,dx$ and $\Phi(x) = e^{-\pi|x|^2}$. Then the family $\{g_\varepsilon\}_{\varepsilon > 0}$ is a multiple of an approximate identity, and by Theorem 1.9.7 (b) we have

$$(\Phi_t * g_\varepsilon)(y) \to c\,\Phi_t(y) \qquad \text{as } \varepsilon \to 0$$

for any $y \in \mathbf{R}^n$ and any $t > 0$, since Φ_t is uniformly continuous on $\mathbf{R}^n$. Then

$$|c|\,|\Phi_t(y)| = \lim_{\varepsilon \to 0}|(\Phi_t * g_\varepsilon)(y)| = \liminf_{\varepsilon \to 0} \frac{1}{\varepsilon^n}\left|(\Phi_{t/\varepsilon} * g)\left(\frac{y}{\varepsilon}\right)\right|,$$

and so for any $x \in \mathbf{R}^n$ and any $t > 0$ we obtain

$$|c| \sup_{t>0}\sup_{y:|y-x|<t} |\Phi_t(y)| \le \liminf_{\varepsilon \to 0} \frac{1}{\varepsilon^n} \sup_{t>0}\sup_{y:|y-x|<t} \left|(\Phi_{t/\varepsilon} * g)\left(\frac{y}{\varepsilon}\right)\right|. \tag{7.2.8}$$

On one hand, the right-hand side of (7.2.8) is

$$\liminf_{\varepsilon \to 0} \sup_{t'>0}\sup_{y':|y'-\frac{x}{\varepsilon}|<t'} \frac{1}{\varepsilon^n}|(\Phi_{t'} * g)(y')| \le \liminf_{\varepsilon \to 0} \frac{1}{\varepsilon^n} M^*(g;\Phi)\left(\frac{x}{\varepsilon}\right).$$

On the other hand the left-hand side of (7.2.8) satisfies

$$|c|\sup_{t>0}\sup_{y:|y-x|<t}|\Phi_t(y)| \ge |c|\,|\Phi_{|x|}(x)| = \frac{|c|e^{-\pi}}{|x|^n}.$$

Suppose $c \ne 0$. Taking L^1 norms and applying Fatou's lemma, we deduce

$$|c|\int_{\mathbf{R}^n} \frac{e^{-\pi}dx}{|x|^n} \le \int_{\mathbf{R}^n} \liminf_{\varepsilon \to 0} M^*(g;\Phi)\left(\frac{x}{\varepsilon}\right)\frac{dx}{\varepsilon^n} \le \liminf_{\varepsilon \to 0}\int_{\mathbf{R}^n} M^*(g;\Phi)\left(\frac{x}{\varepsilon}\right)\frac{dx}{\varepsilon^n} = \|g\|_{H^1}.$$

The quantity on the left equals ∞ but $\|g\|_{H^1} < \infty$, a contradiction. Thus $c = 0$. □

We end this section by showing that Schwartz functions with sufficient vanishing moments[1] lie in H^p.

Theorem 7.2.7. *Let $0 < p \le 1$ and $N = [\frac{n}{p} - n]$. Then every Schwartz function Ψ on $\mathbf{R}^n$ with $\int_{\mathbf{R}^n} \Psi(x)x^\alpha\,dx = 0$ for all multi-indices α with $|\alpha| \le N$, lies in $H^p(\mathbf{R}^n)$.*

---

[1] The αth moment of a function Ψ is $\int_{\mathbf{R}^n} \Psi(x)x^\alpha\,dx$, where α is a multi-index.

Proof. We will make use of (7.1.10) with $\Phi(x) = e^{-\pi|x|^2}$.
(a) If $t < 1$ and $y \in \mathbf{R}^n$ is such that $|y - x| < t$, then

$$2 + |y| \geq 2 + |x| - |x - y| \geq 2 + |x| - t \geq 1 + |x|;$$

Thus the variable y on the right in (7.1.10) can be replaced by x when $|y - x| < t$. Then (7.1.10) yields

$$\sup_{0 < t < 1} \sup_{y: |y-x| < t} |(\Phi_t * \Psi)(y)| \leq \frac{C_M}{(1 + |x|)^M},$$

where M is arbitrarily large.
(b) If $t \geq 1$ and $y \in \mathbf{R}^n$ is such that $|y - x| < t$, then

$$2 + t^{-1}|y| \geq 2 + t^{-1}|x| - t^{-1}|x - y| \geq 2 + t^{-1}|x| - 1 \geq 1 + t^{-1}|x|;$$

thus the variable y on the right in (7.1.10) can also be replaced by x. Then

$$\sup_{t \geq 1} \sup_{y: |y-x| < t} |(\Phi_t * \Psi)(y)| \leq \sup_{t \geq 1} \frac{C_M t^{-n-N-1}}{(1 + t^{-1}|x|)^M} \leq \begin{cases} C_M & \text{if } |x| \leq 1, \\ \frac{C_M}{|x|^{n+N+1}} \sup_{s > 0} \frac{s^{n+N+1}}{(1+s)^M} & \text{if } |x| > 1. \end{cases}$$

Here we changed variables $s = |x|/t$ when $|x| \neq 0$. So choosing $M > n + N + 1$, we obtain

$$M^*(\Psi; \Phi)(x) \leq \frac{C'_M}{(1 + |x|)^{n+N+1}}. \tag{7.2.9}$$

The choice of $N = [\frac{n}{p} - n]$ implies $p(n + N + 1) > n$; hence (7.2.9) yields that the function $M^*(\Psi; \Phi)$ lies in $L^p(\mathbf{R}^n)$. $\square$

Exercises

7.2.1. Recall the translation of a tempered distribution u is defined by $\langle \tau^{x_0} u, \varphi \rangle = \langle u, \tau^{-x_0} \varphi \rangle$, $\varphi \in \mathscr{S}(\mathbf{R}^n)$, $x_0 \in \mathbf{R}^n$, $\tau^{-x_0}\varphi(x) = \varphi(x + x_0)$, $x \in \mathbf{R}^n$. Show that the H^p quasi-norm is *translation-invariant*, meaning that for any $x_0 \in \mathbf{R}^n$, one has $\|\tau^{x_0} u\|_{H^p} = \|u\|_{H^p}$ when $0 < p < \infty$.

7.2.2. Let $0 < p < \infty$. Observe that for $f, g \in H^p$ we have

$$\|f + g\|_{H^p}^{\min(1,p)} \leq \|f\|_{H^p}^{\min(1,p)} + \|g\|_{H^p}^{\min(1,p)}$$

and $\|\lambda f\|_{H^p} = |\lambda| \|f\|_{H^p}$ when $\lambda \in \mathbf{C}$. Conclude that the expression $\|\cdot\|_{H^p}$ is a norm when $p \geq 1$ and a quasi-norm when $0 < p < 1$.

7.2.3. Show that for $0 < p < 1$, the $H^p(\mathbf{R}^n)$ quasi-norm remains invariant under dilations of distributions of the form $u \mapsto \lambda^{n/p} u^\lambda$, $\lambda > 0$, where $\langle u^\lambda, \varphi \rangle = \langle u, \varphi_{1/\lambda} \rangle$

and $\varphi_\lambda(x) = \lambda^{-n}\varphi(\lambda^{-1}x)$, $\varphi \in \mathscr{S}$, $u \in \mathscr{S}'$, $x \in \mathbf{R}^n$. Note that for $1 \le p < \infty$, the $H^p(\mathbf{R}^n)$ norm remains invariant under dilations of functions $f \mapsto f^\lambda$, where $f^\lambda(x) = \lambda^{\frac{n}{p}} f(\lambda x)$, $x \in \mathbf{R}^n$.

7.2.4. Prove that H^p contains the space of Schwartz functions whose Fourier transforms vanish in neighborhoods of the origin.

7.2.5. Let $0 < p < \infty$ and let $u_j \in H^p(\mathbf{R}^n)$, $j = 1, 2, \ldots$. If $u_j \to u$ in $\mathscr{S}'$ as $j \to \infty$, show that
$$\|u\|_{H^p} \le \liminf_{j \to \infty} \|u_j\|_{H^p}.$$

[*Hint:* Use Fatou's lemma.]

7.3 H^p Atoms

We have seen that Schwartz functions Ψ with vanishing moments lie in H^p. A close examination of Theorem 7.2.7 indicates that the only property used of the function Ψ was its decay at infinity. This is certainly the case if Ψ has compact support. This observation motivates the following definition.

Definition 7.3.1. Let $0 < p \le 1$. An H^p *atom*[2] is a function A with the properties:

(i) A supported in a cube Q (with sides parallel to the axes).

(ii) $|A(x)| \le |Q|^{-1/p}$ for all $x \in Q$.

(iii) $\int_Q A(x) x^\alpha dx = 0$ for all multi-indices α with $|\alpha| \le [\frac{n}{p} - n]$.

Condition (iii) is referred to as the *vanishing moment property* and reduces to vanishing integral when $p = 1$. Condition (ii) provides only a natural normalization.

Theorem 7.3.2. H^p *atoms lie in* H^p. *Precisely, there is a constant* $C(n,p)$ *that depends only on* n *and* p *such that* $\|A\|_{H^p} \le C(n,p)$ *for any* H^p *atom* A.

Proof. If the claimed assertion is proven for atoms supported in cubes centered at the origin, then it also holds for all atoms, since translations of atoms are atoms and H^p is translation invariant (Exercise 7.3.1). So we fix an atom A supported in a cube Q of side length $\ell = \ell(Q)$ centered at the origin. We define the function $\Psi(y) = \ell^{\frac{n}{p}} A(\ell y)$. Then Ψ is an atom supported on $[-\frac{1}{2}, \frac{1}{2}]^n$ and is related to A by $A(x) = \ell^{-\frac{n}{p}} \frac{1}{\ell^n} \Psi(\frac{x}{\ell}) = \ell^{n-\frac{n}{p}} \Psi_\ell(x)$. Set $\Phi(x) = e^{-\pi|x|^2}$. Then one has
$$\Phi_t * A = \ell^{n-\frac{n}{p}}(\Phi_t * \Psi_\ell).$$

Moreover, by Definition 7.3.1 (ii) we have

---
[2] Also called an L^∞ atom for H^p.

$$|\Psi(y)| = |\ell^{\frac{n}{p}} A(\ell y)| \le \ell^{\frac{n}{p}} |Q|^{-\frac{1}{p}} \chi_{\frac{1}{\ell}Q}(y) = \chi_{[-\frac{1}{2},\frac{1}{2}]^n}(y) \le \frac{C_K}{(1+|y|)^K}$$

for any $K > 0$ with $C_K = (1+\sqrt{n}/2)^K$. When $t \le \ell$, Theorem 7.1.1 gives

$$\ell^{n-\frac{n}{p}} |(\Phi_t * \Psi_\ell)(y)| \le \frac{C_M \ell^{n-\frac{n}{p}} \ell^{-n}}{(2+\ell^{-1}|y|)^M} \le \frac{C_M \ell^{-\frac{n}{p}}}{(1+\ell^{-1}|x|)^M}, \quad (7.3.1)$$

for any $M > n$, where in the last inequality we took x such that $|y-x| < t \le \ell$.

Notice that in view of property (iii) in Definition 7.3.1, the function $\Psi(y) = \ell^{\frac{n}{p}} A(\ell y)$ has vanishing moments up to and including order $N = [\frac{n}{p} - n]$. When $t > \ell$, applying Theorem 7.1.2, for $|y-x| < t$ and any $L > 0$, we obtain

$$|(\Phi_t * A)(y)| = \ell^{n-\frac{n}{p}} |(\Phi_t * \Psi_\ell)(y)| \le \frac{C_L t^{-n} \ell^{n-\frac{n}{p}} (\ell/t)^{N+1}}{(2+t^{-1}|y|)^L} \le \frac{C_L t^{-N-1-n} \ell^{n-\frac{n}{p}+N+1}}{(1+t^{-1}|x|)^L},$$

where, as before, the last inequality comes from the fact that $|y-x| < t$. This gives

$$\sup_{t>\ell} \sup_{y: |y-x|<t} |(\Phi_t * A)(y)| \le C_L \ell^{-N-1-n} \ell^{n-\frac{n}{p}+N+1} = C_L \ell^{-\frac{n}{p}} \quad (7.3.2)$$

and also, by changing variables $s = |x|/t$ (when $|x| \ne 0$), it also gives

$$\sup_{t>\ell} \sup_{y: |y-x|<t} |(\Phi_t * A)(y)| \le C_L \ell^{-\frac{n}{p}} \left(\frac{\ell}{|x|}\right)^{n+N+1} \sup_{s>0} \frac{s^{N+1+n}}{(1+s)^L}. \quad (7.3.3)$$

Here we chose $L > N+1+n$ and so the supremum in s in (7.3.3) reduces to a constant. Combining (7.3.2) and (7.3.3) yields

$$\sup_{t>\ell} \sup_{y: |y-x|<t} |(\Phi_t * A)(y)| \le \frac{C_L' \ell^{-\frac{n}{p}}}{(1+\ell^{-1}|x|)^{n+N+1}}. \quad (7.3.4)$$

Finally, (7.3.4) and (7.3.1) imply

$$M^*(A; \Phi)(x) \le \frac{C_M \ell^{-\frac{n}{p}}}{(1+\ell^{-1}|x|)^M} + \frac{C_L' \ell^{-\frac{n}{p}}}{(1+\ell^{-1}|x|)^{n+N+1}},$$

and this function has L^p quasi-norm bounded by a constant independent of ℓ. In fact, M and L can be chosen to depend only on n and N (thus on n and p), so the final constant controlling the L^p quasi-norm of $M^*(A; \Phi)$ is bounded by a constant $C(n,p)$ depending only on n and p. $\square$

Example 7.3.3. Every compactly supported and bounded function g with mean value zero lies in $H^1(\mathbf{R}^n)$. Indeed, every such function is supported in a big cube Q and is bounded by $c|Q|^{-1}$, where $c = \|g\|_{L^\infty} |Q|$. Then $g = cA$, where A is an H^1 atom, so g lies in $H^1(\mathbf{R}^n)$.

7.3 H^p Atoms

In the rest of this section we provide a strengthening of Theorem 7.3.2 that can handle even unbounded functions with vanishing moments.

We begin with some observations regarding the derivatives of the Gaussian function $e^{-\pi |x|^2}$. A straightforward calculation via induction gives that the Nth derivative of the function $t \mapsto e^{-\pi t^2}$ equals $p_N(t)e^{-\pi t^2}$, where p_N is a polynomial of degree N. Thus $|\frac{d^N}{dt^N} e^{-\pi t^2}| \leq B_N(1+|t|)^N e^{-\pi t^2}$ for some constant B_N. Extending this to n dimensions (by separation of variables), we obtain that for multi-indices $\beta = (\beta_1, \ldots, \beta_n)$ with size $|\beta| = N$ we have

$$\partial^\beta e^{-\pi |x|^2} = P_\beta(x) e^{-\pi |x|^2}, \qquad (7.3.5)$$

where $x = (x_1, \ldots, x_n)$ and $P_\beta(x) = \prod_{j=1}^n p_{\beta_j}(x_j)$. Thus for any β with $|\beta| = N$, one has

$$|\partial^\beta e^{-\pi |x|^2}| \leq B_N^n (1+|x|)^N e^{-\pi |x|^2}, \qquad x \in \mathbf{R}^n. \qquad (7.3.6)$$

These facts are useful in the calculation of H^p quasi-norms.

Theorem 7.3.4. *Let $0 < p \leq 1$ and $1 < q \leq \infty$. Suppose that g lies in $L^q(\mathbf{R}^n)$ and is supported in a cube Q. Suppose moreover, that $\int_Q g(x) x^\alpha \, dx = 0$ for $|\alpha| \leq [\frac{n}{p} - n]$. Then g lies in $H^p(\mathbf{R}^n)$ and there is a constant $C_{n,p,q}$ such that[3]*

$$\|g\|_{H^p} \leq C_{n,p,q} |Q|^{\frac{1}{p} - \frac{1}{q}} \|g\|_{L^q}.$$

Proof. As H^p is translation invariant, we may assume that Q is centered at the origin. We denote by Q^* the cube Q dilated $2\sqrt{n}$ times. Let us set $\Phi(x) = e^{-\pi |x|^2}$. For this Φ we use (7.2.2) for $x \in Q^*$ to write

$$\|M^*(g; \Phi) \chi_{Q^*}\|_{L^p} \leq C \|M(g) \chi_{Q^*}\|_{L^p} \leq C |Q^*|^{\frac{1}{p} - \frac{1}{q}} \|M(g)\|_{L^q}, \qquad (7.3.7)$$

where in the last step we made use of Hölder's inequality (1.1.4) applied to the functions χ_{Q^*} and $M(g)$ with exponents $(\frac{1}{p} - \frac{1}{q})^{-1}$ and q, respectively. Corollary 1.4.7 yields that the expression on the right in (7.3.7) is bounded by

$$C |Q^*|^{\frac{1}{p} - \frac{1}{q}} \frac{2q}{q-1} 3^{\frac{n}{q}} \|g\|_{L^q} \leq C' |Q|^{\frac{1}{p} - \frac{1}{q}} \|g\|_{L^q}. \qquad (7.3.8)$$

We now turn to the case where $x \notin Q^*$. Let us fix y such that $|y - x| < t$. Exploiting the fact that g has vanishing moments up to order $N = [\frac{n}{p} - n]$, Taylor's expansion in (7.1.7) [and the expression for the remainder (7.1.8)], allow us to write

$$|(\Phi_t * g)(y)|$$
$$= \left| \frac{1}{t^n} \int_Q \left[\Phi\left(\frac{y-z}{t}\right) - \sum_{|\alpha| \leq N} \partial^\alpha \Phi\left(\frac{y}{t}\right) \frac{1}{\alpha!} \left(-\frac{z}{t}\right)^\alpha \right] g(z) \, dz \right|$$

---
[3] The normalized function $|Q|^{-\frac{1}{p} + \frac{1}{q}} g / \|g\|_{L^q}$ is called an L^q atom for H^p.

$$= \frac{N+1}{t^n}\left|\int_Q\left[\sum_{|\beta|=N+1}\int_0^1(1-\theta)^N\partial^\beta\Phi\left(\frac{y}{t}-\theta\frac{z}{t}\right)\frac{1}{\beta!}\left(-\frac{z}{t}\right)^\beta d\theta\right]g(z)\,dz\right|$$

$$\leq \frac{C_N B_{N+1}^n}{t^{n+N+1}}\int_Q\left[\int_0^1(1-\theta)^N\left(1+\left|\frac{y-\theta z}{t}\right|\right)^{N+1}e^{-\pi|\frac{y-\theta z}{t}|^2}d\theta\right]|z|^{N+1}|g(z)|\,dz, \quad (7.3.9)$$

having used (7.3.6). Notice that $x \notin Q^*$ gives that $|x| \geq \sqrt{n}\ell$, where $\ell = \ell(Q)$ is the side length of Q. Also, $z \in Q$ implies $|z| \leq \frac{1}{2}\sqrt{n}\ell$.

Case 1: $|x| \geq 4t$. Then we have

$$|y-\theta z| \geq |y|-|z| \geq |x|-|x-y|-|z| \geq |x|-t-\frac{1}{2}\sqrt{n}\ell \geq \frac{|x|}{2}-t \geq \frac{|x|}{4}$$

and also

$$|y-\theta z| \leq |y|+|z| \leq |x|+|y-x|+|z| \leq |x|+t+\frac{1}{2}\sqrt{n}\ell \leq |x|+\frac{|x|}{4}+\frac{|x|}{2} \leq 2|x|.$$

These two estimates imply

$$\left(1+\left|\frac{y-\theta z}{t}\right|\right)^{N+1}e^{-\pi|\frac{y-\theta z}{t}|^2} \leq \left(1+\frac{2|x|}{t}\right)^{N+1}e^{-\pi(\frac{|x|}{4t})^2}.$$

Inserting this bound in (7.3.9), we obtain

$$|(\Phi_t * g)(y)| \leq \frac{C_N B_{N+1}^n}{t^{n+N+1}}\left(1+\frac{2|x|}{t}\right)^{N+1}e^{-\pi(\frac{|x|}{4t})^2}\int_Q|z|^{N+1}|g(z)|\,dz$$

$$\leq \frac{C_N B_{N+1}^n}{(2|x|)^{n+N+1}}\left(1+\frac{2|x|}{t}\right)^{2N+2+n}e^{-\pi(\frac{|x|}{4t})^2}\int_Q|z|^{N+1}|g(z)|\,dz$$

$$\leq \frac{C_N' B_{N+1}^n}{|x|^{n+N+1}}\int_Q|z|^{N+1}|g(z)|\,dz,$$

as the function $s \mapsto (1+8s)^{2N+2+n}e^{-\pi s^2}$ is bounded.

Case 2: $|x| < 4t$. The function in the brackets in (7.3.9) is bounded, so we obtain

$$|(\Phi_t * g)(y)| \leq \frac{C_N''}{t^{n+N+1}}\int_Q|z|^{N+1}|g(z)|\,dz \leq \frac{C_N'' 4^{n+N+1}}{|x|^{n+N+1}}\int_Q|z|^{N+1}|g(z)|\,dz,$$

which is the same estimate as in Case 1.

So, in both cases 1 and 2, when $x \notin Q^*$ we proved that

$$M^*(g;\Phi)(x) = \sup_{t>0}\sup_{y:|y-x|<t}|(\Phi_t * g)(y)| \leq \frac{C_{n,N}'}{|x|^{n+N+1}}\int_Q|z|^{N+1}|g(z)|\,dz. \quad (7.3.10)$$

But

$$\int_Q|z|^{N+1}|g(z)|\,dz \leq c|Q|^{\frac{N+1}{n}}\int_Q|g(z)|\,dz \leq c|Q|^{\frac{N+1}{n}}|Q|^{\frac{1}{q'}}\|g\|_{L^q},$$

so for $x \notin Q^*$ we obtain

$$M^*(g;\Phi)(x) \leq \frac{C'_{n,N}}{|x|^{n+N+1}}|Q|^{\frac{N+1}{n}}|Q|^{\frac{1}{q}}\|g\|_{L^q}.$$

Since $N = [\frac{n}{p} - n]$, it follows that $p(n+N+1) > n$; hence $|x|^{-(n+N+1)p}$ is integrable over the set $|x| \geq \sqrt{n}\ell(Q)$ and the integral produces a factor of the order of $|Q|^{-(1+\frac{N+1}{n})p+1}$. Thus we derive the estimate

$$\left[\int_{(Q^*)^c} M^*(g;\Phi)^p dx\right]^{\frac{1}{p}} \leq C_{n,p}|Q|^{-(1+\frac{N+1}{n})+\frac{1}{p}}|Q|^{\frac{N+1}{n}}|Q|^{\frac{1}{q}}\|g\|_{L^q} = C_{n,p}|Q|^{\frac{1}{p}-\frac{1}{q}}\|g\|_{L^q}.$$

This estimate, together with (7.3.7) and (7.3.8), provides the required conclusion. □

Exercises

7.3.1. Prove that condition (iii) in Definition 7.3.1 is equivalent to the condition that $\int_Q A(x)(x-x_0)^\alpha dx = 0$ for all $|\alpha| \leq [\frac{n}{p} - n]$ and any $x_0 \in \mathbf{R}^n$. Conclude that translations of atoms are atoms.

7.3.2. Observe that H^q atoms are H^p atoms when $0 < q < p \leq 1$. Then verify that the function

$$\chi_{[0,1]^n}\prod_{j=1}^n(6x_j^2 - 6x_j + 1)$$

is an $H^p(\mathbf{R}^n)$ atom but not an $H^q(\mathbf{R}^n)$ atom when $0 < q \leq \frac{n}{n+2} < p \leq 1$.

7.3.3. Let ϕ be a compactly supported and smooth function and let β be a multi-index with $|\beta| \geq 1$. Prove that $\partial^\beta \phi$ lies in $H^p(\mathbf{R}^n)$ for all p satisfying $\frac{n}{n+|\beta|} < p \leq 1$.

7.3.4. Let $x_1, x_2 \in \mathbf{R}^n$ satisfy $x_1 \neq x_2$. Prove that the difference of Dirac masses $\delta_{x_1} - \delta_{x_2}$ lies in $H^p(\mathbf{R}^n)$ for $\frac{n}{n+1} < p < 1$.

7.3.5. Let $x_1, x_2 \in \mathbf{R}^n$ satisfy $x_1 \neq x_2$ and let α be a multi-index. Show that $\partial^\alpha \delta_{x_1} - \partial^\alpha \delta_{x_2}$ lies in $H^p(\mathbf{R}^n)$ for all p satisfying $\frac{n}{n+1+|\alpha|} < p < \frac{n}{n+|\alpha|}$. [Hint: Use (7.3.5).]

7.4 Grand Maximal Function

Our goal in this section is to show that the definition of Hardy spaces does not depend on the specific choice of the function $\Phi(x) = e^{-\pi|x|^2}$. The following lemma will be crucial in proving this fact.

Lemma 7.4.1. *Let $m \in \mathbf{Z}^+$ and fix Φ in $\mathscr{S}(\mathbf{R}^n)$ with integral equal to 1 [such as $\Phi(x) = e^{-\pi|x|^2}$]. Then there exists a constant $C_0(\Phi, m)$ such that for any function $\Psi \in \mathscr{S}(\mathbf{R}^n)$, there are Schwartz functions $\Theta^{(s)}$, $0 < s \leq 1$, with the properties*

$$\Psi(x) = \int_0^1 (\Theta^{(s)} * \Phi_s)(x) \, ds \tag{7.4.1}$$

and

$$\frac{1}{s^m} \int_{\mathbf{R}^n} (1+|x|)^m |\Theta^{(s)}(x)| \, dx \leq C_0(\Phi, m) \int_{\mathbf{R}^n} (1+|x|)^m \sum_{|\alpha| \leq m+1} |\partial^\alpha \Psi(x)| \, dx. \tag{7.4.2}$$

Proof. We start with a smooth function η on the real line that satisfies $0 \leq \eta \leq 1$, $\eta(s) = 0$ for $s \geq \frac{7}{8}$ and $\eta(s) = 1$ for $s \leq \frac{1}{2}$.

Then we define

$$\zeta(s) = \frac{s^m}{m!} \eta(s)$$

for $s \in \mathbf{R}$. Then ζ lies in $\mathscr{C}^\infty$ and satisfies

$$0 \leq \zeta(s) \leq \frac{s^m}{m!} \quad \text{for all } 0 \leq s \leq 1,$$
$$\zeta(s) = \frac{s^m}{m!} \quad \text{for all } 0 \leq s \leq \frac{1}{2},$$
$$\zeta(s) = 0 \quad \text{for all } s \geq \frac{7}{8}.$$

See Figure 7.1.

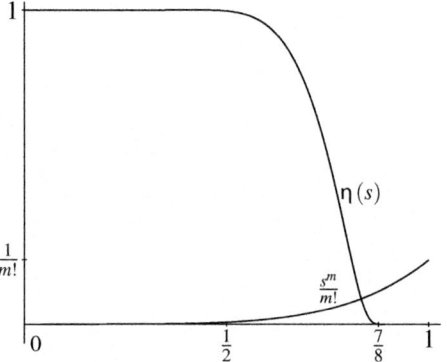

Fig. 7.1 The functions $s^m/m!$ and $\eta(s)$ plotted on $[0, 1]$.

Next, for $s \in (0, 1]$, we define the family of functions

$$\Theta^{(s)} = (-1)^{m+1} \zeta(s) \, \Xi^{(s)} * \Psi - \frac{d^{m+1}\zeta}{ds^{m+1}}(s) \overbrace{\Phi_s * \cdots * \Phi_s}^{m+1 \text{ terms}} * \Psi, \tag{7.4.3}$$

where $\Xi^{(s)}$ is a function chosen so that

$$\frac{d^{m+1}}{ds^{m+1}} \left(\overbrace{\Phi_s * \cdots * \Phi_s}^{m+2 \text{ terms}} \right) = \Xi^{(s)} * \Phi_s \tag{7.4.4}$$

for any $s > 0$. Precisely, $\Xi^{(s)}$ is equal to

$$\sum c_{m, k_1, \ldots, k_{m+1}} \overbrace{\Phi_s * \cdots * \Phi_s}^{m+1-(k_1 + \cdots + k_{m+1}) \text{ terms}} * \overbrace{\frac{d\Phi_s}{ds} * \cdots * \frac{d\Phi_s}{ds}}^{k_1 \text{ terms}} * \cdots * \overbrace{\frac{d^{m+1}\Phi_s}{ds^{m+1}} * \cdots * \frac{d^{m+1}\Phi_s}{ds^{m+1}}}^{k_{m+1} \text{ terms}},$$

7.4 Grand Maximal Function

where the sum is taken over all tuples $(k_1, \ldots, k_{m+1})$ of nonnegative integers such that $1k_1 + \cdots + (m+1)k_{m+1} = m+1$, and

$$c_{m,k_1,\ldots,k_{m+1}} = \frac{(m+2)!}{(m+2-(k_1+\cdots+k_{m+1}))!} \frac{1}{1!^{k_1} \cdots (m+1)!^{k_{m+1}}} \frac{1}{k_1! \cdots k_{m+1}!}.$$

The precise expression of $\Xi^{(s)}$ follows from the Faà di Bruno formula (Appendix F) with $g(t) = t^{m+2}$.

We claim that (7.4.1) holds for this choice of $\Theta^{(s)}$. To verify this assertion, we apply $m+1$ integration by parts to write

$$\int_0^1 \Theta^{(s)} * \Phi_s \, ds = \int_0^1 (-1)^{m+1} \zeta(s) \Xi^{(s)} * \Phi_s * \Psi \, ds - \int_0^1 \frac{d^{m+1}\zeta}{ds^{m+1}}(s) \overbrace{\Phi_s * \ldots * \Phi_s}^{m+2 \text{ terms}} * \Psi \, ds$$

$$= \int_0^1 (-1)^{m+1} \zeta(s) \Xi^{(s)} * \Psi * \Phi_s \, ds$$

$$+ \frac{d^m \zeta}{ds^m}(0) \lim_{s \to 0+} \big(\overbrace{\Phi * \cdots * \Phi}^{m+2 \text{ terms}} \big)_s * \Psi$$

$$- (-1)^{m+1} \int_0^1 \zeta(s) \frac{d^{m+1}}{ds^{m+1}} \Big(\overbrace{\Phi_s * \cdots * \Phi_s}^{m+2 \text{ terms}} \Big) * \Psi \, ds,$$

noting that all the boundary terms vanish except for the term at $s=0$ in the first integration by parts. The first and the third terms in the previous expression on the right add up to zero, while the second term is equal to Ψ, since Φ has integral 1. This implies that the family $\{(\Phi * \cdots * \Phi)_s\}_{s>0}$ is an approximate identity as $s \to 0$ (Example 1.9.2). Therefore, (7.4.1) holds.

We now prove estimate (7.4.2). Let Ω be the $(m+1)$-fold convolution of Φ. For the term after the minus sign in (7.4.3), we note that the $(m+1)$st derivative of $\zeta(s)$ vanishes on $[0, \tfrac{1}{2}]$, so that we may write

$$\int_{\mathbf{R}^n} (1+|x|)^m \left| \frac{d^{m+1}\zeta(s)}{ds^{m+1}} \right| |\Omega_s * \Psi(x)| \, dx$$

$$\leq C_m \chi_{[\frac{1}{2},1]}(s) \int_{\mathbf{R}^n} (1+|x|)^m \left[\int_{\mathbf{R}^n} \tfrac{1}{s^n} |\Omega(\tfrac{x-y}{s})| \, |\Psi(y)| \, dy \right] dx$$

$$\leq C_m \chi_{[\frac{1}{2},1]}(s) \int_{\mathbf{R}^n} \int_{\mathbf{R}^n} (1+|y+sx|)^m |\Omega(x)| \, |\Psi(y)| \, dy \, dx$$

$$\leq C_m \chi_{[\frac{1}{2},1]}(s) \int_{\mathbf{R}^n} \int_{\mathbf{R}^n} (1+|sx|)^m |\Omega(x)| \, (1+|y|)^m |\Psi(y)| \, dy \, dx$$

$$\leq C_m \chi_{[\frac{1}{2},1]}(s) \left(\int_{\mathbf{R}^n} (1+|x|)^m |\Omega(x)| \, dx \right) \left(\int_{\mathbf{R}^n} (1+|y|)^m |\Psi(y)| \, dy \right)$$

$$\leq C_{m,\Phi} s^m \int_{\mathbf{R}^n} (1+|y|)^m |\Psi(y)| \, dy,$$

as $\chi_{[\frac{1}{2},1]}(s) \leq 2^m s^m$. To obtain a similar estimate for the term before the minus sign in (7.4.3), we argue as follows. A generic term in the sum defining $\Xi^{(s)}$ has the form

$$\frac{d^{j_1}\Phi_s}{ds^{j_1}} * \cdots * \frac{d^{j_L}\Phi_s}{ds^{j_L}},$$

where $j_1, \ldots, j_L$ are nonnegative integers satisfying $j_1 + \cdots + j_L = m+1$, in view of the fact that $1k_1 + \cdots + (m+1)k_{m+1} = m+1$. For $j_1 \leq m+1$ we have

$$\int_{\mathbf{R}^n} (1+|x|)^m \left| \frac{d^{j_1}\Phi_s}{ds^{j_1}} * \Psi(x) \right| dx$$

$$= \int_{\mathbf{R}^n} (1+|x|)^m \left| \frac{d^{j_1}}{ds^{j_1}} \int_{\mathbf{R}^n} \Phi(y)\Psi(x-sy)\,dy \right| dx$$

$$= \int_{\mathbf{R}^n} (1+|x|)^m \left| \int_{\mathbf{R}^n} \Phi(y) \frac{d^{j_1}}{ds^{j_1}} \Psi(x-sy)\,dy \right| dx$$

$$\leq \int_{\mathbf{R}^n} (1+|x|)^m \int_{\mathbf{R}^n} |\Phi(y)| \left[\sum_{|\alpha| \leq j_1} |\partial^\alpha \Psi(x-sy)| |y|^{|\alpha|} \right] dy\,dx$$

$$\leq \int_{\mathbf{R}^n} \int_{\mathbf{R}^n} (1+|x+sy|)^m |\Phi(y)| \sum_{|\alpha| \leq j_1} |\partial^\alpha \Psi(x)| (1+|y|)^{j_1} dy\,dx$$

$$\leq \int_{\mathbf{R}^n} (1+|y|)^{j_1} |\Phi(y)| (1+|y|)^m dy \int_{\mathbf{R}^n} (1+|x|)^m \sum_{|\alpha| \leq j_1} |\partial^\alpha \Psi(x)| dx$$

$$\leq C'_{m,\Phi} \int_{\mathbf{R}^n} (1+|x|)^m \sum_{|\alpha| \leq j_1} |\partial^\alpha \Psi(x)| dx,$$

using $j_1 \leq m+1$. We have now proved that for some constant $C'_{m,\Phi} > 0$ we have

$$\int_{\mathbf{R}^n} (1+|x|)^m \left| \frac{d^{j_1}\Phi_s}{ds^{j_1}} * \Psi(x) \right| dx \leq C'_{m,\Phi} \sum_{|\alpha_1| \leq j_1} \int_{\mathbf{R}^n} (1+|x|)^m |\partial^{\alpha_1} \Psi(x)| dx. \quad (7.4.5)$$

Applying (7.4.5) to the function $\frac{d^{j_2}\Phi_s}{ds^{j_2}} * \Psi$ in place of Ψ we obtain

$$\int_{\mathbf{R}^n} (1+|x|)^m \left| \frac{d^{j_1}\Phi_s}{ds^{j_1}} * \frac{d^{j_2}\Phi_s}{ds^{j_2}} * \Psi(x) \right| dx$$

$$\leq C'_{m,\Phi} \sum_{|\alpha_1| \leq j_1} \int_{\mathbf{R}^n} (1+|x|)^m \left| \frac{d^{j_2}\Phi_s}{ds^{j_2}} * \partial^{\alpha_1} \Psi(x) \right| dx$$

$$\leq (C'_{m,\Phi})^2 \sum_{|\alpha_1| \leq j_1} \sum_{|\alpha_2| \leq j_2} \int_{\mathbf{R}^n} (1+|x|)^m |\partial^{\alpha_2} \partial^{\alpha_1} \Psi(x)| dx,$$

where the last estimate follows by another application of (7.4.5). Continuing in this way, we deduce the existence of a positive constant C_m^Φ such that

7.4 Grand Maximal Function

$$\int_{\mathbf{R}^n}(1+|x|)^m\left|\frac{d^{j_1}\Phi_s}{ds^{j_1}}*\cdots*\frac{d^{j_L}\Phi_s}{ds^{j_L}}*\Psi(x)\right|dx\le C_m^\Phi\int_{\mathbf{R}^n}(1+|x|)^m\sum_{|\alpha|\le m+1}|\partial^\alpha\Psi(x)|\,dx,$$

as $j_1+\cdots+j_L=m+1$. Summing these estimates over all terms that appear in the sum defining $\Xi^{(s)}$, we deduce the same estimate for $\Xi^{(s)}*\Psi$.

Keeping in mind that the term before the minus sign in (7.4.3) contains the function $\zeta(s)$, which is pointwise bounded by s^m for $0<s\le 1$, yields the desired estimate. This concludes the proof of (7.4.2). □

Remark 7.4.2. We use the notation of Lemma 7.4.1. A straightforward adaptation of the preceding proof yields that for some constant $C_0'(\Phi,m)$ one has

$$\frac{1}{s^m}\int_{\mathbf{R}^n}(1+|x|)^m\left|s\frac{d}{ds}\Theta^{(s)}(x)\right|dx\le C_0'(\Phi,m)\int_{\mathbf{R}^n}(1+|x|)^m\sum_{|\alpha|\le m+1}|\partial^\alpha\Psi(x)|\,dx.$$

To verify this assertion, we will show that $s\frac{d}{ds}\Theta^{(s)}$ satisfies similar estimates to $\Theta^{(s)}$. We first observe that for any Schwartz function Φ there is another Schwartz function $\check{\Phi}(x)=-n\Phi(x)-\nabla\Phi(x)\cdot x$ such that $s\frac{d}{ds}\Phi_s=\check{\Phi}_s$ for any $s>0$. Notice that if Φ has integral 1, this will not be the case for $\check{\Phi}$. However, in the proof of (7.4.2), we did not make use of the fact that Φ has integral 1, just that Φ was a Schwartz function. Also, the proof did not depend on Φ_s being convolved with itself; it would work with the convolution of distinct Schwartz functions. Applying $s\frac{d}{ds}$ to (7.4.3) we obtain

$$s\frac{d}{ds}\Theta^{(s)}=(-1)^{m+1}s\frac{d\zeta}{ds}(s)\,\Xi^{(s)}*\Psi$$

$$+(-1)^{m+1}\zeta(s)\,s\frac{d\Xi^{(s)}}{ds}*\Psi$$

$$-s\frac{d^{m+2}\zeta}{ds^{m+2}}(s)\,\overbrace{\Phi_s*\cdots*\Phi_s}^{m+1\text{ terms}}*\Psi$$

$$-\frac{d^{m+1}\zeta}{ds^{m+1}}(s)\,(m+1)\,\check{\Phi}_s*\overbrace{\Phi_s*\cdots*\Phi_s}^{m\text{ terms}}*\Psi.$$

We notice that both $s\frac{d}{ds}\zeta(s)$ and $s\frac{d^{m+2}}{ds^{m+2}}\zeta(s)$ are bounded by a constant multiple of s^m, just like $\zeta(s)$ and $\frac{d^{m+1}}{ds^{m+1}}\zeta(s)$ were. Additionally, in $s\frac{d}{ds}\Xi^{(s)}(x)$ one occurrence of Φ_s is replaced by $\check{\Phi}_s$; hence these expressions satisfy similar estimates as $\Xi^{(s)}$.

In the proof of Lemma 7.4.1 the norm-looking quantity

$$\int_{\mathbf{R}^n}(1+|x|)^m\sum_{|\alpha|\le m+1}|\partial^\alpha\Psi(x)|\,dx,\qquad \Psi\in\mathscr{S}(\mathbf{R}^n),$$

appeared. It turns out that this expression plays a crucial role in the theory of Hardy spaces. In particular, it can be used to show that the function $\Phi(x)=e^{-\pi|x|^2}$ in the definition of H^p can be replaced by any other Schwartz function with non vanishing integral.

Definition 7.4.3. For a fixed positive integer N we define the expression

$$\mathfrak{N}_N(\varphi) = \int_{\mathbf{R}^n} (1+|x|)^N \sum_{|\alpha| \leq N+1} |\partial^\alpha \varphi(x)|\, dx \qquad (7.4.6)$$

on Schwartz functions φ. We denote by $\mathscr{F}_N$ the subset of Schwartz functions

$$\mathscr{F}_N = \left\{ \varphi \in \mathscr{S}(\mathbf{R}^n) : \mathfrak{N}_N(\varphi) \leq 1 \right\}. \qquad (7.4.7)$$

Using this we define the *grand maximal function of* $f \in \mathscr{S}'(\mathbf{R}^n)$ *(with respect to N)* by

$$\mathscr{M}_N(f)(x) = \sup_{\varphi \in \mathscr{F}_N} M^*(f;\varphi)(x).$$

Theorem 7.4.4. *Fix* $0 < p < \infty$ *and* $\Phi \in \mathscr{S}(\mathbf{R}^n)$ *with* $\int_{\mathbf{R}^n} \Phi(x)\, dx \neq 0$. *Let* $N \in \mathbf{Z}^+$, $N \geq [\frac{n}{p}]+1$. *Then there is a constant* $C(n,p,\Phi)$ *such that for any* $f \in \mathscr{S}'(\mathbf{R}^n)$

$$\frac{1}{\mathfrak{N}_N(\Phi)} \|M^*(f;\Phi)\|_{L^p} \leq \|\mathscr{M}_N(f)\|_{L^p} \leq C(n,p,\Phi) \|M^*(f;\Phi)\|_{L^p}. \qquad (7.4.8)$$

Proof. Obviously the lower inequality in (7.4.8) holds as the Schwartz function $\Phi/\mathfrak{N}_N(\Phi)$ lies in $\mathscr{F}_N$ and therefore for all $f \in \mathscr{S}'(\mathbf{R}^n)$ we have

$$M^*\!\left(f; \frac{\Phi}{\mathfrak{N}_N(\Phi)}\right) \leq \mathscr{M}_N(f).$$

Let $b > 0$. A tool that will be used in the proof of the upper inequality in (7.4.8) is the following *auxiliary maximal function*

$$M_b^{**}(f;\Phi)(x) = \sup_{t>0} \sup_{y \in \mathbf{R}^n} \frac{|(\Phi_t * f)(x-y)|}{(1+|y|/t)^b}. \qquad (7.4.9)$$

We observe that

$$M^*(f;\Phi) \leq 2^b M_b^{**}(f;\Phi) \qquad (7.4.10)$$

as for any $t > 0$ we have

$$\sup_{y \in \mathbf{R}^n,\, |y|<t} |(\Phi_t * f)(x-y)| \leq \sup_{y \in \mathbf{R}^n} \frac{2^b}{(1+|y|/t)^b} |(\Phi_t * f)(x-y)|.$$

The role of M_b^{**} is apparent in the following assertions:
(A) For every $b > n/p$ and every Φ in $\mathscr{S}(\mathbf{R}^n)$ there exists $C_1(n,p,b) < \infty$ such that for all $f \in \mathscr{S}'(\mathbf{R}^n)$ we have

$$\|M_b^{**}(f;\Phi)\|_{L^p} \leq C_1(n,p,b) \|M^*(f;\Phi)\|_{L^p}. \qquad (7.4.11)$$

7.4 Grand Maximal Function

(B) For every $b > 0$ and every Φ in $\mathscr{S}(\mathbf{R}^n)$ with $\int_{\mathbf{R}^n} \Phi(x)\,dx = 1$ there exists a constant $C_2(b, \Phi) < \infty$ such that if $N \geq [b] + 1$ for all $f \in \mathscr{S}'(\mathbf{R}^n)$ we have

$$\left\|\mathscr{M}_N(f)\right\|_{L^p} \leq C_2(b, \Phi) \left\|M_b^{**}(f; \Phi)\right\|_{L^p}. \tag{7.4.12}$$

We now prove these statements. It follows from the definition of

$$M^*(f; \Phi)(z) = \sup_{t>0} \sup_{|w-z|<t} |(\Phi_t * f)(w)|$$

that

$$|(\Phi_t * f)(x-y)| \leq M^*(f; \Phi)(z) \qquad \text{if } z \in B(x-y, t).$$

But the ball $B(x-y, t)$ is contained in the ball $B(x, |y| + t)$; hence it follows that

$$\begin{aligned}
|(\Phi_t * f)(x-y)|^{\frac{n}{b}} &\leq \frac{1}{|B(x-y,t)|} \int_{B(x-y,t)} M^*(f;\Phi)(z)^{\frac{n}{b}}\,dz \\
&\leq \frac{1}{|B(x-y,t)|} \int_{B(x,|y|+t)} M^*(f;\Phi)(z)^{\frac{n}{b}}\,dz \\
&\leq \left(\frac{|y|+t}{t}\right)^n M\!\left(M^*(f;\Phi)^{\frac{n}{b}}\right)(x),
\end{aligned}$$

from which we conclude that for all $x \in \mathbf{R}^n$ we have

$$M_b^{**}(f;\Phi)(x) \leq \left[M\!\left(M^*(f;\Phi)^{\frac{n}{b}}\right)(x)\right]^{\frac{b}{n}}.$$

Raising to the power p and using the fact that $p > n/b$ and the boundedness of the Hardy–Littlewood maximal operator M on $L^{pb/n}$, we obtain conclusion (7.4.11).

In proving (B) we may replace b by the integer $b_0 = [b] + 1$. Let Φ be a Schwartz function with integral equal to 1. Applying Lemma 7.4.1 with $m = b_0$, we write any function φ in $\mathscr{S}(\mathbf{R}^n)$ as

$$\varphi(y) = \int_0^1 (\Theta^{(s)} * \Phi_s)(y)\,ds$$

for some choice of Schwartz functions $\Theta^{(s)}$. Then we have

$$\varphi_t(y) = \int_0^1 ((\Theta^{(s)})_t * \Phi_{ts})(y)\,ds$$

for all $t > 0$. Let $f \in \mathscr{S}'(\mathbf{R}^n)$. We claim that

$$\varphi_t * f = \int_0^1 (\Theta^{(s)})_t * \Phi_{ts} * f\,ds, \tag{7.4.13}$$

noting that this integral converges absolutely, in view of Theorem 2.7.1. We will prove (7.4.13) at the end.

Assuming (7.4.13), for a fixed $x \in \mathbf{R}^n$ and for y in $B(x,t)$ we write

$$\begin{aligned}
|(\varphi_t * f)(y)| &\leq \int_0^1 \int_{\mathbf{R}^n} |(\Theta^{(s)})_t(z)||(\Phi_{ts} * f)(y-z)|\,dz\,ds \\
&\leq \int_0^1 \int_{\mathbf{R}^n} |(\Theta^{(s)})_t(z)| M_{b_0}^{**}(f;\Phi)(x) \left(\frac{|x-(y-z)|}{st}+1\right)^{b_0} dz\,ds \\
&\leq \int_0^1 s^{-b_0} \int_{\mathbf{R}^n} |(\Theta^{(s)})_t(z)| M_{b_0}^{**}(f;\Phi)(x) \left(\frac{|x-y|}{t}+\frac{|z|}{t}+1\right)^{b_0} dz\,ds \\
&\leq 2^{b_0} M_{b_0}^{**}(f;\Phi)(x) \int_0^1 s^{-b_0} \int_{\mathbf{R}^n} |\Theta^{(s)}(w)|\,(|w|+1)^{b_0} dw\,ds \\
&\leq 2^{b_0} M_{b_0}^{**}(f;\Phi)(x) \int_0^1 s^{-b_0} C_0(\Phi,b_0) s^{b_0} \mathfrak{N}_{b_0}(\varphi)\,ds,
\end{aligned}$$

where we applied conclusion (7.4.2) of Lemma 7.4.1. Setting $N = b_0 = [b]+1$, we obtain for y in $B(x,t)$ and $\varphi \in \mathscr{S}(\mathbf{R}^n)$,

$$|(\varphi_t * f)(y)| \leq 2^{b_0} C_0(\Phi,b_0)\, \mathfrak{N}_{b_0}(\varphi)\, M_{b_0}^{**}(f;\Phi)(x).$$

Taking the supremum over all y in $B(x,t)$, over all $t > 0$, and over all φ in $\mathscr{F}_N$, we obtain the pointwise estimate

$$\mathscr{M}_N(f)(x) \leq 2^{b_0} C_0(\Phi,b_0) M_{b_0}^{**}(f;\Phi)(x), \qquad x \in \mathbf{R}^n,$$

where $N = b_0$. This clearly yields (7.4.12) if we set $C_2 = 2^{b_0} C_0(\Phi,b_0)$.

It remains to prove (7.4.13). Let us set

$$H(s,x) = (\Theta^{(s)} * \Phi_s)(x), \qquad x \in \mathbf{R}^n.$$

Notice that $((\Theta^{(s)})_t * \Phi_{ts})(x) = t^{-n} H(s,x/t)$. It will suffice to show that the partial sums of $\int_0^1 H(s,x/t)\,ds$ converge to this integral in the topology of Schwartz functions. In other words, if $s_i = i/N$, $i = 1, 2, \ldots, N$, we must prove that for fixed $t > 0$ we have

$$\sup_{x \in \mathbf{R}^n} \left| \sum_{i=1}^N \int_{s_{i-1}}^{s_i} [x^\alpha \partial_x^\beta H(s_{i-1}, \tfrac{x}{t}) - x^\alpha \partial_x^\beta H(u, \tfrac{x}{t})]\,du \right| \to 0 \qquad (7.4.14)$$

as $N \to \infty$ for all multi-indices α, β. Using the fundamental theorem of calculus we can express the supremum in (7.4.14) as

$$\sup_{x \in \mathbf{R}^n} \left| \sum_{i=1}^N \int_{s_{i-1}}^{s_i} \left[\int_{s_{i-1}}^u x^\alpha \partial_x^\beta \frac{d}{ds} H(s, \tfrac{x}{t})\,ds \right] du \right|,$$

which is pointwise bounded by

7.4 Grand Maximal Function

$$\frac{1}{N} \sup_{x \in \mathbf{R}^n} \int_0^1 \left| x^\alpha \partial_x^\beta \frac{d}{ds} H\left(s, \frac{x}{t}\right) \right| ds.$$

This expression can be rewritten as

$$\frac{t^{|\alpha|}}{t^{|\beta|} N} \sup_{x \in \mathbf{R}^n} \int_0^1 \left[\left(\frac{x}{t}\right)^\alpha \left(\frac{d}{ds} \Theta^{(s)} * \frac{(\partial_x^\beta \Phi)_s}{s^{|\beta|}} + \Theta^{(s)} * \frac{d}{ds} \frac{(\partial_x^\beta \Phi)_s}{s^{|\beta|}} \right) \left(\frac{x}{t}\right) \right] ds. \quad (7.4.15)$$

Now the quantity inside the square brackets in (7.4.15) can be estimated by

$$\frac{2^{|\alpha|-1}}{s^{|\beta|}} \int_{\mathbf{R}^n} \left| \frac{d}{ds} \Theta^{(s)}(y) \right| |y|^{|\alpha|} \left| (\partial_x^\beta \Phi)_s \left(\frac{x}{t} - y\right) \right| dy$$

$$+ \frac{2^{|\alpha|-1}}{s^{|\beta|}} \int_{\mathbf{R}^n} \left| \frac{d}{ds} \Theta^{(s)}(y) \right| \left| \frac{x}{t} - y \right|^{|\alpha|} \left| (\partial_x^\beta \Phi)_s \left(\frac{x}{t} - y\right) \right| dy$$

$$+ \frac{2^{|\alpha|-1}}{s^{|\beta|}} \int_{\mathbf{R}^n} \left| \Theta^{(s)}(y) \right| |y|^{|\alpha|} \left| \frac{d}{ds} (\partial_x^\beta \Phi)_s \left(\frac{x}{t} - y\right) \right| dy$$

$$+ \frac{|\beta| 2^{|\alpha|-1}}{s^{|\beta|+1}} \int_{\mathbf{R}^n} \left| \Theta^{(s)}(y) \right| |y|^{|\alpha|} \left| (\partial_x^\beta \Phi)_s \left(\frac{x}{t} - y\right) \right| dy$$

$$+ \frac{2^{|\alpha|-1}}{s^{|\beta|}} \int_{\mathbf{R}^n} \left| \Theta^{(s)}(y) \right| \left| \frac{x}{t} - y \right|^{|\alpha|} \left| \frac{d}{ds} (\partial_x^\beta \Phi)_s \left(\frac{x}{t} - y\right) \right| dy$$

$$+ \frac{|\beta| 2^{|\alpha|-1}}{s^{|\beta|+1}} \int_{\mathbf{R}^n} \left| \Theta^{(s)}(y) \right| \left| \frac{x}{t} - y \right|^{|\alpha|} \left| (\partial_x^\beta \Phi)_s \left(\frac{x}{t} - y\right) \right| dy$$

$$\leq \frac{C}{s^{n+|\beta|}} \int_{\mathbf{R}^n} \left| \frac{d}{ds} \Theta^{(s)}(y) \right| (1 + |y|)^m \, dy$$

$$+ \frac{C s^{|\alpha|}}{s^{n+|\beta|}} \int_{\mathbf{R}^n} \left| \frac{d}{ds} \Theta^{(s)}(y) \right| dy$$

$$+ \frac{C}{s^{n+1+|\beta|}} \int_{\mathbf{R}^n} \left| \Theta^{(s)}(y) \right| (1 + |y|)^m \, dy$$

$$+ \frac{C}{s^{n+1+|\beta|}} \int_{\mathbf{R}^n} \left| \Theta^{(s)}(y) \right| (1 + |y|)^m \, dy$$

$$+ \frac{C s^{|\alpha|}}{s^{n+1+|\beta|}} \int_{\mathbf{R}^n} \left| \Theta^{(s)}(y) \right| dy$$

$$+ \frac{C s^{|\alpha|}}{s^{n+1+|\beta|}} \int_{\mathbf{R}^n} \left| \Theta^{(s)}(y) \right| dy$$

for any $m \geq |\alpha|$, where C is a constant depending on $\Phi, \alpha, \beta, m, n$. Recalling (7.4.2) and the estimate in Remark 7.4.2, picking $m \geq |\beta| + n + 2$ and $m \geq |\alpha|$, we obtain that the preceding displayed expression is bounded, so inserting this in (7.4.15) yields a finite integral, and thus (7.4.15) tends to zero as $N \to \infty$. This concludes the proof of (7.4.13). $\square$

Exercises

7.4.1. Let $0 < p \leq 1$ and $N = [\frac{n}{p}] + 1$. Prove that there is a constant $C(p,n)$ such that for any $f \in H^p(\mathbf{R}^n)$ and any $\varphi \in \mathscr{S}(\mathbf{R}^n)$ and all $t > 0$ we have

$$\|\varphi_t * f\|_{L^\infty} \leq C(p,n)\, t^{-\frac{n}{p}}\, \mathfrak{N}_N(\varphi_t)\|f\|_{H^p}$$

and

$$\|\varphi_t * f\|_{L^p} \leq C(p,n)\, \mathfrak{N}_N(\varphi_t)\|f\|_{H^p}.$$

Deduce that for all r satisfying $p \leq r \leq \infty$ one has

$$\|\varphi_t * f\|_{L^r} \leq C(p,n)\, \mathfrak{N}_N(\varphi_t)\, t^{-n(\frac{1}{p}-\frac{1}{r})}\|f\|_{H^p}.$$

7.4.2. Let $0 < p \leq 1$. Show that for all f in $H^p(\mathbf{R}^n)$, the distributional Fourier transform $\widehat{f}$ is a continuous function. Also prove that there exists a constant $C_{n,p}$ such that for all $\xi \neq 0$

$$|\widehat{f}(\xi)| \leq C_{n,p}\, |\xi|^{\frac{n}{p}-n}\|f\|_{H^p}.$$

[*Hint:* Use the preceding exercise with $r = 1$, $t = |\xi|^{-1}$ and $\varphi(x) = e^\pi e^{-\pi|x|^2}$.]

7.4.3. Let $0 < p \leq 1$ and $N = [\frac{n}{p}] + 1$. Prove that for any $f \in H^p(\mathbf{R}^n)$ and any $\varphi \in \mathscr{S}(\mathbf{R}^n)$ one has

$$|\langle f, \varphi \rangle| \leq \mathfrak{N}_N(\varphi) \inf_{|z| \leq 1} \mathscr{M}_N(f)(z).$$

Use this estimate to show the existence of a constant $C_{n,p}$ such that

$$|\langle f, \varphi \rangle| \leq \mathfrak{N}_N(\varphi) C_{n,p} \|f\|_{H^p}.$$

Conclude that if $f_j \to f$ in H^p, then $f_j \to f$ in $\mathscr{S}'$.

7.4.4. (a) Let $x_0 \in \mathbf{S}^{n-1}$. Prove that there is a constant $C_n > 0$ such that the sequence of functions $f_k = \chi_{B(kx_0,1)} - \chi_{B(-kx_0,1)}$, $k \geq 12$, satisfies[4]

$$\|f_k\|_{H^1} \geq C_n \log \tfrac{k}{4}.$$

(b) Show that $f = \sum_{j=2}^{\infty} \frac{1}{j^2} f_{2^{2j}}$ lies in $L^1(\mathbf{R}^n) \setminus H^1(\mathbf{R}^n)$ and has integral zero.
[*Hint:* Fix a $\Phi \in \mathscr{C}^\infty$ supported in $B(0,1)$ and equal to 1 on $\overline{B(0,\frac{1}{2})}$. Prove that $(\Phi_{t_x} * f_k)(x) \geq v_n t_x^{-n}$ for $x \in B(kx_0, \frac{k}{4})$, where $t_x = 2|x - kx_0| + 2$.]

7.4.5. Given A a closed subset of $\mathbf{R}^n$ and $0 < \gamma < 1$, define

$$A_\gamma^* = \left\{ x \in \mathbf{R}^n : \inf_{r>0} \frac{|A \cap B(x,r)|}{|B(x,r)|} \geq \gamma \right\}.$$

---

[4] A weaker assertion is claimed in Exercise 7.8.2 via a different method.

Show that A_γ^* is a closed subset of A and that it satisfies

$$|(A_\gamma^*)^c| \leq \frac{3^n}{1-\gamma}|A^c|.$$

[*Hint:* Show that

$$(A_\gamma^*)^c \subseteq \{x \in \mathbf{R}^n : M(\chi_{A^c})(x) > 1-\gamma\},$$

where M is the Hardy–Littlewood maximal function.]

7.4.6. (a) For a (not necessarily measurable) function F on $\mathbf{R}_+^{n+1}$ and $a > 0$, set

$$F_a^*(x) = \sup_{t>0} \sup_{y:\, |y-x|<at} |F(y,t)|.$$

Let $0 < a < b < \infty$. Prove that for $\lambda > 0$ all sets below are open and satisfy

$$|\{F_a^* > \lambda\}| \leq |\{F_b^* > \lambda\}| \leq 3^n a^{-n}(a+b)^n|\{F_a^* > \lambda\}|.$$

(b) For $f \in \mathscr{S}'(\mathbf{R}^n)$ and $\Phi \in \mathscr{S}(\mathbf{R}^n)$ and $x \in \mathbf{R}^n$ define

$$M_a^*(f; \Phi)(x) = \sup_{t>0} \sup_{y:\, |y-x|<at} |(\Phi_t * f)(y)|.$$

Conclude that $\|M_a^*(f;\Phi)\|_{L^p}$ and $\|M_b^*(f;\Phi)\|_{L^p}$ are comparable for all $a,b,p > 0$. [*Hint:* Apply Exercise 7.4.5 with $\gamma = \frac{(a+b)^n - a^n}{(a+b)^n}$ and $A := \{F_a^* \leq \lambda\}$, and prove that $\{F_b^* > \lambda\} \subseteq (A_\gamma^*)^c$.]

7.5 The Whitney Decomposition of Open Sets

We denote by $\ell(Q)$ the side length of a cube Q. In this section we decompose a proper open subset Ω of $\mathbf{R}^n$ as a union of dyadic cubes with side lengths comparable to their distance to the boundary of Ω.

As usual, we denote by $\mathscr{D}_k$ be the collection of all dyadic cubes of the form

$$\{(x_1,\ldots,x_n) \in \mathbf{R}^n : m_j 2^{-k} \leq x_j < (m_j+1)2^{-k}\},$$

where $m_j \in \mathbf{Z}$. Bisecting each side, we can write each cube in $\mathscr{D}_k$ as a union of 2^n cubes in $\mathscr{D}_{k+1}$. We denote by $\mathscr{D} = \cup_{k\in\mathbf{Z}}\mathscr{D}_k$ the set of all dyadic cubes.

Definition 7.5.1. Two dyadic cubes in $\mathbf{R}^n$ are called *adjacent* if they are disjoint and their boundaries intersect (touch). This intersection could be a point, an edge, or a face of dimension at most $n-1$.

The closures of adjacent dyadic intervals have a common endpoint, but the closures of adjacent dyadic squares in $\mathbf{R}^2$ could share a corner or an edge. In $\mathbf{R}^3$ they may share a corner, an edge or a face.

Theorem 7.5.2. *(**Whitney decomposition**) Let Ω be a nonempty proper open subset of $\mathbf{R}^n$. Then there exists a family of disjoint dyadic cubes $\mathscr{F} = \{Q_j\}_j$ such that*
(a) $\bigcup_j Q_j = \Omega$.
(b) *For every $Q_j \in \mathscr{F}$ we have*

$$\sqrt{n}\,\ell(Q_j) \leq \mathrm{dist}\,(\overline{Q_j}, \Omega^c) < 4\sqrt{n}\,\ell(Q_j).$$

Thus there exists $\xi_j \in \Omega^c$ such that $|\xi_j - \text{center of } Q_j| \leq \frac{9}{2}\sqrt{n}\,\ell(Q_j)$.
(c) *If the cubes Q_j and Q_k in $\mathscr{F}$ are adjacent, then*

$$\frac{1}{4} \leq \frac{\ell(Q_j)}{\ell(Q_k)} \leq 4.$$

(d) *Given Q_j in $\mathscr{F}$ there exist at most $6^n - 4^n$ cubes Q_k in $\mathscr{F}$ adjacent to Q_j.*
(e) *Let $0 < \varepsilon < \frac{2}{5}$. For $Q_j \in \mathscr{F}$ define Q_j^* as the cube with the same center as Q_j and with $\ell(Q_j^*) = (1+\varepsilon)\ell(Q_j)$. Then if Q_j, Q_i in $\mathscr{F}$ are disjoint and not adjacent, we must have that Q_j^* and Q_i^* are disjoint. Moreover, all Q_j^* are contained in Ω and*

$$\chi_\Omega \leq \sum_j \chi_{Q_j^*} \leq 2^n \chi_\Omega.$$

Proof. Write the open set Ω as the union of Ω_k, $k \in \mathbf{Z}$, where

$$\Omega_k = \{x \in \Omega : 2\sqrt{n}\,2^{-k} \leq \mathrm{dist}(x, \Omega^c) < 4\sqrt{n}\,2^{-k}\}.$$

Let

$$\mathscr{G} = \{Q \in \mathscr{D} : \exists\, k \in \mathbf{Z} \text{ such that } Q \in \mathscr{D}_k \text{ and } Q \cap \Omega_k \neq \emptyset\}.$$

We show that the collection $\mathscr{G}$ satisfies property (b). Let $Q \in \mathscr{G} \cap \mathscr{D}_k$ for some k and pick $x \in \Omega_k \cap Q$. Then we have

$$\sqrt{n}\,2^{-k} \leq \mathrm{dist}(x, \Omega^c) - \sqrt{n}\,\ell(Q) \leq \mathrm{dist}(Q, \Omega^c) \leq \mathrm{dist}(x, \Omega^c) < 4\sqrt{n}\,2^{-k}.$$

This proves that the collection $\mathscr{G}$ satisfies the first assertion in (b). Now, given $Q \in \mathscr{G}$ there exists a point $\xi \in \Omega^c$ whose distance from Q is at most $4\sqrt{n}\,\ell(Q)$. Then the distance from ξ to the center of Q is at most

$$4\sqrt{n}\,\ell(Q) + \frac{\sqrt{n}}{2}\ell(Q).$$

This proves that $\mathscr{G}$ satisfies the second assertion in (b) as well.

We note that every Q in $\mathscr{G}$ is contained in Ω since it has distance at least $\sqrt{n}\,\ell(Q)$ (which is strictly positive) from its complement. Thus $\bigcup_{Q \in \mathscr{G}} Q \subseteq \Omega$. Now for every $x \in \Omega$ there exists $k \in \mathbf{Z}$ such that x lies in Ω_k. But this x lies in some cube $Q_x \in \mathscr{D}_k$, hence $Q_x \cap \Omega_k \neq \emptyset$, so $Q_x \in \mathscr{G}$ and thus $x \in \bigcup_{Q \in \mathscr{G}} Q$. Then $\Omega \subseteq \bigcup_{Q \in \mathscr{G}} Q$. Combining these facts we conclude that

$$\Omega = \bigcup_{Q \in \mathscr{G}} Q.$$

7.5 The Whitney Decomposition of Open Sets

The problem is that the cubes in $\mathscr{G}$ may not be disjoint. We then refine $\mathscr{G}$ by eliminating those cubes that are contained in some other cubes in the collection. Two dyadic cubes are disjoint or are related by inclusion. For every cube Q in $\mathscr{G}$ we can therefore consider the unique *maximal* cube $Q^{\max}$ in $\mathscr{G}$ that contains it. Two different such maximal cubes must have disjoint interiors by maximality.

Now set $\mathscr{F} = \{Q^{\max} : Q \in \mathscr{G}\}$. The collection of cubes $\{Q_j\}_j = \mathscr{F}$ satisfies (a) and (b) by construction, and we now turn our attention to the proof of (c). Observe that if Q_j and Q_k in $\mathscr{F}$ are adjacent then

$$\sqrt{n}\,\ell(Q_j) \leq \operatorname{dist}(Q_j, \Omega^c) \leq \operatorname{dist}(Q_j, Q_k) + \operatorname{dist}(Q_k, \Omega^c) < 0 + 4\sqrt{n}\,\ell(Q_k),$$

and, as the roles of Q_j and Q_k could be interchanged, this proves (c).

To prove (d), note that the largest number of dyadic cubes adjacent to a fixed cube $Q_j \in \mathscr{D}_k \cap \mathscr{F}$ is obtained when all the cubes have the smallest possible size, i.e., they lie in $\mathscr{D}_{k+2}$ in view of (c). But $\frac{6}{4}Q_j$ contains 6^n subcubes of side length 2^{-k-2} and Q_j itself contains 4^n such subcubes. This means that $6^n - 4^n$ subcubes of $\frac{6}{4}Q_j$ of side length 2^{-k-2} must be adjacent to Q_j. This yields the assertion in (d). [Here $\frac{6}{4}Q_j$ is concentric with Q_j and $\ell(\frac{6}{4}Q_j) = \frac{6}{4}\ell(Q_j)$.]

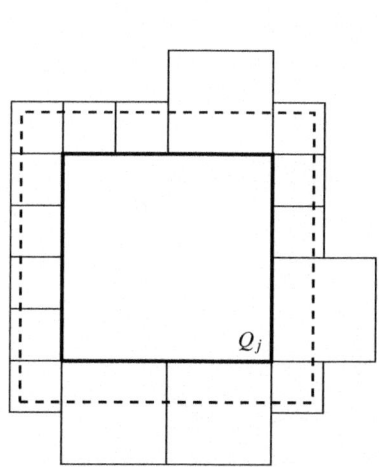

Fig. 7.2 The cube $(1 + \frac{2}{5})Q_j = \frac{7}{5}Q_j$ (shown in dots) is properly contained in the union of Q_j and its adjacent cubes. This is because all adjacent cubes have side lengths at least $\frac{1}{4}\ell(Q_j)$.

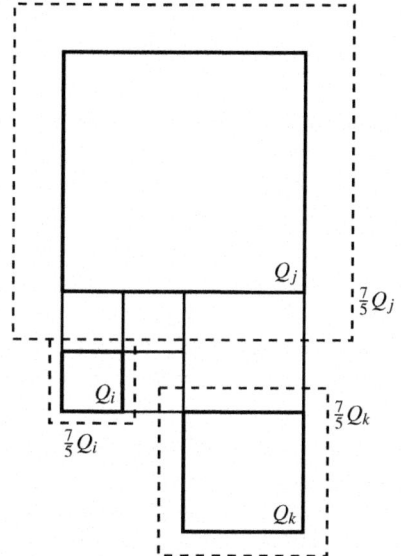

Fig. 7.3 If Q_j is not adjacent to either Q_i or Q_k and $\ell(Q_j) = 4\ell(Q_i) = 2\ell(Q_k)$, then $\frac{7}{5}Q_j \cap \frac{7}{5}Q_k$ is empty. Also $\frac{7}{5}Q_j$ and $\frac{7}{5}Q_i$ are disjoint but their boundaries may touch. Hence, $(1+\varepsilon)$-envelops of non-adjacent cubes in $\mathscr{F}$ are disjoint if $\varepsilon < \frac{2}{5}$.

For part (e) we verify first that if Q_j, Q_i in $\mathscr{F}$ are not adjacent, then Q_j^* and Q_i^* are disjoint. Without loss of generality assume that $\ell(Q_j) \geq \ell(Q_i)$. If $\ell(Q_j) = \ell(Q_i)$, then the distance between Q_j and Q_i is at least $\ell(Q_j)$, which forces Q_j^* and Q_i^* to be disjoint. Also, if $\ell(Q_j) = 2\ell(Q_i)$, then $\frac{7}{5}Q_j$ and $\frac{7}{5}Q_i$ are disjoint. Finally, if $\ell(Q_j) = 4\ell(Q_i)$, then, in the worst case, the disjoint cubes $\frac{7}{5}Q_j$ and $\frac{7}{5}Q_i$ share parts of their boundaries as

$$\frac{1}{5}\ell(Q_j) + \frac{1}{5}\ell(Q_i) = \ell(Q_i) \le \text{dist}(Q_i, Q_j);$$

see Figure 7.3. Thus for $\varepsilon < \frac{2}{5}$, Q_j^* and Q_i^* are always disjoint.

Next we claim that each $Q_j^* = (1+\varepsilon)Q_j$ is contained in Ω if $\varepsilon < \frac{2}{5}$; to see this we observe that Q_j^* is contained in the union of Q_j and its adjacent cubes (see Figure 7.2), since $\frac{1}{5}\ell(Q_j)$ is smaller than the length of any adjacent cube, which is at least $\frac{1}{4}\ell(Q_j)$ by part (b).

The lower inequality in (e) is a consequence of the facts that $Q_j \subseteq Q_j^*$ and that the union of the Q_j equals Ω. For the upper inequality in (e), recall that if Q_j^* and Q_i^* intersect and $i \ne j$, then Q_j and Q_i are adjacent. So we need to find the maximum number of pairwise adjacent dyadic cubes Q_i such that the intersection of the corresponding Q_i^* is non empty. A moment's thought gives that this number is at most 2^n by the construction of dyadic cubes (and this happens exactly when these cubes share a common corner.) This proves the upper inequality in (e). □

Definition 7.5.3. The cubes Q_j obtained in the construction of Theorem 7.5.2 are called the *Whitney cubes* of Ω.

Example 7.5.4. Suppose that our open set is $\Omega = (0,1)$. Then the dyadic intervals $[\frac{1}{4}, \frac{2}{4})$ and $[\frac{2}{4}, \frac{3}{4})$ have distance from Ω^c exactly equal to their side length. And these are the largest dyadic intervals contained in Ω with distance to Ω^c at least their side length. The next generation of dyadic intervals with the same property are $[\frac{1}{8}, \frac{2}{8})$ and $[\frac{6}{8}, \frac{7}{8})$, the next generation are $[\frac{1}{16}, \frac{2}{16})$ and $[\frac{14}{16}, \frac{15}{16})$, etc. All these intervals form a Whitney decomposition of $(0,1)$.

The Whitney decomposition of the unit disc in $\mathbf{R}^2$ is obtained via a similar procedure and is shown in Figure 7.4.

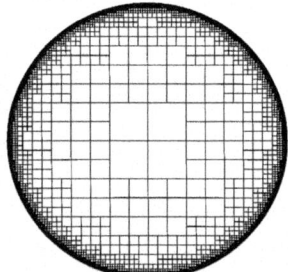

Fig. 7.4 The Whitney decomposition of the unit disk.

Remark 7.5.5. Let Ω, Ω' be nonempty open sets satisfying $\Omega \subseteq \Omega' \subsetneq \mathbf{R}^n$. Then every Whitney cube of Ω is contained in some Whitney cube of Ω'.

To verify this assertion, define $\mathscr{G}'$ and $\mathscr{F}'$ associated with Ω' in the same way that $\mathscr{G}$ and $\mathscr{F}$ are associated with Ω. For $k \in \mathbf{Z}$, we also define Ω_k' analogously. We claim that

$$\Omega_k \subseteq \bigcup_{l \le k} \Omega_l'. \qquad (7.5.1)$$

7.5 The Whitney Decomposition of Open Sets

Indeed, let $x \in \Omega_k$ for some $k \in \mathbf{Z}$. Then $-k$ is the largest integer with the property $2^{-k} \leq \frac{1}{2\sqrt{n}}\text{dist}(x, \Omega^c)$. As $\text{dist}(x, \Omega^c) \leq \text{dist}(x, (\Omega')^c)$, it follows that the largest $-l$ such that $2^{-l} \leq \frac{1}{2\sqrt{n}}\text{dist}(x, (\Omega')^c)$ must satisfy $-l \geq -k$. Then $x \in \Omega'_l$ for some $l \leq k$.

Let Q be a cube in $\mathscr{F}$. Then Q belongs to $\mathscr{G}$ and thus there is a $k \in \mathbf{Z}$ such that $Q \in \mathscr{D}_k$ and $Q \cap \Omega_k \neq \emptyset$. It follows from (7.5.1) that $Q \cap \Omega'_l \neq \emptyset$ for some $l \leq k$. Pick a dyadic ancestor Q' of Q of length 2^{-l}. Then $Q' \in \mathscr{D}_l$ and $Q' \cap \Omega'_l \neq \emptyset$. Hence Q' lies in $\mathscr{G}'$ and is therefore contained in some cube Q'' in $\mathscr{F}'$. Then every Whitney cube Q of Ω is contained in a Whitney cube Q'' of Ω'.

Next we construct a smooth partition of unity adapted to a Whitney decomposition.

Lemma 7.5.6. *Let $\{Q_j\}_j$ be the Whitney decomposition of a nonempty and proper open subset Ω of $\mathbf{R}^n$. Then there are functions $\{\varphi_j\}_j$ of class $\mathscr{C}^\infty$ such that*

(i) $0 \leq \varphi_j \leq 1$ *for all* j.
(ii) φ_j *is supported in* $\frac{9}{8}Q_j$, *which is concentric with Q_j and has length $\frac{9}{8}\ell(Q_j)$.*
(iii) *For any multi-index α there is $C_\alpha > 0$ such that for all j we have*

$$|\partial^\alpha \varphi_j| \leq \frac{C_\alpha}{\ell(Q_j)^{|\alpha|}}.$$

(iv) *For all j we have*

$$\frac{1}{2^n} \leq \frac{1}{|Q_j|}\int_{\mathbf{R}^n} \varphi_j(y)\,dy \leq \left(\frac{9}{8}\right)^n.$$

(v) *The family $\{\varphi_j\}_j$ forms a partition of unity of Ω, i.e.,*

$$\sum_j \varphi_j = \chi_\Omega.$$

(vi) *For any multi-index α there is a constant B_α such that for all i, j we have*

$$|\partial^\alpha(\varphi_i \varphi_j)| \leq \frac{B_\alpha}{\ell(Q_j)^{|\alpha|+n}} \int_{\mathbf{R}^n} \varphi_i(y)\varphi_j(y)\,dy.$$

Proof. Start with an even nonnegative $\mathscr{C}^\infty$ function ω on $\mathbf{R}$ such that

$$\omega(t) = \begin{cases} 0 & |t| \geq \frac{1}{2} + \frac{1}{16}, \\ 1 & |t| \leq \frac{1}{2}, \\ \text{strictly decreasing on} & \left(\frac{1}{2}, \frac{1}{2} + \frac{1}{16}\right). \end{cases} \quad (7.5.2)$$

Such a function exists by Proposition 1.7.3 (b). Now, given a dyadic interval I on the real line, we adapt ω to I by defining

$$\phi_I(t) = \omega\left(\frac{t - c_I}{|I|}\right), \quad (7.5.3)$$

where c_I is the center of I. It follows by construction that for all $k \in \mathbf{Z}^+$ there is a constant C_k such that

$$\left|\phi_I^{(k)}\right| \leq C_k |I|^{-k}. \tag{7.5.4}$$

In fact, $C_k = \|\omega^{(k)}\|_{L^\infty}$. We extend the definition of ϕ_I to higher dimensions as follows: for a dyadic cube $Q = I_1 \times \cdots \times I_n$ we define

$$\Phi_Q(x_1,\ldots,x_n) = \phi_{I_1}(x_1)\cdots\phi_{I_n}(x_n). \tag{7.5.5}$$

In view of (7.5.4), for any multi-index γ there is a constant c_γ such that

$$|\partial^\gamma \Phi_Q| \leq c_\gamma \ell(Q)^{-|\gamma|}. \tag{7.5.6}$$

Let $\{Q_j\}_j$ be the Whitney decomposition of the open set Ω and consider the function Φ_{Q_j} adapted to Q_j as defined in (7.5.5). Notice that $Q_j^* = \frac{9}{8} Q_j$ satisfies $\Phi_{Q_j} \leq \chi_{Q_j^*}$, and property (e) in Theorem 7.5.2 yields

$$\chi_\Omega \leq \sum_j \Phi_{Q_j} \leq 2^n \chi_\Omega. \tag{7.5.7}$$

The family $\{\Phi_{Q_j}\}_j$ is not a partition of unity of Ω, although it is quite close to it. To create a partition of unity from it we define

$$\varphi_j = \begin{cases} \Phi_{Q_j}\left(\sum_s \Phi_{Q_s}\right)^{-1} & \text{on } \Omega, \\ 0 & \text{on } \Omega^c. \end{cases} \tag{7.5.8}$$

We observe that properties (i), (ii), and (v) hold for φ_j by construction.

To prove (iii) we appeal to the Faà di Bruno formula (Appendix F) which says that for a multi-index γ, $\frac{1}{\gamma!} \partial^\gamma \frac{1}{\sum_s \Phi_{Q_s}}$ equals

$$\sum_{\substack{(m_1,\ldots,m_k) \\ (\beta_1,\ldots,\beta_k)}} \frac{(-1)^{m_1+\cdots+m_k}(m_1+\cdots+m_k)!}{(\sum_s \Phi_{Q_s})^{m_1+\cdots+m_k+1}} \frac{\left(\frac{1}{\beta_1!}\partial^{\beta_1}\sum_s \Phi_{Q_s}\right)^{m_1}}{m_1!} \cdots \frac{\left(\frac{1}{\beta_k!}\partial^{\beta_k}\sum_s \Phi_{Q_s}\right)^{m_k}}{m_k!},$$

where the sum is taken over a finite set of $(m_1,\ldots,m_k)$ and $(\beta_1,\ldots,\beta_k)$, where m_j are nonnegative integers and β_j are multi-indices related by $\gamma = m_1\beta_1 + \cdots + m_k\beta_k$ for certain values of k. For a given x that belongs to a fixed Whitney cube Q_r, if Q_s is not Q_r or is not adjacent to Q_r, then Φ_{Q_s} vanishes near x; thus if $\Phi_{Q_s}(x) \neq 0$, then Q_s is comparable to Q_r. In view of this information, the preceding identity gives

$$\left|\partial^\gamma \frac{1}{\sum_s \Phi_{Q_s}}(x)\right| \leq C'_\gamma \sum_{\substack{(m_1,\ldots,m_k) \\ (\beta_1,\ldots,\beta_k)}} \frac{1}{\ell(Q_r)^{m_1|\beta_1|}} \cdots \frac{1}{\ell(Q_r)^{m_k|\beta_k|}} \leq \frac{C''_\gamma}{\ell(Q_r)^{|\gamma|}}. \tag{7.5.9}$$

Applying Leibniz's rule to the function in (7.5.8) and using (7.5.9) and (7.5.6) yields

7.5 The Whitney Decomposition of Open Sets

$$|\partial^\alpha \varphi_j(x)| \leq \sum_{\gamma \leq \alpha} \binom{\alpha}{\gamma} \left|\partial^\gamma \frac{1}{\Sigma_s \Phi_{Q_s}}(x)\right| \left|\partial^{\alpha-\gamma} \Phi_{Q_j}\right| \leq \sum_{\gamma \leq \alpha} \binom{\alpha}{\gamma} \frac{C''_\gamma}{\ell(Q_r)^{|\gamma|}} \frac{c_{\alpha-\gamma}}{\ell(Q_j)^{|\alpha|-|\gamma|}},$$

when $x \in Q_r$. But if φ_j does not vanish near x, then Q_r is Q_j or adjacent to Q_j. Then $\ell(Q_r) \approx \ell(Q_j)$ and this proves the claim in (iii).

We turn our attention to property (iv). Notice that by construction one has

$$1 \leq \frac{1}{|Q_j|} \int_{\mathbf{R}^n} \Phi_{Q_j}(y)\, dy = \prod_{m=1}^n \frac{1}{\ell(Q_j)} \int_{\mathbf{R}^n} \omega\left(\frac{y_m - (c_{Q_j})_m}{\ell(Q_j)}\right) dy_m \leq \left(\frac{9}{8}\right)^n$$

(c_{Q_j} = center of Q_j). These inequalities combined with (7.5.7) yield those in (iv).

To prove (vi) we argue as follows. If $\varphi_i \varphi_j \equiv 0$, then the assertion is trivial. We therefore assume that the function $\varphi_i \varphi_j$ is not identically equal to zero. Then the associated cubes Q_i and Q_j either coincide or are adjacent, and thus they have comparable sizes. Let us assume without loss of generality that $|Q_i| \geq |Q_j|$. Given a multi-index α, we apply Leibniz's rule and the estimates in part (iii) to obtain

$$|\partial^\alpha(\varphi_i \varphi_j)| \leq \sum_{\beta \leq \alpha} \binom{\alpha}{\beta} \frac{C_\beta}{\ell(Q_i)^{|\beta|}} \frac{C_{\alpha-\beta}}{\ell(Q_j)^{|\alpha|-|\beta|}} \leq \frac{B'_\alpha}{\ell(Q_j)^{|\alpha|}}. \tag{7.5.10}$$

It will be sufficient to prove that there is a constant $B'' > 0$ such that

$$\frac{1}{\ell(Q_j)^n} \int_{\mathbf{R}^n} \varphi_i(y) \varphi_j(y)\, dy \geq B''. \tag{7.5.11}$$

Then (vi) would follow with $B_\alpha = B'_\alpha / B''$ by combining (7.5.10) and (7.5.11). The proof of (7.5.11) is contained in the following lemma [part (c)]. □

In the next lemma, φ_j are as in Lemma 7.5.6, while ϕ_I and Φ_Q are introduced in (7.5.3) and (7.5.5), respectively. These are defined in terms of ω given in (7.5.2).

Lemma 7.5.7. (a) *There is a constant B such that for any two dyadic intervals I, J with $\ell(J) \leq 4\ell(I)$, if $\phi_I \phi_J$ is not identically equal to zero, we have*

$$\frac{1}{|J|} \int_{\mathbf{R}} \phi_I(t) \phi_J(t)\, dt \geq B. \tag{7.5.12}$$

(b) *There is a constant B' depending on the dimension such that for any two dyadic cubes Q, R with $\ell(R) \leq 4\ell(Q)$, if $\Phi_Q \Phi_R$ is not the zero function, one has*

$$\frac{1}{|R|} \int_{\mathbf{R}^n} \Phi_Q(y) \Phi_R(y)\, dy \geq B'. \tag{7.5.13}$$

(c) *Let Q_i be a Whitney cube of an open set and let Q_j be a Whitney cube of another open set such that $\ell(Q_j) \leq 4\ell(Q_i)$. If $\varphi_i \varphi_j$ is not the zero function then we have*

$$\frac{1}{|Q_j|} \int_{\mathbf{R}^n} \varphi_i(y) \varphi_j(y)\, dy \geq B'' = 2^{-2n} B'. \tag{7.5.14}$$

Proof. (a) By a translation, we may assume that $J = [0, 2^m)$ for some $m \in \mathbf{Z}$. Then by applying a dilation, we reduce matters to the situation where $J = [0, 1)$ and I is a dyadic interval of size at least $1/4$. Since $\phi_I \phi_J$ is not the zero function, we must have $\frac{9}{8} I \cap \frac{9}{8} J \neq \emptyset$; hence one of the following is true: I is contained in J, or I is adjacent to J, or I contains J. Then the only possibilities for I are

$$[-\tfrac{1}{4}, 0), \; [0, \tfrac{1}{4}), \; [\tfrac{1}{4}, \tfrac{2}{4}), \; [\tfrac{2}{4}, \tfrac{3}{4}), \; [\tfrac{3}{4}, 1), \; [1, \tfrac{5}{4})$$

$$[-\tfrac{1}{2}, 0), \; [0, \tfrac{1}{2}), \; [\tfrac{1}{2}, 1), \; [1, \tfrac{3}{2}),$$

$$[-1, 0), \; [0, 1), \; [1, 2),$$

$$[-2^k, 0), \; [0, 2^k), \qquad \text{for some } k \geq 1.$$

Then it suffices to show that for any interval I above with c_I its center we have

$$\int_{\mathbf{R}} \phi_I(t) \phi_{[0,1)}(t) \, dt = \int_{\mathbf{R}} \omega\Big(\frac{t - c_I}{|I|}\Big) \omega(t - \tfrac{1}{2}) \, dt \geq B. \tag{7.5.15}$$

Exploiting the fact that $\omega(\cdot - \tfrac{1}{2})$ is strictly decreasing and does not vanish on $[1, \tfrac{17}{16})$ indicates that (7.5.15) is valid and in fact, the first of the listed intervals produces the smallest possible constant; thus (7.5.12) holds.

(b) Let $Q = I_1 \times \cdots \times I_n$ and $R = J_1 \times \cdots \times J_n$, where $|I_1| = \cdots = |I_n| = \ell(Q)$ and $|J_1| = \cdots = |J_n| = \ell(R)$. As $\Phi_Q \Phi_R$ is not the zero function, we must have that $\frac{9}{8} Q \cap \frac{9}{8} R \neq \emptyset$ which implies that $\frac{9}{8} I_m \cap \frac{9}{8} J_m \neq \emptyset$ for every $m \in \{1, \ldots, n\}$. Moreover, $\ell(R) \leq 4\ell(Q)$ implies that $|J_m| \leq 4|I_m|$ for every m. The definition of Φ_Q given in (7.5.5) gives that the integral in (7.5.13) is equal to a product of integrals such as those appearing in (7.5.12). So using (7.5.12) we obtain (7.5.13) with $B' = B^n$.

(c) In view of (7.5.7) and (7.5.8) we have that $\varphi_i \geq 2^{-n} \Phi_{Q_i}$ and $\varphi_j \geq 2^{-n} \Phi_{Q_j}$. Then we use (7.5.13) to deduce (7.5.14). □

Exercises

7.5.1. Fix a dyadic cube Q. Show that there exist $3^n - 1$ adjacent dyadic cubes to Q of equal side length, $4^n - 2^n$ adjacent dyadic cubes to Q of half its side length, and $6^n - 4^n$ adjacent dyadic cubes to Q of one quarter its side length,

7.5.2. Let $\{Q_i\}_i$ and $\{Q'_j\}_j$ be two collections of pairwise disjoint dyadic cubes in each collection. Suppose that $\Omega = \cup_i Q_i \subseteq \cup_j Q'_j = \Omega'$ and that the $\{Q_i\}_i$ as well as the $\{Q'_j\}_j$ satisfy property (b) of Theorem 7.5.2. Prove that if $Q \in \{Q_i\}_i$ and $Q' \in \{Q'_j\}_j$ then only one of the following three options is possible: (i) Q and Q' are disjoint, (ii) $Q \subseteq Q'$, or (iii) Q' is a proper subset of Q and $\ell(Q') = \tfrac{1}{2} \ell(Q)$.

7.5.3. Prove the following: There is a constant A_n such that if Q_i and Q_j are adjacent dyadic cubes in the Whitney decomposition of an open set $\Omega \neq \mathbf{R}^n$, then there are at least 2^n dyadic cubes R of side length $\tfrac{1}{2^6} \min(\ell(Q_i), \ell(Q_j))$ such that

$$\varphi_i(x)\varphi_j(x) \geq \frac{A_n}{|Q_j|} \int_{\mathbf{R}^n} \varphi_i(y)\varphi_j(y)\,dy$$

for all $x \in R$. Here $\{\varphi_j\}_j$ is the partition of unity of Lemma 7.5.7 adapted to $\{Q_j\}_j$.

7.5.4. Show that for any multi-index α there is a constant B_α such that for the partition of unity of Lemma 7.5.7 adapted to the Whitney cubes $\{Q_j\}_j$ of an open set $\Omega \neq \mathbf{R}^n$ we have

$$\partial^\beta\left[\varphi_{j_1}\cdots\varphi_{j_r}\right] \leq B_\alpha \max(|Q_{j_1}|,\ldots,|Q_{j_r}|)^{-\frac{|\beta|}{n}}.$$

7.5.5. Prove that there is a constant c_n depending on the dimension such that for any finite collection of distinct Whitney cubes $Q_{j_1},\ldots,Q_{j_r}$ of an open set Ω we have

$$\int_{\mathbf{R}^n} \varphi_{j_1}(y)\cdots\varphi_{j_r}(y)\,dy \geq c_n \min(|Q_{j_1}|,\ldots,|Q_{j_r}|)\varphi_{j_1}\cdots\varphi_{j_r},$$

where $\{\varphi_j\}_j$ is the partition of unity of Lemma 7.5.7 adapted to $\{Q_j\}_j$.
[*Hint:* Note that $r \leq 2^n$. If $Q_j = I_1^j \times \cdots \times I_n^j$ and $\varphi_{j_1}(x)\cdots\varphi_{j_r}(x) \neq 0$, then all intervals in the set $\{I_1^{j_1},\ldots,I_1^{j_r}\}$ have comparable lengths (ratio between $\frac{1}{4}$ and 4) and are either adjacent or related by inclusion. Show that for some $c_n^1 > 0$,

$$\int_{\mathbf{R}} \phi_{I_1^{j_1}}(y)\cdots\phi_{I_1^{j_r}}(y)\,dy \geq c_n^1 \min(|I_1^{j_1}|,\ldots,|I_1^{j_r}|).$$

Reduce matters to the case when one of the largest intervals is equal to $[0,1)$.]

7.6 Atomic Decomposition of H^1

In this section we prove the atomic characterization of $H^1(\mathbf{R}^n)$, which says that every element of H^1 can be expressed as an infinite sum of H^1-atoms. The following theorem provides a precise formulation.

Theorem 7.6.1. *Given a function f in $H^1(\mathbf{R}^n)$ there exists a sequence of H^1 atoms a_s, $s = 1, 2,\ldots$, and a sequence of positive numbers λ_s such that*

$$f = \sum_{s=1}^\infty \lambda_s a_s, \quad \text{a.e.} \tag{7.6.1}$$

Also, there are constants $c_n, C_n > 0$ depending only on the dimension such that

$$c_n \|f\|_{H^1} \leq \sum_{s=1}^\infty \lambda_s \leq C_n \|f\|_{H^1}. \tag{7.6.2}$$

Moreover, the series $\sum_{s=1}^\infty \lambda_s a_s$ converges to f in H^1.

Proof. We fix a nonzero function f in H^1. By Theorem 7.2.4 (b), f lies in $L^1(\mathbf{R}^n)$. We also fix an $N \geq n+1$ and consider the grand maximal function $\mathcal{M}_N$. If there is a $k \in \mathbf{Z}$ such that $\Omega^k = \{x \in \mathbf{R}^n : \mathcal{M}_N(f)(x) > 2^k\}$ is empty, we let k_{00} be the smallest such integer k, otherwise we set $k_{00} = \infty$. We observe that for each $k < k_{00}$, Ω^k is a nonempty and proper open subset of $\mathbf{R}^n$. In the remainder of this proof, all all indices k that appear will be tacitly assumed to be strictly less than k_{00}.

For $k \in \mathbf{Z}$ apply Theorem 7.5.2 to write

$$\Omega^k = \{x \in \mathbf{R}^n : \mathcal{M}_N(f)(x) > 2^k\} = \bigcup_{i=1}^{\infty} Q_i^k,$$

where $\{Q_i^k\}_{i=1}^{\infty}$ are the dyadic Whitney cubes of Ω^k. Let ℓ_i^k be the length of Q_i^k and $\{\varphi_i^k\}_i$ be the partition of unity adapted to the Whitney cubes Q_i^k, according to Lemma 7.5.6. Then we set

$$m_i^k = \frac{1}{\|\varphi_i^k\|_{L^1}} \int_{\mathbf{R}^n} f \varphi_i^k \, dy$$

and

$$g_k = f \chi_{(\Omega^k)^c} + \sum_{i=1}^{\infty} m_i^k \varphi_i^k. \tag{7.6.3}$$

We claim that there is a constant $C_1(n)$, that depends on the dimension, such that

$$|m_i^k| \leq C_1(n) 2^k \tag{7.6.4}$$

for all i and k. To see this, by Theorem 7.5.2 (b) we pick a $\xi_i^k \notin \Omega^k$ such that $|\xi_i^k - c_i^k| \leq \frac{9}{2}\sqrt{n}\,\ell_i^k$, where c_i^k is the center of Q_i^k. Define the function

$$\Phi_i^k(x) = \varphi_i^k(\xi_i^k - \ell_i^k x)$$

and notice that it is supported in the set $\{x \in \mathbf{R}^n : |\xi_i^k - \ell_i^k x - c_i^k| \leq \frac{1}{2} \cdot \frac{9}{8}\sqrt{n}\,\ell_i^k\}$, which is contained in the ball $B(0, \frac{82}{16}\sqrt{n})$. Moreover, by Lemma 7.5.6 (iii), Φ_i^k has derivatives bounded by a constant independent of i and k. These observations yield the existence of a constant $C_1'(n) > 0$ such that $\frac{1}{C_1'(n)}\Phi_i^k \in \mathscr{F}_N$ for all i and k. Then

$$|m_i^k| = \frac{1}{\|\varphi_i^k\|_{L^1}} \left| \int_{\mathbf{R}^n} f(y) \overbrace{\Phi_i^k\left(\frac{\xi_i^k - y}{\ell_i^k}\right)}^{\varphi_i^k(y)} dy \right|$$

$$= C_1'(n) \frac{(\ell_i^k)^n}{\|\varphi_i^k\|_{L^1}} \left| \left(\left(\frac{\Phi_i^k}{C_1'}\right)_{\ell_i^k} * f \right)(\xi_i^k) \right|$$

$$\leq C_1'(n) 2^n \mathcal{M}_N(f)(\xi_i^k) \qquad \text{[Lemma 7.5.6 (iv)]}$$

$$\leq C_1(n) 2^k \qquad \text{as } (\xi_i^k \notin \Omega^k). \tag{7.6.5}$$

7.6 Atomic Decomposition of H^1

Next we claim that there is a constant $C_2(n)$ such that $|g_k| \leq C_2(n)2^k$. Indeed, in view of (7.6.3) and (7.6.4) this assertion holds on Ω^k. Additionally, picking ϕ in $\mathscr{F}_N$ with $c_\phi = \int \phi\, dy \neq 0$, Theorem 2.5.5 yields that $\phi_t * f \to c_\phi f$ as $t \to 0$. But $|\phi_t * f| \leq \mathscr{M}_N(f)$, so $|f| \leq 2^k/c_\phi$ on $(\Omega^k)^c$. Hence the assertion $|g_k| \leq C_2(n)2^k$ also holds on $(\Omega^k)^c$. Consequently, $g_k \to 0$ as $k \to -\infty$.

On the other hand,

$$f - g_k = \sum_{i=1}^{\infty}(f - m_i^k)\varphi_i^k$$

is supported in Ω^k, which tends to the empty set as $k \to \infty$. This implies that the support of $f - g_k$ tends to the empty set; in other words, $g_k \to f$ a.e. as $k \to \infty$. These observations allow us to conclude that

$$f = \sum_{k=-\infty}^{\infty}(g_{k+1} - g_k) \qquad \text{a.e.} \qquad (7.6.6)$$

We now write

$$g_{k+1} - g_k = (f - g_k) - (f - g_{k+1}) = \sum_{i=1}^{\infty}(f - m_i^k)\varphi_i^k - \sum_{j=1}^{\infty}(f - m_j^{k+1})\varphi_j^{k+1}.$$

At this point one is tempted to consider multiples of the functions $\{(f - m_i^k)\varphi_i^k\}_{i,k}$ as the atoms appearing in the decomposition of f. These functions have integral zero and are supported in cubes, but the problem is that they may be unbounded. So we introduce a further decomposition to fix this issue. Recalling that each cube Q_j^{k+1} is contained in some cube Q_i^k (Remark 7.5.5), we group together all φ_j^{k+1} whose support intersects a given φ_i^k. Precisely, if $\varphi_i^k \varphi_j^{k+1}$ is the zero function, we define $m_{i,j}^{k,k+1} = 0$, while if $\varphi_i^k \varphi_j^{k+1}$ is not the zero function we set

$$m_{i,j}^{k,k+1} = \frac{1}{\int \varphi_i^k \varphi_j^{k+1}\, dy}\int_{\mathbf{R}^n} f(y)\varphi_i^k(y)\varphi_j^{k+1}(y)\, dy.$$

Then we introduce the functions

$$\alpha_i^k = (f - m_i^k)\varphi_i^k - \left(\sum_{j=1}^{\infty}(f - m_{i,j}^{k,k+1})\varphi_j^{k+1}\right)\varphi_i^k \qquad (7.6.7)$$

$$= \left[f\chi_{(\Omega^{k+1})^c} - \left(m_i^k - \sum_{j=1}^{\infty}m_{i,j}^{k,k+1}\varphi_j^{k+1}\right)\right]\varphi_i^k, \qquad (7.6.8)$$

$$\beta_j^{k+1} = -(f - m_j^{k+1})\varphi_j^{k+1} + \left(\sum_{i=1}^{\infty}(f - m_{i,j}^{k,k+1})\varphi_i^k\right)\varphi_j^{k+1} \qquad (7.6.9)$$

$$= \left[-f\chi_{(\Omega^k)^c} + \left(m_j^{k+1} - \sum_{i=1}^{\infty}m_{i,j}^{k,k+1}\varphi_i^k\right)\right]\varphi_j^{k+1}. \qquad (7.6.10)$$

In view of the definitions of m_i^k, m_j^{k+1}, and $m_{i,j}^{k,k+1}$, (7.6.7) and (7.6.9) give that α_i^k and β_j^{k+1} have integral zero. Moreover, summing over i in (7.6.7) and over j in (7.6.9) yields the identity

$$g_{k+1} - g_k = \sum_{i=1}^{\infty} \alpha_i^k + \sum_{j=1}^{\infty} \beta_j^{k+1}. \tag{7.6.11}$$

We claim that $|m_{i,j}^{k,k+1}| \leq C_3(n) 2^k$ for a constant $C_3(n)$ that only depends on the dimension. Indeed, if $\varphi_i^k \varphi_j^{k+1} \neq 0$, then $\frac{9}{8}Q_i^k \cap \frac{9}{8}Q_j^{k+1} \neq \emptyset$ which implies that either Q_j^{k+1} is contained in Q_i^k or in another Whitney cube $Q_{i'}^k$ adjacent to Q_i^k (Remark 7.5.5). In either case we have $\ell_j^{k+1} \leq 4\ell_i^k$. We pick a point $\xi_j^{k+1} \in (\Omega^{k+1})^c$ within $\frac{9}{2}\sqrt{n}\,\ell_j^{k+1}$ units from the center of Q_j^{k+1} and we define

$$\Psi_{i,j}^{k,k+1}(x) = \varphi_i^k(\xi_j^{k+1} - \ell_j^{k+1} x)\varphi_j^{k+1}(\xi_j^{k+1} - \ell_j^{k+1} x).$$

As $\ell_j^{k+1} \leq 4\ell_i^k$, by Leibniz's rule and Lemma 7.5.6 (iii), all derivatives of $\Psi_{i,j}^{k,k+1}$ are bounded above by a constant independent of i,j,k; moreover, $\Psi_{i,j}^{k,k+1}$ is supported in $B(0, \frac{82}{16}\sqrt{n})$. So, there is a constant $C_3'(n)$ such that $\frac{1}{C_3'(n)}\Psi_{i,j}^{k,k+1} \in \mathscr{F}_N$ uniformly in i,j and k. Using an argument similar to that leading to (7.6.5) and the following fact [(7.5.14) in Lemma 7.5.7]

$$\frac{(\ell_j^{k+1})^n}{\int \varphi_i^k \varphi_j^{k+1} dy} \leq \frac{1}{B''},$$

we deduce that $|m_{i,j}^{k,k+1}| \leq C_3(n) 2^k$.

Additionally, we pick a Schwartz function $\phi \in \mathscr{F}_N$ with $c_\phi = \int \phi\, dy \neq 0$, and we notice that $\phi_t * f \to c_\phi f$ as $t \to 0$ (Theorem 2.5.5), but $|\phi_t * f| \leq \mathscr{M}_N(f)$, so $|f| \leq 2^k/c_\phi$ whenever $\mathscr{M}_N(f) \leq 2^k$, i.e. on $(\Omega^k)^c$ (and likewise $|f| \leq 2^{k+1}/c_\phi$ on $(\Omega^{k+1})^c$). These estimates inserted in (7.6.8) and (7.6.10) provide

$$|\alpha_i^k| \leq C_4(n) 2^k \qquad |\beta_j^{k+1}| \leq C_4(n) 2^{k+1}$$

for some constant $C_4(n) > 0$ depending only on the dimension. Moreover, we notice that α_i^k and β_j^{k+1} are supported in the cubes $\frac{9}{8}Q_i^k$ and $\frac{9}{8}Q_j^{k+1}$, respectively, and have mean value zero. So suitable normalizations of them are H^1 atoms.

To create atoms, we define the constants

$$\mu_i^k = C_4(n) 2^k |\tfrac{9}{8}Q_i^k|, \qquad \mu_j^{k+1} = C_4(n) 2^{k+1} |\tfrac{9}{8}Q_j^{k+1}|$$

and the following normalizations of α_i^k and β_j^{k+1}:

7.6 Atomic Decomposition of H^1

$$A_i^k = \frac{\alpha_i^k}{\mu_i^k}, \qquad B_j^{k+1} = \frac{\beta_j^{k+1}}{\mu_j^{k+1}}.$$

Then A_i^k, B_j^{k+1} are H^1 atoms. In view of (7.6.11), (7.6.6) we have

$$f = \sum_{k=-\infty}^{\infty} \sum_{i=1}^{\infty} \left(\mu_i^k A_i^k + \mu_i^{k+1} B_i^{k+1} \right) \qquad \text{a.e.} \qquad (7.6.12)$$

We estimate the ℓ^1 norm of the sequences of coefficients as follows:

$$\sum_{k=-\infty}^{\infty} \sum_{i=1}^{\infty} (\mu_i^k + \mu_i^{k+1}) = (\tfrac{9}{8})^n C_4(n) \sum_{k=-\infty}^{\infty} \sum_{i=1}^{\infty} (2^k |Q_i^k| + 2^{k+1} |Q_i^{k+1}|)$$

$$= (\tfrac{9}{8})^n C_4(n) \sum_{k=-\infty}^{\infty} (2^k |\Omega^k| + 2^{k+1} |\Omega^{k+1}|)$$

$$= 2(\tfrac{9}{8})^n C_4(n) \sum_{k=-\infty}^{\infty} 2^k |\Omega^k|$$

$$= 4(\tfrac{9}{8})^n C_4(n) \sum_{k=-\infty}^{\infty} \int_{2^{k-1}}^{2^k} |\{x \in \mathbf{R}^n : \mathcal{M}_N(f)(x) > 2^k\}| \, d\lambda$$

$$\leq 4(\tfrac{9}{8})^n C_4(n) \int_0^{\infty} |\{x \in \mathbf{R}^n : \mathcal{M}_N(f)(x) > \lambda\}| \, d\lambda \qquad (7.6.13)$$

$$= 4(\tfrac{9}{8})^n C_4(n) \|\mathcal{M}_N(f)\|_{L^1}$$

$$\leq C_n \|f\|_{H^1}.$$

This proves the upper inequality in (7.6.2) with a_s being the sequence combining A_i^k and B_j^{k+1} and λ_s being the sequence combining μ_i^k and μ_j^{k+1}.

For the lower inequality in (7.6.2) we use that if $f = \sum_{s=1}^{\infty} \lambda_s a_s$ a.e. with $\lambda_s > 0$ and a_s being H^1 atoms, then

$$M^*(f; \Phi) \leq \sum_{s=1}^{\infty} \lambda_s M^*(a_s; \Phi),$$

where $\Phi(x) = e^{-\pi |x|^2}$. Applying the H^1 norm and using Theorem 7.3.2 we deduce

$$\|f\|_{H^1} = \|M^*(f; \Phi)\|_{L^1} \leq \sum_{s=1}^{\infty} \lambda_s \|M^*(a_s; \Phi)\|_{L^1} \leq C(n, 1) \sum_{s=1}^{\infty} \lambda_s.$$

The lower inequality in (7.6.2) now follows with $c_n = 1/C(n, 1)$.

Finally, we turn to the assertion that the series $\sum_{s=1}^{\infty} \lambda_s a_s$ converges to f in H^1. The fact that $f \in H^1$ and the upper inequality in (7.6.2) gives that

$$\sum_{k \in \mathbf{Z}} \sum_{i=1}^{\infty} (\mu_i^k + \mu_i^{k+1}) = \sum_{i=1}^{\infty} \sum_{k \in \mathbf{Z}} (\mu_i^k + \mu_i^{k+1}) < \infty.$$

This implies that

$$\lim_{M\to\infty} \sum_{|k|>M} \sum_{i=1}^{\infty} (\mu_i^k + \mu_i^{k+1}) = 0 \quad \text{and} \quad \lim_{M\to\infty} \sum_{k\in\mathbf{Z}} \sum_{i=M+1}^{\infty} (\mu_i^k + \mu_i^{k+1}) = 0.$$

Setting $S_M = \sum_{|k|\leq M} \sum_{i=1}^{M} (\mu_i^k A_i^k + \mu_i^{k+1} B_i^{k+1})$ we have

$$S_M - f = \sum_{|k|>M} \left(\sum_{i=1}^{\infty} \mu_i^k A_i^k + \mu_i^{k+1} B_i^{k+1} \right) + \sum_{|k|\leq M} \left(\sum_{i>M} \mu_i^k A_i^k + \mu_i^{k+1} B_i^{k+1} \right).$$

We apply the H^1 norm on these expressions. Using the lower inequality in (7.6.2), we obtain

$$\|S_M - f\|_{H^1} \leq \frac{1}{c_n} \left[\sum_{|k|>M} \sum_{i=1}^{\infty} (\mu_i^k + \mu_i^{k+1}) + \sum_{k\in\mathbf{Z}} \sum_{i>M} (\mu_i^k + \mu_i^{k+1}) \right],$$

and this converges to zero as $M \to \infty$.

We finally define $\{\lambda_s\}_{s=1}^{\infty}$ to be an enumeration of $\{\mu_i^k, \mu_i^{k+1}\}_{i,k}$ and analogously we let $\{a_s\}_{s=1}^{\infty}$ be an enumeration of $\{A_i^k, B_i^{k+1}\}_{i,k}$. □

Corollary 7.6.2. *For every $f \in H^1(\mathbf{R}^n)$ we have*

$$\|f\|_{H^1} \approx \inf\left\{ \sum_{j=1}^{\infty} \lambda_j : f = \sum_{j=1}^{\infty} \lambda_j a_j \text{ a.e., } a_j \text{ are } H^1 \text{ atoms, } \lambda_j > 0, \text{ and } \sum_{j=1}^{\infty} \lambda_j < \infty \right\}.$$

Proof. Given $f \in H^1$, by Theorem 7.6.1 we know that there exist H^1 atoms a_j and $\lambda_j > 0$ such that $f = \sum_{j=1}^{\infty} \lambda_j a_j$ a.e. and moreover,

$$\sum_{j=1}^{\infty} \lambda_j \leq C_n \|f\|_{H^1}.$$

The infimum over all such expressions $\sum_{j=1}^{\infty} \lambda_j$ is even smaller. This gives

$$\inf\left\{ \sum_{j=1}^{\infty} \lambda_j : f = \sum_{j=1}^{\infty} \lambda_j a_j, a_j \text{ are } H^1 \text{ atoms, } \lambda_j > 0, \text{ and } \sum_{j=1}^{\infty} \lambda_j < \infty \right\} \leq C_n \|f\|_{H^1}.$$

For the other direction, we notice that for any representation of f as $\sum_{j=1}^{\infty} \lambda_j a_j$ we have

$$\|f\|_{H^1} = \Big\| \sum_{j=1}^{\infty} \lambda_j a_j \Big\|_{H^1} \leq \sum_{j=1}^{\infty} \lambda_j \|a_j\|_{H^1} \leq c_n \sum_{j=1}^{\infty} \lambda_j$$

in view of Theorem 7.3.2. Taking the infimum over all such representations we obtain the other direction of the equivalence. □

Exercises

7.6.1. Provide another proof of Theorem 7.2.6 using the atomic decomposition. Precisely, use Theorem 7.6.1 to show that if $f \in H^1(\mathbf{R}^n)$, then $\int_{\mathbf{R}^n} f(x)\,dx = 0$.

7.6.2. Let $\varepsilon > 0$. Express the function $|x|^{-1+\varepsilon}\operatorname{sgn}(x)\chi_{0<|x|\le 1}$ defined on the real line as a sum of the form in (7.6.1).

7.6.3. Let $\varepsilon > 0$. Show that the function $h(x) = \frac{1}{x}\left(\log\frac{1}{|x|}\right)^{-1-\varepsilon}\chi_{|x|<1/2}$ lies in the Hardy space $H^1(\mathbf{R})$ although

$$\int_{-1/2}^{1/2} |h(t)|\log|h(t)|\,dt = \infty.$$

[*Hint:* For $j = 1,2,\dots$ define atoms $a_j = cj^{1+\varepsilon}(h\chi_{R_j} - h_{R_j})$ supported in $R_j = (2^{-j}, 2^{-j+1})$ and $b_j = cj^{1+\varepsilon}(h\chi_{L_j} - h_{L_j})$ supported in $L_j = (-2^{-j+1}, -2^{-j})$ for a suitable constant $c > 0$ independent of j. Write $h = \sum_{j=1}^{\infty} \frac{1}{cj^{1+\varepsilon}}(a_j + b_j)$.]

7.6.4. (Calderón–Zygmund decomposition on H^1) Fill in the steps below to obtain the following result related to the atomic decomposition of H^1: Prove that there exist constants A_n, B_n, C_n such that for any $f \in H^1(\mathbf{R}^n)$ and $\alpha > 0$ there exist functions g and b on $\mathbf{R}^n$ and a collection of disjoint dyadic cubes $\{Q_j\}_j$ such that

(1) $f = g + b$.

(2) $\|g\|_{L^\infty} \le A_n \alpha$.

(3) $b = \sum_j b_j$, where each b_j is supported in $\frac{9}{8}Q_j$.

(4) $\int_{Q_j} b_j(x)\,dx = 0$.

(5) $\|b_j\|_{H^1} \le B_n \int_{Q_j^*} \mathcal{M}_N(f)\,dx$.

(6) $\|b\|_{H^1} \le 2^n B_n \int_{\{\mathcal{M}_N(f)>\alpha\}} \mathcal{M}_N(f)\,dx \le C_n \|f\|_{H^1}$.

[*Hint:* For fixed $N \ge n+1$, write $\Omega = \{\mathcal{M}_N(f) > \alpha\}$ as a union of Whitney cubes $\{Q_j\}_j$ according to Theorem 7.5.2. Let φ_j be the associated partition of unity according to Lemma 7.5.7. Define $b_j = (f - m_j)\varphi_j$ where

$$m_j = \frac{1}{\int \varphi_j\,dy}\int_{\mathbf{R}^n} f(y)\varphi_j(y)\,dy$$

and

$$g = f\chi_{\Omega^c} + \sum_j m_j \varphi_j.$$

Then $b = \sum_j b_j$ and properties **(1)**, **(3)**, and **(4)** hold by construction.

Property **(2)**: It will be sufficient to prove that $|m_j| \leq C_n^1 \alpha$ for a constant C_n^1. Pick a $\xi_j \notin \Omega$ such that $|z - \xi_j| < 6\sqrt{n}\ell_j$ for all $z \in Q_j$ [cf. Theorem 7.5.2 (b)]. Here $\ell_j = \ell(Q_j)$. Notice that $\mathscr{M}(f)(\xi_j) \leq \alpha$. Then write $m_j = ((\Phi_j)_{\ell_j} * f)(\xi_j)$ where

$$\Phi_j(y) = \frac{\ell_j^n}{\|\varphi_j\|_{L^1}} \varphi_j(\xi_j - \ell_j y), \qquad y \in \mathbf{R}^n.$$

Verify that $C_n^2 \Phi_j \in \mathscr{F}_N$ for all j for a fixed constant C_n^2.

Property **(5)**: Let $\Phi \in \mathscr{S}$ have integral equal to 1 and be supported in $B(0,1)$. Let $Q_j^* = \frac{9}{8} Q_j$ and by a translation assume that Q_j is centered at the origin. Prove

$$\int_{Q_j^*} M^*(b_j; \Phi)(x)\,dx + \int_{(Q_j^*)^c} M^*(b_j; \Phi)(x)\,dx \leq B_n \int_{Q_j^*} \mathscr{M}_N(f)(x)\,dx.$$

Case 1: Fix $x \in Q_j^*$ and y with $|y - x| < t$. Estimate the first integral above by writing

$$\int_{\mathbf{R}^n} \Phi_t(y-z) b_j(z)\,dz = \frac{q^n}{t^n} \int_{\mathbf{R}^n} \Psi_q(y'-z) f(z)\,dz - m_j \int_{\mathbf{R}^n} \Phi_t(y-z)\varphi_j(z)\,dz,$$

where $q = \min(\ell_j, t)$. Here $\Psi(z) = \Phi(z)\varphi_j(y - tz)$ and $y' = y$ when $t \leq \ell_j$, while

$$\Psi(z) = \Phi\Big(\frac{\ell_j}{t} z - \frac{\xi_j - y}{t}\Big) \varphi_j(\xi_j - \ell_j z) \qquad \text{and } y' = \xi_j$$

when $\ell_j < t$ and ξ_j is as in Theorem 7.5.2 (b). In both cases, show that there is a constant C_n^4 such that $C_n^4 \Psi$ lies in $\mathscr{F}_N$ and use that $|m_j| \leq C_n^1 \alpha$.

Case 2: Fix $x \notin Q_j^*$, $t > 0$, and y with $|y - x| < t$. Then $|x| \geq \sqrt{n}\ell_j$. Notice that $(\Phi_t * b_j)(y)$ vanishes if $|x| > 4t$. So we may suppose that $\sqrt{n}\ell_j \leq |x| \leq 4t$. Write

$$(\Phi_t * b_j)(y) = \frac{1}{t^{n+1}} \sum_{k=1}^n \int_{\mathbf{R}^n} \int_0^1 \partial_k \Phi\Big(\frac{y - \theta z}{t}\Big) \varphi_j(z)(f(z) - m_j) z_k\,d\theta dz,$$

using the mean value theorem and the vanishing integral property of b_j. The goal is to prove that the preceding expression is bounded by $C(n)\ell_j^{n+1}|x|^{-n-1}$. For the term containing m_j this is straightforward. For the term containing $f(z)$ define

$$R(z) = \varphi_j(z) \sum_{k=1}^n z_k \int_0^1 \partial_k \Phi\Big(\frac{y - \theta z}{t}\Big) d\theta, \qquad \Omega(z) = \ell_j^n R(\xi_j - \ell_j z),$$

and notice $R(z) = \frac{1}{\ell_j^n} \Omega\big(\frac{\xi_j - z}{\ell_j}\big)$. Then matters reduce to estimating $\frac{1}{t^{n+1}}(\Omega_{\ell_j} * f)(\xi_j)$. Now show $\mathfrak{N}_N(\Omega) \leq C_n' \ell_j^{n+1}$ for some constant C_n' and use that $1/t \leq 4/|x|$.

Property **(6)**: Use Theorem 7.5.2 (e).]

7.7 Singular Integrals on the Hardy Space H^1

Singular integrals map L^p to L^p for $1 < p < \infty$ but this is not the case when $p = 1$. In this section we show that singular integrals map the Hardy space H^1 to L^1.

We begin by reviewing the definition of singular integrals. Let K be a function defined on $\mathbf{R}^n \setminus \{0\}$ that satisfies the size estimate (3.3.3), the smoothness estimate (3.3.4), and the cancellation estimate (3.3.5) for some $A_1, A_2, A_3 < \infty$. The cancellation condition (3.3.5) implies that there exists a sequence $\delta_j \downarrow 0$ as $j \to \infty$ such that the following limit exists:

$$\lim_{j \to \infty} \int_{\delta_j \leq |x| \leq 1} K(x)\, dx = L_0.$$

This gives that for a smooth and compactly supported function φ on $\mathbf{R}^n$, the limit

$$\lim_{j \to \infty} \int_{|x-y| > \delta_j} K(x-y)\varphi(y)\, dy = T(\varphi)(x) \qquad (7.7.1)$$

exists and defines a linear operator T on $\mathscr{C}_0^\infty$. This operator T is called a *singular integral* and is given by convolution with a tempered distribution W that coincides with the function K on $\mathbf{R}^n \setminus \{0\}$.

We know that such an operator T, initially defined on $\mathscr{C}_0^\infty(\mathbf{R}^n)$, admits an extension that is L^p bounded for all $1 < p < \infty$ and is also of weak type $(1,1)$. All these norms are bounded above by constant multiple of the quantity $A_1 + A_2 + A_3$; in particular, by Corollary 3.4.3, the L^2 norm of T is bounded by $9\omega_{n-1}A_1 + A_2 + A_3$. Therefore, such a T is well defined on $L^1(\mathbf{R}^n)$ and in particular on $H^1(\mathbf{R}^n)$, which is contained in $L^1(\mathbf{R}^n)$.

Theorem 7.7.1. *Let K satisfy (3.3.3), (3.3.4), and (3.3.5), and let T be defined as in (7.7.1). Then there is a constant C_n such that for all f in $H^1(\mathbf{R}^n)$ we have*

$$\|T(f)\|_{L^1} \leq C_n(A_1 + A_2 + A_3)\|f\|_{H^1}. \qquad (7.7.2)$$

Proof. We start by checking the validity of (7.7.2) on H^1 atoms. Since T is a convolution operator (i.e., it commutes with translations), it suffices to take the atom f supported in a cube Q centered at the origin. Let $f = a$ be such an atom, supported in cube Q centered at zero, and let

$$Q^* = 2\sqrt{n}\, Q$$

be a concentric cube with side length $2\sqrt{n}$ times that of Q. We write

$$\int_{\mathbf{R}^n} |T(a)(x)|\, dx = \int_{Q^*} |T(a)(x)|\, dx + \int_{(Q^*)^c} |T(a)(x)|\, dx \qquad (7.7.3)$$

and we estimate each term separately. We have

$$\int_{Q^*} |T(a)(x)|\,dx \le |Q^*|^{\frac{1}{2}} \left(\int_{Q^*} |T(a)(x)|^2\,dx \right)^{\frac{1}{2}}$$
$$\le (9\omega_{n-1}A_1 + A_2 + A_3)|Q^*|^{\frac{1}{2}} \left(\int_Q |a(x)|^2\,dx \right)^{\frac{1}{2}}$$
$$\le 9\omega_{n-1}(A_1 + A_2 + A_3)|Q^*|^{\frac{1}{2}}|Q|^{\frac{1}{2}}|Q|^{-1}$$
$$= 9\omega_{n-1}(2\sqrt{n})^{\frac{n}{2}}(A_1 + A_2 + A_3),$$

where we used the L^2 boundedness of T (Corollary 3.4.3) and property (ii) of atoms in Definition 7.3.1.

Let Q be a cube centered at the origin. We claim that if $x \notin Q^*$ and $y \in Q$, then $|x| \ge 2|y|$ and $x - y$ stays away from zero; thus $K(x-y)$ is well defined. To see this assertion we note that
$$|x| \ge \ell(Q)\sqrt{n}$$
and
$$|y| \le \frac{1}{2}\ell(Q)\sqrt{n}$$
imply
$$|x| \ge \ell(Q)\sqrt{n} \ge 2|y|.$$

See Figure 7.5.

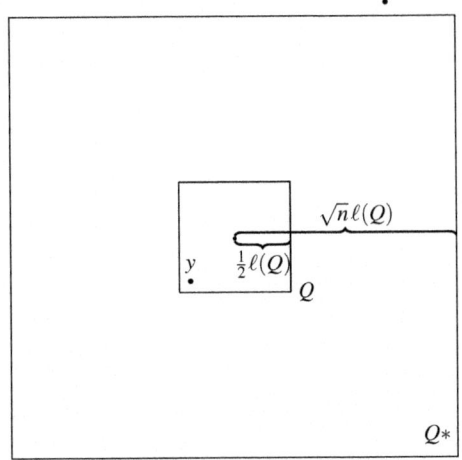

Fig. 7.5 The situation where $x \notin Q^*$ and $y \in Q$.

Moreover, in this case $T(a)(x)$ can be expressed as an absolutely convergent integral of a against $K(x - \cdot)$. Exploiting the fact that atoms have mean value zero we write

$$\int_{(Q^*)^c} |T(a)(x)|\,dx = \int_{(Q^*)^c} \left| \int_Q K(x-y)a(y)\,dy \right| dx$$
$$= \int_{(Q^*)^c} \left| \int_Q (K(x-y) - K(x))a(y)\,dy \right| dx$$
$$\le \int_Q \int_{(Q^*)^c} |K(x-y) - K(x)|\,dx\,|a(y)|\,dy$$
$$\le \int_Q \int_{|x| \ge 2|y|} |K(x-y) - K(x)|\,dx\,|a(y)|\,dy$$
$$\le A_2 \int_Q |a(y)|\,dy$$
$$\le A_2.$$

7.7 Singular Integrals on the Hardy Space H^1

Combining this calculation with the previous one and inserting the final conclusions in (7.7.3) we deduce that H^1 atoms a satisfy

$$\|T(a)\|_{L^1} \leq C'_n(A_1+A_2+A_3), \tag{7.7.4}$$

where $C'_n = 9\omega_{n-1}(2\sqrt{n})^{\frac{n}{2}}$. We now prove (7.7.2) for another constant C_n. In view of Theorem 7.6.1 and Corollary 7.6.2 we can write $f \in H^1$ as $f = \sum_{j=1}^{\infty} \lambda_j a_j$, where $\lambda_j > 0$, the series converges in L^1, the a_j are H^1 atoms, and

$$\sum_{j=1}^{\infty} \lambda_j \leq C''_n \|f\|_{H^1}. \tag{7.7.5}$$

Since T maps L^1 to $L^{1,\infty}$ (Theorem 3.6.1), $T(f)$ is already a well-defined $L^{1,\infty}$ function. We claim that

$$T(f) = \sum_{j=1}^{\infty} \lambda_j T(a_j) \quad \text{a.e.,} \tag{7.7.6}$$

noting that the series in (7.7.6) converges in L^1 and produces a well-defined integrable function.

To prove (7.7.6), we make use of the fact that T is of weak type $(1,1)$. For a given $\delta > 0$ we have

$$\left|\left\{\left|T(f) - \sum_{j=1}^{\infty} \lambda_j T(a_j)\right| > \delta\right\}\right|$$

$$\leq \left|\left\{\left|T(f) - \sum_{j=1}^{N} \lambda_j T(a_j)\right| > \delta/2\right\}\right| + \left|\left\{\left|\sum_{j=N+1}^{\infty} \lambda_j T(a_j)\right| > \delta/2\right\}\right|$$

$$\leq \frac{2}{\delta} \|T\|_{L^1 \to L^{1,\infty}} \left\|f - \sum_{j=1}^{N} \lambda_j a_j\right\|_{L^1} + \frac{2}{\delta} \left\|\sum_{j=N+1}^{\infty} \lambda_j T(a_j)\right\|_{L^1}$$

$$\leq \frac{2}{\delta} \|T\|_{L^1 \to L^{1,\infty}} \left\|\sum_{j=N+1}^{\infty} \lambda_j a_j\right\|_{L^1} + \frac{2}{\delta} C'_n (A_1+A_2+A_3) \sum_{j=N+1}^{\infty} \lambda_j.$$

Obviously $\|\sum_{j=N+1}^{\infty} \lambda_j a_j\|_{L^1} \leq \sum_{j=N+1}^{\infty} \lambda_j$, so both terms converge to zero as $N \to \infty$. We conclude that

$$\left|\left\{x \in \mathbf{R}^n : \left|T(f)(x) - \sum_{j=1}^{\infty} \lambda_j T(a_j)(x)\right| > \delta\right\}\right| = 0$$

for all $\delta > 0$, which implies (7.7.6).

Now that (7.7.6) is established, we deduce (7.7.2) with $C_n = C'_n C''_n$ by taking L^1 norms in (7.7.6) and using (7.7.4) and (7.7.5). $\square$

Exercises

7.7.1. Suppose that T is a sublinear operator that maps $L^1(\mathbf{R}^n)$ to $L^{1,\infty}(\mathbf{R}^n)$. Given f in $H^1(\mathbf{R}^n)$ written in atomic decomposition as $f = \sum_{j=1}^{\infty} \lambda_j a_j$, $\lambda_j > 0$, prove that

$$|T(f)| \leq \sum_{j=1}^{\infty} \lambda_j |T(a_j)| \qquad \text{a.e.}$$

7.7.2. Let T be an operator as in the statement of Theorem 7.7.1. Let $\Psi \in \mathscr{S}(\mathbf{R}^n)$ have Fourier transform supported in an annulus of the form $0 < c_1 < |\xi| < c_2 < \infty$ and consider the Littlewood–Paley operator Δ_j^{Ψ} associated with Ψ. Prove the existence of a constant C_n (depending on the dimension) such that for any $f \in L^1$ we have

$$\left\|\Delta_j^{\Psi} T(f)\right\|_{L^1} \leq C_n (A_1 + A_2 + A_3) \left\|\Delta_j^{\Psi}(f)\right\|_{L^1}.$$

[*Hint:* Pick $\Omega \in \mathscr{S}$ so that its Fourier transform is equal to 1 on the annulus $c_1 < |\xi| < c_2$ and vanishes off the annulus $\frac{1}{2}c_1 < |\xi| < 2c_2$. Then Ω lies in H^1 by Theorem 7.2.7 and we can apply Theorem 7.7.1.]

7.7.3. Fix $0 < A, B < \infty$. Let $\{K_j\}_{j=1}^{N}$ be a sequence of functions on $\mathbf{R}^n \setminus \{0\}$ that satisfies

$$\left(\sum_{j=1}^{N} |K_j(x)|^2\right)^{\frac{1}{2}} \leq A|x|^{-n}, \qquad x \neq 0,$$

$$\sup_{y \neq 0} \int_{|x| \geq 2|y|} \left(\sum_{j=1}^{N} |K_j(x-y) - K_j(x)|^2\right)^{\frac{1}{2}} \leq A,$$

$$\sup_{1 \leq j \leq N} \sup_{0 < \varepsilon < R} \left|\int_{\varepsilon < |y| < R} K_j(y)\, dy\right| \leq A.$$

Define T_j by $T_j(\varphi)(x) = \text{p.v.} \int_{\mathbf{R}^n} K_j(x-y)\varphi(y)\,dy$ when $\varphi \in \mathscr{C}_0^{\infty}$ and suppose that $\vec{T} = \{T_j\}_{j=1}^{N}$ admits a bounded extension that maps $L^2(\mathbf{R}^n)$ to $L^2(\mathbf{R}^n, \ell_N^2)$ with bound B. Show that $\vec{T}$ also maps $H^1(\mathbf{R}^n)$ to $L^1(\mathbf{R}^n, \ell_N^2)$ with norm bounded by $C_n(A+B)$, where C_n depends only on the dimension n (and is independent of N). [*Hint:* Use Theorem 4.1.1 and the idea of the proof of Theorem 7.7.1.]

7.7.4. Let B, δ and Ψ be as in Theorem 4.4.2. Use the previous exercise to show that the Littlewood–Paley square function of Theorem 4.4.2 satisfies

$$\left\|\left(\sum_{j \in \mathbf{Z}} |\Delta_j^{\Psi}(f)|^2\right)^{\frac{1}{2}}\right\|_{L^1} \leq C_{n,\delta} B \|f\|_{H^1}$$

for all $f \in H^1(\mathbf{R}^n)$. The constant $C_{n,\delta}$ depends only on the indicated parameters.

7.8 Duality Between H^1 and BMO

In this section we obtain that BMO is the dual space of H^1. A crucial element in this assertion is the completeness of H^1 which implies the weak* compactness of the unit ball of $(H^1)^*$.

Proposition 7.8.1. *Every Cauchy sequence in H^1 converges; thus H^1 is a Banach space.*

Proof. Let $\{f_k\}_{k=1}^{\infty}$ be a Cauchy sequence in H^1. Then f_k is Cauchy in L^1 and thus it converges in L^1 to an integrable function f. Let $\Phi(x) = e^{-\pi|x|^2}$. Then, as $\Phi \in L^{\infty}$, we have for any $t > 0$ any $y \in \mathbf{R}^n$ and any $m \in \mathbf{Z}^+$

$$|(f_k - f_m) * \Phi_t(y)| \to |(f - f_m) * \Phi_t(y)|.$$

Consequently, for $|y - x| < t$ one has

$$|(f - f_m) * \Phi_t(y)| = \liminf_{k \to \infty} |(f_k - f_m) * \Phi_t(y)| \leq \liminf_{k \to \infty} M^*(f_k - f_m; \Phi)(x).$$

So taking the supremum over all y with $|y - x| < t$ and $t > 0$ we obtain

$$M^*(f - f_m; \Phi)(x) \leq \liminf_{k \to \infty} M^*(f_k - f_m; \Phi)(x)$$

for all $x \in \mathbf{R}^n$. Fatou's lemma now gives

$$\left\| M^*(f - f_m; \Phi) \right\|_{L^1} \leq \liminf_{k \to \infty} \left\| M^*(f_k - f_m; \Phi) \right\|_{L^1}. \tag{7.8.1}$$

As $\{f_k\}_k$ is Cauchy in H^1, the expression on the right is finite and we deduce from (7.8.1) that $f - f_m$ lies in H^1, hence so does f. Finally, it follows from (7.8.1) that

$$\limsup_{m \to \infty} \left\| M^*(f - f_m; \Phi) \right\|_{L^1} \leq \limsup_{m \to \infty} \limsup_{k \to \infty} \left\| M^*(f_k - f_m; \Phi) \right\|_{L^1} = 0,$$

where the equality is a consequence of the fact that $\{f_k\}_k$ is a Cauchy sequence in H^1. We conclude that $f_m \to f$ in H^1 as $m \to \infty$. $\square$

Definition 7.8.2. We denote by $H_0^1(\mathbf{R}^n)$ the space of all finite linear combinations of $H^1(\mathbf{R}^n)$ atoms. By Theorem 7.6.1, $H_0^1(\mathbf{R}^n)$ is dense in $H^1(\mathbf{R}^n)$. Fix $b \in BMO(\mathbf{R}^n)$. Given $g \in H_0^1$ we define a linear functional

$$L_b(g) = \int_{\mathbf{R}^n} g(x) b(x) \, dx \tag{7.8.2}$$

as an absolutely convergent integral. Observe that the integral in (7.8.2) and thus the definition of L_b on H_0^1 remain the same if b is replaced by $b + c$, where c is an additive constant, as $H^1(\mathbf{R}^n)$ atoms have mean value zero.

Moreover, notice that if the function b happens to lie in L^{∞}, then L_b is well-defined on the entire H^1, not only on H_0^1.

Proposition 7.8.3. *There is a constant C_n such that for any function $b \in L^\infty$ we have*

$$\|L_b\|_{H^1 \to \mathbf{C}} \leq C_n \|b\|_{BMO}. \tag{7.8.3}$$

Proof. Let b be a bounded BMO function. Let C_n be a constant that satisfies

$$\inf\left\{\sum_{j=1}^\infty \lambda_j : f = \sum_{j=1}^\infty \lambda_j a_j \text{ a.e.}, a_j \text{ are } H^1 \text{atoms}, \lambda_j > 0, \text{ and } \sum_{j=1}^\infty \lambda_j < \infty\right\} \leq C_n \|f\|_{H^1}$$

for any $f \in H^1(\mathbf{R}^n)$, as stated in Corollary 7.6.2. Given f in H^1, find a sequence of H^1 atoms $\{a_k\}_k$ supported in cubes Q_k and $\lambda_k > 0$, $k = 1, \ldots$, such that

$$f = \sum_{k=1}^\infty \lambda_k a_k \qquad \text{a.e.} \tag{7.8.4}$$

and

$$\sum_{k=1}^\infty \lambda_k \leq C'_n \|f\|_{H^1},$$

where C'_n is any constant strictly bigger than C_n. Since the series in (7.8.4) converges in H^1, it must converge in L^1, and then we have

$$|L_b(f)| = \left|\int_{\mathbf{R}^n} f(x) b(x) \, dx\right|$$

$$= \left|\sum_{k=1}^\infty \lambda_k \int_{Q_k} a_k(x)(b(x) - b_{Q_k}) \, dx\right|$$

$$\leq \sum_{k=1}^\infty \lambda_k \frac{1}{|Q_k|} \int_{Q_k} |b(x) - b_{Q_k}| \, dx$$

$$\leq C'_n \|f\|_{H^1} \|b\|_{BMO}.$$

As $C'_n > C_n$ is arbitrary, this proves (7.8.3) for b in L^∞. $\square$

Having established Proposition 7.8.3, we turn to the goal of extending the definition of L_b on the entire H^1 for functions b in BMO that are not necessarily bounded. To achieve this, we fix $b \in BMO$ and let $b_M(x) = b\chi_{|b| \leq M}$ for $M = 1, 2, 3, \ldots$. Since $\|b_M\|_{BMO} \leq \frac{9}{4}\|b\|_{BMO}$ (Exercise 6.1.2), the sequence of linear functionals $\{L_{b_M}\}_M$ lies in a multiple of the unit ball of $(H^1)^*$ and by the Banach–Alaoglu theorem there is a subsequence $M_j \to \infty$ as $j \to \infty$ such that $L_{b_{M_j}}$ converges weakly* to a bounded linear functional $\widetilde{L}_b$ on H^1. In other words, for all f in $H^1(\mathbf{R}^n)$ we have

$$L_{b_{M_j}}(f) \to \widetilde{L}_b(f) \qquad \text{as } j \to \infty.$$

If a^Q is a fixed H^1 atom supported in a cube Q then

$$L_{b_{M_j}}(a^Q) = \int_Q b_{M_j} a^Q \, dy = \int_Q (b_{M_j} - (b_{M_j})_Q) a^Q \, dy$$

7.8 Duality Between H^1 and BMO

and analogously for $L_b(a^Q)$, so

$$|L_{b_{M_j}}(a^Q) - L_b(a^Q)| \leq \|a^Q\|_{L^\infty}\left[\|(b_{M_j} - (b_{M_j})_Q) - (b - b_Q)\|_{L^1(Q)}\right]$$

$$\leq \frac{1}{|Q|}\|b_{M_j} - b\|_{L^1(Q)} + |(b_{M_j} - b)_Q|.$$

But both terms tend to zero as $j \to \infty$ by the Lebesgue dominated convergence theorem. The same conclusion holds for any finite linear combination of the a^Q. Thus for all $g \in H_0^1$ we have $L_{b_{M_j}}(g) \to L_b(g)$, and consequently, $L_b(g) = \widetilde{L}_b(g)$ for all $g \in H_0^1$. Since H_0^1 is dense in H^1 and L_b and $\widetilde{L}_b$ coincide on H_0^1, it follows that $\widetilde{L}_b$ is the unique bounded extension of L_b on H^1. This process provides an extension of L_b on the entire space H^1 as a weak limit of bounded linear functionals.

These arguments prove that every BMO function b gives rise to a bounded linear functional $\widetilde{L}_b$ on $H^1(\mathbf{R}^n)$ (henceforth denoted by L_b) that satisfies

$$\|L_b\|_{H^1 \to \mathbf{C}} \leq C_n \|b\|_{BMO}. \tag{7.8.5}$$

The main contribution of the next theorem is the converse assertion.

Theorem 7.8.4. *There exist finite constants C_n and C'_n such that the following statements are valid:*
(a) *Given $b \in BMO(\mathbf{R}^n)$, the linear functional L_b lies in $(H^1(\mathbf{R}^n))^*$ and has norm at most $C_n\|b\|_{BMO}$. Moreover, the mapping $b \mapsto L_b$ from BMO to $(H^1)^*$ is injective.*
(b) *For every bounded linear functional L on H^1 there exists a BMO function b such that*

$$\|b\|_{BMO} \leq C'_n \|L\|_{H^1 \to \mathbf{C}}$$

and such that $L(f) = L_b(f)$ for all functions $f \in H_0^1(\mathbf{R}^n)$.

Proof. (a) We already showed in (7.8.5) that for all $b \in BMO(\mathbf{R}^n)$, L_b lies in $(H^1(\mathbf{R}^n))^*$ and has norm at most $C_n\|b\|_{BMO}$. To show that the embedding $b \mapsto L_b$ is injective, we need to prove that if $L_b = 0$, then b is a constant function. But this is a consequence of Exercise 7.8.5. So we focus attention on assertion (b). Fix a bounded linear functional L on $H^1(\mathbf{R}^n)$ and also fix a cube Q. Consider the space $L^2(Q)$ of all square integrable functions supported in Q with norm

$$\|g\|_{L^2(Q)} = \left(\int_Q |g(x)|^2 dx\right)^{1/2}.$$

We denote by $L_0^2(Q)$ the closed subspace of $L^2(Q)$ consisting of all functions in $L^2(Q)$ with mean value zero. In view of Theorem 7.3.4 every function g in $L_0^2(Q)$ lies in $H^1(\mathbf{R}^n)$ and satisfies the norm estimate

$$\|g\|_{H^1} \leq c_n |Q|^{\frac{1}{2}} \|g\|_{L^2}. \tag{7.8.6}$$

Since $L_0^2(Q)$ is a subspace of H^1, it follows from (7.8.6) that the linear functional L on H^1 is also a linear functional on $L_0^2(Q)$ and its norm satisfies

$$\|L\|_{L_0^2(Q)\to\mathbf{C}} \leq c_n |Q|^{\frac{1}{2}} \|L\|_{H^1\to\mathbf{C}}. \tag{7.8.7}$$

We extend L to a linear functional $\widetilde{L}$ on $L^2(Q)$ by setting

$$\widetilde{L}(h) = L(h - h_Q), \qquad h \in L^2(Q).$$

We notice that for $h \in L^2(Q)$ we have

$$|\widetilde{L}(h)| \leq \|L\|_{L_0^2(Q)\to\mathbf{C}} \|h - h_Q\|_{L^2(Q)}$$
$$\leq \|L\|_{L_0^2(Q)\to\mathbf{C}} \left[\|h\|_{L^2(Q)} + |h_Q| |Q|^{\frac{1}{2}}\right]$$
$$\leq 2\|L\|_{L_0^2(Q)\to\mathbf{C}} \|h\|_{L^2(Q)},$$

thus

$$\|\widetilde{L}\|_{L^2(Q)\to\mathbf{C}} \leq 2 \|L\|_{L_0^2(Q)\to\mathbf{C}}. \tag{7.8.8}$$

Then $\widetilde{L}$ lies in $(L^2(Q))^*$ and by the Riesz representation theorem for the Hilbert space $L^2(Q)$, there is a function $F^Q \in L^2(Q)$ such that

$$\widetilde{L}(h) = \int_Q h F^Q \, dx \qquad \text{for all } h \in L^2(Q),$$

and

$$\|F^Q\|_{L^2} = \|\widetilde{L}\|_{L^2(Q)\to\mathbf{C}}. \tag{7.8.9}$$

Restricting to $L_0^2(Q)$ we can write

$$L(g) = \int_Q g F^Q \, dx, \qquad \text{for all } g \in L_0^2(Q). \tag{7.8.10}$$

Combining (7.8.8) and (7.8.9) we obtain.

$$\|F^Q\|_{L^2(Q)} \leq 2 \|L\|_{L_0^2(Q)\to\mathbf{C}}. \tag{7.8.11}$$

Thus for any cube Q in $\mathbf{R}^n$, there is square integrable function F^Q supported in Q such that (7.8.10) is satisfied.

We now show that if a cube Q is contained in another cube Q', then F^Q differs a.e. from $F^{Q'}$ by a constant on Q. Indeed, for all $g \in L_0^2(Q)$ we have

$$\int_Q F^{Q'}(x) g(x) \, dx = L(g) = \int_Q F^Q(x) g(x) \, dx$$

7.8 Duality Between H^1 and BMO

and thus
$$\int_Q (F^{Q'}(x) - F^Q(x)) g(x)\, dx = 0.$$

The result of Exercise 7.8.5 implies that $F^{Q'} - F^Q$ is equal to a constant a.e. on Q.

Let
$$Q_m = \left[-\frac{m}{2}, \frac{m}{2}\right]^n$$

for $m = 1, 2, \ldots$. Then $|Q_1| = 1$. We define $b \in L^1_{\text{loc}}(\mathbf{R}^n)$ by setting

$$b(x) = F^{Q_m}(x) - \frac{1}{|Q_1|}\int_{Q_1} F^{Q_m}(y)\, dy \tag{7.8.12}$$

whenever $x \in Q_m$. We check that this definition is unambiguous. Let $1 \leq \ell < m$. Then for $x \in Q_\ell$, $b(x)$ is also defined as in (7.8.12) with ℓ in the place of m. The difference of these two functions is $F^{Q_m} - F^{Q_\ell} - (F^{Q_m} - F^{Q_\ell})_{Q_1}$, and we claim this is zero a.e., since $F^{Q_m} - F^{Q_\ell}$ is constant a.e. on Q_ℓ (which is contained in Q_m) and thus it coincides with its integral over Q_1 (recall $|Q_1| = 1$).

Next we claim that for any cube Q there is a constant C_Q such that

$$F^Q = b - C_Q \qquad \text{on } Q. \tag{7.8.13}$$

Indeed, given a cube Q pick the smallest $m = m(Q)$ such that Q is contained in Q^m. Then we write

$$F^Q = \underbrace{F^Q - F^{Q_m}}_{\text{constant on } Q} + \underbrace{F^{Q_m} - (F^{Q_m})_{Q_1}}_{b(x)} + \underbrace{(F^{Q_m})_{Q_1}}_{\text{constant on } Q}$$

and let $-C_Q$ be the sum of the first and third expressions (constants on Q) above.

We have constructed a locally integrable function b such that for all cubes Q and all $g \in L^2_0(Q)$ we have

$$\int_Q b(x) g(x)\, dx = \int_Q (F^Q(x) + C_Q) g(x)\, dx = \int_Q F^Q(x) g(x)\, dx = L(g), \tag{7.8.14}$$

as follows from (7.8.10) and (7.8.13). We next show that b lies in $BMO(\mathbf{R}^n)$. By (7.8.13), (7.8.11), and (7.8.7) we can write

$$\begin{aligned}
\sup_Q \frac{1}{|Q|}\int_Q |b(x) - C_Q|\, dx &= \sup_Q \frac{1}{|Q|}\int_Q |F^Q(x)|\, dx \\
&\leq \sup_Q |Q|^{-1} |Q|^{\frac{1}{2}} \|F^Q\|_{L^2(Q)} \\
&\leq 2 \sup_Q |Q|^{-\frac{1}{2}} \|L\|_{L^2_0(Q) \to \mathbf{C}} \\
&\leq c_n \|L\|_{H^1 \to \mathbf{C}} \\
&< \infty.
\end{aligned}$$

In view of Proposition 6.1.3 we deduce that $b \in BMO$ and

$$\|b\|_{BMO} \leq 2c_n \|L\|_{H^1 \to \mathbf{C}}.$$

Finally, (7.8.14) implies that for all $g \in H_0^1(\mathbf{R}^n)$ one has

$$L(g) = \int_{\mathbf{R}^n} b(x)g(x)\,dx = L_b(g).$$

This proves that the linear functional L coincides with L_b on the dense subspace H_0^1 of H^1. Consequently, $L = L_b$, and this concludes the proof of part (b). $\square$

Exercises

7.8.1. Let $u_k \in H^p$ have uniformly bounded quasi-norms and suppose that $u_k \to u$ in $\mathscr{S}'(\mathbf{R}^n)$. Prove that $u \in H^p$. [*Hint:* Modify the proof of Proposition 7.8.1.]

7.8.2. Let $x_0 \in \mathbf{R}^n \setminus \{0\}$. Let $B(x, \delta)$ denote the ball of radius $\delta > 0$ centered at x. Prove that the sequence of functions $f_k = \chi_{B(kx_0,1)} - \chi_{B(-kx_0,1)}$ satisfies

$$\|f_k\|_{H^1} \to \infty \qquad \text{as } k \to \infty.$$

[*Hint:* Notice that $k^n f_k(k\,\cdot) \to \delta_{x_0} - \delta_{-x_0}$ as $k \to \infty$. Use Exercises 7.8.1 and 7.2.3.]

7.8.3. Show that H^p is a complete quasi-normed space for any $0 < p \leq 1$.
[*Hint:* Modify the proof of Proposition 7.8.1 and use Exercises 7.4.3 and 7.8.1.]

7.8.4. Let $f_k \in H^1(\mathbf{R}^n)$ satisfy $|f_k| \leq F$ for all k, where F is integrable over $\mathbf{R}^n$. Suppose that $f_k \to f$ a.e. Prove that

$$\|f\|_{H^1} \leq \liminf_{k \to \infty} \|f_k\|_{H^1}.$$

7.8.5. Suppose that $u \in L^1_{\text{loc}}(\mathbf{R}^n)$ is supported in a cube Q and has the property

$$\int_Q u(x)g(x)\,dx = 0$$

for all bounded functions g on Q with mean value zero. Show that u is almost everywhere equal to a constant. [*Hint:* If u were not a.e. constant on Q there would exist real numbers $c < d$ such that the sets $E = \{w \leq c\} \cap Q$ and $F = \{w \geq d\} \cap Q$ have positive measure, where w is either $\operatorname{Re} u$ or $\operatorname{Im} u$. Consider the bounded function $g = -|F|$ on E, $g = |E|$ on F, and $g = 0$ elsewhere on Q.]

Chapter 8
Weighted Inequalities

8.1 Appearance of Weights

Weights are positive functions that produce useful absolutely continuous measures. Weights, in particular, are intricately connected with the theory of the Hardy–Littlewood maximal operator. We motivate our discussion on weights by obtaining a weighted version of the weak-type $(1,1)$ inequality of the Hardy–Littlewood maximal operator. First we adapt Lemma 1.4.5 to general measures. We recall that a positive Borel measure is a positive measure defined on the Borel sets, that is, the smallest σ-algebra containing the open sets in $\mathbf{R}^n$. Such a measure is called *regular* if for any Borel measurable set $A \subseteq \mathbf{R}^n$ the following properties hold:

$$\mu(A) = \sup\{\mu(K) : K \text{ is compact subset of } A\},$$
$$\mu(A) = \inf\{\mu(G) : G \text{ is open set containing } A\}.$$

Lemma 8.1.1. (*Covering lemma for general measures*) *Let $\{B_1, B_2, \ldots, B_N\}$ be a finite collection of open balls in $\mathbf{R}^n$. Then there exists a finite subcollection $\{B_{j_1}, \ldots, B_{j_L}\}$ of pairwise disjoint balls ($L \leq N$) such that*

$$v\left(\bigcup_{i=1}^{N} B_i\right) \leq \sum_{r=1}^{L} v(3B_{j_r}) \tag{8.1.1}$$

for any positive Borel measure v. Moreover, the same result is valid if B_j are open cubes with sides parallel to the axes.

Proof. The proof of Lemma 1.4.5 provides a collection of balls $\{B_{j_1}, \ldots, B_{j_L}\}$ that are pairwise disjoint and such that each ball B_k in the original collection is contained in $3B_{j_i}$ for some $i \in \{1, \ldots, L\}$. In other words,

$$\bigcup_{i=1}^{N} B_i \subseteq \bigcup_{r=1}^{L} 3B_{j_r},$$

and (8.1.1) follows by the subadditivity of v.

Finally, we note that the same argument applies if each B_j is a cube. For if two cubes intersect, then the smaller one is contained in the triple of the larger cube. $\square$

Theorem 8.1.2. *(Fefferman–Stein inequality) Let M be the uncentered Hardy–Littlewood operator with respect to balls. Let $u \geq 0$ be a measurable function. Then for any measurable function f on $\mathbf{R}^n$ we have*

$$\int_{\{M(f)>\lambda\}} u(y)\,dy \leq \frac{3^n}{\lambda} \int_{\{M(f)>\lambda\}} |f(x)| M(u)(x)\,dx \qquad (8.1.2)$$

and when $1 < p < \infty$ we have

$$\left(\int_{\mathbf{R}^n} M(f)(x)^p u(x)\,dx\right)^{\frac{1}{p}} \leq 2\left(\frac{3^n p}{p-1}\right)^{\frac{1}{p}} \left(\int_{\mathbf{R}^n} |f(x)|^p M(u)(x)\,dx\right)^{\frac{1}{p}}. \qquad (8.1.3)$$

The same estimates are valid if M is replaced by M_c, the Hardy–Littlewood operator with respect to cubes.

Proof. We introduce two positive measures $\mu = M(u)\,dx$ and $\nu = u\,dx$. The claimed inequalities (8.1.2) and (8.1.3) can be restated in terms of μ and ν as follows:

$$\nu(\{M(f) > \lambda\}) \leq \frac{3^n}{\lambda} \int_{\{M(f)>\lambda\}} |f|\,d\mu \qquad (8.1.4)$$

and, for $1 < p < \infty$,

$$\left(\int_{\mathbf{R}^n} M(f)^p\,d\nu\right)^{\frac{1}{p}} \leq 2\left(\frac{3^n p}{p-1}\right)^{\frac{1}{p}} \left(\int_{\mathbf{R}^n} |f|^p\,d\mu\right)^{\frac{1}{p}} \qquad (8.1.5)$$

for a measurable function f. We first prove these inequalities under the assumption that u is bounded and has compact support.

We consider the set $E_\lambda = \{x \in \mathbf{R}^n : M(f)(x) > \lambda\}$. In the proof of Theorem 1.4.6 this set was shown to be open. Let K be a compact subset of E_λ. For each $x \in K$ there exists an open ball B_x containing the point x such that

$$\int_{B_x} |f(y)|\,dy > \lambda |B_x|. \qquad (8.1.6)$$

Observe that $B_x \subset E_\lambda$ for all x. By compactness there exists a finite subcover $\{B_{x_1}, \ldots, B_{x_N}\}$ of K. Using Lemma 8.1.1 we find a subcollection of pairwise disjoint balls $B_{x_{j_1}}, \ldots, B_{x_{j_L}}$ such that (8.1.1) holds. Then by (8.1.6) we obtain

$$\nu(K) \leq \nu\left(\bigcup_{i=1}^{N} B_{x_i}\right)$$
$$\leq \sum_{i=1}^{L} \nu(3 B_{x_{j_i}})$$

8.1 Appearance of Weights

$$= 3^n \sum_{i=1}^{L} \frac{v(3B_{x_{j_i}})}{|3B_{x_{j_i}}|} |B_{x_{j_i}}|$$

$$\leq \frac{3^n}{\lambda} \sum_{i=1}^{L} \frac{v(3B_{x_{j_i}})}{|3B_{x_{j_i}}|} \int_{B_{x_{j_i}}} |f|\,dy$$

$$\leq \frac{3^n}{\lambda} \sum_{i=1}^{L} M(u)(y_i) \int_{B_{x_{j_i}}} |f|\,dy,$$

where y_i is an arbitrary point in $3B_{x_{j_i}}$. Taking the infimum over all such y_i we deduce that $v(K)$ is bounded by

$$\frac{3^n}{\lambda} \sum_{i=1}^{L} \left(\inf_{3B_{x_{j_i}}} M(u)\right) \int_{B_{x_{j_i}}} |f|\,dy \leq \frac{3^n}{\lambda} \sum_{i=1}^{L} \left(\inf_{B_{x_{j_i}}} M(u)\right) \int_{B_{x_{j_i}}} |f|\,dy$$

$$\leq \frac{3^n}{\lambda} \sum_{i=1}^{L} \int_{B_{x_{j_i}}} |f| M(u)\,dy.$$

Thus

$$v(K) \leq \frac{3^n}{\lambda} \sum_{i=1}^{L} \int_{B_{x_{j_i}}} |f| M(u)\,dy \leq \frac{3^n}{\lambda} \int_{E_\lambda} |f|\,d\mu,$$

as the balls $B_{x_{j_i}}$ are disjoint and contained in E_λ. Taking the supremum over all compact subsets K of E_λ we derive (8.1.4) using the regularity of the measure v, which follows from the fact that u is compactly supported and bounded (Exercise 8.1.1).

We now prove (8.1.5). Fix $1 < p < \infty$. We split $f = f_0 + f_\infty$, where $f_0 = f\chi_{|f|\leq \lambda/2}$ and $f_\infty = f\chi_{|f|>\lambda/2}$. Then

$$\{M(f) > \lambda\} \subseteq \{M(f_0) > \lambda/2\} \cup \{M(f_\infty) > \lambda/2\} = \{M(f_\infty) > \lambda/2\},$$

since the first set is empty, by the definition of f_0 which forces $M(f_0) \leq \lambda/2$.

The pth power of the left-hand side of (8.1.5) can be written as

$$\int_0^\infty p\lambda^{p-1} v(\{M(f) > \lambda\})\,d\lambda \leq \int_0^\infty p\lambda^{p-1} v(\{M(f_\infty) > \lambda/2\})\,d\lambda$$

$$\leq 3^n \int_0^\infty p\lambda^{p-1} \frac{2}{\lambda} \int_{\mathbf{R}^n} |f_\infty|\,d\mu\,d\lambda$$

$$= 3^n 2 \int_0^\infty p\lambda^{p-2} \int_{\{|f|>\lambda/2\}} |f|\,d\mu\,d\lambda$$

$$= 3^n 2 \int_{\mathbf{R}^n} \int_0^{2|f(x)|} p\lambda^{p-2} |f(x)|\,d\lambda\,d\mu(x)$$

$$= 3^n 2^p \frac{p}{p-1} \int_{\mathbf{R}^n} |f(x)|^p\,d\mu(x).$$

This yields (8.1.5). The use of Tonelli's theorem in the penultimate equality is based on the fact that $d\mu$ is a σ-finite measure (if g is is measurable and satisfies $0 \leq g < \infty$ a.e., then gdx is a σ-finite measure on $\mathbf{R}^n$).

For a general nonnegative measurable function u, we introduce

$$u_N = u\chi_{u \leq N}\chi_{B(0,N)}$$

and measures

$$dv_N = u_N dx, \qquad d\mu_N = M(u_N)dx.$$

Then (8.1.4) and (8.1.5) hold for v_N and μ_N. Letting $N \to \infty$ and applying the LMCT then yields (8.1.4) and (8.1.5) in general.

Finally, we note that a repetition of this proof yields the same inequalities for M_c (in place of M), as Lemma 8.1.1 is also valid for cubes in place of balls. □

Remark 8.1.3. We discuss an alternative proof of (8.1.5) under the assumption that $u > 0$ on a set of positive measure. Note that as $M(u)$ never vanishes, the measures μ and dx are mutually absolutely continuous. This gives $\|f\|_{L^\infty(\mu)} = \|f\|_{L^\infty}$. Now we note that

$$\inf\left\{D > 0 : \nu(\{M(f) > D\}) = 0\right\} \leq \|f\|_{L^\infty}$$

as $\{M(f) > \|f\|_{L^\infty}\}$ is a set of Lebesgue measure zero and hence of ν measure zero. (Thus $\|f\|_{L^\infty}$ is one of the D that appear in the set.) This discussion leads to the inequality

$$\|M(f)\|_{L^\infty(\nu)} \leq \|f\|_{L^\infty} = \|f\|_{L^\infty(\mu)}. \qquad (8.1.7)$$

By considering the truncations μ_N and ν_N as in the proof of Theorem 8.1.2, we may assume that μ and ν are σ-finite measures. Interpolating between (8.1.7) and (8.1.4) using Theorem 1.3.3, we obtain that (8.1.5) is a direct consequence of (1.3.6). We note, however, that a slight improvement of the constant in (8.1.5) can be obtained via the technique suggested in Exercise 8.1.2.

This discussion motivates the study of general estimates of the form

$$\|T(f)\|_{L^p(udx)} \leq C\|f\|_{L^p(wdx)}, \qquad (8.1.8)$$

where u, w are a.e. positive functions and T is an operator, such as the Hardy–Littlewood maximal operator. If the functions u, w are locally integrable, then they are called *weights* and estimates of the type (8.1.8) are called *weighted inequalities*. In the remaining sections we focus on weighted inequalities of the form (8.1.8) only when $u = w$, and we seek to characterize the functions u for which such estimates hold. Note that if $M(u) \leq cu$, (i.e., if u is an A_1 weight according to Definition 8.2.5), then Theorem 8.1.2 addresses this situation but does not provide a characterization.

One may wonder if there exists a weak-type estimate for the Hardy–Littlewood maximal function with respect to a measure that is not absolutely continuous with respect to Lebesgue measure. This is not the case, as the following result indicates.

Proposition 8.1.4. *Let μ be a regular positive Borel measure with the property $\mu(K) < \infty$ whenever K is a compact subset of $\mathbf{R}^n$. Suppose that there is a constant*

8.1 Appearance of Weights

$C > 0$ and there exists p in $[1, \infty)$ such that for all $f \in L^p(\mathbf{R}^n)$ one has

$$\sup_{\lambda > 0} \lambda \mu(\{M(f) > \lambda\})^{\frac{1}{p}} \leq C \|f\|_{L^p(\mathbf{R}^n, d\mu)}. \tag{8.1.9}$$

Then μ is absolutely continuous with respect to Lebesgue measure.

Proof. Fix a subset E of $\mathbf{R}^n$ with $|E| = 0$ and $\varepsilon > 0$. Let K be an arbitrary compact subset of E. As $\mu(K) < \infty$, by the regularity of μ, there is an open subset U of $\mathbf{R}^n$ such that $K \subset U$ and $\mu(U \setminus K) < \varepsilon$. Consider the function $\chi_{U \setminus K}$ which is equal to χ_U a.e. with respect to Lebesgue measure. Then $M(\chi_U) = M(\chi_{U \setminus K})$ everywhere and $M(\chi_U)(x) = 1$ for all $x \in U$. Applying hypothesis (8.1.9), with $\lambda = \frac{1}{2}$, we obtain

$$\tfrac{1}{2}\mu(K)^{\frac{1}{p}} \leq \tfrac{1}{2}\mu(\{M(\chi_U) > \tfrac{1}{2}\})^{\frac{1}{p}}$$
$$= \tfrac{1}{2}\mu(\{M(\chi_{U \setminus K}) > \tfrac{1}{2}\})^{\frac{1}{p}}$$
$$\leq C\mu(U \setminus K)^{\frac{1}{p}}$$
$$\leq C\varepsilon^{\frac{1}{p}}.$$

Letting $\varepsilon \to 0$ we obtain $\mu(K) = 0$. Taking the supremum over all compact subsets K of E and using the regularity of μ, we conclude $\mu(E) = 0$.

We have now proved that $|E| = 0$ implies $\mu(E) = 0$, and hence μ is absolutely continuous with respect to Lebesgue measure. $\square$

Exercises

8.1.1. Show that for any nonnegative integrable function u on $\mathbf{R}^n$ the measure $u\,dx$ is regular.

8.1.2. Fix $1 < p < \infty$ and let $u \geq 0$ be a measurable function on $\mathbf{R}^n$. Show that inequality (8.1.3) of Theorem 8.1.2 can be improved to

$$\left(\int_{\mathbf{R}^n} M(f)^p u\,dx\right)^{\frac{1}{p}} \leq (3^n p)^{\frac{1}{p}} \frac{p}{p-1} \left(\int_{\mathbf{R}^n} |f|^p M(u)\,dx\right)^{\frac{1}{p}}.$$

[*Hint:* Split $f = f_0 + f_\infty$, where $f_0 = f\chi_{|f| \leq \varepsilon \lambda}$ and $f_\infty = f\chi_{|f| > \varepsilon \lambda}$. Then use that $\{M(f) > \lambda\}$ is a.e. contained in $\{M(f_\infty) > (1-\varepsilon)\lambda\}$ and optimize over $\varepsilon \in (0,1)$.]

8.1.3. Let $v = u\,dx$ and $\mu = M(u)\,dx$ be as in the proof of Theorem 8.1.2. Prove that for $1 < p < \infty$ and f measurable we have

$$\|M(f)\|_{L^p(\mathbf{R}^n, dv)}^p \leq \frac{3^n p}{p-1} \|M(f)\|_{L^p(\mathbf{R}^n, d\mu)}^{p-1} \|f\|_{L^p(\mathbf{R}^n, d\mu)}.$$

8.1.4. Let $u \geq 0$ be a measurable function on $\mathbf{R}^n$. Define the measures $d\nu = u\,dx$ and $d\mu = M_c(u)\,dx$. Prove that for any measurable function f and any $0 < \varepsilon < 1$ we have
$$\nu(\{M_c(f) > \lambda\}) \leq \frac{3^n}{(1-\varepsilon)\lambda} \int_{\{|f| > \varepsilon\lambda\}} |f|\,d\mu.$$
[*Hint:* Split $f = f_0 + f_\infty$, where $f_\infty = f\chi_{\{|f| > \varepsilon\lambda\}}$ and use the version of (8.1.2) for the maximal function M_c.]

8.2 The A_p Condition

We begin with the formal definition of a weight.

Definition 8.2.1. A *weight* is a nonnegative locally integrable function on $\mathbf{R}^n$ that vanishes or takes the value ∞ only on a set of measure zero. Given a weight w and a measurable set E, we denote the $w\,dx$-measure of E by
$$w(E) = \int_E w(x)\,dx.$$
Since weights are locally integrable functions, we have $w(E) < \infty$ for all sets E contained in some ball. Moreover $w(E) > 0$ for any set E with positive Lebesgue measure. The *weighted L^p spaces* are denoted by $L^p(\mathbf{R}^n, w)$, or simply $L^p(w)$, and consist of all measurable functions g that satisfy
$$\|g\|_{L^p(w)} = \left(\int_{\mathbf{R}^n} |g|^p w\,dx\right)^{\frac{1}{p}} < \infty.$$
Here $0 < p < \infty$. Analogously one may define the weighted weak L^p spaces, which are denoted by $L^{p,\infty}(w)$.

It follows from this definition that the reciprocal of a weight is another weight if and only if it is locally integrable. It is often useful to work with a dense subspace of $L^p(w)$, and the next proposition identifies one.

Proposition 8.2.2. *Let $0 < p < \infty$. Bounded functions with compact support are dense in $L^p(w)$ for any weight w on $\mathbf{R}^n$.*

Proof. Given $f \in L^p(w)$ consider the functions $f_k = f\chi_{|f| \leq k}\chi_{B(0,k)}$ for $k \in \mathbf{Z}^+$. The assertion follows by noticing that $|f_k - f|^p w \to 0$ in L^1 by the LDCT. $\square$

Let M_c be the Hardy–Littlewood maximal operator associated with cubes.[1] Motivated by the discussion in the previous section, we seek for weights w such that for $1 < p < \infty$ there is a constant $C_p = C_p(w)$ with the property

---
[1] Cubes in this text have sides parallel to the axes (unless indicated otherwise).

8.2 The A_p Condition

$$\int_{\mathbf{R}^n} M_c(f)^p w \, dx \leq C_p^p \int_{\mathbf{R}^n} |f|^p w \, dx, \qquad \text{when } f \in L^p(w). \tag{8.2.1}$$

Let us fix $1 < p < \infty$ and suppose that (8.2.1) is valid for some weight w and all $f \in L^p(w)$. Applying (8.2.1) to the function $f \chi_Q$ and using that[2] $|f|_Q \leq M_c(f \chi_Q)(x)$ for all $x \in Q$, we obtain

$$w(Q) |f|_Q^p \leq \int_Q M_c(f \chi_Q)^p w \, dx \leq C_p^p \int_Q |f|^p w \, dx. \tag{8.2.2}$$

It follows that

$$\left(\frac{1}{|Q|} \int_Q |f| \, dt \right)^p \leq \frac{C_p^p}{w(Q)} \int_Q |f|^p w \, dx \tag{8.2.3}$$

for all cubes Q and all functions f. We could obtain a condition involving w by eliminating the integrals in (8.2.3). We achieve this by choosing a function f such that the two integrals are equal; indeed, pick $f = w^{-p'/p}$, which gives $f^p w = w^{-p'/p}$. If we knew that $\inf_Q w > 0$ for all cubes Q, it would follow from (8.2.3) that

$$\sup_{Q \text{ cubes}} \left(\frac{1}{|Q|} \int_Q w \, dx \right) \left(\frac{1}{|Q|} \int_Q w^{-\frac{1}{p-1}} \, dx \right)^{p-1} \leq C_p^p. \tag{8.2.4}$$

This condition involves only the weight w and arbitrary cubes Q.

Now, if $\inf_Q w = 0$ for certain cubes Q, we take $f = (w + \varepsilon)^{-p'/p}$ to obtain

$$\left(\frac{1}{|Q|} \int_Q w \, dx \right) \left(\frac{1}{|Q|} \int_Q (w + \varepsilon)^{-\frac{p'}{p}} \, dx \right)^p \left(\frac{1}{|Q|} \int_Q \frac{w \, dx}{(w + \varepsilon)^{p'}} \right)^{-1} \leq C_p^p \tag{8.2.5}$$

for all $\varepsilon > 0$. Replacing $w \, dx$ by $(w + \varepsilon) \, dx$ in the last integral of (8.2.5) we obtain

$$\left(\frac{1}{|Q|} \int_Q w \, dx \right) \left(\frac{1}{|Q|} \int_Q (w + \varepsilon)^{-\frac{p'}{p}} \, dx \right)^{p-1} \leq C_p^p, \tag{8.2.6}$$

from which we can still deduce (8.2.4) by letting $\varepsilon \to 0$ by appealing to the Lebesgue monotone convergence theorem. This condition motivates the following definition.

Definition 8.2.3. Let $1 < p < \infty$ and w be a weight. The expression

$$[w]_{A_p} = \sup_{Q \text{ cubes in } \mathbf{R}^n} \left(\frac{1}{|Q|} \int_Q w \, dx \right) \left(\frac{1}{|Q|} \int_Q w^{-\frac{1}{p-1}} \, dx \right)^{p-1} \tag{8.2.7}$$

is called the A_p *Muckenhoupt characteristic constant of* w, or simply the A_p characteristic constant of w.

A weight w is said to be *of class* A_p if $[w]_{A_p} < \infty$. The class of A_p weights is

$$A_p = \{ w \text{ weights on } \mathbf{R}^n \text{ such that } [w]_{A_p} < \infty \}.$$

---

[2] Recall that $|f|_Q$ denotes the average of $|f|$ over a cube Q.

Remark 8.2.4. It follows from (8.2.7) that if $w \in A_p$ then $\sigma = w^{-\frac{1}{p-1}} = w^{1-p'}$ is locally integrable, and thus it is also a weight. But then $w = \sigma^{-\frac{1}{p'-1}}$, thus (8.2.7) transforms to

$$[w]_{A_p}^{p'-1} = \sup_{Q \text{ cubes in } \mathbf{R}^n} \left(\frac{1}{|Q|}\int_Q \sigma^{-\frac{1}{p'-1}}dx\right)^{p'-1}\left(\frac{1}{|Q|}\int_Q \sigma\, dx\right)$$

as $(p-1)(p'-1) = 1$ or, equivalently, to

$$[\sigma]_{A_{p'}} = [w]_{A_p}^{\frac{1}{p-1}} = [w]_{A_p}^{p'-1}$$

when $1 < p < \infty$. The weight σ is often called the *dual weight* to w. Notice that condition (8.2.7) can be restated in a more symmetric form as

$$[w]_{A_p}^{\frac{1}{p}} = [\sigma]_{A_{p'}}^{\frac{1}{p'}} = \sup_{Q \text{ cubes in } \mathbf{R}^n} \frac{w(Q)^{\frac{1}{p}}\sigma(Q)^{\frac{1}{p'}}}{|Q|} = \sup_{Q \text{ cubes in } \mathbf{R}^n} (w_Q)^{\frac{1}{p}}(\sigma_Q)^{\frac{1}{p'}} < \infty,$$

where w_Q is the average of w over Q and σ_Q the average of σ over Q.

The argument leading to (8.2.4) does not capture the case $p = 1$, which we examine separately. Assume that for some weight w there is a constant $C_1 < \infty$ such that

$$w(\{M_c(f) > \alpha\}) \leq \frac{C_1}{\alpha}\int_{\mathbf{R}^n}|f|w\,dx, \qquad \text{when } f \in L^1(w). \tag{8.2.8}$$

Since $M_c(f)(x) \geq |f|_Q$ for all $x \in Q$, it follows from (8.2.8) that for all $\alpha < |f|_Q$ we have

$$w(Q) \leq w(\{M_c(f) > \alpha\}) \leq \frac{C_1}{\alpha}\int_{\mathbf{R}^n}|f|w\,dx. \tag{8.2.9}$$

Replacing f by $f\chi_Q$ in (8.2.9) we deduce that

$$\frac{1}{|Q|}\int_Q|f|\,dy \leq \frac{C_1}{w(Q)}\int_Q|f|w\,dy \tag{8.2.10}$$

for all $f \in L^1(w)$ and all cubes Q. Testing this condition on characteristic functions, $f = \chi_S$, we obtain

$$\frac{|S|}{|Q|} \leq C_1 \frac{w(S)}{w(Q)}, \tag{8.2.11}$$

where S is any measurable subset of the cube Q.

Recall that the *essential infimum* of a function w over a set E is defined as

$$\operatorname*{ess.inf}_E(w) = \inf\{b > 0 : |\{x \in E : w(x) < b\}| > 0\}.$$

Then for every $a > \text{ess.inf}_Q(w)$ there exists a subset S_a of Q with positive measure such that $w(x) < a$ for all $x \in S_a$. Applying (8.2.11) to the set S_a, we obtain

8.2 The A_p Condition

$$\frac{1}{|Q|}\int_Q w\,dy \leq \frac{C_1}{|S_a|}\int_{S_a} w\,dy \leq C_1 a, \qquad (8.2.12)$$

which implies that for all cubes Q and almost all $x \in Q$

$$\frac{1}{|Q|}\int_Q w\,dy \leq C_1 w(x). \qquad (8.2.13)$$

This means that for every cube Q there exists a null set $N(Q)$ such that (8.2.13) holds for all x in $Q\setminus N(Q)$. Let N be the union of all the null sets $N(Q)$ for all cubes Q with centers in $\mathbf{Q}^n$ and rational lengths. Then N is a null set and for every x in $Q\setminus N$, (8.2.13) holds for all cubes Q with centers in $\mathbf{Q}^n$ and rational lengths. By density, (8.2.13) must also hold for all cubes Q that contain a fixed x in $\mathbf{R}^n\setminus N$. It follows that for $x \in \mathbf{R}^n\setminus N$ we have

$$M_c(w)(x) = \sup_{Q\ni x}\frac{1}{|Q|}\int_Q w\,dy \leq C_1 w(x). \qquad (8.2.14)$$

Therefore the assumption (8.2.8) leads to the conclusion

$$M_c(w) \leq C_1 w \quad \text{a.e. in } \mathbf{R}^n, \qquad (8.2.15)$$

where C_1 is the same constant as in (8.2.13).

Definition 8.2.5. A nonnegative function w is called an A_1 *weight* if there is a constant $C_1 \in (0,\infty)$ such that

$$M_c(w) \leq C_1 w \quad \text{a.e.} \qquad (8.2.16)$$

We define the class A_1 as follows:

$$A_1 = \{w \text{ weights on } \mathbf{R}^n \text{ such that (8.2.16) is valid for some constant } C_1\}.$$

If w is a weight, then the finite quantity

$$[w]_{A_1} = \sup_{Q \text{ cubes in } \mathbf{R}^n}\left(\frac{1}{|Q|}\int_Q w\,dy\right)\|w^{-1}\|_{L^\infty(Q)} \qquad (8.2.17)$$

is called the A_1 *Muckenhoupt characteristic constant* of w, or simply the A_1 *characteristic constant* of w. Note that the smallest constant C_1 in (8.2.16) is exactly $[w]_{A_1}$, and in this case

$$\frac{1}{|Q|}\int_Q w\,dy \leq [w]_{A_1} \operatorname*{ess.inf}_{y \in Q} w(y) \qquad (8.2.18)$$

for all cubes Q in $\mathbf{R}^n$. Also $[w]_{A_1}$ is the smallest constant such that (8.2.18) is valid.

Remark 8.2.6. We define the A_1 characteristic constant (with respect to balls) by

$$[w]_{A_1}^{\text{balls}} = \sup_{B \text{ balls in } \mathbf{R}^n}\left(\frac{1}{|B|}\int_B w\,dy\right)\|w^{-1}\|_{L^\infty(B)}. \qquad (8.2.19)$$

We leave as an exercise the fact that $[w]_{A_1}^{\text{balls}}$ and $[w]_{A_1}$ are in fact equivalent.

We also define $[w]_{A_p}^{\text{balls}}$ as in (8.2.7) with balls in place of cubes. Then we have

$$\left(v_n 2^{-n}\right)^p \leq \frac{[w]_{A_p}}{[w]_{A_p}^{\text{balls}}} \leq \left(n^{n/2} v_n 2^{-n}\right)^p, \tag{8.2.20}$$

as cubes have size comparable to those of the inscribed and circumscribed balls. See Figure 8.1 and Exercise 8.2.1.

We have now shown that the validity of (8.2.1) implies that the weight is of class A_p and that (8.2.8) implies that the weight is of class A_1. In the next result we prove the converse when $p = 1$ and a weaker version of the converse when $p > 1$. The full version of the converse for $p > 1$ is shown later in Theorem 8.4.3.

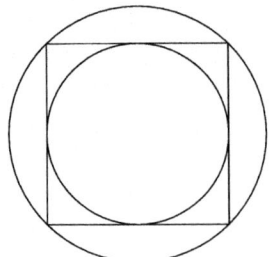

Fig. 8.1 The measure of a cube is comparable to those of the inscribed and circumscribed balls.

Theorem 8.2.7. *Let M_c be the Hardy–Littlewood maximal operator (with respect to cubes). Let $w \in A_p$ and $1 \leq p < \infty$. Then we have*

$$\left\|M_c\right\|_{L^p(w) \to L^{p,\infty}(w)} \leq 3^n [w]_{A_p}^{\frac{1}{p}}. \tag{8.2.21}$$

Proof. This result is essentially contained in Theorem 8.1.2. For instance, when $p = 1$ we restate (8.1.2) (with M_c in place of M) as

$$w(\{M_c(f) > \lambda\}) \leq \frac{3^n}{\lambda} \int_{\{M_c(f) > \lambda\}} |f| M_c(w) \, dx \tag{8.2.22}$$

and as $M_c(w) \leq [w]_{A_1} w$ a.e., we deduce (8.2.21).

When $1 < p < \infty$ we do not restate (8.1.3) but we revisit its proof. The dual weight $\sigma = w^{1-p'}$ plays a key role in this inequality. Let $E_\lambda = \{x \in \mathbf{R}^n : M_c(f)(x) > \lambda\}$ and let K be a compact subset of E_λ. For each $x \in K$ select an open cube Q^x such that $|Q^x|^{-1} \int_{Q^x} |f| \, dx > \lambda$. By the compactness of K, we can work with only finitely many such cubes Q^x.

We let $\{Q_j\}_j$ be the finite subcollection of the finite collection of Q^x, given by Lemma 8.1.1 (adapted to cubes). Then we write

$$\lambda |Q_j| < \int_{Q_j} |f| \, dx = \int_{Q_j} \left(|f| w^{\frac{1}{p}}\right) \left(w^{\frac{1-p'}{p}}\right) dx \leq \left(\int_{Q_j} |f|^p w \, dx\right)^{\frac{1}{p}} \sigma(Q_j)^{\frac{1}{p'}},$$

8.2 The A_p Condition

in view of Hölder's inequality. This estimate in conjunction with Lemma 8.1.1 yields

$$\lambda^p w(K) \leq \sum_j w(3Q_j)\lambda^p$$

$$= 3^{np} \sum_j \frac{w(3Q_j)}{|3Q_j|^p} (\lambda |Q_j|)^p$$

$$\leq 3^{np} \sum_j \frac{w(3Q_j)}{|3Q_j|^p} \left(\int_{Q_j} |f|\, dx \right)^p$$

$$\leq 3^{np} \sum_j \frac{w(3Q_j)}{|3Q_j|^p} \left(\int_{Q_j} |f|^p w\, dx \right) \sigma(Q_j)^{\frac{p}{p'}}$$

$$\leq 3^{np} \left[\sup_j \frac{w(3Q_j)}{|3Q_j|^p} \sigma(3Q_j)^{\frac{p}{p'}} \right] \sum_j \int_{Q_j} |f|^p w\, dx.$$

But notice that

$$\sup_j \frac{w(3Q_j)}{|3Q_j|^p} \sigma(3Q_j)^{\frac{p}{p'}} = \left[\sup_j \frac{w(3Q_j)^{\frac{1}{p}}}{|3Q_j|} \sigma(3Q_j)^{\frac{1}{p'}} \right]^p \leq [w]_{A_p}.$$

Combining these facts, we deduce

$$\lambda^p w(K) \leq 3^{np} [w]_{A_p} \int_{\mathbf{R}^n} |f|^p w\, dx,$$

and from this we derive (8.2.21) after taking the supremum over all compact subsets K of E_λ. □

Remark 8.2.8. Theorem 8.2.7 is also valid for the maximal operators M, $\mathcal{M}$, and $\mathcal{M}_c$ which are pointwise equivalent to M_c.

Exercises

8.2.1. Verify the validity of (8.2.20).

8.2.2. Suppose $Aw \leq u \leq Bw$, where A, B are positive constants and u, w are weights. Show that

$$\frac{A}{B} \leq \frac{[w]_{A_p}}{[u]_{A_p}} \leq \frac{B}{A}.$$

8.2.3. Suppose that $0 < \delta < 1$ and $w \in A_q$ where $1 \leq q < \infty$. Let $p - 1 = \delta(q - 1)$. Show that $w^\delta \in A_p$ and

$$[w^\delta]_{A_p} \leq [w]_{A_q}^\delta.$$

8.2.4. Show that the weight $u(x) = (1 + |x|)^{-n}$ does not lie in A_p for any $p \geq 1$.

8.2.5. Let v be a real-valued locally integrable function on $\mathbf{R}^n$ and let $1 < p < \infty$.
(a) If e^v is an A_p weight, show that

$$\max\left\{\sup_{Q \text{ cubes}} \frac{1}{|Q|}\int_Q e^{v-v_Q}\,dx,\ \sup_{Q \text{ cubes}}\left(\frac{1}{|Q|}\int_Q e^{-(v-v_Q)\frac{1}{p-1}}\,dx\right)^{p-1}\right\} \le [e^v]_{A_p}.$$

(b) Conversely, if

$$C_1 = \sup_{Q \text{ cubes}} \frac{1}{|Q|}\int_Q e^{v-v_Q}\,dx < \infty, \qquad C_2 = \sup_{Q \text{ cubes}} \frac{1}{|Q|}\int_Q e^{-(v-v_Q)\frac{1}{p-1}}\,dx < \infty,$$

then e^v lies in A_p and $[e^v]_{A_p} \le C_1 C_2^{p-1}$. [*Hint:* Use Jensen's inequality.]

8.2.6. (a) Show that if $\varphi \in A_2$, then $\log \varphi \in BMO$ and $\|\log \varphi\|_{BMO} \le [\varphi]_{A_2}$.
(b) Conclude that if $\varphi \in A_p$ with $p > 2$, then $\log \varphi \in BMO$ and

$$\|\log \varphi\|_{BMO} \le (p-1)[\varphi]_{A_p}^{\frac{1}{p-1}}.$$

[*Hint:* Part (a) Use Exercise 8.2.5 (a) with $p=2$. Part (b) Use that $\varphi^{-\frac{1}{p-1}} \in A_{p'}$.]

8.2.7. Let $f \in BMO$ be nonconstant and let $c = 2^{-n-1}\|f\|_{BMO}^{-1}$. Show that

$$\left[e^{cf}\right]_{A_2} \le (1+e)^2.$$

[*Hint:* Use Exercise 8.2.5 (b) and Corollary 6.2.2 with $\gamma = \frac{1}{2}\frac{1}{2^n e}$.]

8.3 Properties of A_p Weights

Cubes in this text have sides parallel to the axes, unless indicated otherwise. Given a cube Q, we denote by λQ the cube with the same center as Q and side length λ times the side length of Q.

Definition 8.3.1. A positive Borel measure μ is called *doubling* if there is constant $A \in (0,\infty)$ such that

$$\mu(2Q) \le A\mu(Q) \tag{8.3.1}$$

for all cubes Q in $\mathbf{R}^n$. In this case, the constant A is called the *doubling constant* of the measure μ.

Lebesgue measure is certainly doubling but the measure $e^{|x|}dx$ is not doubling on any Euclidean space $\mathbf{R}^n$. Next we summarize some basic properties of A_p weights and among them we show that the measure $w\,dx$ is doubling whenever $w \in A_p$.

8.3 Properties of A_p Weights

Proposition 8.3.2. *Let* $w \in A_p$ *for some* $1 \le p < \infty$ *and* $\lambda > 0$. *Then*

(1) $[w^\lambda]_{A_p} = [w]_{A_p}$, *where* $w^\lambda(x) = w(\lambda x)$.

(2) $[\tau^z w]_{A_p} = [w]_{A_p}$, *where* $\tau^z w(x) = w(x-z)$, $z \in \mathbf{R}^n$.

(3) $[\lambda w]_{A_p} = [w]_{A_p}$ *for all* $\lambda > 0$.

(4) $[w]_{A_p} \ge 1$ *for all* $w \in A_p$. *Equality holds if and only if* w *is a constant.*

(5) *For* $p > 1$, *the* A_p *characteristic constant of* w *can be expressed as:*

$$[w]_{A_p} = \sup_{\substack{Q \text{ cubes} \\ \text{in } \mathbf{R}^n}} \sup_{\substack{f \in L^p(Q, w dy) \\ \int_Q |f|^p w dy > 0}} \left\{ \frac{\left(\frac{1}{|Q|} \int_Q |f| dy\right)^p}{\frac{1}{w(Q)} \int_Q |f|^p w dy} \right\}.$$

(6) *The measure* $w dx$ *is doubling. More generally, for* $\lambda > 1$ *and cubes* Q *we have*

$$w(\lambda Q) \le \lambda^{np} [w]_{A_p} w(Q).$$

Proof. The proofs of (1), (2), and (3) are left as exercises. Property (4) follows by an application of Hölder's inequality with exponents p and p' as in

$$1 = \frac{1}{|Q|} \int_Q dx = \frac{1}{|Q|} \int_Q w^{\frac{1}{p}} w^{-\frac{1}{p}} dx \le [w]_{A_p}^{\frac{1}{p}},$$

with equality holding only when $w^{\frac{1}{p}} = c w^{-\frac{1}{p}}$ a.e. for some $c > 0$ (i.e., when w is a.e. equal to a constant). To prove (5), we apply Hölder's inequality with exponents p and p' to obtain

$$\begin{aligned}
\left(\frac{1}{|Q|} \int_Q |f| dx\right)^p &= \left(\frac{1}{|Q|} \int_Q |f| w^{\frac{1}{p}} w^{-\frac{1}{p}} dx\right)^p \\
&\le \frac{1}{|Q|^p} \left(\int_Q |f|^p w dx\right) \left(\int_Q w^{-\frac{p'}{p}} dx\right)^{\frac{p}{p'}} \\
&= \left(\frac{1}{w(Q)} \int_Q |f|^p w dx\right) \left(\frac{1}{|Q|} \int_Q w dx\right) \left(\frac{1}{|Q|} \int_Q w^{-\frac{1}{p-1}} dx\right)^{p-1} \\
&\le [w]_{A_p} \left(\frac{1}{w(Q)} \int_Q |f|^p w dx\right).
\end{aligned}$$

This argument proves the $\ge$ inequality in (5) when $p > 1$. The reverse inequality follows by taking $f = (w+\varepsilon)^{-p'/p}$ as in (8.2.5) and letting $\varepsilon \to 0$.

To prove (6), we apply (5) to the function $f = \chi_Q$ and place λQ in the place of Q in (5). We obtain

$$w(\lambda Q) \le \lambda^{np} [w]_{A_p} w(Q),$$

which implies that $w dx$ is a doubling measure with doubling constant $A = 2^{np} [w]_{A_p}$. This proves (6). □

Two additional properties of the A_p characteristic constants are listed below; note that the only assumption needed in the following proposition is that w is a weight.

Proposition 8.3.3. *For any weight w we have*
$$[w]_{A_q} \le [w]_{A_p} \tag{8.3.2}$$
when $1 \le p < q < \infty$. Moreover,
$$\lim_{q \to 1^+} [w]_{A_q} = [w]_{A_1}. \tag{8.3.3}$$

Proof. The hypothesis $1 \le p < q < \infty$ can be expressed as $0 < q'-1 < p'-1 \le \infty$. Then statement (8.3.2) is a consequence of
$$\|w^{-1}\|_{L^{q'-1}(Q, \frac{dx}{|Q|})} \le \|w^{-1}\|_{L^{p'-1}(Q, \frac{dx}{|Q|})}, \tag{8.3.4}$$

which is a consequence of Hölder's inequality.

We now turn to (8.3.3). The sequence $[w]_{A_q}$ is increasing as q decreases, so it has a limit, which could be infinite.

We consider first the case $[w]_{A_1} < \infty$. Given $\varepsilon > 0$, there is a cube Q such that
$$[w]_{A_1} - \varepsilon < w_Q \|w^{-1}\|_{L^\infty(Q)}.$$

As $q \downarrow 1$, we have $q' - 1 \uparrow \infty$ and thus, by Exercise 1.1.7, we have
$$w_Q \|w^{-1}\|_{L^{q'-1}(Q)} \to w_Q \|w^{-1}\|_{L^\infty(Q)},$$

where the expressions on the left are monotonically increasing by (8.3.4). Then there is a $q_0(Q) > 1$ such that for all q satisfying $1 < q \le q_0(Q)$ one has
$$[w]_{A_1} - \varepsilon < w_Q \|w^{-1}\|_{L^{q'-1}(Q)}.$$

As the expression on the right is bounded by $[w]_{A_q}$, it follows that
$$[w]_{A_1} \le \liminf_{q \to 1}[w]_{A_q}.$$

The reverse inequality
$$\limsup_{q \to 1}[w]_{A_q} \le [w]_{A_1}$$

is a consequence of the monotonicity of the characteristic constants, i.e., (8.3.2).

Now if $[w]_{A_1} = \infty$, then given $M > 0$ there is a cube Q such that
$$M < w_Q \|w^{-1}\|_{L^\infty(Q)}.$$

Then, by Exercise 1.1.7, there is $q_1(Q) > 1$ such that when $1 < q \le q_1(Q)$ one has
$$M < w_Q \|w^{-1}\|_{L^{q'-1}(Q)} \le [w]_{A_q}.$$

This shows that $[w]_{A_q} \to \infty$ as $q \downarrow 1$. $\square$

8.3 Properties of A_p Weights

Corollary 8.3.4. *Suppose that* $\sup_{q>1}[w]_{A_q} < \infty$. *Then* $w \in A_1$.

Proof. This assertion follows by combining (8.3.2) and (8.3.3). □

Example 8.3.5. Let $1 < p < \infty$ and $a > -n$. We investigate for which real numbers a, the locally integrable function $|x|^a$ is an A_p weight, that is, when

$$[|\cdot|^a]_{A_p}^{\text{balls}} = \sup_{B \text{ balls}} \left(\frac{1}{|B|} \int_B |x|^a \, dx\right) \left(\frac{1}{|B|} \int_B |x|^{-a\frac{p'}{p}} \, dx\right)^{\frac{p}{p'}} < \infty. \tag{8.3.5}$$

We split balls $B = B(x_0, R)$ in $\mathbf{R}^n$ into two categories: of type I which means $|x_0| \geq 3R$ and type II which means $|x_0| < 3R$. If $B = B(x_0, R)$ is of type I, then for x satisfying $|x - x_0| \leq R$ we must have

$$\frac{2}{3}|x_0| \leq |x_0| - R \leq |x| \leq |x_0| + R \leq \frac{4}{3}|x_0|,$$

thus the expression inside the supremum in (8.3.5) is comparable to

$$|x_0|^a \left(|x_0|^{-a\frac{p'}{p}}\right)^{\frac{p}{p'}} = 1.$$

If $B(x_0, R)$ is a ball of type II, then $B(0, 5R)$ has size comparable to $B(x_0, R)$ and contains it. Since the measure $|x|^a \, dx$ is doubling, the integrals of the function $|x|^a$ over $B(x_0, R)$ and over $B(0, 5R)$ are comparable. It suffices therefore to estimate the expression inside the supremum in (8.3.5), in which we have replaced $B(x_0, R)$ by $B(0, 5R)$. But this is

$$\left(\frac{1}{v_n(5R)^n} \int_{B(0,5R)} |x|^a \, dx\right) \left(\frac{1}{v_n(5R)^n} \int_{B(0,5R)} |x|^{-a\frac{p'}{p}} \, dx\right)^{\frac{p}{p'}}$$

$$= \left(\frac{n}{(5R)^n} \int_0^{5R} r^{a+n-1} \, dr\right) \left(\frac{n}{(5R)^n} \int_0^{5R} r^{-a\frac{p'}{p}+n-1} \, dr\right)^{\frac{p}{p'}},$$

which is seen easily to be finite and independent of R exactly when $-n < a < n\frac{p}{p'}$. We conclude that $|x|^a$ is an A_p weight, $1 < p < \infty$, if and only if $-n < a < n(p-1)$.

The previous proof can be suitably modified to include the case $p = 1$; in this case we obtain that $|x|^a$ is an A_1 weight if and only if $-n < a \leq 0$.

Thus, given $1 \leq p, q < \infty$ with $q > p$, the weight $|x|^a$ lies in A_q but not in A_p when $n(p-1) < a < n(q-1)$.

Exercises

8.3.1. Let $1 < p < \infty$ and $a \geq n(p-1)$. Show that the measure $|x|^a \, dx$ is doubling but $|x|^a$ is not in A_p.

8.3.2. (a) Let $w_1, w_2 \in A_p$ where $1 \leq p < \infty$. Prove that
$$[w_1 + w_2]_{A_p} \leq [w_1]_{A_p} + [w_2]_{A_p}.$$

(b) If $w_1 \in A_{p_1}$ and $w_2 \in A_{p_2}$ then
$$[w_1 + w_2]_{A_p} \leq [w_1]_{A_{p_1}} + [w_2]_{A_{p_2}},$$
where $1 \leq p_1, p_2 < \infty$ and $p = \max(p_1, p_2)$. [*Hint:* Part (b). Use (8.3.2).]

8.3.3. Let $w_1 \in A_{p_1}$ and $w_2 \in A_{p_2}$ where $1 \leq p_1, p_2 < \infty$ and let $p = \max(p_1, p_2)$.
(a) Prove that
$$[\max(w_1, w_2)]_{A_p} \leq [w_1]_{A_{p_1}} + [w_2]_{A_{p_2}}.$$

(b) Let $c_p = 1$ when $p \leq 2$ and $c_p = 2^{p-2}$ when $p > 2$. Prove that
$$[\min(w_1, w_2)]_{A_p} \leq c_p\big([w_1]_{A_{p_1}} + [w_2]_{A_{p_2}}\big)$$

[*Hint:* Consider first the case where $p_1 = p_2 = p$. Then use (8.3.2).]

8.3.4. (a) Let $1 \leq p < \infty$, $w_1 \in A_p$ and $w_2 \in A_p$ and $\theta_j \geq 0$ satisfy $\theta_1 + \theta_2 = 1$. Prove that
$$[w_1^{\theta_1} w_2^{\theta_2}]_{A_p} \leq [w_1]_{A_p}^{\theta_1} [w_2]_{A_p}^{\theta_2}.$$

(b) Use (8.3.2) to conclude that
$$[w_1^{\theta_1} w_2^{\theta_2}]_{A_p} \leq [w_1]_{A_{p_1}}^{\theta_1} [w_2]_{A_{p_2}}^{\theta_2}$$
where $1 \leq p_1, p_2 < \infty$ and $p = \max(p_1, p_2)$.

8.3.5. Show that the function
$$u(x) = \begin{cases} \log \frac{1}{|x|} & \text{when } |x| < \frac{1}{e}, \\ 1 & \text{otherwise} \end{cases}$$
is an A_1 weight on $\mathbf{R}^n$. [*Hint:* Consider balls of type I and II (Example 8.3.5).]

8.3.6. ([19]) Let $w \in A_p$ and $v \in A_{p'}$. Prove that for any cube Q we have
$$\left(\frac{1}{|Q|} \int_Q w\, dx\right)^{\frac{1}{p}} \left(\frac{1}{|Q|} \int_Q v\, dx\right)^{\frac{1}{p'}} \leq [w]_{A_p}^{\frac{1}{p}} [v]_{A_{p'}}^{\frac{1}{p'}} \left(\frac{1}{|Q|} \int_Q w^{\frac{1}{p}} v^{\frac{1}{p'}}\, dx\right).$$

[*Hint:* Prove $\left(\frac{1}{|Q|} \int_Q w\, dx\right)^{\frac{1}{p}} \left(\frac{1}{|Q|} \int_Q v\, dx\right)^{\frac{1}{p'}} \left(\frac{1}{|Q|} \int_Q w^{-\frac{1}{p}} v^{-\frac{1}{p'}}\, dx\right) \leq [w]_{A_p}^{\frac{1}{p}} [v]_{A_{p'}}^{\frac{1}{p'}}.$]

8.4 Strong-Type A_p Estimates

In this section we obtain a version of Theorem 8.2.7 in which the weak L^p quasi-norm (with respect to w) is replaced by the strong L^p norm. To achieve this goal we work with a maximal function associated with a general positive Borel measure μ.

Definition 8.4.1. Let f be a measurable function on $\mathbf{R}^n$. We define the *centered maximal function with respect to cubes associated with μ* of f as follows:

$$\mathscr{M}_c^\mu(f)(x) = \sup_{\varepsilon>0} \frac{1}{\mu(Q(x,\varepsilon))} \int_{Q(x,\varepsilon)} |f(y)|\, d\mu(y),$$

where $Q(x,\varepsilon)$ is the open cube centered at x of side length 2ε. Likewise, the *uncentered maximal function with respect to cubes associated with μ* is defined by

$$M_c^\mu(f)(x) = \sup_{Q \ni x} \frac{1}{\mu(Q)} \int_Q |f(y)|\, d\mu(y),$$

where the supremum is taken over all open cubes containing the given point x. Here f is a μ-measurable function.

Theorem 8.4.2. *Let f be a measurable function and let $1 < p < \infty$.*
(a) Let μ be a doubling regular positive Borel measure and let $A < \infty$ be a constant such that $\mu(3Q) \le A\mu(Q)$ for all cubes Q in $\mathbf{R}^n$. Then we have

$$\sup_{\lambda>0} \lambda \mu(\{M_c^\mu(f) > \lambda\}) \le A \int_{\{M_c^\mu(f) > \lambda\}} |f|\, d\mu. \tag{8.4.1}$$

Additionally, if μ is σ-finite, then

$$\|M_c^\mu(f)\|_{L^p(\mu)} \le 2 \left(\frac{Ap}{p-1}\right)^{\frac{1}{p}} \|f\|_{L^p(\mu)}. \tag{8.4.2}$$

(b) Let μ be a regular positive Borel measure with the properties $\mu(K) < \infty$ for all compact sets K in $\mathbf{R}^n$ and $\mu(G) > 0$ for all nonempty open sets G in $\mathbf{R}^n$. Then we have

$$\sup_{\lambda>0} \lambda \mu(\{\mathscr{M}_c^\mu(f) > \lambda\}) \le 4^n \int_{\mathbf{R}^n} |f|\, d\mu \tag{8.4.3}$$

and

$$\|\mathscr{M}_c^\mu(f)\|_{L^p(\mu)} \le 2 \left(\frac{4^n p}{p-1}\right)^{\frac{1}{p}} \|f\|_{L^p(\mu)}. \tag{8.4.4}$$

Proof. (a) We obtain (8.4.1) by modifying the proof leading to (8.1.4). We consider the open set $E_\lambda = \{x \in \mathbf{R}^n : M_c^\mu(f)(x) > \lambda\}$. Let K be a compact subset of E_λ. For each $x \in K$ there exists an open cube Q_x containing the point x such that

$$\int_{Q_x} |f|\, d\mu > \lambda \mu(Q_x). \tag{8.4.5}$$

Observe that $Q_x \subset E_\lambda$ for all x. By compactness there exists a finite subcover $\{Q_{x_1}, \ldots, Q_{x_N}\}$ of K. Lemma 8.1.1 yields a subcollection of pairwise disjoint cubes $Q_{x_{j_1}}, \ldots, Q_{x_{j_L}}$ such that (8.1.1) holds. Using (8.4.5) we obtain

$$\mu(K) \leq \mu\left(\bigcup_{i=1}^{N} Q_{x_i}\right) \leq \sum_{i=1}^{L} \mu(3Q_{x_{j_i}}) \leq A \sum_{i=1}^{L} \mu(Q_{x_{j_i}}) \leq \frac{A}{\lambda} \sum_{i=1}^{L} \int_{Q_{x_{j_i}}} |f| \, d\mu \leq \frac{A}{\lambda} \int_{E_\lambda} |f| \, d\mu$$

as the cubes $Q_{x_{j_i}}$ are disjoint and contained in E_λ. Taking the supremum over all compact subsets K of E_λ and using the regularity of μ, we derive (8.4.1).

We now turn to assertion (8.4.2). Fix $1 < p < \infty$. We split $f = f_0 + f_\infty$, where $f_0 = f\chi_{|f| \leq \lambda/2}$ and $f_\infty = f\chi_{|f| > \lambda/2}$. Then

$$\{M_c^\mu(f) > \lambda\} \subseteq \{M_c^\mu(f_0) > \lambda/2\} \cup \{M_c^\mu(f_\infty) > \lambda/2\} = \{M_c^\mu(f_\infty) > \lambda/2\},$$

since the first set is empty by the definition of f_0. We prove this estimate just as we proved (8.1.5). The pth power of the left-hand side of (8.4.2) is

$$\int_0^\infty p\lambda^{p-1} \mu(\{M_c^\mu(f) > \lambda\}) \, d\lambda \leq \int_0^\infty p\lambda^{p-1} \mu(\{M_c^\mu(f_\infty) > \lambda/2\}) \, d\lambda$$

$$\leq A \int_0^\infty p\lambda^{p-1} \frac{2}{\lambda} \int_{\mathbf{R}^n} |f_\infty| \, d\mu \, d\lambda$$

$$= 2A \int_0^\infty p\lambda^{p-2} \int_{\{|f| > \lambda/2\}} |f| \, d\mu \, d\lambda$$

$$= 2A \int_{\mathbf{R}^n} \int_0^{2|f(x)|} p\lambda^{p-2} |f(x)| \, d\lambda \, d\mu(x)$$

$$= 2^p A \frac{p}{p-1} \int_{\mathbf{R}^n} |f(x)|^p \, d\mu(x).$$

The σ-finiteness of μ was used in the application of Tonelli's theorem.

(b) We first prove (8.4.3) for $f \in L^1(\mathbf{R}^n, \mu)$. We begin by showing that the set

$$E_\lambda = \{\mathcal{M}_c^\mu(f) > \lambda\}$$

is open. If we knew that for any fixed $r > 0$ the function

$$x \mapsto \frac{1}{\mu(Q(x,r))} \int_{Q(x,r)} |f| \, d\mu(y) \tag{8.4.6}$$

were continuous, then $\mathcal{M}_c^\mu(f)$ would be lower semicontinuous as the supremum of continuous functions and then E_λ would be open. To establish the continuity, if $x_n \to x_0$, then $\mu(Q(x_n, r)) \to \mu(Q(x_0, r))$ and also $\int_{Q(x_n,r)} |f| \, d\mu(y) \to \int_{Q(x_0,r)} |f| \, d\mu(y)$ by the Lebesgue dominated convergence theorem. As $\mu(Q(x_n, r)) \neq 0 \neq \mu(Q(x_0, r))$, it follows that the function in (8.4.6) is continuous.

Given a compact subset K of E_λ, for any $x \in K$ select an open cube $Q_x = Q(x, \delta_x)$ of length $2\delta_x$ centered at x such that

8.4 Strong-Type A_p Estimates

$$\frac{1}{\mu(Q_x)} \int_{Q_x} |f| d\mu(y) > \lambda.$$

Applying Lemma G.0.1 (Appendix G) we extract a sequence of points $\{x_j\}_{j=1}^m$ in K and a subfamily $\{Q_{x_j}\}_{j=1}^m$ of the family of the cubes $\{Q_x : x \in K\}$ such that

$$K \subseteq \bigcup_{j=1}^m Q_{x_j}$$

and that for all $y \in \mathbf{R}^n$ we have

$$\sum_{j=1}^m \chi_{Q_{x_j}}(y) \leq 4^n. \tag{8.4.7}$$

Then

$$\mu(K) \leq \sum_{j=1}^m \mu(Q_{x_j}) \leq \sum_{j=1}^m \frac{1}{\lambda} \int_{Q_{x_j}} |f| d\mu(y) \leq \frac{4^n}{\lambda} \int_{\mathbf{R}^n} |f| d\mu(y),$$

where the last inequality is a consequence of (8.4.7). Taking the supremum over all compact subsets K of E_λ and using the regularity of $d\mu$ we deduce (8.4.3).

Now we remove the assumption that $f \in L^1(\mathbf{R}^n, \mu)$. To do so, given a general measurable function f we define the sequence

$$f_N = f \chi_{|f| \leq N} \chi_{B(0,N)}, \qquad N = 1, 2, \ldots.$$

Then $f_N \in L^1(\mathbf{R}^n, \mu)$ and thus (8.4.3) holds for f_N in place of f. The LMCT yields the conclusion for a general f.

Finally assertion (8.4.4) is obtained just like (8.4.2). $\square$

Theorem 8.4.3. *Let $w \in A_p(\mathbf{R}^n)$ for some $1 < p < \infty$. Then we have*

$$\|\mathcal{M}_c\|_{L^p(w) \to L^p(w)} \leq 3^{(n+1)p'} \left(16^n p p'\right)^{\frac{1}{p}} [w]_{A_p}^{\frac{1}{p-1}}. \tag{8.4.8}$$

Remark 8.4.4. If one opted to use conclusion (a) instead of (b) in Theorem 8.4.2, the constant 4^n in (8.4.8) should be replaced by $3^{np}[w]_{A_p}$ [cf. Proposition 8.3.2 (6)].

Proof. Recall the dual weight $\sigma = w^{1-p'}$ which lies in $A_{p'}$. (Equivalently we have $w = \sigma^{1-p}$.)

We fix a measurable function f on $\mathbf{R}^n$ and a cube Q with sides parallel to the axes. For every $x \in Q$, we denote by Q^x the smallest cube (also with sides parallel to the axes) centered at x, contained in $3Q$, and containing Q. See Figure 8.2.

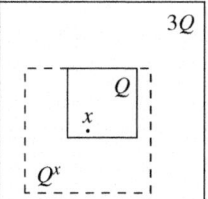

Fig. 8.2 The cubes Q, Q^x, and $3Q$.

Then for any $x \in Q$ we have

$$\frac{1}{|Q|}\int_Q |f|\,dy \le \frac{\sigma(Q^x)}{|Q|}\frac{1}{\sigma(Q^x)}\int_{Q^x}|f|\sigma^{-1}\sigma\,dy \le \frac{\sigma(3Q)}{|Q|}\mathscr{M}_c^\sigma(f\sigma^{-1})(x).$$

As this estimate is valid for all $x \in Q$, we average it over Q after raising it to the power $p-1$. This gives

$$\left(\frac{1}{|Q|}\int_Q |f|\,dy\right)^{p-1} \le \left(\frac{\sigma(3Q)}{|Q|}\right)^{p-1}\frac{1}{|Q|}\int_Q (\mathscr{M}_c^\sigma(f\sigma^{-1}))^{p-1}\,dy$$

$$= \left(\frac{\sigma(3Q)}{|Q|}\right)^{p-1}\frac{w(Q)}{|Q|}\frac{1}{w(Q)}\int_Q (\mathscr{M}_c^\sigma(f\sigma^{-1}))^{p-1}w^{-1}w\,dy$$

$$\le \left(\frac{\sigma(3Q)}{|Q|}\right)^{p-1}\frac{w(3Q)}{|Q|}\mathscr{M}_c^w\!\left((\mathscr{M}_c^\sigma(f\sigma^{-1}))^{p-1}w^{-1}\right)(c_Q),$$

where c_Q is the center of Q. Expressing $|Q|$ in terms of $|3Q|$ we deduce the estimate

$$\left(\frac{1}{|Q|}\int_Q |f|\,dy\right)^{p-1} \le 3^{np}[w]_{A_p}\mathscr{M}_c^w\!\left((\mathscr{M}_c^\sigma(f\sigma^{-1}))^{p-1}w^{-1}\right)(c_Q)$$

which implies, for any $x \in \mathbf{R}^n$,

$$\mathscr{M}_c(f)(x) \le 3^{np'}[w]_{A_p}^{\frac{1}{p-1}}\left(\mathscr{M}_c^w\!\left((\mathscr{M}_c^\sigma(f\sigma^{-1}))^{p-1}w^{-1}\right)(x)\right)^{\frac{1}{p-1}}.$$

It follows that

$$\|\mathscr{M}_c(f)\|_{L^p(w)} \le 3^{np'}[w]_{A_p}^{\frac{1}{p-1}}\left[\int_{\mathbf{R}^n}\left(\mathscr{M}_c^w\!\left((\mathscr{M}_c^\sigma(f\sigma^{-1}))^{p-1}w^{-1}\right)\right)^{p'}w\,dx\right]^{\frac{1}{p'}\frac{p'}{p}}.$$

We now apply (8.4.4) in Theorem 8.4.2 first with respect to the measure $w\,dx$ on $L^{p'}(w)$ and then with respect to the measure $\sigma\,dx$ on $L^p(\sigma)$. We obtain

$$\|\mathscr{M}_c(f)\|_{L^p(w)} \le 3^{np'}[w]_{A_p}^{\frac{1}{p-1}}[2(4^n p)^{\frac{1}{p'}}]^{\frac{p'}{p}}\left[\int_{\mathbf{R}^n}\left((\mathscr{M}_c^\sigma(f\sigma^{-1}))^{p-1}w^{-1}\right)^{p'}w\,dx\right]^{\frac{1}{p'}\frac{p'}{p}}$$

$$= 3^{np'}[w]_{A_p}^{\frac{1}{p-1}}[2(4^n p)^{\frac{1}{p'}}]^{\frac{p'}{p}}\left[\int_{\mathbf{R}^n}(\mathscr{M}_c^\sigma(f\sigma^{-1}))^p\sigma\,dx\right]^{\frac{1}{p}}$$

$$\le 3^{np'}[w]_{A_p}^{\frac{1}{p-1}}[2(4^n p)^{\frac{1}{p'}}]^{\frac{p'}{p}}[2(4^n p')^{\frac{1}{p}}]\left[\int_{\mathbf{R}^n}(f\sigma^{-1})^p\sigma\,dx\right]^{\frac{1}{p}}$$

$$= 3^{np'}2^{p'}(16^n pp')^{\frac{1}{p}}[w]_{A_p}^{\frac{1}{p-1}}\|f\|_{L^p(w)}.$$

This proves (8.4.8). $\square$

Exercises

8.4.1. Let $w \in A_p$ for some $1 < p < \infty$. Let μ be a regular positive Borel measure with the property that the space $(\mathbf{R}^n, \mu)$ is σ-finite. Show that the sublinear operator

$$f \mapsto \mathcal{M}_c^\mu(w^{\frac{1}{p}}|f|)w^{-\frac{1}{p}}$$

maps $L^p(w, d\mu)$ to itself with norm bounded by $2\left(\frac{4^n p}{p-1}\right)^{1/p}$.

8.4.2. Let $w \in A_p$ for some $1 < p < \infty$ and let $0 < q < \infty$. Show that the operator

$$f \mapsto \left(M(|f|^q w) w^{-1}\right)^{\frac{1}{q}}$$

maps $L^{p'q}(w)$ to itself with norm bounded by $C_{n,p,q}[w]_{A_p}^{1/q}$, $C_{n,p,q}$ being a constant.

8.4.3. Let μ be a regular doubling positive Borel measure such that $(\mathbf{R}^n, \mu)$ is σ-finite and let $A < \infty$ be a constant such that $\mu(3Q) \le A\mu(Q)$ for all cubes Q in $\mathbf{R}^n$. Prove that M_c^μ maps $L^{p,\infty}(\mu)$ to itself with norm at most $\frac{Ap}{p-1}$, $1 < p < \infty$.
[*Hint:* Use (8.4.1) and Theorem 1.2.10.]

8.5 The Jones Factorization of Weights

We begin by building an A_p weight from two A_1 weights.

Proposition 8.5.1. *Let w_1, w_2 be two A_1 weights and let $1 < p < \infty$. Then $w_1 w_2^{1-p}$ is an A_p weight which satisfies*

$$[w_1 w_2^{1-p}]_{A_p} \le [w_1]_{A_1} [w_2]_{A_1}^{p-1}. \tag{8.5.1}$$

Proof. For any cube Q we have

$$\left(\frac{1}{|Q|}\int_Q w_1 w_2^{1-p} dx\right)\left(\frac{1}{|Q|}\int_Q (w_1 w_2^{1-p})^{-\frac{1}{p-1}} dx\right)^{p-1}$$

$$\le \left(\frac{1}{|Q|}\int_Q w_1 \|w_2^{-1}\|_{L^\infty(Q)}^{p-1} dx\right)\left(\frac{1}{|Q|}\int_Q \|w_1^{-1}\|_{L^\infty(Q)}^{\frac{1}{p-1}} w_2 \, dx\right)^{p-1}$$

$$= \left(\frac{1}{|Q|}\int_Q w_1 \, dx\right) \|w_1^{-1}\|_{L^\infty(Q)} \left(\frac{1}{|Q|}\int_Q w_2 \, dx\right)^{p-1} \|w_2^{-1}\|_{L^\infty(Q)}^{p-1}.$$

Taking the supremum over all cubes Q, we deduce (8.5.1). $\square$

It is rather remarkable that the converse of Proposition 8.5.1 is also valid.

Theorem 8.5.2. *Suppose that w is an A_p weight for some $1 < p < \infty$. Then there exist A_1 weights w_1 and w_2 such that*

$$w = w_1 w_2^{1-p}$$

and moreover, there exist finite constants $c_1(n,p)$, $c_2(n,p)$ such that

$$[w_1]_{A_1} \leq c_1(n,p)[w]_{A_p}, \qquad [w_2]_{A_1} \leq c_2(n,p)[w]_{A_p}^{\frac{1}{p-1}}. \tag{8.5.2}$$

Proof. We are seeking weights w_1 and w_2 such that $w = w_1 w_2^{1-p}$. Setting $u = w_2^{p-1}$, we express w_1 and w_2 in terms of u as follows: $w_1 = wu$ and $w_2 = u^{p'-1}$. We are therefore looking for a weight u such that both wu and $u^{p'-1}$ are A_1 weights, i.e., they satisfy

$$\mathcal{M}_c(wu) \leq Cuw \qquad \text{and} \qquad \mathcal{M}_c(u^{p'-1}) \leq Cu^{p'-1},$$

for some constant C, or equivalently,

$$w^{-1}\mathcal{M}_c(wu) \leq Cu \qquad \text{and} \qquad \mathcal{M}_c(u^{p'-1})^{p-1} \leq Cu. \tag{8.5.3}$$

We would like to find a *sub-eigenvector* of a positive operator T, which means a positive function u that satisfies $T(u) \leq Cu$. In our case, T could be one of the operators

$$f \mapsto w^{-1}\mathcal{M}_c(wf)$$

or

$$f \mapsto \mathcal{M}_c(f^{p'-1})^{p-1}.$$

But as these are positive operators, it will be sufficient to take T to be their sum, i.e.,

$$T(f) = w^{-1}\mathcal{M}_c(wf) + \mathcal{M}_c(f^{p'-1})^{p-1}.$$

At this point, we make the assumption $1 < p \leq 2$. This restriction on p is crucial in making the positive operator $f \mapsto \mathcal{M}_c(f^{p'-1})^{p-1} = \mathcal{M}_c(f^{p'-1})^{\frac{1}{p'-1}}$ sublinear and countably subadditive.[3] Moreover, the operator T maps $L^{p'}(w)$ to itself. To see this, let

$$C(n,p) = 3^{np'}2^p\left(4^n\sqrt{pp'}\right)^{\frac{2}{p}}$$

be the constant that appears in (8.4.8). Then

$$\left(\int_{\mathbf{R}^n}[w^{-1}\mathcal{M}_c(wf)]^{p'}w\,dx\right)^{\frac{1}{p'}} \leq C(n,p')[w^{1-p'}]_{A_{p'}}^{\frac{1}{p'-1}}\|f\|_{L^{p'}(w)}$$

and

---

[3] A positive operator T is called countably subadditive if $T(\sum_{j=0}^\infty f_j) \leq \sum_{j=0}^\infty T(f_j)$ when $f_j \geq 0$.

8.5 The Jones Factorization of Weights

$$\left(\int_{\mathbf{R}^n}(\mathcal{M}_c(|f|^{p'-1})^{p-1})^{p'}w\,dx\right)^{\frac{1}{p'}} \leq \left[C(n,p)[w]_{A_p}^{\frac{1}{p-1}}\left(\int_{\mathbf{R}^n}(|f|^{p'-1})^p w\,dx\right)^{\frac{1}{p}}\right]^{\frac{p}{p'}},$$

and a combination of these estimates yields

$$B = \|T\|_{L^{p'}(w) \to L^{p'}(w)} \leq \left[C(n,p') + C(n,p)^{\frac{p}{p'}}\right][w]_{A_p} < \infty.$$

We now fix a nonzero function f in $L^{p'}(w)$ and we define the function

$$u = \sum_{k=0}^{\infty} \frac{T^k(f)}{(2B)^k}$$

which is positive everywhere. This series converges in $L^{p'}(w)$ and so it defines a function u in $L^{p'}(w)$. Moreover, this construction ensures that u is a sub-eigenvector of T, in the aforementioned sense, as

$$\begin{aligned} T(u) &= T\left(\sum_{k=0}^{\infty} \frac{T^k(f)}{(2B)^k}\right) \\ &\leq \sum_{k=0}^{\infty} \frac{T^{k+1}(f)}{(2B)^k} \\ &= 2B\sum_{k=1}^{\infty} \frac{T^k(f)}{(2B)^k} \\ &\leq 2Bu, \end{aligned}$$

having used that $T(\lambda f) = |\lambda| T(f)$ for $\lambda \in \mathbf{C}$. So now we have found a function u such that both inequalities in (8.5.3) are valid with $C = 2B$. These inequalities imply that

$$[wu]_{A_1} \leq 2B \quad \text{and} \quad [u^{p'-1}]_{A_1} \leq (2B)^{\frac{1}{p'-1}}$$

and directly translate into the estimates claimed in (8.5.2), given that $w_1 = wu$ and $w_2 = u^{p'-1}$ and that B is bounded by a constant multiple of $[w]_{A_p}$.

We now turn to the case $p > 2$. Given a weight $w \in A_p$ for $p > 2$, we factor its dual weight $\sigma = w^{1-p'}$, which lies in $A_{p'}$ and $p' < 2$. By the previous case we are able to write

$$\sigma = v_1 v_2^{1-p'}, \quad \text{where} \quad [v_1]_{A_1} \leq c_1(n,p)[\sigma]_{A_{p'}}, \quad [v_2]_{A_1} \leq c_2(n,p)[\sigma]_{A_{p'}}^{\frac{1}{p'-1}}.$$

Then $w = \sigma^{1-p}$ satisfies

$$w = v_1^{1-p} v_2, \quad \text{where} \quad [v_1]_{A_1} \leq c_1(n,p)[w]_{A_p}^{\frac{1}{p-1}}, \quad [v_2]_{A_1} \leq c_2(n,p)[w]_{A_p},$$

and the claimed conclusion follows with $w_1 = v_2$, $w_2 = v_1$ and the constants $c_1(n,p)$ and $c_2(n,p)$ interchanged. $\square$

We have managed to obtain a description of A_p weights in terms of A_1 weights. But it remains to understand the structure of A_1 weights. In particular, we would like to have ways to build A_1 weights. We implicitly saw such a way in the proof of Theorem 8.5.2. Let us recap it. Let $f \in L^p(\mathbf{R}^n, \mu)$ for some $1 < p < \infty$. Define

$$u = \sum_{k=0}^{\infty} \frac{M^k(f)}{(2\|M\|)^k},$$

where $\|M\| < \infty$ is the norm of the Hardy–Littlewood maximal operator M on $L^p(\mathbf{R}^n, \mu)$ (Theorem 8.4.2). Then u is a well-defined function in $L^p(\mathbf{R}^n, \mu)$ and satisfies

$$\begin{aligned}
M(u) &= M\Big(\sum_{k=0}^{\infty} \frac{M^k(f)}{(2\|M\|)^k} \Big) \\
&\leq \sum_{k=0}^{\infty} \frac{M^{k+1}(f)}{(2\|M\|)^k} \\
&= 2\|M\| \sum_{k=1}^{\infty} \frac{M^k(f)}{(2\|M\|)^k} \\
&\leq 2\|M\| \, u,
\end{aligned}$$

so u is an A_1 weight.

This process of construction is more or less abstract, but there is a more concrete way to build A_1 weights.

Theorem 8.5.3. *Let f be a measurable function such that $M_c(f) < \infty$ a.e. Then for $0 \leq \varepsilon < 1$, the function $M_c(f)^\varepsilon$ is an A_1 weight that satisfies*

$$\big[M_c(f)^\varepsilon\big]_{A_1} \leq \frac{9^n + 3^n}{1 - \varepsilon}. \tag{8.5.4}$$

Proof. Fix $x \in \mathbf{R}^n$. Given a measurable function f such that $M_c(f) < \infty$ a.e. and a cube Q that contains x, we will show that

$$\frac{1}{|Q|} \int_Q M_c(f)(y)^\varepsilon \, dy \leq \frac{9^n}{1-\varepsilon} M_c(f)(x)^\varepsilon + 3^n M_c(f)(x)^\varepsilon. \tag{8.5.5}$$

To prove (8.5.5), we fix a cube Q and split $f = f\chi_{3Q} + f\chi_{(3Q)^c}$. Then

$$\begin{aligned}
\frac{1}{|Q|} \int_Q M_c(f\chi_{3Q})(y)^\varepsilon \, dy &\leq \frac{1}{|Q|} \frac{|Q|^{1-\varepsilon}}{1-\varepsilon} \|M_c(f\chi_{3Q})\|_{L^{1,\infty}}^\varepsilon \\
&\leq \frac{1}{1-\varepsilon} \frac{1}{|Q|^\varepsilon} \Big(3^n \int_{3Q} |f(y)| \, dy \Big)^\varepsilon \\
&\leq \frac{3^{n\varepsilon} 3^{n\varepsilon}}{1-\varepsilon} M_c(f)(x)^\varepsilon, \tag{8.5.6}
\end{aligned}$$

8.5 The Jones Factorization of Weights

where we made use of the inequality of Exercise 1.2.6 (a) (Kolmogorov's inequality) and the weak-type $(1,1)$ boundedness of M_c (with constant 3^n).

For $f\chi_{(3Q)^c}$ we only need to notice that for all y in Q

$$M_c(f\chi_{(3Q)^c})(y) \leq 3^n M_c(f)(x). \tag{8.5.7}$$

The reason for the validity of (8.5.7) is that any cube R that contains the point y in Q and meets $(3Q)^c$ must have side length at least that of Q, hence $3R$ contains x; see Figure 8.3. Thus

$$\frac{1}{|R|} \int_R |f|\chi_{(3Q)^c} \, dy$$
$$\leq \frac{3^n}{|3R|} \int_{3R} |f| \, dy$$
$$\leq 3^n M_c(f)(x).$$

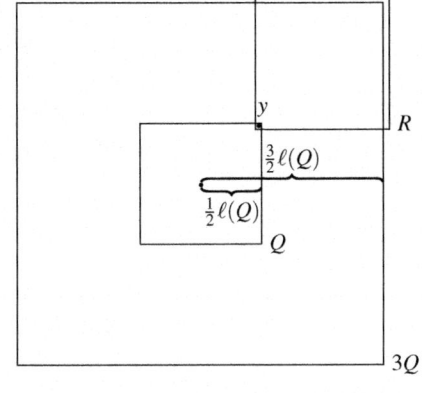

Fig. 8.3 A cube R that contains a point $y \in Q$ and meets $(3Q)^c$.

Averaging the εth power of (8.5.7) over Q, we obtain

$$\frac{1}{|Q|} \int_Q M_c(f\chi_{(3Q)^c})(y)^\varepsilon \, dy \leq 3^{n\varepsilon} M_c(f)(x)^\varepsilon.$$

Combining the estimates for $f\chi_{(3Q)^c}$ and $f\chi_{3Q}$ and using the subadditivity property $M_c(f)^\varepsilon \leq M_c(f\chi_{(3Q)^c})^\varepsilon + M_c(f\chi_{3Q})^\varepsilon$, we deduce (8.5.5). □

We show in Theorem 8.6.5 that Theorem 8.5.3 can essentially be reversed.

Exercises

8.5.1. Let $0 \leq \alpha, \beta < 1$ and $a, b \in \mathbf{R}^n$. Prove that the weight

$$\frac{(1+|x-a|)^{n\alpha(p-1)}}{(1+|x-b|)^{n\beta}}$$

lies in A_p for $1 < p < \infty$. In particular, $(1+|x|)^{n\gamma} \in A_2$ when $-1 < \gamma < 1$. [*Hint:* Use Proposition 8.5.1 and Theorem 8.5.3.]

8.5.2. Find a measurable bounded function f on $\mathbf{R}^n$ and a constant c_n such that

$$[M(f)^\varepsilon]_{A_1} \geq \frac{c_n}{1-\varepsilon}$$

for any $0 < \varepsilon < 1$, i.e., the reverse inequality to (8.5.4) holds.

8.5.3. Let $1 \leq p_0 < p < \infty$. If $w_1 \in A_{p_0}$ and $w_2 \in A_1$, then prove that $w_1 w_2^{p_0-p}$ lies in A_p and

$$[w_1 w_2^{p_0-p}]_{A_p} \leq [w_1]_{A_{p_0}} [w_2]_{A_1}^{p-p_0}.$$

8.5.4. Let $1 \leq p_0 < p < \infty$. Given a weight $w \in A_p$ obtain the existence of weights $u_1 \in A_{p_0}$ and $u_2 \in A_1$ such that the factorization $w = u_1 u_2^{p_0-p}$ holds and, moreover,

$$[u_1]_{A_{p_0}} \leq c_1(n,p) c_2(n,p)^{p_0-1} [w]_{A_p}^{\frac{p_0-1}{p-1}+1}, \qquad [u_2]_{A_1} \leq c_2(n,p) [w]_{A_p}^{\frac{1}{p-1}},$$

where $c_1(n,p)$, $c_2(n,p)$ are the constants in Theorem 8.5.2. [*Hint:* Pick $u_2 = w_2$, where w_2 is the weight obtained in the factorization of w in Theorem 8.5.2.]

8.5.5. Let $1 < p < p_0 < \infty$. If $w_1 \in A_{p_0}$ and $w_2 \in A_1$, then prove that the weight $(w_1^{p-1} w_2^{p_0-p})^{\frac{1}{p_0-1}}$ lies in A_p and

$$\left[(w_1^{p-1} w_2^{p_0-p})^{\frac{1}{p_0-1}}\right]_{A_p} \leq [w_1]_{A_{p_0}}^{\frac{p-1}{p_0-1}} [w_2]_{A_1}^{\frac{p_0-p}{p_0-1}}.$$

8.6 Reverse Hölder Property of A_p Weights

Hölder's inequality on the probability space $(Q, \frac{1}{|Q|} dx)$ yields

$$\frac{1}{|Q|} \int_Q w\, dx \leq \left(\frac{1}{|Q|} \int_Q w^r\, dx\right)^{\frac{1}{r}}$$

for $1 < r < \infty$. The interesting fact is that this inequality can be reversed for A_p weights. In fact one has the so-called reverse-Hölder property

$$\left(\frac{1}{|Q|} \int_Q w^r\, dx\right)^{\frac{1}{r}} \leq \frac{K}{|Q|} \int_Q w\, dx, \qquad (8.6.1)$$

for some constant K uniformly over all cubes Q.

Definition 8.6.1. Let $1 < r < \infty$. We define RH_r to be the space of all weights w on $\mathbf{R}^n$ that satisfy (8.6.1). For $w \in RH_r$ we set $[w]_{RH_r}$ to be the smallest constant K such that (8.6.1) holds for all cubes Q.

8.6 Reverse Hölder Property of A_p Weights

Before we embark into the proof of the reverse Hölder property, we discuss the *reverse weak-type* $(1,1)$ inequality for the Hardy–Littlewood maximal operator.

Theorem 8.6.2. *For each $f \in L^1(\mathbf{R}^n)$, cube[4] Q, and $\lambda > 0$ we have*

$$\frac{1}{2^n \lambda} \int_{\{x \in Q: |f| > \lambda\}} |f| \, dx \leq |\{x \in Q : M_c(f) > \lambda\}|. \tag{8.6.2}$$

Proof. Although (8.6.2) is stated in terms of M_c it will be convenient to work with its dyadic counterpart M_d. So we fix a function $f \in L^1(\mathbf{R}^n)$ and consider the set $E_\lambda = \{x \in Q : M_d(f)(x) > \lambda\}$. For each $x \in E_\lambda$ there is a maximal dyadic cube Q^x such that

$$\frac{1}{|Q^x|} \int_{Q^x} |f| \, dy > \lambda.$$

As the dyadic ancestor of Q^x is not maximal, this implies that

$$\frac{1}{|Q^x|} \int_{Q^x} |f| \, dy \leq 2^n \lambda.$$

We denote by $\{Q_j\}_j$ the collection of all maximal dyadic cubes Q^x, where $x \in E_\lambda$. Then the Q_j are pairwise disjoint and $\cup_j Q_j = E_\lambda$. As $|f| \leq M_d(f)$ a.e., it follows that

$$\begin{aligned}
\frac{1}{2^n \lambda} \int_{\{x \in Q: |f| > \lambda\}} |f| \, dx &\leq \frac{1}{2^n \lambda} \int_{\{x \in Q: M_d(f) > \lambda\}} |f| \, dx \\
&= \frac{1}{2^n \lambda} \sum_j \int_{Q_j} |f| \, dx \\
&\leq \sum_j |Q_j| \\
&= |E_\lambda| \\
&\leq |\{x \in Q : M_c(f) > \lambda\}|,
\end{aligned}$$

since $M_d \leq M_c$. Then (8.6.2) is proved. $\square$

Theorem 8.6.3. *For any $w \in A_1$, any $K > 1$, and any r satisfying*

$$1 \leq r \leq 1 + \frac{K-1}{K} \frac{1}{e^{1/e} 2^n [w]_{A_1}} \tag{8.6.3}$$

we have $[w]_{RH_r} \leq K$.

Proof. Applying (8.6.2) to $f = w$ we obtain

$$\frac{1}{2^n \lambda} \int_{\{x \in Q: w > \lambda\}} w \, dx \leq |\{x \in Q : M_c(w) > \lambda\}|.$$

---

[4] The cube Q could have infinite side length or could be the whole space.

It follows from this that

$$\frac{1}{2^n\lambda} w(\{x \in Q : w > \lambda\}) \leq |\{x \in Q : w > \lambda/[w]_{A_1}\}|,$$

as w is an A_1 weight. In view of (1.2.3), we write

$$\int_Q w^r \, dx = \int_Q w^{r-1} w \, dx$$

$$= (r-1) \int_0^\infty \lambda^{r-2} w(\{x \in Q : w(x) > \lambda\}) \, d\lambda$$

$$= (r-1) \left[\int_0^a + \int_a^\infty \right] \lambda^{r-2} w(\{x \in Q : w(x) > \lambda\}) \, d\lambda$$

$$\leq a^{r-1} w(Q) + (r-1) \int_a^\infty \lambda^{r-2} 2^n \lambda \left| \{x \in Q : w(x) > \lambda/[w]_{A_1}\} \right| d\lambda$$

$$\leq a^{r-1} w(Q) + 2^n [w]_{A_1}^r (r-1) \int_0^\infty \lambda^{r-1} |\{x \in Q : w(x) > \lambda\}| \, d\lambda$$

$$= a^{r-1} w(Q) + 2^n [w]_{A_1}^r \frac{r-1}{r} \int_Q w^r \, dx.$$

Choosing $a = w_Q = \frac{1}{|Q|} \int_Q w \, dx$ we obtain

$$\left(1 - 2^n [w]_{A_1}^r \frac{r-1}{r}\right) \frac{1}{|Q|} \int_Q w^r \, dx \leq \left(\frac{1}{|Q|} \int_Q w \, dx\right)^r.$$

We now examine for which $r \geq 1$ we have

$$1 - 2^n [w]_{A_1}^r \frac{r-1}{r} \geq \frac{1}{K^r},$$

or equivalently,

$$2^{\frac{n}{r}} \left(\frac{r-1}{r}\right)^{\frac{1}{r}} \leq \frac{(K^r - 1)^{\frac{1}{r}}}{K} \frac{1}{[w]_{A_1}}. \tag{8.6.4}$$

It may be difficult to exactly determine these r, but as $K > 1$, note that (8.6.4) is a consequence of

$$0 \leq 2^n e^{\frac{1}{e}} (r-1) \leq \frac{K-1}{K} \frac{1}{[w]_{A_1}}, \tag{8.6.5}$$

in view of the inequality $(1 - 1/r)^{1/r} \leq e^{1/e} (r-1)$ for $r \geq 1$. This last inequality can be obtained by setting $s = 1 - 1/r$ and observing that the minimum of s^s over $[0,1]$ is $e^{-1/e}$. Now as (8.6.5) is equivalent to (8.6.3), (8.6.1) is proven. □

Remark 8.6.4. The constant $e^{1/e}$ in (8.6.3) could be replaced by the smallest constant c that satisfies the inequality $(1 - 1/t)^{1/t} \leq c(t-1)$ for all $t > 1$.

A consequence of the reverse Hölder property of A_p weights is the following characterization of A_1 weights. This provides a converse to Theorem 8.5.3.

8.6 Reverse Hölder Property of A_p Weights

Theorem 8.6.5. *Let w be an A_1 weight. Then there exist $0 < \varepsilon < 1$, a nonnegative measurable function g such that $g, 1/g \in L^\infty$, and a nonnegative locally integrable function f that satisfies $M_c(f) < \infty$ a.e. such that*

$$w = g M_c(f)^\varepsilon \qquad \text{a.e.} \tag{8.6.6}$$

Proof. By Theorem 8.6.3, there is $\gamma > 0$ such that the reverse Hölder condition

$$\left(\frac{1}{|Q|} \int_Q w^{1+\gamma} dy \right)^{\frac{1}{1+\gamma}} \le \frac{2}{|Q|} \int_Q w\, dy \tag{8.6.7}$$

holds for all cubes Q. We set

$$\varepsilon = \frac{1}{1+\gamma} \qquad \text{and} \qquad f = w^{1+\gamma}$$

so that $f^\varepsilon = w$.

There is a null subset E_Q of each cube Q such that

$$\frac{2}{|Q|} \int_Q w\, dy \le 2[w]_{A_1} w(x) \qquad \text{for all } x \in Q \setminus E_Q.$$

Let N be the union of E_Q over all Q with rational side lenghts and centers in $\mathbf{Q}^n$. It follows from this and from (8.6.7) that M_c satisfies

$$M_c(f)(x) \le 2^{1+\gamma} [w]_{A_1}^{1+\gamma} f(x) \qquad \text{for } x \in \mathbf{R}^n \setminus N.$$

That is, $M_c(f) \le 2^{1+\gamma}[w]_{A_1}^{1+\gamma} f$ a.e. or $M_c(f)^\varepsilon \le 2[w]_{A_1} w < \infty$ a.e. But the Lebesgue differentiation theorem also gives $M_c(f)^\varepsilon \ge f^\varepsilon = w$ a.e. We now set

$$g = \frac{w}{M_c(f)^\varepsilon},$$

and we observe that $\frac{1}{2[w]_{A_1}} \le g \le 1$ a.e. Finally, we point out that (8.6.7) implies that $f = w^{1+\gamma}$ is locally integrable, so the functions f and g are as claimed. $\square$

Lemma 8.6.6. *Let $1 < p < \infty$, $w \in A_p$, and $0 < \delta < 1$. Then $[w^\delta]_{RH_{1/\delta}} \le [w]_{A_p}^\delta$, i.e.,*

$$\frac{1}{|Q|} \int_Q w\, dy \le [w]_{A_p} \left(\frac{1}{|Q|} \int_Q w^\delta dy \right)^{\frac{1}{\delta}} \tag{8.6.8}$$

for any cube Q in $\mathbf{R}^n$.

Proof. Let $\alpha > 0$ and $r \in (1, \infty)$ to be determined later. By Hölder's inequality we have

$$1 = \frac{1}{|Q|} \int_Q w^\alpha w^{-\alpha} dy \le \left(\frac{1}{|Q|} \int_Q w^{\alpha r} dy \right)^{\frac{1}{r}} \left(\frac{1}{|Q|} \int_Q w^{-\alpha r'} dy \right)^{\frac{1}{r'}}. \tag{8.6.9}$$

We now pick α and r satisfying $\alpha r = \delta$ and $\alpha r' = \frac{1}{p-1}$. Solving the system we find the values $r = \delta(p-1) + 1$ and $\alpha = \frac{\delta}{\delta(p-1)+1}$. Then (8.6.9) transforms to

$$1 \leq \left(\frac{1}{|Q|}\int_Q w^\delta \, dy\right)^{\frac{1}{\delta(p-1)+1}} \left(\frac{1}{|Q|}\int_Q w^{-\frac{1}{p-1}} \, dy\right)^{\frac{\delta(p-1)}{\delta(p-1)+1}}$$

or equivalently to

$$1 \leq \left(\frac{1}{|Q|}\int_Q w^\delta \, dy\right)^{\frac{1}{\delta}} \left(\frac{1}{|Q|}\int_Q w^{-\frac{1}{p-1}} \, dy\right)^{p-1}.$$

Multiplying by $\frac{1}{|Q|}w(Q)$ and using the definition of $[w]_{A_p}$ we obtain (8.6.8). $\square$

Remark 8.6.7. By Hölder's inequality, the reverse inequality to (8.6.8) trivially holds with constant 1. So condition (8.6.8) is in fact a reverse Hölder property. Then (8.6.8) essentially says that the normalized L^1 norm of an A_p weight over a cube is comparable to its normalized L^δ quasi-norm over the same cube.

Next we derive another consequence of Lemma 8.6.6.

Lemma 8.6.8. *Let $1 < p < \infty$. Then for any $\delta \in (0,1)$, any weight $w \in A_p$, any cube Q, and any $x \in Q$ we have*

$$M_c(w\chi_Q)(x) \leq [w]_{A_p}\left(M_c(w^\delta \chi_Q)(x)\right)^{1/\delta}. \tag{8.6.10}$$

Proof. Fix a cube Q and a point $x \in Q$. Then

$$M_c(w\chi_Q)(x) = \sup_{\substack{Q' \text{ cube} \\ Q' \ni x}} \frac{1}{|Q'|}\int_{Q'} w\chi_Q \, dy, \tag{8.6.11}$$

where the supremum is taken over all cubes Q' containing x. Now if Q' is not contained in Q, then the estimate

$$\frac{1}{|Q'|}\int_{Q'} w\chi_Q \, dy = \frac{|Q \cap Q'|}{|Q'|} \frac{1}{|Q \cap Q'|}\int_{Q \cap Q'} w\chi_Q \, dy \leq \frac{1}{|Q \cap Q'|}\int_{Q \cap Q'} w\chi_Q \, dy$$

shows that the average of $w\chi_Q$ over Q' is smaller than the average of $w\chi_Q$ over the cube $Q \cap Q'$ which is contained in Q and contains x. Thus the supremum in (8.6.11) can be restricted to cubes Q' contained in Q. Now it follows from (8.6.8) that

$$\frac{1}{|Q'|}\int_{Q'} w \, dy \leq [w]_{A_p}\left(\frac{1}{|Q'|}\int_{Q'} w^\delta \, dy\right)^{\frac{1}{\delta}} \tag{8.6.12}$$

for all cubes Q' contained in Q that contain the point x. For all such cubes Q' we can then replace w by $w\chi_Q$ in (8.6.12) and obtain

8.6 Reverse Hölder Property of A_p Weights

$$\frac{1}{|Q'|}\int_{Q'} w\chi_Q\, dy \leq [w]_{A_p}\left(\frac{1}{|Q'|}\int_{Q'} w^\delta \chi_Q\, dy\right)^{\frac{1}{\delta}} \leq [w]_{A_p}(M_c(w^\delta \chi_Q)(x))^{\frac{1}{\delta}}. \quad (8.6.13)$$

Taking the supremum on the left side of (8.6.13) over all cubes Q' that contain the fixed point x and are contained in Q, and using the observation that the supremum in (8.6.11) can be restricted to such cubes Q', we deduce the validity of (8.6.10). □

We now suitably adapt the argument in the proof of Theorem 8.6.3 to obtain the reverse Hölder property of A_p weights.

Theorem 8.6.9. *For any $w \in A_p$, $1 < p < \infty$, and $K > 1$, and any r satisfying*

$$1 \leq r \leq 1 + \frac{K-1}{K}\frac{1}{2^{3/2}e^{1/e}6^n[w]_{A_p}} \quad (8.6.14)$$

we have $[w]_{RH_r} \leq K$, i.e., for any cube Q in $\mathbf{R}^n$,

$$\left(\frac{1}{|Q|}\int_Q w^r\, dx\right)^{\frac{1}{r}} \leq \frac{K}{|Q|}\int_Q w\, dx. \quad (8.6.15)$$

Proof. Fix a cube Q and $0 < \delta < 1$. Applying first (8.6.2) to $f = w\chi_Q$ we obtain

$$\frac{1}{2^n \lambda} w(\{x \in Q : w(x) > \lambda\}) \leq |\{x \in Q : M_c(w\chi_Q)(x) > \lambda\}|.$$

Next we apply (8.6.10) to write

$$|\{x \in Q : M_c(w\chi_Q)(x) > \lambda\}| \leq |\{x \in Q : M_c(w^\delta \chi_Q)(x) > \lambda^\delta/[w]_{A_p}^\delta\}|$$

$$\leq 2 \cdot 3^n \frac{[w]_{A_p}^\delta}{\lambda^\delta} w^\delta(\{x \in Q : w^\delta(x) > \lambda^\delta/2[w]_{A_p}^\delta\}),$$

where the second inequality is a consequence of Exercise 8.1.4 (with $\varepsilon = 1/2$, $f = w^\delta \chi_Q$, and $u = 1$). Thus we have

$$w(\{x \in Q : w(x) > \lambda\}) \leq 2^{n+1} 3^n \frac{[w]_{A_p}^\delta}{\lambda^{\delta-1}} w^\delta(\{x \in Q : w^\delta(x) > \lambda^\delta/2[w]_{A_p}^\delta\}).$$

Let $r \geq 1$. We use (1.2.3) and the preceding estimate to write

$$\int_Q w^{r-1} w\, dx$$

$$= (r-1)\left[\int_0^{w_Q} + \int_{w_Q}^\infty\right] \lambda^{r-2} w(\{x \in Q : w(x) > \lambda\})\, d\lambda$$

$$\leq w_Q^{r-1} w(Q) + (r-1)\int_{w_Q}^\infty [w]_{A_p}^\delta \frac{\lambda^{r-2}}{\lambda^{\delta-1}} 2^{n+1} 3^n w^\delta\left(\{x \in Q : w^\delta(x) > \frac{\lambda^\delta}{2[w]_{A_p}^\delta}\}\right) d\lambda$$

$$\leq w_Q^{r-1} w(Q) + 2^{n+\frac{r}{\delta}} 3^n [w]_{A_p}^r \frac{1}{\delta}(r-1) \int_0^\infty t^{\frac{r}{\delta}-2} w^\delta(\{x \in Q : w^\delta(x) > t\}) \, dt$$

$$= |Q| \left(\frac{w(Q)}{|Q|}\right)^r + 2^{\frac{r}{\delta}} 6^n [w]_{A_p}^r \frac{1}{\delta} \frac{r-1}{\frac{r}{\delta}-1} \int_Q (w^\delta)^{\frac{r}{\delta}-1} w^\delta \, dx,$$

where we made use of the change of variables $t = \lambda^\delta / 2[w]_{A_p}^\delta$. Thus

$$\left(1 - 2^{\frac{r}{\delta}} 6^n [w]_{A_p}^r \frac{r-1}{r-\delta}\right) \frac{1}{|Q|} \int_Q w^r \, dx \leq \left(\frac{1}{|Q|} \int_Q w \, dx\right)^r.$$

We examine for which $r \geq 1$ we have

$$1 - 2^{\frac{r}{\delta}} 6^n [w]_{A_p}^r \frac{r-1}{r-\delta} \geq \frac{1}{K^r},$$

or equivalently,

$$6^{\frac{n}{r}} 2^{\frac{1}{\delta}} \left(\frac{r-1}{r-\delta}\right)^{\frac{1}{r}} \leq \frac{(K^r - 1)^{\frac{1}{r}}}{K} \frac{1}{[w]_{A_p}}. \tag{8.6.16}$$

Making use of the inequality below (8.6.5), we see that when $0 < \delta < 1 \leq r$ one has

$$\left(\frac{r-1}{r-\delta}\right)^{\frac{1}{r}} = \left(\frac{r-1}{r}\right)^{\frac{1}{r}} \left(\frac{r}{r-\delta}\right)^{\frac{1}{r}} \leq e^{\frac{1}{e}}(r-1) \frac{1}{1-\delta},$$

since $\left(\frac{r}{r-\delta}\right)^{1/r}$ is decreasing in r on $[1, \infty)$. Then (8.6.16) is a consequence of

$$0 \leq 6^n \frac{2^{\frac{1}{\delta}}}{1-\delta} e^{\frac{1}{e}} (r-1) \leq \frac{K-1}{K} \frac{1}{[w]_{A_p}}, \tag{8.6.17}$$

which is derived in (8.6.5). Obviously $\frac{2^{1/\delta}}{1-\delta}$ tends to infinity as $\delta \to 0^+$ or $\delta \to 1^-$, so it is advantageous to choose an intermediate value of δ, such as $\delta = 1/2$. For this choice of δ, (8.6.17) is equivalent to (8.6.14), and so the assertion of the theorem is proved. □

Taking $f = \chi_S$ in Proposition 8.3.2 (5) yields that

$$\frac{w(S)}{w(Q)} \geq \frac{1}{[w]_{A_p}} \left(\frac{|S|}{|Q|}\right)^p \tag{8.6.18}$$

for any cube Q and any measurable subset S of Q. The reverse Hölder inequality allows us to essentially reverse this inequality.

Corollary 8.6.10. *There is a constant c_n such that for any $1 < p < \infty$, any $w \in A_p$, any cube Q, and any measurable subset S of Q we have*

$$\frac{w(S)}{w(Q)} \leq 2 \left(\frac{|S|}{|Q|}\right)^{\frac{1}{1+c_n[w]_{A_p}}}. \tag{8.6.19}$$

8.6 Reverse Hölder Property of A_p Weights

Proof. An application of Hölder's inequality gives

$$w(S) = \int_Q w\chi_S\, dx \le \Big(\int_Q w^r dx\Big)^{\frac{1}{r}} |S|^{\frac{r-1}{r}} = \Big(\frac{1}{|Q|}\int_Q w^r dx\Big)^{\frac{1}{r}} |Q|^{\frac{1}{r}} |S|^{\frac{r-1}{r}} \qquad (8.6.20)$$

for any $r > 1$. We choose $r = 1 + \frac{1}{c_n [w]_{A_p}}$ where $c_n = 2^{5/2} e^{1/e} 6^n$. Then (8.6.15) in Theorem 8.6.9 (with $K = 2$) combined with (8.6.20) yields

$$w(S) \le 2\Big(\frac{1}{|Q|}\int_Q w\, dx\Big) |Q|^{\frac{1}{r}} |S|^{\frac{r-1}{r}},$$

which is a restatement of (8.6.19). $\square$

Corollary 8.6.11. *Let $1 < p < \infty$. Given $w \in A_p$ there is a $q \in (1, p)$ such that $w \in A_q$. Moreover, q satisfies $p - 1 = (q-1)\big(1 + (c_n[w]_{A_p}^{p'-1})^{-1}\big)$, where $c_n = 2^{5/2} e^{1/e} 6^n$, and*

$$[w]_{A_q} \le 2^{p-1} [w]_{A_p}. \qquad (8.6.21)$$

Proof. Let $w \in A_p$. Then $w^{1-p'} \in A_{p'}$. The reverse Hölder condition on the weight $w^{1-p'}$ yields

$$\Big(\frac{1}{|Q|}\int_Q w^{(1-p')r} dx\Big)^{\frac{1}{r}} \le \frac{2}{|Q|}\int_Q w^{1-p'} dx \qquad (8.6.22)$$

where $r = 1 + 1/c_n[w]_{A_p}^{p'-1}$. Let us set $(p'-1)r = q'-1$ for some $1 < q < p$. Then $p - 1 = r(q-1)$ and (8.6.22) translates to

$$\Big(\frac{1}{|Q|}\int_Q w^{1-q'} dx\Big)^{q-1} \le \Big(\frac{2}{|Q|}\int_Q w^{1-p'} dx\Big)^{p-1}. \qquad (8.6.23)$$

Multiplying by w_Q and taking the supremum over cubes Q, we derive (8.6.21). $\square$

Corollary 8.6.12. *For any $1 < p < \infty$ we have $A_p = \cup_{1 \le q < p} A_q$.*

Proof. As the classes A_p are increasing, it follows that A_p contains all the classes A_q with $q < p$. Conversely, if $w \in A_p$, then by Corollary 8.6.11 there is a $q \in (1, p)$ such that $w \in A_q$. This shows that A_p is contained in $\cup_{1 \le q < p} A_q$. $\square$

Exercises

8.6.1. Show that the RH_r classes decrease as r increases. Precisely, show that when $1 \le s \le r$ we have $[w]_{RH_s} \le [w]_{RH_r}$ for $w \in RH_r$.

8.6.2. Let $1 < p < \infty$. If $w \in A_p$ prove that $1/w \in RH_{p'-1}$ and

$$[1/w]_{RH_{p'-1}} \le [w]_{A_p}.$$

8.6.3. Let $w \in A_p$, where $p \geq 1$. Show that there is an $s > 1$ (depending on $[w]_{A_p}$) such that $w^s \in A_p$. [*Hint:* Use Theorems 8.5.2, 8.5.3, 8.6.5, and Proposition 8.5.1.]

8.6.4. ([14]) Let $1 \leq p < \infty$. Given $w \in A_p$ and a cube Q show that there is a measurable subset E_Q of Q (that also depends on w) such that $|E_Q| \geq \frac{1}{2}|Q|$ and

$$\frac{1}{|Q|}\int_Q w\,dx \leq 2^{p-1}[w]_{A_p} w(y) \qquad \text{for all } y \in E_Q.$$

[*Hint:* When $p > 1$ choose $E_Q = \{w \geq \beta w_Q\} \cap Q$, where $\beta^{-1} = 2^{p-1}[w]_{A_p}$. Show that $|Q \setminus E_Q| \leq \frac{1}{2}|Q|$ using the definition of $[w]_{A_p}$. When $p = 1$ choose $E_Q = Q$.]

8.6.5. ([102]) Let $1 < p, q, s < \infty$ be related by $q - 1 = s(p - 1)$.
(a) Let $w \in A_p \cap RH_s$. Show that $w^s \in A_q$ and

$$[w^s]_{A_q} \leq [w]_{A_p}^s [w]_{RH_s}^s.$$

(b) Conversely, given $w^s \in A_q$ prove that $w \in A_p \cap RH_s$; precisely show that

$$[w]_{A_p} \leq [w^s]_{A_q}^{\frac{1}{s}} \qquad \text{and} \qquad [w]_{RH_s} \leq 2^p [w^s]_{A_q}^{\frac{1}{s}}.$$

[*Hint:* Part (b). Apply Exercises 8.2.3 and 8.6.4 to w^s.]

8.7 Weighted Estimates for Singular Integral Operators

We begin by verifying the necessity of the A_p condition for basic singular integrals.

Theorem 8.7.1. *Let w be a weight in $\mathbf{R}^n$, $n \geq 2$, and let $1 \leq p < \infty$. Suppose that each of the Riesz transforms R_j is of weak type (p,p) with respect to w. Then w must be an A_p weight.*

Proof. Let Q be a cube and let f be a nonnegative function on $\mathbf{R}^n$ supported in Q that satisfies $f_Q > 0$. Let Q' be the cube that shares a corner with Q, has the same length as Q, and satisfies $x_j \geq y_j$ for all $1 \leq j \leq n$ whenever $x \in Q'$ and $y \in Q$. Then for $x \in Q'$ we have

$$\left|\sum_{j=1}^n R_j(f)(x)\right| = \frac{\Gamma(\frac{n+1}{2})}{\pi^{\frac{n+1}{2}}} \sum_{j=1}^n \int_Q \frac{x_j - y_j}{|x-y|^{n+1}} f(y)\,dy \geq \frac{\Gamma(\frac{n+1}{2})}{\pi^{\frac{n+1}{2}}} \int_Q \frac{f(y)}{|x-y|^n}\,dy.$$

But if $x \in Q'$ and $y \in Q$ we must have that $|x - y| \leq 2\sqrt{n}\,\ell(Q)$, which implies that $|x-y|^{-n} \geq (2\sqrt{n})^{-n}|Q|^{-1}$. Let $C_n = \Gamma(\frac{n+1}{2})(2\sqrt{n})^{-n}\pi^{-\frac{n+1}{2}}$. It follows that for all $0 < \alpha < C_n f_Q$ we have

$$Q' \subseteq \left\{x \in \mathbf{R}^n : \left|\sum_{j=1}^n R_j(f)(x)\right| > \alpha\right\}.$$

8.7 Weighted Estimates for Singular Integral Operators

Since the operator $\sum_{j=1}^n R_j$ is of weak type (p,p) with respect to w (with constant C), we must have

$$w(Q') \leq \frac{C^p}{\alpha^p}\int_Q f^p w\,dx$$

for all $\alpha < C_n f_Q$, which implies that

$$f_Q^p \leq \frac{C_n^{-p}C^p}{w(Q')}\int_Q f^p w\,dx. \tag{8.7.1}$$

We observe that we can reverse the roles of Q and Q' and obtain

$$g_{Q'}^p \leq \frac{C_n^{-p}C^p}{w(Q)}\int_{Q'} g^p w\,dx \tag{8.7.2}$$

for all g supported in Q'. In particular, taking $g = \chi_{Q'}$ in (8.7.2) gives that

$$w(Q) \leq C_n^{-p}C^p w(Q').$$

Using this estimate and (8.7.1), we obtain

$$f_Q^p \leq \frac{(C_n^{-p}C^p)^2}{w(Q)}\int_Q f^p w\,dx. \tag{8.7.3}$$

Using the characterization of $[w]_{A_p}$ in Proposition 8.3.2 (5), it follows that

$$[w]_{A_p} \leq (C_n^{-p}C^p)^2 < \infty;$$

hence $w \in A_p$. □

We now show that a singular integral operator is bounded from $L^p(w)$ to itself when $1 < p < \infty$ and $w \in A_p$. We need two lemmas to achieve this. Recall the operators $M_c^\#$ of Definition 6.4.1 and $M_d^\#$ of Definition 6.4.2.

Lemma 8.7.2. *Let $0 < A_1, A_2, A_3 < \infty$ and suppose that K is defined on $\mathbf{R}^n \setminus \{0\}$ and satisfies the size condition $|K(x)| \leq A_1|x|^{-n}$, $x \neq 0$, the smoothness condition*

$$|K(x-y) - K(x)| \leq A_2|y|^\delta |x|^{-n-\delta}, \qquad |x| \geq 2|y| \tag{8.7.4}$$

and the cancellation condition (3.3.5). Suppose that W is a tempered distribution on $\mathbf{R}^n$ defined in terms of (3.3.8) and let T be the operator given by convolution with W. Then for any $s > 1$ there is a constant $C_{n,s,\delta}$ such that

$$M_c^\#(T(f)) \leq C_{n,s,\delta}(A_1 + A_2 + A_3)M_c(|f|^s)^{\frac{1}{s}}$$

for any bounded function f on $\mathbf{R}^n$ with compact support.

Proof. Let us fix a bounded function f on $\mathbf{R}^n$ with compact support, a cube Q, and $x \in Q$. It will be enough to show that there is a constant C_Q such that

$$\frac{1}{|Q|} \int_Q |T(f) - C_Q| \, dy \leq C_{n,s,\delta} M_c(|f|^s)^{\frac{1}{s}}(x). \tag{8.7.5}$$

As we have done on various occasions, we estimate separately the contribution of f near the cube and far from it. So we write $f = f_0 + f_\infty$, where $f_0 = f\chi_{Q^*}$ and $Q^* = (4\sqrt{n}+1)Q$. We chose the constant $C_Q = T(f_\infty)(x)$. Then

$$\frac{1}{|Q|}\int_Q|T(f)(y)-C_Q|\,dy \leq \frac{1}{|Q|}\int_Q|T(f_0)(y)|\,dy + \frac{1}{|Q|}\int_Q|T(f_\infty)(y)-T(f_\infty)(x)|\,dy.$$

By the boundedness of T on L^s (Corollary 3.6.2) the first term on the right satisfies

$$\frac{1}{|Q|}\int_Q|T(f_0)(y)|\,dy \leq \left(\frac{1}{|Q|}\int_Q|T(f_0)(y)|^s\,dy\right)^{\frac{1}{s}}$$

$$\leq C'_{n,s}(A_1+A_2+A_3)\left(\frac{1}{|Q|}\int_{Q^*}|f(y)|^s\,dy\right)^{\frac{1}{s}}$$

$$\leq C''_{n,s}M_c(|f|^s)^{\frac{1}{s}}(x).$$

Now observe that if $x,y \in Q$ and $z \in (Q^*)^c$, then

$$|x-z| \geq \frac{4\sqrt{n}+1}{2}\ell(Q) - \frac{1}{2}\ell(Q) = 2\sqrt{n}\,\ell(Q) \geq 2|x-y|,$$

and hence (8.7.4) implies

$$|K(y-z) - K(x-z)| = |K((x-z)-(x-y)) - K(x-z)| \leq \frac{A_2|x-y|^\delta}{|x-z|^{n+\delta}}.$$

Moreover, if $w \in \{x,y\}$, then $T(f_\infty)(w) = \int_{\mathbf{R}^n} K(w-z)f_\infty(z)\,dz$ by (3.3.3) and the LDCT. Thus we may write

$$\frac{1}{|Q|}\int_Q|T(f_\infty)(y)-T(f_\infty)(x)|\,dy \leq \frac{1}{|Q|}\int_Q\int_{(Q^*)^c}|K(y-z)-K(x-z)||f(z)|\,dz\,dy$$

$$\leq \frac{1}{|Q|}\int_Q\int_{(Q^*)^c}\frac{A_2|x-y|^\delta|f(z)|}{|x-z|^{n+\delta}}\,dz\,dy$$

$$\leq c_{n,\delta}A_2 \int_{\mathbf{R}^n}\frac{\ell(Q)^\delta|f(z)|}{(|x-z|+\ell(Q))^{n+\delta}}\,dz$$

$$\leq c'_{n,\delta}A_2 M_c(f)(x)$$

$$\leq c'_{n,\delta}A_2 M_c(|f|^s)(x)^{\frac{1}{s}},$$

having made use of Corollary 2.5.2. Thus (8.7.5) holds with $C_{n,s,\delta} = C''_{n,s} + c'_{n,\delta}$. $\square$

Lemma 8.7.3. *Suppose that $1 < p_0 < p < \infty$ and $w \in A_p$. Then for every function f in $L^1_{\mathrm{loc}}(\mathbf{R}^n)$ with the property*

8.7 Weighted Estimates for Singular Integral Operators

$$\sup_{0<\lambda\leq N} \lambda^{p_0} w(\{M_d(f) > \lambda\}) < \infty \quad \text{for every } N > 0 \tag{8.7.6}$$

we have

$$\left(\int_{\mathbf{R}^n} M_d(f)^p w\, dx\right)^{\frac{1}{p}} \leq 2^{1+\frac{1}{p}+(n+p+2)(1+c_n[w]_{A_p})} \left(\int_{\mathbf{R}^n} M_c^{\#}(f)^p w\, dx\right)^{\frac{1}{p}}. \tag{8.7.7}$$

Proof. Let $\Omega_\lambda = \{M_d(f) > \lambda\}$ and let Q_j be the maximal dyadic cubes that appear in the proof of Theorem 6.4.5 which satisfy $\cup_j Q_j = \Omega_\lambda$. Estimate (6.4.4) says

$$|S_j| \leq 2^n \gamma |Q_j|$$

where $S_j = \{x \in Q_j : M_d(f)(x) > 2\lambda,\ M_d^{\#}(f)(x) \leq \gamma\lambda\}$ and $\gamma > 0$. Combining this fact with (8.6.19), we obtain

$$\frac{w(S_j)}{w(Q_j)} \leq 2\left(\frac{|S_j|}{|Q_j|}\right)^{\frac{1}{1+c_n[w]_{A_p}}} \leq 2(2^n\gamma)^{\frac{1}{1+c_n[w]_{A_p}}} \leq 2^{n+1}\gamma^{\frac{1}{1+c_n[w]_{A_p}}}.$$

Multiplying by $w(Q_j)$ and adding over j (using the notation of Theorem 6.4.5), we arrive at the estimate

$$w(\{M_d(f) > 2\lambda,\ M_d^{\#}(f) \leq \gamma\lambda\})$$
$$= \sum_j w(\{x \in Q_j : M_d(f)(x) > 2\lambda,\ M_d^{\#}(f)(x) \leq \gamma\lambda\})$$
$$\leq 2^{n+1}\gamma^{\frac{1}{1+c_n[w]_{A_p}}} \sum_j w(Q_j)$$
$$= 2^{n+1}\gamma^{\frac{1}{1+c_n[w]_{A_p}}} w(\{M_d(f) > \lambda\}). \tag{8.7.8}$$

We now adapt the proof of (6.4.10) to account for the presence of the weight w. For a positive real number N we set

$$I_N = \int_0^N p\lambda^{p-1} w(\{x \in \mathbf{R}^n : M_d(f)(x) > \lambda\})\, d\lambda.$$

We note that I_N is finite, as it is bounded by

$$\left(\int_0^N p\lambda^{p-p_0-1}\, d\lambda\right) \sup_{0<\lambda\leq N} \lambda^{p_0} w(\{M_d(f) > \lambda\}) < \infty.$$

We now write

$$I_N = 2^p \int_0^{\frac{N}{2}} p\lambda^{p-1} w(\{x \in \mathbf{R}^n : M_d(f)(x) > 2\lambda\})\, d\lambda$$

and we use (8.7.8) to obtain the following sequence of inequalities:

$$I_N \leq 2^p \int_0^{\frac{N}{2}} p\lambda^{p-1} w\big(\{x \in \mathbf{R}^n : M_d(f)(x) > 2\lambda,\, M_d^\#(f)(x) \leq \gamma\lambda\}\big)\, d\lambda$$

$$+ 2^p \int_0^{\frac{N}{2}} p\lambda^{p-1} w\big(\{x \in \mathbf{R}^n : M_d^\#(f)(x) > \gamma\lambda\}\big)\, d\lambda$$

$$\leq 2^p 2^{n+1} \gamma^{\frac{1}{1+c_n[w]_{A_p}}} \int_0^{\frac{N}{2}} p\lambda^{p-1} w\big(\{x \in \mathbf{R}^n : M_d(f)(x) > \lambda\}\big)\, d\lambda$$

$$+ 2^p \int_0^{\frac{N}{2}} p\lambda^{p-1} w\big(\{x \in \mathbf{R}^n : M_d^\#(f)(x) > \gamma\lambda\}\big)\, d\lambda$$

$$\leq 2^p 2^{n+1} \gamma^{\frac{1}{1+c_n[w]_{A_p}}} I_N + \frac{2^p}{\gamma^p} \int_0^{\frac{N\gamma}{2}} p\lambda^{p-1} w\big(\{x \in \mathbf{R}^n : M_d^\#(f)(x) > \lambda\}\big)\, d\lambda.$$

We pick γ so that $2^p 2^{n+1} \gamma^{\frac{1}{1+c_n[w]_{A_p}}} = \frac{1}{2}$. Since I_N is finite, we subtract from both sides of the inequality the number $\frac{1}{2} I_N$ to obtain

$$\frac{I_N}{2} \leq 2^p 2^{(p+n+2)p(1+c_n[w]_{A_p})} \int_0^{\frac{N\gamma}{2}} p\lambda^{p-1} w\big(\{x \in \mathbf{R}^n : M_d^\#(f)(x) > \lambda\}\big)\, d\lambda.$$

From this we deduce (8.7.7) by letting $N \to \infty$ and using that $M_d^\# \leq M_c^\#$. $\square$

Theorem 8.7.4. *Let T be a singular integral operator as in Lemma 8.7.2 and let $w \in A_p$ for some $1 < p < \infty$. Then there is a constant $C(n,p,w)$ such that for any bounded and compactly supported function f we have*

$$\big\|T(f)\big\|_{L^p(w)} \leq C(n,p,w)(A_1 + A_2 + A_3)\big\|f\big\|_{L^p(w)}. \tag{8.7.9}$$

Thus T admits a unique bounded extension on $L^p(w)$ (by Proposition 8.2.2).

Proof. Fix $1 < p < \infty$ and $w \in A_p$. Let f be bounded and compactly supported. By Corollary 8.6.11 we pick q such that $1 < q < p$ and $w \in A_q$. We claim that $T(f)$ lies in $L^q(w)$. Assuming this claim, by the boundedness of M_d on $L^q(w)$, it follows that $M_d(T(f)) \in L^q(w)$. Then (8.7.6) holds with $p_0 = q$; then Lemma 8.7.3 gives

$$\left(\int_{\mathbf{R}^n} |T(f)|^p w\, dx\right)^{\frac{1}{p}} \leq \left(\int_{\mathbf{R}^n} M_d(T(f))^p w\, dx\right)^{\frac{1}{p}} \leq C \left(\int_{\mathbf{R}^n} M_c^\#(T(f))^p w\, dx\right)^{\frac{1}{p}}.$$

An application of Lemma 8.7.2 now yields that

$$\left(\int_{\mathbf{R}^n} M_c^\#(T(f))^p w\, dx\right)^{\frac{1}{p}} \leq C_{n,s,\delta}(A_1+A_2+A_3)\left(\int_{\mathbf{R}^n} M_c(|f|^s)^{\frac{p}{s}} w\, dx\right)^{\frac{1}{p}}$$

for any $s > 1$. Picking $s \in (1, p)$ and applying Theorem 8.4.3 (with p/s in place of p), we obtain

8.7 Weighted Estimates for Singular Integral Operators

$$\left[\left(\int_{\mathbf{R}^n} M_c(|f|^s)^{\frac{p}{s}} w\,dx\right)^{\frac{s}{p}}\right]^{\frac{1}{s}} \leq \left[C'(n,p,w)\big\|\,|f|^s\big\|_{L^{\frac{p}{s}}(w)}\right]^{\frac{1}{s}} = C''(n,p,w)\|f\|_{L^p(w)}.$$

This concludes the proof of (8.7.9) and it remains to establish the claim that $T(f)$ lies in $L^q(w)$. To achieve this, let f be bounded and supported in the ball $B(0,R)$. Then for some r with $1 < r < q$, Hölder's inequality gives

$$\int_{B(0,2R)} |T(f)|^q w\,dx \leq \left(\int_{\mathbf{R}^n} |T(f)|^{qr'} dx\right)^{\frac{1}{r'}} \left(\int_{B(0,2R)} w^r dx\right)^{\frac{1}{r}},$$

and this expression is finite by the boundedness of T on $L^{qr'}$ (Corollary 3.6.2), provided r is chosen small enough so that the reverse Hölder property (8.6.15) is valid. We now show the finiteness of the $L^q(w)$ norm outside $B(0,2R)$. First we note that for $|x| \geq 2R$ one has $|x - y| \geq \frac{1}{2}|x|$ when $|y| \leq R$ and so

$$|T(f)(x)| \leq \int_{B(0,R)} \frac{A_1 |f(y)|}{|x-y|^n}\,dy \leq A_1 |B(0,R)| \frac{2^n \|f\|_{L^\infty}}{|x|^n} = \frac{C_f}{|x|^n}.$$

Then

$$\int_{B(0,2R)^c} |T(f)|^q w\,dx \leq \sum_{k=0}^\infty \int_{2R2^k \leq |x| \leq 2R2^{k+1}} \frac{C_f^q w(x)}{|x|^{nq}}\,dx \leq C_f^q \sum_{k=0}^\infty \frac{w(B(0,R2^{k+2}))}{(2^{k+1}R)^{nq}}.$$

Notice that the cube $Q(0,R2^{k+3})$ contains $B(0,R2^{k+2})$. Appealing to (8.6.18), we write

$$w(B(0,R2^{k+2})) \leq w(B(0,R))[w]_{A_\rho} \left(\frac{|Q(0,R2^{k+3})|}{|B(0,R)|}\right)^\rho = \text{constant } 2^{kn\rho},$$

where $\rho \in (1,q)$ is suitably picked so that $w \in A_\rho$. In view of this estimate the series

$$\sum_{k=0}^\infty (2^{k+1}R)^{-nq} w(B(0,R2^{k+2}))$$

converges and so $\|T(f)\|_{L^q(w)} < \infty$. □

Exercises

8.7.1. The *transpose* T^t of a linear operator T is defined by

$$\langle T(f), g \rangle = \langle f, T^t(g) \rangle$$

for all f, g in a range subspace of the domain and range of T, respectively. Suppose that T is a linear operator that maps $L^p(\mathbf{R}^n, v\,dx)$ to itself for some $1 < p < \infty$ and some $v \in A_p$. Show that the transpose operator T^t maps $L^{p'}(\mathbf{R}^n, w\,dx)$ to itself with the same norm, where $w = v^{1-p'} \in A_{p'}$.

8.7.2. Let K, T, and $T^{(*)}$ be as in Theorem 3.8.1 and let $1 < p < \infty$. Prove that $T^{(*)}$ is bounded from $L^p(w)$ to itself for any $w \in A_p$.

8.7.3. Let $\alpha > 0$, $w \in A_1$, and $f \in L^1(\mathbf{R}^n, w) \cap L^1(\mathbf{R}^n)$. Let $f = g + b$ be the Calderón–Zygmund decomposition of f at height $\alpha > 0$ given in Theorem 3.5.2, such that $b = \sum_j b_j$, where each b_j is supported in a dyadic cube Q_j, $\int_{Q_j} b_j(x)\,dx = 0$, and Q_j and Q_k have disjoint interiors when $j \neq k$. Prove that

(a) $\|g\|_{L^1(w)} \leq [w]_{A_1} \|f\|_{L^1(w)}$ and $\|g\|_{L^\infty(w)} = \|g\|_{L^\infty} \leq 2^n \alpha$,

(b) $\|b_j\|_{L^1(w)} \leq (1 + [w]_{A_1}) \|f\|_{L^1(Q_j, w)}$ and $\|b\|_{L^1(w)} \leq (1 + [w]_{A_1}) \|f\|_{L^1(w)}$,

(c) $\sum_j w(Q_j) \leq \frac{[w]_{A_1}}{\alpha} \|f\|_{L^1(w)}$.

8.7.4. Let δ, K, and T be as Lemma 8.7.2. Prove that there is a constant $C_{n,\delta}$ such that for any weight w, for any cube Q, and any function $F \in L^1(w)$ supported in Q and with $\int_Q F(x)\,dx = 0$ we have

$$\int_{(Q^*)^c} |T(F)(x)|\,w(x)\,dx \leq C_{n,\delta} A_2 \int_{\mathbf{R}^n} |F(y)|\,M(w)(y)\,dy.$$

Here $Q^* = (4\sqrt{n} + 1)Q$ is a fixed concentric multiple of Q.

8.7.5. Let δ, K, and T be as in Lemma 8.7.2. Fix $w \in A_1$. Suppose that T maps $L^2(w)$ to $L^2(w)$ with bound $B_w > 0$. Prove that there is a constant $C_{n,\delta}$ such that

$$\|T\|_{L^1(w) \to L^{1,\infty}(w)} \leq C_{n,\delta}(A + B_w)[w]_{A_1}^2.$$

[*Hint:* Apply the idea of the proof of Theorem 3.6.1 using the Calderón–Zygmund decomposition $f = g + b$ of Exercise 8.7.3 at height $\gamma\alpha$ for a suitable γ. The bad function can be handled via Exercise 8.7.4.]

Historical Notes

The weak L^p spaces were introduced in [63], [64] as natural endpoints of the scale of Lorentz spaces. An early treatment of Lorentz spaces appeared in the article by Hunt [41]. The normability of the weak spaces L^p for $1 < p \leq \infty$ can be traced back to general principles obtained by Kolmogorov [52]. Theorem 1.3.3 first appeared without proof in Marcinkiewicz's note [67] and was reintroduced by Zygmund in [124].

Theorem 2.4.1 (Riesz–Thorin interpolation theorem) can be traced back to Riesz [84] in the context of bilinear forms. Riesz's student Thorin [104], [105] developed an approach in the study of this result based on the maximum modulus principle. Tamarkin and Zygmund [103] provided a more efficient approach to Thorin's method. The one-dimensional maximal function originated in the work of Hardy and Littlewood [34]. Its n-dimensional analog was introduced and shown to be bounded by Wiener [119].

The Fourier transform can be traced back to Fourier [27]. The theory of distributions was developed by Schwartz [90], [91]. For a concise introduction to this theory one may consult Hörmander [40]. Lemma 2.9.1 is due to van der Corput [114].

The L^p boundedness of the Hilbert transform for $1 < p < \infty$ is due to M. Riesz but was obtained for the related conjugate function [83], [85] and was based on interpolation [84]. The weak-type $(1,1)$ property of the Hilbert transform is due to Kolmogorov [51]. The inequality in Exercise 3.8.3 is due to Cotlar [17]. Operators of the kind T_Ω and the stopping-time argument of Theorem 3.5.2 are due to Calderón and Zygmund [9]. In the same article, Calderón and Zygmund used this decomposition to prove Theorem 3.6.1 for T_Ω when Ω is a Lipschitz function on the unit sphere. The more general condition (3.3.4) first appeared in Hörmander [39]. The method of rotations (Corollary 3.6.4) appeared in the article of Calderón and Zygmund [10]. Example 3.3.2 is taken from Muckenhoupt [74].

The development of the theory of singular integrals in the vector-valued setting originated in the article of Benedek, Calderón, and Panzone [3]. This reference contains a general theorem which covers both Theorems 4.1.1 and 4.3.2. Theorem 4.3.3 is due to Fefferman and Stein [25]. Early versions of Theorem 4.4.2 can be found in [60], [61], [62]. These works depend on complex-analysis techniques and contain one-dimensional results. The real-variable treatment of the Littlewood–Paley theorem which allowed its higher-dimensional extension was pioneered by Stein [95].

The one-dimensional version of the Riesz potentials appeared in work of Weyl [117], but they were later systematically studied by Riesz [86] on $\mathbf{R}^n$. The Bessel potentials were introduced by Aronszajn and Smith [2] and by Calderón [7]. The strong type estimates in Theorem 5.1.3 were obtained by Hardy and Littlewood [33] in one dimension and by Sobolev [92] in higher dimensions, while the weak-type estimate first appeared in Zygmund [124]. The proof of Theorem 5.1.3 is taken from Hedberg [35]. Theorem 5.4.9 is due to Sobolev [92] when s is a positive integer. Theorem 5.3.2 is due to Mikhlin [73] and Theorem 5.3.6 to Hörmander [39], although both references contain slightly different formulations. A version of Theorem 5.7.3 in the context of one-dimensional Fourier series can be found in Marcinkiewicz's article [68]. Calderón and Torchinsky [8] obtained Theorem 5.6.5; the underlying Theorem 5.6.4 is also due to them. Interpolation of analytic families of operators (Theorem 5.5.3) is due to Stein [94]; the critical Proposition C.0.2 was previously established by Hirschman [38]. Estimates of the type that appear in Exercises 5.5.2, 5.5.3 can be traced back to the article of Kato and Ponce [47].

The pioneering article of Fefferman and Stein [26] provided the foundation of the theory of Hardy spaces. The decomposition of open sets given in Theorem 7.5.2 is due to Whitney [118]. The one-dimensional atomic decomposition of Hardy spaces is due to Coifman [13] and its higher-dimensional extension to Latter [56]. A simplification of some of the technical details in Latter's proof was subsequently obtained by Latter and Uchiyama [57].

The space of functions of bounded mean oscillation and Theorem 6.2.1 first appeared in the work of John and Nirenberg [43]. The duality of H^1 and BMO (Theorem 7.8.4) was announced by Fefferman in [24], but its first proof appeared in the article of Fefferman and Stein [26]. The proof of Theorem 7.8.4 is based on the atomic decomposition of H^1, which was obtained subsequently. Dyadic BMO is studied in Garnett and Jones [30]. The sharp maximal function was introduced by Fefferman and Stein [26] in interpolation when one endpoint space is BMO. Theorem 6.5.4 was independently obtained by Peetre [78], Spanne [93], and Stein [96].

The A_p condition first appeared in a paper of Rosenblum [87] in a somewhat different form. The characterization of A_p when $n = 1$ in terms of the boundedness of the Hardy–Littlewood maximal operator was obtained by Muckenhoupt [75]. The estimate in (8.4.8) can also be reversed, as shown by Buckley [6]. The proof of Theorem 8.2.7 is based on that in Lerner [58]. The particular version of Lemma G.0.1 is adapted from that in de Guzmán [21]. The fact that A_p weights satisfy the reverse Hölder condition is due to Coifman and Fefferman [14]. The characterization of A_1 weights (Theorem 8.5.3) is due to Coifman and Rochberg [15]. The necessity and sufficiency of the A_p condition for the boundedness of the Hilbert transform on weighted L^p spaces was obtained by Hunt, Muckenhoupt, and Wheeden [42]. Weighted L^p estimates controlling Calderón–Zygmund operators by the Hardy–Littlewood maximal operator were obtained by Coifman [12]. The factorization of A_p weights was conjectured by Muckenhoupt and proved by Jones [45]. The proof in the text is one of several ones given afterwards, based on the so-called Rubio de Francia algorithm, i.e., the series of iterates of an operator that create a sub-eigenvector of it. Parts of the exposition in Chapter 8 was based on the notes of Duoandikoetxea [23].

General reference texts on Fourier Analysis include: Duoandikoetxea [22], García-Cuerva and Rubio de Francia [29], Grafakos [31], [32], Katznelson [48], Körner [53], Meyer [70], [71], Meyer and Coifman [72], Muscalu and Schlag [76], [77], Pereyra and Ward [81], Pinsky [82], Stein [97], [98], Stein and Shakarchi [99], Stein and Weiss [100], Torchinsky [107], Wolff [122], Zygmund [125]. More specialized books include: de Guzmán [20], [21] on covering lemmas; Bennett and Sharpley [4], Bergh and Löfström [5], Kislyakov and Kruglyak [49], Krein, Petunin, and Semenov [55] on interpolation; Cruz-Uribe, Martell, and Pérez [18], Kokilashvili [50], Strömberg and Torchinsky [101], Wilson [121] on weighted estimates; Frazier, Jawerth, and Weiss [28], Peetre [79], Sawano [88], Schmeisser and Triebel [89], Triebel [109], [110], [111], [112], Yuan, Sickel, and Yang [120] on different types of function spaces; Adams and Fournier [1], Lieb and Loss [59], Maz'ya [69], Ziemer [123] for topics on Sobolev spaces; Lu [65], Uchiyama [113], Weisz [116] on Hardy spaces; Christ [11], Journé [46], Lu, Ding, and Yan [66], Tolsa [106], Torres [108], Volberg [115] on singular integrals. The proof of Besicovitch's lemma in Appendix G follows the exposition in Jones [44].

Finally, several aspects of dyadic harmonic analysis can be found in Pereyra [80]. Many topics in this book can be studied in terms of wavelets; on this the reader may consult Hérnandez and Weiss [36].

Appendix A
Orthogonal Matrices

An $n \times n$ matrix A is called orthogonal if $AA^t = I$. This implies that $A^{-1} = A^t$. It follows that $\det A = 1$ or $\det A = -1$. Let

$$A = \begin{bmatrix} a_{11} & a_{12} & \cdots & a_{1n} \\ a_{21} & a_{22} & \cdots & a_{2n} \\ \vdots & \vdots & \cdots & \vdots \\ a_{n1} & a_{n2} & \cdots & a_{nn} \end{bmatrix}; \quad \text{then} \quad A^t = \begin{bmatrix} a_{11} & a_{21} & \cdots & a_{n1} \\ a_{12} & a_{22} & \cdots & a_{n2} \\ \vdots & \vdots & \cdots & \vdots \\ a_{1n} & a_{2n} & \cdots & a_{nn} \end{bmatrix}.$$

In view of the property $AA^t = I$, it follows that

$$a_{j1}^2 + a_{j2}^2 + \cdots + a_{jn}^2 = 1, \quad j \in \{1,2,\ldots,n\},$$

$$a_{j1}a_{k1} + a_{j2}a_{k2} + \cdots + a_{jn}a_{kn} = 0, \quad k \neq j.$$

In other words, the set of rows of A is an orthonormal basis of $\mathbf{R}^n$, and so is the set of columns of A.

Then for $x = (x_1, \ldots, x_n) \in \mathbf{R}^n$ we have $|Ax| = |x|$. Indeed,

$$|Ax|^2 = \sum_{j=1}^{n} (a_{j1}x_1 + \cdots + a_{jn}x_n)^2$$

$$= \sum_{j=1}^{n} \sum_{k=1}^{n} \sum_{l=1}^{n} a_{jk}x_k a_{jl}x_l$$

$$= \sum_{j=1}^{n} \sum_{k=1}^{n} a_{jk}^2 x_k^2 + \sum_{j=1}^{n} \sum_{1 \leq k \neq l \leq n} a_{jk}a_{jl}x_k x_l$$

$$= \sum_{k=1}^{n} x_k^2 \sum_{j=1}^{n} a_{jk}^2 + \sum_{1 \leq k \neq l \leq n} x_k x_l \sum_{j=1}^{n} a_{jk}a_{jl}$$

$$= \sum_{k=1}^{n} x_k^2 \cdot 1 + \sum_{1 \leq k \neq l \leq n} x_k x_l \cdot 0 = |x|^2.$$

© The Editor(s) (if applicable) and The Author(s), under exclusive license to Springer Nature Switzerland AG 2024
L. Grafakos, *Fundamentals of Fourier Analysis*, Graduate Texts in Mathematics 302, https://doi.org/10.1007/978-3-031-56500-7

Appendix B
Subharmonic Functions

A locally integrable function f on an open subset O of $\mathbf{R}^n$ with values in $[-\infty,\infty)$ is called subharmonic if it is upper semicontinuous, which means
$$\limsup_{y\to x} f(y) \le f(x)$$
for every $x \in O$ and
$$f(x) \le \frac{1}{|B(x,r)|} \int_{B(x,r)} f(y)\,dy \tag{B.0.1}$$
for any $x \in O$ and every $r > 0$ such that $B(x,r) \subset O$. If $f \in \mathscr{C}^2$, then the above condition is equivalent to $\Delta f \ge 0$. A function is called log-subharmonic if it is nonnegative and its logarithm is subharmonic.

An interesting property of subharmonic functions is the following maximum modulus principle. For simplicity we state this result only in the case $n = 2$ using complex number notation.

Lemma B.0.1. *Let O be an open connected subset of the complex plane with compact closure $\overline{O}$. Let V be a subharmonic function on O and U be a harmonic function on O. Assume that for every $z_0 \in \overline{O}\setminus O$ we have*
$$\limsup_{z\to z_0} \big(V(z) - U(z)\big) \le 0. \tag{B.0.2}$$
Then for all $z \in O$ we have
$$V(z) - U(z) \le 0.$$

Proof. Let $M = \sup_{z\in O}(V(z)-U(z))$. Suppose that $M > 0$ to obtain a contradiction. Let z_k be a sequence in O such that $(V-U)(z_k) \to M$. In view of (B.0.2), z_k cannot accumulate near the boundary of O, thus $\{z_k\}_k$ has a limit point $z \in O$. By the semicontinuity we must have $(V-U)(z) = M$. Then the set
$$E = \{w \in O : (V-U)(w) = M\}$$

is nonempty. But for an upper semicontinuous function g, sets of the form $\{g \geq \alpha\}$ are closed; thus E must be closed by taking $\alpha = M$.

To show that E is open, let $w \in E$. The mean value property (B.0.1) yields that $V - U = M$ a.e. on an open disk $B(w,r)$ contained in O. Hence E is dense in $B(w,r)$, but E is closed, implying that $B(w,r) \subseteq E$. This shows that E is also open. As O is connected, we must have $E = O$. So, unless $U = V$, in which case the claim is obvious, we have a contradiction. Thus M cannot be positive, which means that we should have $M \leq 0$. □

Lemma B.0.2. *The sum of two log-subharmonic functions is log-subharmonic.*

Proof. Let $\varphi(x,y) = \log(e^x + e^y)$ defined on $\mathbf{R}^2$. Then φ is obviously increasing in each variable. Also φ is a convex function of both variables, i.e., it satisfies

$$\varphi((1-\theta)(x_1,y_1) + \theta(x_2,y_2)) \leq (1-\theta)\varphi(x_1,y_1) + \theta\varphi(x_2,y_2)$$

for all (x_1,y_1), (x_2,y_2) in $\mathbf{R}^2$ and all $\theta \in [0,1]$. Indeed, writing

$$\varphi(x,y) = x + \log(1 + e^{y-x})$$

and using the convexity of the function $t \mapsto \log(1+e^t)$ on the real line, we can easily obtain this assertion.

Suppose that F, G are subharmonic functions on $\mathbf{R}^n$. Then the fact that φ is increasing in each variable and Jensen's inequality (which can be used since φ is convex) gives

$$\varphi(F(x), G(x)) \leq \varphi\left(\frac{1}{|B(x,r)|} \int_{B(x,r)} F(y)\,dy\,,\,\frac{1}{|B(x,r)|} \int_{B(x,r)} G(y)\,dy\right)$$
$$\leq \frac{1}{|B(x,r)|} \int_{B(x,r)} \varphi(F(y), G(y))\,dy,$$

which implies that $\varphi(F(x), G(x))$ is subharmonic.

Now let f, g be log-subharmonic functions. Writing $f = e^F$ and $g = e^G$, then $\log(f+g) = \varphi(F,G)$. But $\varphi(F,G)$ was shown to be subharmonic, thus $\log(f+g)$ is also subharmonic. □

We need a few facts from the theory of analytic functions with values in Banach spaces. Let $\mathscr{B}$ be a Banach space with norm $\|\cdot\|_{\mathscr{B}}$ and let $\mathbf{f}$ be a mapping from an open subset U of $\mathbf{C}$ to $\mathscr{B}$. We say that $\mathbf{f}$ is analytic from U to $\mathscr{B}$ if for every z_0 in U there is an element $\mathbf{f}'(z_0)$ in $\mathscr{B}$ such that

$$\lim_{z \to z_0} \left\| \mathbf{f}'(z_0) - \frac{\mathbf{f}(z) - \mathbf{f}(z_0)}{z - z_0} \right\|_{\mathscr{B}} = 0.$$

Theorem B.0.3. *Let U be an open subset of $\mathbf{C}$ and let $\mathbf{f}$ be a mapping from U to a Banach space $\mathscr{B}$. Then $\mathbf{f}$ is analytic if and only if for every bounded linear functional Λ on $\mathscr{B}$ we have*

B Subharmonic Functions

$$\lim_{z \to z_0} \Lambda \left(\frac{\mathbf{f}(z) - \mathbf{f}(z_0)}{z - z_0} \right)$$

exists in **C**.

Theorem B.0.4. *Let U be an open subset of **C**, let $B(z_0, r)$ be a disk contained in U, and let $\mathbf{f}$ be an analytic mapping from U to $\mathcal{B}$. Then $\mathbf{f}$ has a unique power series expansion*

$$\mathbf{f}(z) = \sum_{n=0}^{\infty} \mathbf{a}_n (z - z_0)^n,$$

where $\mathbf{a}_n \in \mathcal{B}$ and the series converges in the norm of $\mathcal{B}$ for any $|z - z_0| < r$ and uniformly in the norm of $\mathcal{B}$ on any subdisk $|z - z_0| \leq r'$, where $r' < r$.

We denote by **S** the open unit strip, i.e., the set of all points z in the plane with $0 < \operatorname{Re} z < 1$.

Lemma B.0.5. *Let (X, μ) be a measure space with $\mu(X) < \infty$ and let V be a complex-valued function defined on $X \times \mathbf{S}$ such that the mapping $z \mapsto V(\cdot, z)$ from **S** to the Banach space $L^1(X)$ is analytic. Then the function*

$$z \mapsto F(z) = \int_X |V(x, z)|^q \, d\mu(x)$$

is log-subharmonic for any $0 < q \leq 1$.

Proof. Given z_0 in **S**, there is an $r > 0$ such that the closed disk $\overline{B(z_0, r)}$ of radius r centered at z_0 is contained in **S** and there exist functions $a_{k,z_0}(x)$ such that

$$V(x, z) = \sum_{k=0}^{\infty} a_{k,z_0}(x)(z - z_0)^k,$$

where the series converges in $L^1(X)$ uniformly in $z \in \overline{B(z_0, r)}$. We claim that

$$F_N(z) = \int_X \left| \sum_{k=0}^{N} a_{k,z_0}(x)(z - z_0)^k \right|^q d\mu$$

converges uniformly [in $z \in \overline{B(z_0, r)}$] to $F(z)$ as $N \to \infty$. Indeed, using the inequality $||a|^q - |b|^q| \leq |a - b|^q$ we obtain

$$|F_N(z) - F(z)| \leq \int_X \left| \sum_{k=N+1}^{\infty} a_{k,z_0}(x)(z - z_0)^k \right|^q d\mu$$

$$\leq \mu(X)^{1-q} \left(\int_X \left| \sum_{k=N+1}^{\infty} a_{k,z_0}(x)(z - z_0)^k \right| d\mu \right)^q,$$

and this tends to zero (as $N \to \infty$) uniformly in $z \in \overline{B(z_0, r)}$.

To show that $\log|F(z)|$ is subharmonic in $B(z_0,r)$, it will suffice to prove that $\log|F_N(z)|$ are subharmonic in $B(z_0,r)$, as $\log|F_N(z)|$ converge to $\log|F(z)|$ uniformly on $\overline{B(z_0,r)}$.

For each k there is a sequence of simple functions $\{a_{k,z_0}^j\}_{j=1}^\infty$ such that $\|a_{k,z_0} - a_{k,z_0}^j\|_{L^1(X)} \to 0$ as $j \to \infty$. Let

$$G_N^j(z) = \int_X \left| \sum_{k=0}^N a_{k,z_0}^j(x)(z-z_0)^k \right|^q d\mu.$$

Then, as before we have

$$|F_N(z) - G_N^j(z)| \le \int_X \left| \sum_{k=0}^N (a_{k,z_0}(x) - a_{k,z_0}^j(x))(z-z_0)^k \right|^q d\mu$$

$$\le \sum_{k=0}^N |z-z_0|^{kq} \int_X |a_{k,z_0}(x) - a_{k,z_0}^j(x)|^q d\mu$$

and this tends to zero as $j \to \infty$ uniformly in $z \in \overline{B(z_0,r)}$. We may now replace each $F_N(z)$ by a suitable $G_N^{j_N}$. It will now suffice to show that $\log G_N^j$ is subharmonic for any N and j. To achieve this we write

$$\sum_{k=0}^N a_{k,z_0}^j(x)(z-z_0)^k = \sum_i^m p_i(z)\chi_{E_i},$$

where $p_i(z)$ are polynomials and E_i are pairwise disjoint measurable subsets of X. Then

$$\int_X \left| \sum_{k=0}^N a_{k,z_0}^j(x)(z-z_0)^k \right|^q d\mu(x) = \sum_i^m |p_i(z)|^q \mu(E_i)$$

and as each function $|p_i(z)|^q$ is log-subharmonic, the same is true for the finite sum of these functions by Lemma B.0.2. □

Appendix C
Poisson Kernel on the Unit Strip

We denote by **S** the unit strip $\mathbf{S} = \{x + iy : 0 < x < 1, y \in \mathbf{R}\}$ and by **D** the unit disk $\{x + iy \in \mathbf{C} : x^2 + y^2 < 1\}$. Consider the conformal mapping

$$\varphi(w) = \frac{i - e^{\pi i w}}{1 - i e^{\pi i w}}$$

from **S** to **D** which has a continuous bijective extension from $\overline{\mathbf{S}}$ to $\overline{\mathbf{D}} \setminus \{i, -i\}$. Notice that $\varphi(0) = -1$, $\varphi(1) = 1$, $\varphi(\frac{1}{2}) = 0$ and the image of the line $\{\frac{1}{2} + it : -\infty < t < \infty\}$ is the open segment $\{is : s \in (-1, 1)\}$, preserving orientation. Consider the Poisson kernel $\operatorname{Re} \frac{1+z}{1-z}$, defined for $z \in \overline{\mathbf{D}} \setminus \{1\}$, composed with φ, i.e., the function

$$(x, y) \mapsto \operatorname{Re} \frac{1 + \varphi(x + iy)}{1 - \varphi(x + iy)} = \frac{\sin(\pi x)}{\cosh(\pi y) + \cos(\pi x)}$$

defined on $\overline{\mathbf{S}} \setminus \{1\}$. Being the harmonic image of a conformal mapping, this function is harmonic on **S**; i.e., for all $(x, y) \in \mathbf{S}$ we have

$$\frac{\partial^2}{\partial x^2}\left(\frac{\sin(\pi x)}{\cosh(\pi y) + \cos(\pi x)}\right) + \frac{\partial^2}{\partial y^2}\left(\frac{\sin(\pi x)}{\cosh(\pi y) + \cos(\pi x)}\right) = 0.$$

As the boundary of **S** has two disjoint pieces and integration over each piece will be written separately, we introduce the "half" Poisson kernel Ω on $\overline{\mathbf{S}} \setminus \{1\}$ via

$$\Omega(x, y) = \frac{1}{2} \frac{\sin(\pi x)}{\cosh(\pi y) + \cos(\pi x)} = \frac{1}{2} \frac{\cot(\frac{\pi x}{2})}{\left[\cot^2(\frac{\pi x}{2}) + \tanh^2(\frac{\pi y}{2})\right] \cosh^2(\frac{\pi y}{2})}, \quad \text{(C.0.1)}$$

where $0 \leq x \leq 1$ and $-\infty < y < \infty$ but $(x, y) \neq (1, 0)$. This function is nonnegative and satisfies

$$\int_{-\infty}^{+\infty} \Omega(x, t) \, dt = x \qquad \text{(C.0.2)}$$

for all $0 \leq x < 1$; thus

© The Editor(s) (if applicable) and The Author(s), under exclusive license to Springer Nature Switzerland AG 2024
L. Grafakos, *Fundamentals of Fourier Analysis*, Graduate Texts in Mathematics 302, https://doi.org/10.1007/978-3-031-56500-7

$$\int_{-\infty}^{+\infty} \Omega(x,t)\,dt + \int_{-\infty}^{+\infty} \Omega(1-x,t)\,dt = x + (1-x) = 1, \qquad 0 < x < 1.$$

Proposition C.0.1. *Suppose that M_0 and M_1 are continuous functions defined on the real line that satisfy*

$$\int_{-\infty}^{+\infty} |M_0(t)|\,e^{-\pi|t|}\,dt + \int_{-\infty}^{+\infty} |M_1(t)|\,e^{-\pi|t|}\,dt < \infty. \tag{C.0.3}$$

Then the function

$$u(x,y) = \int_{-\infty}^{+\infty} \Omega(1-x, y-t) M_0(t)\,dt + \int_{-\infty}^{+\infty} \Omega(x, y-t) M_1(t)\,dt$$

is harmonic on the unit strip and satisfies $u(x,y) \to M_0(y_0)$ as $(x,y) \to (0^+, y_0)$ and $u(x,y) \to M_1(y_1)$ as $(x,y) \to (1^-, y_1)$.

Proof. The harmonicity of u is verified by passing the derivatives inside the integral via the Lebesgue dominated convergence theorem, taking into account (C.0.3). We show that $u(x,y) \to M_0(y_0)$ as $(x,y) \to (0^+, y_0)$, for $y_0 \in \mathbf{R}$. An analogous argument works for the other boundary line.

Let us fix $y_0 \in \mathbf{R}$. Given $\varepsilon > 0$ there is a $\delta > 0$ such that

$$|t - y_0| < \delta \implies |M_0(t) - M_0(y_0)| < \varepsilon. \tag{C.0.4}$$

Using (C.0.2) we write

$$|u(x,y) - M_0(y_0)|$$
$$\leq \int_{-\infty}^{+\infty} \Omega(1-x, y-t)\,|M_0(t) - M_0(y_0)|\,dt + \int_{-\infty}^{+\infty} \Omega(x, y-t)\,|M_1(t) - M_0(y_0)|\,dt.$$

We now take $(x,y) \in \overline{\mathbf{S}}$ with

$$|x - 0| + |y - y_0| < \delta' = \min\left(\frac{\delta}{2}, \frac{1}{3}, \varepsilon, \frac{\varepsilon}{C(\delta)}\right),$$

where

$$C(\delta) = \frac{\cosh(\frac{\pi\delta}{2})}{\cosh(\frac{\pi\delta}{2}) - 1}.$$

The integral

$$\int_{-\infty}^{+\infty} \Omega(x, y-t)\,|M_1(t) - M_0(y_0)|\,dt$$

is at most a constant multiple of ε, as we can verify from the observation

$$|\Omega(x, y-t)| \leq \frac{1}{2} \frac{|\sin(\pi x)|}{\cosh(\pi|y-t|) + \frac{1}{2}} \leq \frac{|\sin(\pi x)|}{e^{\pi|y-t|}} \leq \pi|x| \frac{e^{\pi|y|}}{e^{\pi|t|}} \leq \varepsilon\pi \frac{e^{\pi|y_0|} e^{\pi/3}}{e^{\pi|t|}}$$

and hypothesis (C.0.3). For the other integral we note that

$$\int_{|t-y_0|<\delta} \Omega(1-x,y-t)|M_0(t)-M_0(y_0)|\,dt \leq \varepsilon(1-x) < \varepsilon,$$

in view of (C.0.2) and (C.0.4). Now, if $|t-y_0| \geq \delta$ then $|t-y| \geq \delta/2$ (since $|y-y_0| < \delta/2$); hence

$$|\Omega(1-x,y-t)| \leq \frac{\frac{1}{2}|\sin(\pi x)|}{\cosh(\pi|y-t|)-1} \leq C(\delta)\frac{\frac{1}{2}|\sin(\pi x)|}{\cosh(\pi|y-t|)} \leq \frac{C(\delta)|\sin(\pi x)|}{e^{\pi|y-t|}}.$$

Thus, one has

$$\int_{|t-y_0|\geq \delta} \Omega(1-x,y-t)|M_0(t)-M_0(y_0)|\,dt$$

$$\leq C(\delta)|x|\,\pi\, e^{\pi|y_0|}e^{\pi/3} \int_{\mathbb{R}} e^{-\pi|t|}(|M_0(t)|+|M_0(y_0)|)\,dt,$$

and this expression is at most a constant multiple of ε in view of the choice of δ'. □

Proposition C.0.2. *Fix $C, a > 0$ with $a < \pi$. Let F be a continuous function on the closed unit strip $\overline{S}$ whose logarithm is subharmonic in S and satisfies*

$$\sup_{0\leq x\leq 1} \log|F(x+iy)| \leq Ce^{a|y|}, \qquad -\infty < y < \infty. \tag{C.0.5}$$

If M_0, M_1 are continuous functions on the line that satisfy $M_0(y) \geq \log|F(iy)|$ and $M_1(y) \geq \log|F(1+iy)|$ for all $y \in (-\infty, \infty)$ and also

$$|M_0(y)| \leq Ce^{a|y|}, \qquad -\infty < y < \infty,$$
$$|M_1(y)| \leq Ce^{a|y|}, \qquad -\infty < y < \infty.$$

Then for any $\theta \in (0,1)$ we have

$$\log|F(\theta)| \leq \int_{-\infty}^{+\infty} \Omega(1-\theta,t)M_0(t)\,dt + \int_{-\infty}^{+\infty} \Omega(\theta,t)M_1(t)\,dt. \tag{C.0.6}$$

Proof. Consider the rectangle $D_T = (0,1) \times (-T,T)$ for some $T > 0$ and choose $a' \in (a,\pi)$ and $\varepsilon > 0$. For $\kappa \in \{0,1\}$ define continuous functions

$$M_\kappa^T(y) = \begin{cases} M_\kappa(y) & \text{if } |y| \leq T, \\ L_\kappa(y) & \text{if } T < |y| \leq T+1, \\ 0 & \text{if } |y| > T+1, \end{cases}$$

where $L_\kappa(y)$ is a line segment joining $M_\kappa(T)$ to 0 for $y \in [T, T+1]$ and 0 to $M_\kappa(-T)$ for $y \in [-T-1, -T]$.

Now define a harmonic function U_T on D_T by setting

$$U_T(x,y) = \int_{-\infty}^{+\infty} \Omega(1-x,y-t)M_0^T(t)dt + \int_{-\infty}^{+\infty} \Omega(x,y-t)M_1^T(t)dt$$
$$+ \varepsilon \cosh(a'y)\cos\left(a'\left(x-\frac{1}{2}\right)\right)$$

for $(x,y) \in D_T$. Identifying z with $x+iy$, we claim that for T sufficiently large, the function
$$\log|F(z)| - U_T(z)$$
is negative on the boundary of D_T. To prove this assertion we first notice that by Proposition C.0.1, $U_T(x,y)$ is harmonic on D_T and on $\{iy : |y| \leq T\}$ it coincides with
$$M_0(y) + \varepsilon \cosh(a'y)\cos\left(\frac{a'}{2}\right),$$
which is bigger than $\log|F(iy)|$, while on $\{1+iy : |y| \leq T\}$ it coincides with
$$M_1(y) + \varepsilon \cosh(a'y)\cos\left(\frac{a'}{2}\right),$$
which is bigger than $\log|F(1+iy)|$; hence the assertion is valid on the vertical parts of the boundary of D_T. On the horizontal boundary pieces of D_T, i.e., for $z = x \pm iT$, first notice that
$$\cosh(a'y)\cos\left(a'\left(x-\frac{1}{2}\right)\right) \geq \frac{1}{2}e^{a'T}\cos\left(a'\left(x-\frac{1}{2}\right)\right) \geq \frac{1}{2}e^{a'T}\cos\left(\frac{a'}{2}\right)$$
and then observe that
$$U_T(z) \geq \int_{-T}^{T} \Omega(1-x,y-t)M_0(t)\,dt + \int_{-T}^{T} \Omega(x,y-t)M_1(t)\,dt + \frac{\varepsilon}{2}e^{a'T}\cos\left(\frac{a'}{2}\right)$$
$$- \int_{T\leq|t|\leq T+1} [\Omega(1-x,y-t) + \Omega(x,y-t)]\max\left(|M_0(\pm T)|,|M_1(\pm T)|\right)dt$$
$$\geq \int_{-\infty}^{+\infty} \Omega(1-x,y-t)(-Ce^{aT})\,dt + \int_{-\infty}^{+\infty} \Omega(x,y-t)(-Ce^{aT})\,dt$$
$$- Ce^{aT}(1-x) - Ce^{aT}x + \frac{\varepsilon}{2}e^{a'T}\cos\left(\frac{a'}{2}\right)$$
$$= -2Ce^{aT}(1-x) - 2Ce^{aT}x + \frac{\varepsilon}{2}e^{a'T}\cos\left(\frac{a'}{2}\right)$$
$$= -2Ce^{aT} + \frac{\varepsilon}{2}e^{a'T}\cos\left(\frac{a'}{2}\right)$$
$$> Ce^{aT}$$
$$\geq \log|F(z)|,$$

where the strict inequality holds for all $T \geq T_0(a',\varepsilon)$, where $T_0(a',\varepsilon)$ depends on $\varepsilon > 0$ and on $a' \in (a,\pi)$. Thus $U_T - \log|F|$ is positive on the horizontal boundary pieces of D_T. Consequently, $U_T - \log|F|$ is positive on the boundary of D_T.

By Proposition C.0.1, the function U_T is harmonic on D_T and by assumption $\log|F|$ is subharmonic on **S**. Applying the maximum principle for subharmonic functions (Lemma B.0.1) to the functions $V = \log|F|$ and $U = U_T$ on the domain D_T (which has compact closure) we obtain that

$$\chi_{D_T}(z) \log|F(z)| \leq \chi_{D_T}(z) U_T(z).$$

We now take the limit as $T \to \infty$ on both sides of this inequality and we use the LDCT to prove (C.0.6) with the extra term $\varepsilon \cosh(a'y)\cos(a'(x-\frac{1}{2}))$ on the right. As $\varepsilon > 0$ was arbitrary, the conclusion follows by letting $\varepsilon \to 0$. $\square$

Corollary C.0.3. *Let F be a bounded continuous function on the closed unit strip $\overline{\mathbf{S}}$ such that $\log|F|$ is subharmonic in **S**. Suppose that for some $B_0, B_1 > 0$ we have $|F(it)| \leq B_0$ and $|F(1+it)| \leq B_1$ for all t real. Then for any $\theta \in (0,1)$ we have*

$$|F(\theta)| \leq B_0^{1-\theta} B_1^{\theta}.$$

*In particular, this is the case if F is continuous and bounded on $\overline{\mathbf{S}}$ and analytic in **S**.*

Proof. This is a straightforward consequence of Proposition C.0.2 with $M_0 = \log B_0$, $M_1 = \log B_1$ and identity (C.0.2). Notice that condition (C.0.5) is obviously satisfied in this case.

Corollary C.0.4. *Let F be a continuous function on the closed unit strip $\overline{\mathbf{S}}$ that satisfies (C.0.5) and such that $\log|F|$ is subharmonic in **S**. Suppose that for some positive constants M, B_0, B_1 we have $|F(it)| \leq B_0(1+|t|)^M$ for all t real and also $|F(1+it)| \leq B_1(1+|t|)^M$ for all t real. Then for any $\theta \in (0,1)$ we have*

$$|F(\theta)| \leq C_{M,\theta} B_0^{1-\theta} B_1^{\theta},$$

where $C_{M,\theta}$ is a positive constant that depends only on M and θ.

Obviously, the subharmonicity of $\log|F|$ is satisfied if F is analytic in **S**.

Appendix D
Density for Subadditive Operators

It is well known that it suffices to obtain quantitative estimates for linear operators on a dense subspace of its domain. Something analogous is valid for subadditive operators.

Theorem D.0.1. *Suppose that T is a positive symmetric subadditive operator defined on a dense subspace V of a quasi-Banach space X that takes values in the space of measurable functions on a measure space Y. This means that $T(\varphi)$ is real for all $\varphi \in V$,*
$$T(-\varphi) = T(\varphi) \geq 0 \qquad \text{for all } \varphi \in V$$
and
$$T(\varphi + \psi) \leq T(\varphi) + T(\psi) \qquad \text{for all } \varphi, \psi \in V.$$
Let $0 < p < \infty$. Suppose that for some $C \in (0, \infty)$ we are given the estimate
$$\|T(\varphi)\|_{L^p} \leq C\|\varphi\| \qquad \text{for all } \varphi \in V.$$
Then there is a unique positive subadditive operator $\overline{T}$ defined on X such that $\overline{T}$ coincides with T on V and satisfies
$$\|\overline{T}(f)\|_{L^p} \leq C\|f\|$$
for all $f \in X$.

Proof. Given $f \in X$, let φ_n be a Cauchy sequence converging to f in X. Then
$$T(\varphi_n) - T(\varphi_m) \leq T(\varphi_n - \varphi_m)$$
and likewise
$$T(\varphi_m) - T(\varphi_n) \leq T(\varphi_m - \varphi_n) = T(\varphi_n - \varphi_m).$$
It follows that
$$|T(\varphi_n) - T(\varphi_m)| \leq T(\varphi_n - \varphi_m)$$
and consequently

$$\left\|T(\varphi_n) - T(\varphi_m)\right\|_{L^p} \leq \left\|T(\varphi_n - \varphi_m)\right\|_{L^p} \leq C\left\|\varphi_n - \varphi_m\right\|,$$

which indicates that the sequence $\{T(\varphi_n)\}_n$ is Cauchy in $L^p(Y)$ and thus it converges to an element which we call $\overline{T}(f)$. We note that the definition of $\overline{T}(f)$ does not depend on φ_n. Indeed, if ψ_n is another sequence from V that converges to f in X, then the preceding argument gives

$$\left\|T(\varphi_n) - T(\psi_n)\right\|_{L^p} \leq C\left\|\varphi_n - \psi_n\right\| \to 0;$$

thus the sequences $\{T(\varphi_n)\}_n$ and $\{T(\psi_n)\}_n$ tend to the same limit in L^p. Note that $\overline{T}(f)$ is positive and symmetric since it is the L^p limit of positive and symmetric operators. Moreover, $\overline{T}(f)$ is subadditive as it is a limit of subadditive operators. Finally, $\overline{T}$ is an extension of T as it coincides with T on V, for given $\varphi \in V$ we take the constant sequence in V converging to itself.

We need to show that $\overline{T}(f)$ is bounded. Given f pick $\varphi_n \to f$ in X and write

$$\left\|\overline{T}(f)\right\|_{L^p}^{\min(1,p)} \leq \left\|\overline{T}(f) - \overline{T}(\varphi_n)\right\|_{L^p}^{\min(1,p)} + \left\|T(\varphi_n)\right\|_{L^p}^{\min(1,p)}$$
$$\leq \left\|\overline{T}(f) - \overline{T}(\varphi_n)\right\|_{L^p}^{\min(1,p)} + C^{\min(1,p)}\left\|\varphi_n\right\|^{\min(1,p)},$$

so letting $n \to \infty$ we obtain that

$$\left\|\overline{T}(f)\right\|_{L^p} \leq C\|f\|$$

as $\varphi_n \to f$ in X and $\overline{T}(f) - \overline{T}(\varphi_n) \to 0$ in L^p.

Note: The same argument works if the range is $L^{p,\infty}$ equipped with a norm under which it is q-normable for some q. For instance, $L^{p,\infty}$ is normable if $p > 1$, p-normable if $p < 1$, and $(1-\varepsilon)$-normable if $p = 1$ $(0 < \varepsilon < 1)$.

Appendix E
Transposes and Adjoints of Linear Operators

The notion of the transpose of a linear operator is compatible with that of the transpose of a matrix. The transpose A^t of an $n \times n$ matrix A has the fundamental property

$$x \cdot Ay = A^t x \cdot y,$$

where x, y are column vectors in $\mathbf{R}^n$ and $x \cdot y$ is the usual inner product on $\mathbf{R}^n$. Replacing the inner product by an integral and x, y by functions essentially yields the definition of the transpose of a linear operator. Let $1 \leq p, q \leq \infty$. For a bounded linear operator T from $L^p(X, \mu)$ to $L^q(Y, \nu)$ we define the *transpose* of T as the unique linear operator T^t that satisfies

$$\langle T(f), g \rangle = \int_Y T(f) g \, d\nu = \int_X f T^t(g) \, d\mu = \langle f, T^t(g) \rangle$$

for all $f \in L^p(X, \mu)$ and all $g \in L^{q'}(Y, \nu)$. We notice that the real inner product $(f, g) \mapsto \langle f, g \rangle$ also coincides with the action of the distribution f on the function g (if g is a Schwartz function) or vice versa.

Looking at matrices again, we notice that the complex inner product $(z, w) \mapsto z \cdot \overline{w}$ satisfies the identity

$$z \cdot \overline{Aw} = A^* z \cdot \overline{w},$$

where A^* is the conjugate transpose of A, i.e., the matrix whose coefficients are the complex conjugates of the coefficients of A^t. Analogously, for f, g measurable functions on $\mathbf{R}^n$, we consider the *complex inner product*

$$(f, g) \mapsto \int_{\mathbf{R}^n} f(x) \overline{g(x)} \, dx,$$

whenever the integral converges absolutely.

Let $1 \leq p, q \leq \infty$. For a bounded linear operator T from $L^p(X, \mu)$ to $L^q(Y, \nu)$ we define the *adjoint operator* T^* of T as the unique linear operator that satisfies

$$\int_Y T(f) \overline{g} \, d\nu = \int_X f \overline{T^*(g)} \, d\mu \qquad \text{for all } f \in L^p(X, \mu), g \in L^{q'}(Y, \nu).$$

Examples.
(a) If T is an integral operator of the form

$$T(f)(x) = \int_X K(x,y) f(y) \, d\mu(y),$$

then T^* and T^t are also integral operators with kernels $K^*(x,y) = \overline{K(y,x)}$ and $K^t(x,y) = K(y,x)$, respectively.

(b) If $T(f) = T_m(f) = (\widehat{f}m)^\vee$, then $T_m^t = T_{\widetilde{m}}$. This was essentially shown in the proof of Proposition 2.8.6. We also have $T_m^* = T_{\overline{m}}$. To verify this for f, g in $\mathscr{S}(\mathbf{R}^n)$ we write

$$\begin{aligned}
\int_{\mathbf{R}^n} f \, \overline{T^*(g)} \, dx &= \int_{\mathbf{R}^n} T(f) \, \overline{g} \, dx \\
&= \int_{\mathbf{R}^n} \widehat{T(f)} \, \overline{\widehat{g}} \, d\xi \\
&= \int_{\mathbf{R}^n} \widehat{f} \, \overline{\overline{m} g} \, d\xi \\
&= \int_{\mathbf{R}^n} f \, \overline{(\overline{m} \widehat{g})^\vee} \, dx.
\end{aligned}$$

Consequently, if $m(\xi)$ is real-valued, then T_m is *self-adjoint* (i.e., $T_m = T_m^*$) while if $m(\xi)$ is even, then T_m is *self-transpose* (i.e., $T_m = T_m^t$).

Appendix F
Faà di Bruno Formula

This formula provides an identity for a high-order derivative of the composition of two functions.

Suppose that $f \in \mathscr{C}^N$ is defined on an open subset U of the line and takes values in another open subset V of $\mathbf{R}$. Let $g \in \mathscr{C}^N$ be complex-valued function defined on V. Let N be a positive integer. Then for $x \in U$ we have

$$\frac{(g \circ f)^{(N)}(x)}{N!} = \sum_{(m_1,\ldots,m_N)} g^{(m_1+\cdots+m_N)}(f(x)) \frac{\left(\frac{1}{1!}f^{(1)}(x)\right)^{m_1}}{m_1!} \cdots \frac{\left(\frac{1}{N!}f^{(N)}(x)\right)^{m_N}}{m_N!}$$

where the sum is taken over all decompositions of $N = m_1 \cdot 1 + \cdots + m_N \cdot N$, where m_j are nonnegative integers. Here $f^{(j)}$ indicates the jth derivative of f. For a proof of this we refer to [54].

The higher-dimensional extension of this formula is a bit more involved. Suppose that $F : U \to V$, where U is an open subset of $\mathbf{R}^n$ and V is an open subset of $\mathbf{R}$. Let g be complex-valued function defined on V. Suppose that both functions are of class $\mathscr{C}^N$, $N \in \mathbf{Z}^+$. Then for any multi-index α with $1 \leq |\alpha| \leq N$ and all $x \in U$ one has

$$\frac{\partial^\alpha (g \circ F)(x)}{\alpha!} = \sum_{\mathscr{F}} g^{(m_1+\cdots+m_k)}(F(x)) \frac{\left(\frac{1}{\beta_1!}\partial^{\beta_1} F(x)\right)^{m_1}}{m_1!} \cdots \frac{\left(\frac{1}{\beta_k!}\partial^{\beta_k} F(x)\right)^{m_k}}{m_k!},$$

where the sum is taken over the following finite set:

$$\mathscr{F} = \Big\{(m_1,\ldots,m_k;\beta_1,\ldots,\beta_k) \in (\mathbf{Z}^+ \cup \{0\})^k \times ((\mathbf{Z}^+ \cup \{0\})^n)^k : \quad k = |\alpha|$$

such that there is an $s \in \mathbf{Z}^+$ with $1 \leq s \leq k$ so that
$m_i = 0$ and $\beta_i = 0$ for all i with $1 \leq i \leq k-s$
and $m_i > 0$ for all i with $k-s+1 \leq i \leq k$
and $0 \prec \beta_{k-s+1} \prec \cdots \prec \beta_k$, $1 \leq \sum_{i=1}^k m_i \leq k$, $\sum_{i=1}^k m_i \beta_i = \alpha\Big\}.$

We explain the meaning of the ordering $\prec$ on multi-indices. We say that two multi-indices $\alpha = (\alpha_1, \ldots, \alpha_n)$ and $\beta = (\beta_1, \ldots, \beta_n)$ satisfy $\alpha \prec \beta$ if and only if

(i) $|\alpha| < |\beta|$; or
(ii) $|\alpha| = |\beta|$ and $\alpha_1 < \beta_1$; or
(iii) $|\alpha| = |\beta|$ and there is a k such that $\alpha_i = \beta_i$ when $i \leq k$ while $\alpha_{k+1} < \beta_{k+1}$.

This version of the multivariate version of the Faà di Bruno formula can be found in [16, Corollary 2.10].

Appendix G
Besicovitch Covering Lemma

Lemma G.0.1. *Let K be a bounded set in $\mathbf{R}^n$ and suppose that for every $x \in K$ there is an open cube Q_x centered at x with sides parallel to the axes. Then there is an $m \in \mathbf{Z}^+ \cup \{\infty\}$ and there exists a sequence of points $\{x_j\}_{j=1}^m$ in K such that*

$$K \subseteq \bigcup_{j=1}^m Q_{x_j} \tag{G.0.1}$$

and for all $y \in \mathbf{R}^n$ one has

$$\sum_{j=1}^m \chi_{Q_{x_j}}(y) \leq 4^n. \tag{G.0.2}$$

Proof. Let $s_0 = \sup\{\ell(Q_x) : x \in K\}$. If $s_0 = \infty$, then there exists $x_1 \in K$ such that $\ell(Q_{x_1}) > 4L$, where $[-L, L]^n$ contains K. Then K is contained in Q_{x_1} and the statement of the lemma is valid with $m = 1$.

Suppose now that $s_0 < \infty$. Select $x_1 \in K$ such that $\ell(Q_{x_1}) > s_0/2$. Then define

$$K_1 = K \setminus Q_{x_1}, \qquad s_1 = \sup\{\ell(Q_x) : x \in K_1\},$$

and select $x_2 \in K_1$ such that $\ell(Q_{x_2}) > s_1/2$. Next define

$$K_2 = K \setminus (Q_{x_1} \cup Q_{x_2}), \qquad s_2 = \sup\{\ell(Q_x) : x \in K_2\},$$

and select $x_3 \in K_2$ such that $\ell(Q_{x_3}) > s_2/2$. Continue until the first integer m is found such that K_m is an empty set. If no such integer exists, continue this process indefinitely and set $m = \infty$.

We claim that for all $i \neq j$ we have $\frac{1}{3}Q_{x_i} \cap \frac{1}{3}Q_{x_j} = \emptyset$. Indeed, suppose that $i > j$. Then $x_i \in K_{i-1} = K \setminus (Q_{x_1} \cup \cdots \cup Q_{x_{i-1}})$; thus $x_i \notin Q_{x_j}$. Also $x_i \in K_{i-1} \subseteq K_{j-1}$, which implies that $\ell(Q_{x_i}) \leq s_{j-1} < 2\ell(Q_{x_j})$. Since $x_i \notin Q_{x_j}$ and $\ell(Q_{x_j}) > \frac{1}{2}\ell(Q_{x_i})$, it easily follows that $\frac{1}{3}Q_{x_i} \cap \frac{1}{3}Q_{x_j} = \emptyset$.

Next we claim that $\ell(Q_{x_j}) \to 0$ as $j \to \infty$. Indeed, if this was not the case, then there would be an $\varepsilon_0 > 0$ and a subsequence $\{j_r\}_{r=1}^\infty$ of the positive integers such that $\ell(Q_{x_{j_r}}) \geq \varepsilon_0$ for all $r = 1, 2, \ldots$. The cubes $\frac{1}{3}Q_{x_{j_r}}$, $r = 1, 2, \ldots$, are infinitely many and disjoint. But all of these cubes are contained in a bounded set, as their centers lie in K (which is bounded) and their side lengths are at most $s_0 < \infty$, and this is a contradiction. This shows that $\ell(Q_{x_j}) \to 0$ and also shows that $s_j \to 0$ as $j \to \infty$, since $s_j < 2\ell(Q_{x_{j+1}})$ for all j.

We now prove (G.0.1). If $m < \infty$, then $K_m = \emptyset$ and therefore $K \subseteq \bigcup_{j=1}^m Q_{x_j}$. If $m = \infty$, then there is an infinite number of selected cubes Q_{x_j}. As shown, the sequence of their lengths converges to zero. If there exists a $y \in K \setminus \bigcup_{j=1}^\infty Q_{x_j}$, this y would belong to all K_j, $j = 1, 2, \ldots$, and then $s_j \geq \ell(Q_y)$ for all j. But as $s_j \to 0$, then it must be that $\ell(Q_y) = 0$, which would force the open cube Q_y to be empty, a contradiction. Thus (G.0.1) holds.

We now prove (G.0.2) via a general argument concerning a sequence of open cubes Q_{x_i}, $i = 1, 2, \ldots$, satisfying two properties (valid in our setting):

P1 $j < i \implies \ell(Q_{x_i}) < 2\ell(Q_{x_j})$.
P2   If $j < i$, then x_i (the center of Q_{x_i}) is not contained in Q_{x_j}.

We claim that under properties P1 and P2, no point in $\mathbf{R}^n$ belongs to more than 4^n of these cubes. This certainly implies (G.0.2).

To prove this claim we argue by contradiction. Suppose that some point in $\mathbf{R}^n$ belongs to more than 4^n of the cubes Q_{x_i}. By translating all the cubes we may assume that this point is the origin. Extracting from the sequence $Q_{x_1}, Q_{x_2}, \ldots$ those cubes which contain the origin and renumbering the remaining, we may assume that

$$0 \in \bigcap_{i=1}^{4^n+1} Q_{x_i}.$$

We now write $\mathbf{R}^n$ as a union of 2^n higher-dimensional closed quadrants each characterized by the signs of the coordinates of the points it contains. One of these quadrants must contain more than $(4^n + 1)/2^n$ of the $x_1, \ldots, x_{4^n+1}$. By a change of notation we may assume that the quadrant $E = [0, \infty) \times \cdots \times [0, \infty)$ that contains at least $2^n + 1$ of the x_i. We now renumber the Q_{x_i} whose centers belong to E in such a way so that P1 and P2 are preserved; then we may suppose that

$$0 \in \bigcap_{i=1}^{2^n+1} Q_{x_i}, \qquad x_1, \ldots, x_{2^n+1} \in E.$$

In the sequel we use the notation $|z|_{\ell^\infty} = \sup_{1 \leq i \leq n} |z_i|$ for points $z = (z_1, \ldots, z_n)$ in $\mathbf{R}^n$. For simplicity, we also denote by $\ell_i = \ell(Q_{x_i})/2$ half the side length of Q_{x_i}. Then the cube Q_{x_i} equals the set $\{y \in \mathbf{R}^n : |y - x_i|_{\ell^\infty} < \ell_i\}$.

Next, we prove the following facts. All indices i, j below lie in $\{2, 3, \ldots, 2^n + 1\}$.

(A) $E \cap [0, \ell_1)^n$ is contained in Q_{x_1}.

Indeed, as $0 \in Q_{x_1}$, it follows that $|x_1|_{\ell^\infty} < \ell_1$. Thus, if $y \in E \cap [0, \ell_1)^n$, then $|y_1|_{\ell^\infty} < \ell_1$ and as all coordinates of y and x_1 are nonnegative, it follows that $|y - x_1|_{\ell^\infty} < \ell_1$, i.e., y lies in Q_{x_1}.

(B) If $2 \leq i \leq 2^n+1$, then $\ell_i > \ell_1$.

Since x_i does not belong to Q_{x_1} by property P2 and $x_i \in E$, it follows from fact (A) that $x_i \notin [0,\ell_1)^n$. But $0 \in Q_{x_i}$ and this gives $\ell_1 \leq |x_i|_{\ell^\infty} < \ell_i$.

(C) If $2 \leq i \leq 2^n+1$, then $x_i \in [0,2\ell_1)^n \setminus [0,\ell_1)^n$.

The fact that $|x_i|_{\ell^\infty} < \ell_i < 2\ell_1$ implies that $x_i \in (-2\ell_1, 2\ell_1)^n$ but as $x_i \in E$ we have $x_i \in [0,2\ell_1)^n$. Also in (B) it was proved that $\ell_i > \ell_1$ which implies $x_i \notin (-\ell_1,\ell_1)^n$ thus $x_i \notin [0,\ell_1)^n$.

(D) Let $R_1,\ldots,R_{2^n-1}$ be disjoint cubes of length ℓ_1 obtained by bisecting all of the sides of $[0,2\ell_1)^n$ and ignoring the cube $[0,\ell_1)^n$. Then there exist indices i,j such that $1 < j < i \leq 2^n+1$ and such that x_i, x_j belong to the same cube R_k.

We subdivide the cube $[0,2\ell_1)^n$ into 2^n disjoint subcubes by bisecting all of its sides. Removing the cube $[0,\ell_1)^n$ we are left with 2^n-1 subcubes of $[0,2\ell_1)^n$ each of length ℓ_1. These cubes are named $R_1,\ldots,R_{2^n-1}$ and each one of them has the form $[a_1,b_1) \times \cdots \times [a_n,b_n)$, where $[a_k,b_k) \in \{[0,\ell_1),[\ell_1,2\ell_1)\}$ but not all $a_k = 0$. A point belongs to one R_k if and only if all of its coordinates are in $[0,2\ell_1)$ and at least one of them is in $[\ell_1,2\ell_1)$. Fact (C) gives that the centers $x_2,x_3,\ldots,x_{2^n+1}$ lie in the union of R_k. By the pigeonhole principle, one R_k must contain at least two points among the $x_2,x_3,\ldots,x_{2^n+1}$. Thus there exist i,j with $1 < j < i \leq 2^n+1$ such that x_i, x_j belong to the same R_k.

(E) If $1 < j < i \leq 2^n+1$ and $x_i, x_j \in R_k$ (for the same k), then R_k is contained in Q_{x_j}.

Suppose that $y \in R_k$. Then for each κ, the κth coordinate of both y and x_j lie in the same interval $[0,\ell_1)$ or $[\ell_1, 2\ell_1)$. This implies that $|y-x_j|_{\ell^\infty} < \ell_1 < \ell_j$, where the last inequality follows by part (B). Thus y lies in Q_{x_j}.

Having established facts (A)–(E), we now arrive at a contradiction by noting that (E) yields that $x_i \in Q_{x_j}$ which refutes P2. $\square$

Glossary

LMCT	Lebesgue monotone convergence theorem
LDCT	Lebesgue dominated convergence theorem
$A \subseteq B$	A is a subset of B (also denoted by $A \subseteq B$)
$A \subsetneq B$	A is a proper subset of B
$A \supset B$	B is a proper subset of A
A^c	the complement of a set A
χ_E	the characteristic function of the set E
D_f	the distribution function of a function f
$f_n \uparrow f$	f_n increases monotonically to a function f
$\mathbf{Z}$	the set of all integers
$\mathbf{Z}^+$	the set of all positive integers $\{1, 2, 3, \dots\}$
$\mathbf{Z}^n$	the n-fold product of the integers
$\mathbf{R}$	the set of real numbers
$\mathbf{R}^+$	the set of positive real numbers
$\mathbf{R}^n$	the Euclidean n-space
$\mathbf{Q}$	the set of rationals
$\mathbf{C}$	the set of complex numbers
S	the unit strip $\{z \in \mathbf{C} : 0 < \operatorname{Re} z < 1\}$
$\lvert x \rvert$	$\sqrt{\lvert x_1 \rvert^2 + \cdots + \lvert x_n \rvert^2}$ when $x = (x_1, \dots, x_n) \in \mathbf{R}^n$
$\mathbf{S}^{n-1}$	the unit sphere $\{x \in \mathbf{R}^n : \lvert x \rvert = 1\}$
e_j	the vector $(0, \dots, 0, 1, 0, \dots, 0)$ with 1 in the jth entry and 0 elsewhere
$\log t$	the logarithm with base e of $t > 0$
$[t]$	the largest integer less than or equal to a real number t

$[[t]]$	the largest integer strictly less than a real number t
$x \cdot y$	the inner product $\sum_{j=1}^{n} x_j y_j$ when $x = (x_1,\ldots,x_n)$, $y = (y_1,\ldots,y_n)$
$B(x,R)$	the ball of radius R centered at x in $\mathbf{R}^n$
ω_{n-1}	the surface area of the unit sphere $\mathbf{S}^{n-1}$
v_n	the volume of the unit ball $\{x \in \mathbf{R}^n : |x| < 1\}$
$|A|$	the Lebesgue measure of the set $A \subseteq \mathbf{R}^n$
dx	Lebesgue measure
f_B	the average $\frac{1}{|B|}\int_B f(x)\,dx$ of f over the set B
$\langle f,g\rangle$	the real inner product $\int_{\mathbf{R}^n} f(x)g(x)\,dx$
$\langle u,f\rangle$	the action of a distribution u on a function f
p'	the number $p/(p-1)$, whenever $1 < p < \infty$
$1'$	the number ∞
∞'	the number 1
$f = O(g)$	means $|f(x)| \le M|g(x)|$ for some M for x near x_0
$f = o(g)$	means $|f(x)||g(x)|^{-1} \to 0$ as $x \to x_0$
A^t	the transpose of the matrix A
A^*	the conjugate transpose of a complex matrix A
A^{-1}	the inverse of the matrix A
$O(n)$	the space of real matrices satisfying $A^{-1} = A^t$ (orthogonal matrices)
$\|T\|_{X\to Y}$	the norm of the (bounded) operator $T : X \to Y$
$A \approx B$	means that there exists a $c > 0$ such that $c^{-1} \le \frac{B}{A} \le c$
$|\alpha|$	indicates the size $|\alpha_1| + \cdots + |\alpha_n|$ of a multi-index $\alpha = (\alpha_1,\ldots,\alpha_n)$
$\partial_j^m f$	the mth partial derivative of $f(x_1,\ldots,x_n)$ with respect to x_j
$\partial^\alpha f$	$\partial_1^{\alpha_1}\cdots\partial_n^{\alpha_n} f$
$\mathscr{C}^k$	the space of functions f with $\partial^\alpha f$ continuous for all $|\alpha| \le k$
$\mathscr{C}^\infty$	the space of smooth functions $\bigcap_{k=1}^{\infty} \mathscr{C}^k$
$\mathscr{C}_0^\infty$	the space of smooth functions with compact support
$\mathscr{C}_0$	the space of continuous functions with compact support
$\mathscr{C}_{00}$	the space of continuous functions that tend to zero at infinity
$\mathscr{S}$	the space of Schwartz functions
$\mathscr{S}_0$	the space of Schwartz functions φ with the property $\int_{\mathbf{R}^n} x^\gamma \varphi(x)\,dx = 0$ for all multi-indices γ.
$\mathscr{S}'(\mathbf{R}^n)$	the space of tempered distributions on $\mathbf{R}^n$
$\ell(Q)$	the side length of a cube Q in $\mathbf{R}^n$

Glossary

∂Q	the boundary of a cube Q in $\mathbf{R}^n$
$L^p(X,\mu)$	the Lebesgue space over the measure space (X,μ)
$L^p(\mathbf{R}^n)$	the space $L^p(\mathbf{R}^n, |\cdot|)$, $0 < p \leq \infty$
$L^{p,\infty}(X,\mu)$	the weak L^p space over the measure space (X,μ)
$L^1_{\mathrm{loc}}(\mathbf{R}^n)$	the space of functions that lie in $L^1(K)$ for any compact set K in $\mathbf{R}^n$
$|\mu|$	the variation (measure) of a signed Borel measure μ on $\mathbf{R}^n$
$\|\mu\|$	the total variation of a signed Borel measure μ on $\mathbf{R}^n$, i.e., $\int_{\mathbf{R}^n} d|\mu|$.
$\mathscr{M}(\mathbf{R}^n)$	the space of all signed Borel measures on $\mathbf{R}^n$ with finite total variation.
$\mathscr{M}_p(\mathbf{R}^n)$	the space of L^p Fourier multipliers, $1 < p < \infty$
$\mathcal{M}$	the centered Hardy–Littlewood maximal operator with respect to balls
M	the uncentered Hardy–Littlewood maximal operator with respect to balls
$\mathcal{M}_c$	the centered Hardy–Littlewood maximal operator with respect to cubes
M_c	the uncentered Hardy–Littlewood maximal operator with respect to cubes
$\mathcal{M}_\mu$	the centered maximal operator with respect to a measure μ
M_μ	the uncentered maximal operator with respect to a measure μ
M_s	the strong maximal operator
M_d	the dyadic maximal operator
$M_c^\#$	the sharp maximal operator
$M_d^\#$	the dyadic sharp maximal operator

References

1. Adams, Robert A.; Fournier, John J. F., *Sobolev spaces. Second edition.* Pure and Applied Mathematics (Amsterdam), 140. Elsevier/Academic Press, Amsterdam, 2003. xiv+305 pp.
2. Aronszajn, N.; Smith, K. T., *Theory of Bessel potentials, I.* Ann. Inst. Fourier (Grenoble) **11** (1961), 385–475.
3. Benedek; A., Calderón; A.-P., Panzone, R., *Convolution operators on Banach-space valued functions.* Proc. Nat. Acad. Sci. U.S.A. **48** (1962), 356–365.
4. Bennett, Colin; Sharpley, Robert, *Interpolation of Operators.* Pure and Applied Mathematics, 129. Academic Press, Inc., Boston, MA, 1988. xiv+469 pp.
5. Bergh, Jöran; Löfström, Jörgen, *Interpolation Spaces. An Introduction.* Grundlehren der Mathematischen Wissenschaften, No. 223. Springer-Verlag, Berlin-New York, 1976. x+207 pp.
6. Buckley, S. M., *Estimates for operator norms on weighted spaces and reverse Jensen inequalities.* Trans. Amer. Math. Soc. **340** (1993), no. 1, 253–272.
7. Calderón, A. P., *Lebesgue spaces of differentiable functions and distributions.* 1961 Proc. Sympos. Pure Math., Vol. IV pp. 33–49, American Mathematical Society, Providence, R.I.
8. Calderón, A. P.; Torchinsky, A., *Parabolic maximal functions associated with a distribution, II.* Advances in Math. **24** (1977), no. 2, 101–171.
9. Calderón, A. P.; Zygmund, A., *On the existence of certain singular integrals.* Acta Math. **88** (1952), no. 1, 85–139.
10. Calderón, A. P.; Zygmund, A., *On singular integrals.* Amer. J. Math. **78** (1956), 289–309.
11. Christ, Michael, *Lectures on singular integral operators.* CBMS Regional Conference Series in Mathematics, 77. Published for the Conference Board of the Mathematical Sciences, Washington, DC; by the American Mathematical Society, Providence, RI, 1990. x+132 pp.
12. Coifman, R. R., *Distribution function inequalities for singular integrals.* Proc. Nat. Acad. Sci. U.S.A. **69** (1972), 2838–2839.
13. Coifman, Ronald R., *A real variable characterization of H^p.* Studia Math. **51** (1974), 269–274.
14. Coifman, R. R.; Fefferman, C., *Weighted norm inequalities for maximal functions and singular integrals.* Studia Math. **51** (1974), 241–250.
15. Coifman, R. R.; Rochberg, R., *Another characterization of BMO*, Proc. Amer. Math. Soc. **79** (1980), no. 2, 249–254.
16. Constantine, G. M.; Savits, T. H. *A multivariate Faà di Bruno formula with applications.* Trans. Amer. Math. Soc. **348** (1996), no. 2, 503–520.
17. Cotlar, Mischa, *A unified theory of Hilbert transforms and ergodic theorems.* Rev. Mat. Cuyana **1** (1955), 105–167.

18. Cruz-Uribe, David V.; Martell, José Maria; Pérez, Carlos, *Weights, extrapolation and the theory of Rubio de Francia*. Operator Theory: Advances and Applications, 215. Birkhäuser/Springer Basel AG, Basel, 2011. xiv+280
19. Cruz-Uribe, David SFO; Neugebauer C. J., *The structure of the reverse Hölder classes*. Trans. Amer. Math. Soc. **347** (1995), no. 8, 2941–2960.
20. de Guzmán, Miguel, *Real Variable Methods in Fourier Analysis*. Notas de Matemática [Mathematical Notes], 75. North-Holland Mathematics Studies, 46. North-Holland Publishing Co., Amsterdam-New York, 1981. xiii+392 pp.
21. de Guzmán, M., *Differentiation of Integrals in R^n*. With appendices by Antonio Córdoba, and Robert Fefferman, and two by Roberto Moriyón. Lecture Notes in Mathematics, Vol. 481. Springer-Verlag, Berlin-New York, 1975. xii+266 pp.
22. Duoandikoetxea, Javier, *Fourier Analysis*. Translated and revised from the 1995 Spanish original by David Cruz-Uribe. Graduate Studies in Mathematics, 29. American Mathematical Society, Providence, RI, 2001. xviii+222 pp.
23. Duoandikoetxea, Javier, *Forty Years of Muckenhoupt Weights*. Function Spaces and Inequalities, Lecture Notes Paseky nad Jizerou 2013 (J. Lukeš, L. Pick ed.), pp. 23–75. Matfyzpress, Praga, 2013, Czech Republic.
24. Fefferman, Charles, *Characterizations of bounded mean oscillation*. Bull. Amer. Math. Soc. **77** (1971), 587–588.
25. Fefferman, C.; Stein, E. M., *Some maximal inequalities*. Amer. J. Math. **93** (1971), 107–115.
26. Fefferman, C.; Stein, E. M., *H^p spaces of several variables*. Acta Math. **129** (1972), no. 3–4, 137–193.
27. Fourier, Joseph, *Théorie analytique de la chaleur* [Analytical theory of heat]. Reprint of the 1822 original. Éditions Jacques Gabay, Paris, 1988. xxii+644 pp.
28. Frazier, Michael; Jawerth, Björn; Weiss, Guido, *Littlewood-Paley theory and the study of function spaces*. CBMS Regional Conference Series in Mathematics, 79. Published for the Conference Board of the Mathematical Sciences, Washington, DC; by the American Mathematical Society, Providence, RI, 1991. viii+132 pp.
29. García-Cuerva, José; Rubio de Francia, José L., *Weighted Norm Inequalities and Related Topics*. North-Holland Mathematics Studies, 116. Notas de Matemática [Mathematical Notes], 104. North-Holland Publishing Co., Amsterdam, 1985. x+604 pp.
30. Garnett, John B.; Jones, Peter W., *BMO from dyadic BMO*. Pacific J. Math. **99** (1982), no. 2, 351–371.
31. Grafakos, Loukas, *Classical Fourier analysis*. Third edition. Graduate Texts in Mathematics, 249. Springer, New York, 2014. xviii+638 pp.
32. Grafakos, Loukas, *Modern Fourier analysis*. Third edition. Graduate Texts in Mathematics, 250. Springer, New York, 2014. xvi+624 pp.
33. Hardy, G. H.; Littlewood, J. E., *Some properties of fractional integrals I*. Math. Z. **27** (1927), no. 1, 565–606.
34. Hardy, G. H.; Littlewood, J. E., *A maximal theorem with function-theoretic applications*. Acta Math. **54** (1930), no. 1, 81–116.
35. Hedberg, Lars Inge, *On certain convolution inequalities*. Proc. Amer. Math. Soc. **36** (1972), 505–510.
36. Hérnandez, Eugenio; Weiss, Guido, *A First Course on Wavelets*. With a foreword by Yves Meyer. Studies in Advanced Mathematics. CRC Press, Boca Raton, FL, 1996. xx+489 pp.
37. Hewitt, Edwin; Stromberg, Karl, *Real and Abstract Analysis*. A modern treatment of the theory of functions of a real variable. Springer-Verlag, New York 1965 vii+476 pp.
38. Hirschman, I. I. Jr., *A convexity theorem for certain groups of transformations*. J. Analyse Math. **2** (1953), 209–218.
39. Hörmander, Lars, *Estimates for translation invariant operators in L^p spaces*. Acta Math. **104** (1960), 93–140.
40. Hörmander, Lars, *The Analysis of Linear Partial Differential Operators I*. Distribution theory and Fourier analysis. Second edition. Grundlehren der mathematischen Wissenschaften [Fundamental Principles of Mathematical Sciences], 256. Springer-Verlag, Berlin, 1990. xii+440 pp.

References

41. Hunt, Richard, A., *On $L(p,q)$ spaces*. Einseign.Math. (2) **12** (1966), 249–276.
42. Hunt, Richard; Muckenhoupt, Benjamin; Wheeden, Richard, *Weighted norm inequalities for the conjugate function and the Hilbert transform*. Trans. Amer. Math. Soc. **176** (1973), 227–251.
43. John, F.; Nirenberg, L., *On functions of bounded mean oscillation*. Comm. Pure Appl. Math. **14** (1961), 415–426.
44. Jones, Frank, *Lebesgue integration on Euclidean space*. Jones and Bartlett Publishers, Boston, MA, 1993. xvi+588 pp.
45. Jones, Peter W., *Factorization of A_p weights*. Ann. of Math. (2nd Ser.) **111** (1980), no. 3, 511–530.
46. Journé, Jean-Lin, *Calderón–Zygmund operators, pseudodifferential operators and the Cauchy integral of Calderón*. Lecture Notes in Mathematics, 994. Springer-Verlag, Berlin, 1983. vi+128 pp.
47. Kato, Tosio; Ponce, Gustavo, *Commutator estimates and the Euler and Navier-Stokes equations*. Comm. Pure App. Math. **41** (1988), 891–907.
48. Katznelson, Yitzhak, *An introduction to harmonic analysis. Third edition.* Cambridge Mathematical Library. Cambridge University Press, Cambridge, 2004. xviii+314 pp.
49. Kislyakov, Sergey; Kruglyak, Natan, *Extremal problems in interpolation theory, Whitney–Besicovitch coverings, and singular Integrals*. Instytut Matematyczny Polskiej Akademii Nauk. Monografie Matematyczne (New Series) [Mathematics Institute of the Polish Academy of Sciences. Mathematical Monographs (New Series)], 74. Birkhäuser/Springer Basel AG, Basel, 2013. x+316 pp.
50. Kokilashvili, V. M., *Singular integral operators in weighted spaces*. Functions, series, operators, Vol. I, II (Budapest, 1980), 707–714, Colloq. Math. Soc. János Bolyai, 35, North-Holland, Amsterdam, 1983.
51. Kolmogorov, A. N., *Sur les fonctions harmoniques conjuguées et les séries de Fourier*. Fund. Math. **7** (1925), 23–28.
52. Kolmogorov, A. N., *Zur Normierbarkeit eines topologischen Raumes*. Studia Math. **5** (1934), 29–33.
53. Körner, T. W., *Fourier Analysis. Second edition.* Cambridge University Press, Cambridge, 1989. xii+591 pp.
54. Krantz, Steven G.; Parks, Harold R., *A primer of real analytic functions*. Basler Lehrbücher [Basel Textbooks], 4. Birkhäuser Verlag, Basel, 1992. x+184 pp.
55. Kreĭn, S. G.; Petunin, Yu. I.; Semënov, E. M., *Interpolation of linear operators*. Translated from the Russian by J. Szücs. Translations of Mathematical Monographs, 54. American Mathematical Society, Providence, R.I., 1982. xii+375 pp.
56. Latter, Robert H., *A characterization of $H^p(\mathbf{R}^n)$ in terms of atoms*. Studia Math. **62** (1978), no. 1, 92–101.
57. Latter, Robert H.; Uchiyama, Akihito, *The atomic decomposition for parabolic H^p spaces*. Trans. Amer. Math. Soc. **253** (1979), 391–398.
58. Lerner, Andrei K., *An elementary approach to several results on the Hardy–Littlewood maximal operator*. Proc. Amer. Math. Soc., **136** (2008), no. 8, 2829–2833.
59. Lieb, Elliott H.; Loss, Michael, *Analysis. Second edition.* Graduate Studies in Mathematics, 14. American Mathematical Society, Providence, RI, 2001. xxii+346 pp.
60. Littlewood, J. E.; Paley, R. E. A. C., *Theorems on Fourier series and power series*. J. London Math. Soc. **6** (1931), no. 3, 230–233.
61. Littlewood, J. E.; Paley, R. E. A. C., *Theorems on Fourier series and power series (II)*. Proc. London Math. Soc. **42** (1936), no. 1, 52–89.
62. Littlewood, J. E.; Paley, R. E. A. C., *Theorems on Fourier series and power series (III)*. Proc. London Math. Soc. **43** (1937), no. 2, 105–126.
63. Lorentz, G. G., *Some new functional spaces*. Ann. of Math. (2nd Ser.) **51** (1950), no. 1, 37–55.
64. Lorentz, G. G., *On the theory of spaces Λ*. Pacific. J. Math. **1** (1951), 411–429.
65. Lu, Shan Zhen, *Four Lectures on Real H^p Spaces*. World Scientific Publishing Co., Inc., River Edge, NJ, 1995. viii + 217 pp.

66. Lu, Shanzhen; Ding, Yong; Yan, Dunyan, *Singular integrals and related topics*. World Scientific Publishing Co. Pte. Ltd., Hackensack, NJ, 2007. viii+272 pp.
67. Marcinkiewicz, J., *Sur l'interpolation d'operations*, C. R. Acad. Sci. Paris **208** (1939), 1272–1273.
68. Marcinkiewicz, J., *Sur les multiplicateurs des séries de Fourier*, Studia Math. **8** (1939), 78–91.
69. Maz'ja, Vladimir G. *Sobolev spaces*. Translated from the Russian by T. O. Shaposhnikova. Springer Series in Soviet Mathematics. Springer-Verlag, Berlin, 1985. xix+486 pp.
70. Meyer, Yves, *Ondelettes et opérateurs. I.* (French) [Wavelets and operators. I] Ondelettes. [Wavelets] Actualités Mathématiques. [Current Mathematical Topics] Hermann, Paris, 1990. xii+215 pp.
71. Meyer, Yves, *Ondelettes et opérateurs. II.* (French) [Wavelets and operators. II] Opérateurs de Calderón-Zygmund. [Calderón-Zygmund operators] Actualités Mathématiques. [Current Mathematical Topics] Hermann, Paris, 1990. pp. i–xii and 217–384.
72. Meyer, Yves; Coifman, R. R., *Ondelettes et opérateurs. III.* (French) [Wavelets and operators. III] Opérateurs multilinéaires. [Multilinear operators] Actualités Mathématiques. [Current Mathematical Topics] Hermann, Paris, 1991. pp. i–xii and 383–538.
73. Mihlin, S. G., *On the multipliers of Fourier integrals*. [Russian], Dokl. Akad. Nauk. SSSR (N.S.) **109** (1956), 701–703.
74. Muckenhoupt, Benjamin, *On certain singular integrals*. Pacific J. Math. **10** (1960), 239–261.
75. Muckenhoupt, Benjamin, *Weighted norm inequalities for the Hardy maximal function*. Trans. Amer. Math. Soc. **165** (1972), 207–226.
76. Muscalu, Camil; Schlag, Wilhelm, *Classical and multilinear harmonic analysis. Vol. I.* Cambridge Studies in Advanced Mathematics, 137. Cambridge University Press, Cambridge, 2013. xviii+370 pp.
77. Muscalu, Camil; Schlag, Wilhelm, *Classical and multilinear harmonic analysis. Vol. II.* Cambridge Studies in Advanced Mathematics, 138. Cambridge University Press, Cambridge, 2013. xvi+324 pp.
78. Peetre, Jaak, *On convolution operators leaving $L^{p,\lambda}$ spaces invariant*. Ann. Mat. Pura Appl. (4) **72** (1966), 295–304.
79. Peetre, Jaak, *New thoughts on Besov spaces*. Duke University Mathematics Series, No. 1. Duke University, Mathematics Department, Durham, N.C., 1976. vi+305 pp.
80. Pereyra, María Cristina, *Lecture notes on dyadic harmonic analysis*. Second Summer School in Analysis and Mathematical Physics (Cuernavaca, 2000), 1–60, Contemp. Math., 289, Amer. Math. Soc., Providence, RI, 2001.
81. Pereyra, María Cristina; Ward, Lesley A., *Harmonic analysis. From Fourier to wavelets*. Student Mathematical Library, 63. IAS/Park City Mathematical Subseries. American Mathematical Society, Providence, RI; Institute for Advanced Study (IAS), Princeton, NJ, 2012. xxiv+410 pp.
82. Pinsky, Mark A., *Introduction to Fourier analysis and wavelets*. Reprint of the 2002 original. Graduate Studies in Mathematics, 102. American Mathematical Society, Providence, RI, 2009. xx+376 pp.
83. Riesz, Marcel, *Les fonctions conjuguées et les séries de Fourier*. C. R. Acad. Sci. Paris **178** (1924), 1464–1467.
84. Riesz, Marcel, *Sur les maxima des formes bilinéaires et sur les fonctionnelles linéaires*. Acta Math. **49** (1927), no. 3-4, 465–497.
85. Riesz, Marcel, *Sur les fonctions conjuguées*. Math. Z. **27** (1928), no. 1, 218–244.
86. Riesz, Marcel, *L' intégrale de Riemann–Liouville et le problème de Cauchy*. (French) Acta Math. **81** (1949), 1–222.
87. Rosenblum, Marvin, *Summability of Fourier series in $L^p(d\mu)$*. Trans. Amer. Math. Soc. **105** (1962), 32–42.
88. Sawano, Yoshihiro, *Theory of Besov spaces*. Developments in Mathematics, 56. Springer, Singapore, 2018. xxiii+945 pp.
89. Schmeisser, Hans-Jürgen; Triebel, Hans, *Topics in Fourier analysis and function spaces*. A Wiley-Interscience Publication. John Wiley & Sons, Ltd., Chichester, 1987. 300 pp.

90. Schwartz, L. *Théorie des distributions. Tome I.* (French) Publ. Inst. Math. Univ. Strasbourg, 9. Actualités Scientifiques et Industrielles [Current Scientific and Industrial Topics], No. 1091 Hermann & Cie, Paris, 1950. 148 pp.
91. Schwartz, Laurent, *Théorie des distributions. Tome II.* (French) Publ. Inst. Math. Univ. Strasbourg, 10. Actualités Scientifiques et Industrielles [Current Scientific and Industrial Topics], No. 1122 Hermann & Cie, Paris, 1951. 169 pp.
92. Sobolev, S. L., *On a theorem in functional analysis.* [Russian], Mat. Sob. **46** (1938), 471–497.
93. Spanne, Sven, *Sur l' interpolation entre les espaces $\mathscr{L}_k^{p\Phi}$*. Ann. Scuola Norm. Sup. Pisa (3) **20** (1966), 625–648.
94. Stein, Elias M., *Interpolation of linear operators.* Trans. Amer. Math. Soc. **83** (1956), 482–492.
95. Stein, E. M., *On the functions of Littlewood–Paley, Lusin, and Marcinkiewicz.* Trans. Amer. Math. Soc. **88** (1958), 430–466.
96. Stein, E. M., *Singular integrals, harmonic functions, and differentiability properties of functions of several variables.* Singular integrals (Proc. Sympos. Pure Math., Chicago, Ill., 1966), pp. 316–335. Amer. Math. Soc., Providence, R.I., 1967.
97. Stein, Elias M., *Singular integrals and differentiability properties of functions.* Princeton Mathematical Series, No. 30 Princeton University Press, Princeton, N.J. 1970 xiv+290 pp.
98. Stein, Elias M., *Harmonic analysis, real variable methods, orthogonality, and oscillatory integrals.* With the assistance of Timothy S. Murphy. Princeton Mathematical Series, 43. Monographs in Harmonic Analysis, III. Princeton University Press, Princeton, NJ, 1993. xiv+695 pp.
99. Stein, Elias M.; Shakarchi, Rami, *Fourier analysis. An introduction.* Princeton Lectures in Analysis, 1. Princeton University Press, Princeton, NJ, 2003. xvi+311 pp.
100. Stein, Elias M., Weiss, Guido, *Introduction to Fourier Analysis on Euclidean Spaces.* Princeton Mathematical Series, No. 32. Princeton University Press, Princeton, N.J., 1971. x+297 pp.
101. Strömberg, Jan-Olov; Torchinsky, Alberto, *Weighted Hardy spaces.* Lecture Notes in Mathematics, 1381. Springer-Verlag, Berlin, 1989. vi+193 pp.
102. Strömberg, Jan-Olov; Wheeden, Richard L., *Fractional integrals on weighted H^p and L^p spaces.* Trans. Amer. Math. Soc. 287 (1985), no. 1, 293–321.
103. Tamarkin, J. D., Zygmund, A., *Proof of a theorem of Thorin.* Bull. Amer. Math. Soc. **50** (1944), 279–282.
104. Thorin, G. O., *An extension of a convexity theorem due to M. Riesz.* Fys. Säellsk. Förh. **8** (1938), No. 14.
105. Thorin, G. O., *Convexity theorems generalizing those of M. Riesz and Hadamard with some applications.* Comm. Sém. Math. Univ. Lund [Medd. Lunds Univ. Mat. Sem.] 9 (1948), 1–58.
106. Tolsa, Xavier, *Analytic capacity, the Cauchy transform, and non-homogeneous Calderón-Zygmund theory.* Progress in Mathematics, 307. Birkhäuser/Springer, Cham, 2014. xiv+396 pp.
107. Torchinsky, Alberto, *Real-variable methods in harmonic analysis.* Pure and Applied Mathematics, 123. Academic Press, Inc., Orlando, FL, 1986. xii+462 pp.
108. Torres, Rodolfo H., *Boundedness results for operators with singular kernels on distribution spaces.* Mem. Amer. Math. Soc. 90 (1991), no. 442, viii+172 pp.
109. Triebel, Hans, *Theory of function spaces*, Monographs in Mathematics, 78, Birkhäuser Verlag, Basel, 1983.
110. Triebel, Hans, Theory of function spaces. II. Monographs in Mathematics, 84. Birkhäuser Verlag, Basel, 1992. viii+370 pp.
111. Triebel, Hans, Theory of function spaces. III. Monographs in Mathematics, 100. Birkhäuser Verlag, Basel, 2006. xii+426 pp.
112. Triebel, Hans, Theory of function spaces. IV. Monographs in Mathematics, 107. Birkhäuser/Springer, Cham, [2020], ©2020. 160 pp.
113. Uchiyama, Akihito, *Hardy spaces on the Euclidean space.* With a foreword by Nobuhiko Fujii, Akihiko Miyachi and Kozo Yabuta and a personal recollection of Uchiyama by Peter W. Jones. Springer Monographs in Mathematics. Springer-Verlag, Tokyo, 2001. xiv+305 pp.

114. van der Corput, J. G., *Zahlentheoretische Abschätzungen*. [German] Math. Ann. **84** (1921), no. 1–2, 53–79.
115. Volberg, Alexander, *Calderón-Zygmund capacities and operators on nonhomogeneous spaces*. CBMS Regional Conference Series in Mathematics, 100. Published for the Conference Board of the Mathematical Sciences, Washington, DC; by the American Mathematical Society, Providence, RI, 2003. iv+167 pp.
116. Weisz, Ferenc, *Summability of multi-dimensional Fourier series and Hardy spaces*. Mathematics and its Applications, 541. Kluwer Academic Publishers, Dordrecht, 2002. xvi+332 pp.
117. Weyl, H., *Bemerkungen zum Begriff der Differentialquotienten gebrochener Ordnung*. Viertel Natur. Gesellschaft Zürich **62** (1917), 296–302.
118. Whitney, Hassler, *Analytic extensions of differentiable functions defined in closed sets*. Trans. Amer. Math. Soc. **36** (1934), no. 1, 63–89.
119. Wiener, Norbert, *The ergodic theorem*. Duke Math. J. **5** (1939), no. 1, 1–18.
120. Yuan, Wen; Sickel, Winfried; Yang, Dachun, *Morrey and Campanato meet Besov, Lizorkin and Triebel*. Lecture Notes in Mathematics, 2005. Springer-Verlag, Berlin, 2010. xii+281 pp.
121. Wilson, Michael, *Weighted Littlewood-Paley theory and exponential-square integrability*. Lecture Notes in Mathematics, 1924. Springer, Berlin, 2008. xiv+224 pp.
122. Wolff, Thomas H., *Lectures on harmonic analysis*. With a foreword by Charles Fefferman and a preface by Izabella Łaba. Edited by Łaba and Carol Shubin. University Lecture Series, 29. American Mathematical Society, Providence, RI, 2003. x+137 pp.
123. Ziemer, William P., *Weakly differentiable functions*. Sobolev spaces and functions of bounded variation. Graduate Texts in Mathematics, 120. Springer-Verlag, New York, 1989. xvi+308 pp.
124. Zygmund, A., *On a theorem of Marcinkiewicz concerning interpolation of operators*. J. Math. Pures Appl. (9) **35** (1956), 223–248.
125. Zygmund, A., Trigonometric series. Vol. I, II. Third edition. With a foreword by Robert A. Fefferman. Cambridge Mathematical Library. Cambridge University Press, Cambridge, 2002. xii; Vol. I: xiv+383 pp.; Vol. II: viii+364 pp.

Index

BMO, 247
  dyadic, 263
  properties of, 249
H^1, 285, 311
  atomic decomposition of, 311, 316
H^p atom, 289
L^∞, 2
L^p, 2
L^p Fourier multiplier, 207
σ-finite measure space, 1

A
adjacent dyadic cubes, 303
adjoint of a linear operator, 385
adjoint operator, 111
admissible growth, 222
affinely independent, 163
analytic family of operators, 222
ancestor of a dyadic cube, 260
anti-self-adjoint operator, 111
A_1 condition, 337
A_p condition, 335
  necessity of, 362
approximate identity, 45
associative property of convolution, 32
atom in H^p, 289
atomic decomposition of H^1, 311, 316
average of a function, 247

B
Besicovitch covering lemma, 389
Bessel potential, 201, 206
  one-dimensional, 207
Borel measure, 1, 83, 285, 329
Borel sets, 1
Borel–Cantelli lemma, 82

bounded mean oscillation, 247
  dyadic, 263

C
Calderón–Torchinsky multiplier theorem, 236
Calderón–Zygmund decomposition, 128
  on H^1, 317
  on L^q, 132
  weighted, 368
  with bounded overlap, 133
cancellation condition
  for a kernel, 120
centered Hardy–Littlewood maximal function, 22
characteristic constant of a weight
  A_1, 337
  A_p, 335
characterization
  of A_1 weights, 357
Chebyshev's inequality, 12
class of A_1 weights, 337
class of A_p weights, 335
commutative property of convolution, 32
complete measure space, 1
completeness of L^p, $0 < p < \infty$, 6
conjugate harmonic, 140
conjugate Poisson kernel, 140, 144
continuous Fejér kernel, 49
continuously differentiable function
  of order N, 37
convex function, 9
convolution, 32
  of a function with a tempered distribution, 87
Cotlar's inequality, 145

simpler form, 149
counting measure, 8
covering lemma, 24, 329, 389

D
derivative
　of a distribution, 85
　of a function (partial), 37
descendant of a dyadic cube, 260
difference operator, 200
dilation
　of a function, 86
　of a tempered distribution, 86, 288
dilation invariance of H^p, 288
Dirac mass, 83
directional Hilbert transform, 137
distribution
　homogeneous, 89
distributional derivative, 84
distributional inequality
　for the sharp maximal function, 268
distribution function, 11
doubling constant, 340
doubling measure, 340
dual exponent, 2
duality H^1-BMO, 325
dual weight, 336
dyadic average operator, 266
dyadic bounded mean oscillation, 263
dyadic child, 260
dyadic cube, 128, 260
dyadic decomposition, 188
dyadic John–Nirenberg theorem, 265
dyadic maximal function, 261
　sharp, 267
dyadic sharp maximal function, 267

E
essential infimum, 336
essentially bounded function, 2
exponential integrability, 257, 260, 265
exponentially integrable function, 254

F
factorization of A_p weights, 349
Fatou's lemma, 5
　for weak L^p spaces, 15
Fefferman–Stein inequality, 330
Fefferman–Stein vector-valued maximal function inequality, 169
Fejér kernel, 49
　continuous, 49
finitely simple function, 2
Fourier coefficient, 51

Fourier inversion, 56
　for distributions, 86
　on L^2, 64
Fourier transform
　of a Schwartz function, 51
　on $L^1 + L^2$, 66
　on L^2, 62
　properties of, 53, 62, 63
fractional integration by parts, 206, 220
fractional integration theorem, 197
fractional maximal function, 200
Fubini's theorem, 7

G
gamma function, 202
generalized functions, 82
generalized Lebesgue dominated convergence theorem, 6
good lambda inequality
　for the sharp maximal function, 268
gradient, 37
grand maximal function, 298

H
H^1-BMO duality, 325
Hölder's inequality, 2
　for weak L^p spaces, 15
Hörmander multiplier theorem, 212
Hörmander's integral smoothness condition, 120
Hardy space, 284
Hardy space characterizations, 298
Hardy–Littlewood maximal function
　centered, 22
　uncentered, 23
Hardy–Littlewood–Sobolev theorem, 197
harmonic polynomial, 185
Hausdorff–Young inequality, 74
Hilbert transform, 107
　maximal, 143
　smoothly truncated, 144
　truncated, 107, 109, 142
Hilbert transform identity, 143
homogeneous distribution, 89

I
infinitely differentiable function, 37
inhomogeneous singular integral, 120
inner product
　complex, 385
integrable function, 2
integration by parts
　fractional, 206
　for Sobolev spaces, 220

Index

interpolation
 between Sobolev spaces, 229
 Marcinkiewicz theorem, 16
  vector-valued, 21
 of analytic families of operators, 222
 Riesz–Thorin, 69
 using BMO, 273
 using BMO_d, 275
inverse Fourier transform, 56

J
Jensen's inequality, 9
John–Nirenberg theorem, 254
Jumping hat identity, 56, 64

K
Kato–Ponce inequality, 229

L
Laplace transform, 69
Laplace's equation, 185
Laplacian, 118, 195
 powers of, 195
Lebesgue differentiation theorem, 30
 one dimensional, 30
Lebesgue dominated convergence theorem, 6
Lebesgue measure, 7
Lebesgue monotone convergence theorem, 5
 for weak L^p spaces, 14
Lebesgue points, 29
Lebesgue set, 29
Leibniz rule, 38
 for distributions, 92
linear functional, 82
linear operator, 16
Lipschitz condition, 122, 148
Littlewood–Paley operator, 172, 322
 sharp cutoff, 187
Littlewood–Paley square function, 172
Littlewood–Paley theorem, 172, 322
locally integrable functions, 13

M
Marcinkiewicz function, 171
Marcinkiewicz interpolation theorem, 16
Marcinkiewicz multiplier theorem, 240
martingale maximal function, 266
maximal function
 auxiliary M_b^{**}, 298
 centered
  with respect to cubes, 27
 dyadic, 261
 dyadic sharp, 267
 fractional, 200
 grand, 298, 312
 Hardy–Littlewood centered, 22
 Hardy–Littlewood uncentered, 23
 nontangential, 283
 sharp, 266
 sharp with respect to balls, 271
 uncentered
  with respect to cubes, 26
maximal function associated with a measure, 345
maximal Hilbert transform, 143, 147
maximal Riesz transform, 147
maximal singular integral, 113, 149
maximal singular integral operator, 145
maximal singular integrals
 weighted bounds, 368
mean of a function, 247
mean oscillation, 247
Mikhlin multiplier theorem, 209
Minkowski convolution inequality, 34
Minkowski inequality, 4
Minkowski integral inequality, 7
Muckenhoupt characteristic constant
 of an A_1 weight, 337
 of an A_p weight, 335
multi-index, 37
multiplier theorem
 of Calderón and Torchinsky, 236
 of Hörmander, 212
 of Marcinkiewicz, 240
 of Mikhlin, 209

N
n-simplex, 163
necessity of A_p condition, 362
nonsmooth Littlewood–Paley theorem, 188
nontangential maximal function, 283
null set, 1

O
Operator
 of weak type (p,p), 16
Oscillation of a function, 77, 247

P
Parseval's identity, 56, 64
partial derivative, 37
partition of unity, 41, 307
Plancherel's identity, 64
Poincaré inequality, 41
Poisson kernel, 45, 59, 139
 conjugate, 140

polarization, 64
polynomial
    harmonic, 185
positive Borel measure, 1, 329
potential
    Bessel $\mathcal{J}_z$, 201
    Riesz I_s, 196
power weights, 343
powers of Laplacian, 195
principal value integral, 107
product of a function and a tempered distribution, 87

Q

quasi-normed space, 12, 14
quasi-subadditive operator, 16
quasi-subadditivity, 8

R

radial function, 54
radially decreasing function, 76
reflection, 179
    of a function, 53
    of a tempered distribution, 86
regular measure, 329
reverse-Hölder property, 354
reverse wea-type $(1,1)$ inequality
    for the Hardy–Littlewood maximal operator, 355
Riemann–Lebesgue lemma, 57
Riesz potential, 271
Riesz potential operator $\mathcal{I}_s$, 196
Riesz representation theorem, 7
Riesz transform, 113, 116
    truncated, 144
Riesz–Thorin interpolation theorem, 69

S

Schur's test, 74
Schwartz function, 42
Schwartz seminorm, 42
self-adjoint operator, 386
self-transpose operator, 386
semigroup property, 195
sharp maximal function, 266
sharp maximal function with respect to balls, 271
simple function, 2
simplex, 163
singular integral, 319
size condition
    for a kernel, 119
smooth function, 37
smooth function with compact support, 37

smoothly truncated Hilbert transform, 144
smoothly truncated maximal singular integral, 149
smoothly truncated singular integral, 149
smoothness condition
    for a kernel, 119
Sobolev embedding theorem, 218
Sobolev space, 214
    of product type, 240
space
    BMO, 247
    H^1, 285, 311
    L^∞, 2
    L^p, 2
    $L^{p,\infty}$, 12
    $\mathcal{C}^N$, 37
    $\mathcal{C}^\infty$, 37
    $\mathcal{C}_0^\infty$, 37
    $\mathcal{M}_p(\mathbf{R}^n)$, 96
    dyadic BMO, 263
square function, 172
Stein's interpolation theorem for analytic families, 222
step function, 8
stopping-time argument, 128
strong maximal function, 171
subadditive operator, 16
subordination identity, 58
support of a distribution, 88

T

Taylor formula, 281
tempered distributions, 82
tempered function, 83
    at infinity, 83
Tonelli's theorem, 7
translation
    of a function, 53
    of a tempered distribution, 86, 288
translation invariance of H^p, 288
transpose of a linear operator, 367, 385
truncated Hilbert transform, 107, 109, 142
truncated Riesz transform, 144
truncated singular integral operator, 113, 145, 149

U

uncentered Hardy–Littlewood maximal function, 23

V

Van der Corput lemma, 102
vanishing moment property, 289
vector-valued

Hardy–Littlewood maximal inequality, 169, 171
vector-valued inequalities, 158, 159, 162, 322
vector-valued Marcinkiewicz interpolation, 21
vector-valued singular integral, 151

W
weak L^p, 12
weak type $(1,1)$, 25
weak type (p,p), 16

weight, 332, 334
  of class A_1, 337
  of class A_p, 335
weighted L^p space, 334
weighted bounds for maximal singular integrals, 368
weighted inequalities, 332
Whitney cube, 306
Whitney decomposition, 304

Y
Young's inequality, 72

Printed by Printforce, United Kingdom